(Bolyai Janos Matematikai Tarsulat)

COLLOQUIA MATHEMATICA
SOCIETATIS JÁNOS BOLYAI, 38.

RADICAL THEORY

Edited by:

L. MÁRKI
and
R. WIEGANDT

NORTH-HOLLAND
AMSTERDAM – OXFORD – NEW YORK

Budapest, Hungary, 1985

ISBN North-Holland: 0 444 86765 1
ISBN Bolyai: 963 8021 68 3
ISSN Bolyai: 0139 3383

Joint edition published by

JÁNOS BOLYAI MATHEMATICAL SOCIETY
and
ELSEVIER SCIENCE PUBLISHERS B. V.
P. O. Box 1991
1000 BZ Amsterdam, The Netherlands

In the U. S. A. and Canada:

ELSEVIER SCIENCE PUBLISHING COMPANY INC.

52 Vanderbilt Avenue
New York, N. Y. 10017
U. S. A.

Printed in Hungary
Szegedi Nyomda, Szeged

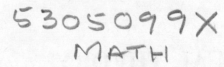

PREFACE

The notion of radical was introduced by Köthe in 1930 for rings; his aim was to determine the structure of rings with zero radical. Later further radicals were defined for rings and the most powerful of them turned out to be the Jacobson radical. In parallel, various notions of nilpotence were introduced for groups and quasigroups and were used in structural investigations. The notion of general radicals arose in works of three mathematicians: at first, in the early forties, Bruck came up with general nilpotences for quasigroups, then, in the early fifties, apparently unaware of Bruck's theory, Amitsur and Kurosh (also independently of each other) introduced general radicals for rings and outlined a generalization to categories. Whereas concrete radicals have been studied more or less intensively ever since, the theory of general radicals of rings started a fast development only after Anderson, Divinsky and Suliński had proved in the mid sixties their fundamental theorem on the hereditariness of semisimple classes. Now the theory of concrete and general radicals is gaining popularity within ring theory; the counterpart of radicals, called torsion theories, constitutes a flourishing part of module theory; furthermore, radical theoretic concepts and aspects have infiltrated also into other branches of algebra, topology and graph theory.

Taking account of this development, some Hungarian algebraists felt that the time had come to organize a conference devoted entirely to radical theory. Upon their initiative, the very first conference of this kind was hosted by the János Bolyai Mathematical Society and held in Eger (Hungary), August 1-7, 1982. As well as Hungarian members of the Organizing Committee, Professors L. A. Bokut' (Novosibirsk) and L. A. Skorniǐakov (Moscow) were very helpful in preparing the colloquium, and we should like to take this opportunity to express our gratitude for their activity.

Realizing the hopes of the organizers, the conference provided a most suitable place for an encounter of scientists from both the East and the West. In our eyes it turned out to be a substantial international conference with 64 algebraists from 22 countries representing all the five continents. The scientific program consisted of 9 invited talks of 30-45 minutes and 34 short communications. Besides a sightseeing tour of the historic town of Eger, an excursion to the near-by Bükk mountains and a banquet gave opportunity for informal discussions, welcomed by people who had known each other only by publications.

Most of the contents of the present volume are written and often expanded versions of talks delivered at the colloquium. Some of them are survey articles by invited speakers, the rest consists of research papers with full proofs, including some by colleagues who would have liked to attend the conference but could not do so for various reasons. The volume also contains unsolved problems presented at the problem session of the colloquium. All contributions have been refereed, and it is a pleasure for us to express our appreciation to the referees as well as the authors for their help and cooperation.

The editors.

CONTENTS

LIST OF PARTICIPANTS

ALMKVIST, Gert, PL 500, S-24300 Höör, Sweden

AMITSUR, Shimshon Avraham, 43 Harlap Str., Jerusalem, Israel

ÁNH, Pham Ngoc, Mathematical Institute, Hungarian Academy of Sciences, H-1053 Budapest, Reáltanoda u. 13-15, Hungary

BACCELLA, Giuseppe, Istituto Matematico, Università dell'Aquila, Via Roma 33, 67100 L'Aquila, Italy

BECKMANN, Peter, Karl-Marx-Universität, Sektion Mathematik, DDR-7010 Leipzig, Karl-Marx-Platz, GDR

BEĬDAR, Konstantin I., USSR, Moskva 117192, Vinnickaya 9, kv. 16

BETSCH, Gerhard, Mathematisches Institut der Universität, Auf der Morgenstelle 10, D-7400 Tübingen 1, GFR

BOKUT', Leonid Arkad'evič, Institute of Mathematics, Siberian Branch of the Academy, 630090 Novosibirsk, Universitetskiĭ pr. 4, USSR

CHEHATA, Chehata Gouda, Faculty of Science, University of Alexandria, Moharrem Bey, Alexandria, Egypt

DIMITRIĆ, Radoslav, Tulane University, Department of Mathematics, New Orleans, LA 70118, USA

DIVINSKY, Nathan, University of British Columbia, Mathematics Department, Vancouver, B. C., V6T 1Y4, Canada

DUBROVIN, Nikolaĭ Ivanovič, Vladimir, 600014, Lakina 133-A, kv. 71, USSR

FUCHS, Laszlo, Tulane University, Department of Mathematics, New Orleans, LA 70118, USA

GARDNER, Barry J., University of Tasmania, Department of Mathematics, POB 252-C, Hobart, Tas. 7001, Australia

GERASIMOV, Viktor Nikolaevič, Institute of Mathematics, Siberian Branch of the Academy, 630090 Novosibirsk, Universitetskiĭ pr. 4, USSR

GOODAIRE, Edgar G., Department of Mathematics and Statistics, Memorial University, St. John's, Newfoundland, A1B 3X7, Canada

GOLAN, Jonathan, Department of Mathematics, University of Haifa, 31999 Haifa, Israel

HÁCH, Tran Lam, Mathematical Institute, Hungarian Academy of Sciences, H-1053 Budapest, Reáltanoda u. 13-15, Hungary

HUYNH, Dinh Van, Institute of Mathematics, Hanoi, Vietnam Hộpthư 631 - Bờ Hồ HANOI

IVANOVA, Tatǐana Velkova, Bulgarian Academy of Sciences, Institute of Mathematics, ul. Akad. G. Bončev 8, 1113 Sofia, Bulgaria

JACOBSON, Nathan, Department of Mathematics, Yale University, New Haven, CT 06520, USA

JENKINS, Terry L., Mathematics Department, University of Wyoming, 220 Ross Hall, Laramie, WY 82070, USA

JESPERS, Eric, Katholieke Universiteit Leuven, Departement Wiskunde, Celestijnenlaan 200 B, B-3030 Leuven (Heverlee), Belgium

KAARLI, Kalle, USSR 202400, Tartu, Tartu State University, Department of Mathematics

KAUTSCHITSCH, Hermann, Institut für Mathematik, UBW Klagenfurt, Universitätsstraße 65-67, A-9010 Klagenfurt, Austria

KEGEL, Otto H., Mathematisches Institut, Albert-Ludwigs-Universität, Albertstraße 23/b, D-7800 Freiburg, GFR

KISS, Emil W., Mathematical Institute, Hungarian Academy of Sciences, H-1053 Budapest, Reáltanoda u. 13-15, Hungary

KMET', František, 94901 Nitra, Lúčna 2, Czechoslovakia

KOǏFMAN, Lazar' Abramovič, USSR, Moskva, 129642, proezd Dežniova, 9-1-51

KOLIBIAR, Milan, Komenský University, Mathematics Department, Mlynská dolina, 816 31-Bratislava, Czechoslovakia

KREMPA, Jan, Institute of Mathematics, University of Warsaw, 00-901 Warsaw, Poland

KUZMIN, Evgeniĭ Nikiforovič, Institute of Mathematics, Siberian Branch of the Academy, 630090 Novosibirsk, Universitetskiĭ pr. 4, USSR

van LEEUWEN, Leo C. A., Rijksuniversiteit te Groningen, Postbus 800, Groningen, The Netherlands

LEX, Wilfried, Institut für Informatik der Universität Clausthal, Erzstr. 1, D-3392 Clausthal-Zellerfeld, GFR

LOI, Nguyen Van, Mathematical Institute, Hungarian Academy of Sciences, H-1053 Budapest, Reáltanoda u. 13-15, Hungary

LOONSTRA, Frans, Haviklaan 25, 2566 XB - Den Haag, The Netherlands

MÁRKI, László, Mathematical Institute, Hungarian Academy of Sciences, H-1053 Budapest, Reáltanoda u. 13-15, Hungary

MARTYNOV, Leonid Matveevič, USSR 644099 Omsk, Pedagogicheskiĭ Institut, Kafedra Algebry

McCRIMMON, Kevin, Department of Mathematics, Mathematics-Astronomy Building, University of Virginia, Charlottesville, VA 22903, USA

MEDVED'EV, Juriĭ A., Institute of Mathematics, Siberian Branch of the Academy of Sciences, 630090 Novosibirsk, Universitetskiĭ pr. 4, USSR

MLITZ, Rainer, Institut für Angewandte und Numerische Mathematik, Technische Universität Wien, Gußhausstraße 27-29, A-1040 Wien, Austria

MORSY, Monir S., Pure Mathematics Department, Faculty of Sciences, Ain-Shams University, Abbasia - Cairo, Egypt

NIEMENMAA, Markku, Department of Mathematics, University of Oulu, SF-90570 Oulu, Finland

OSŁOWSKI, Bogdan J., Zakład Matematyki, Wyžsza Szkoła Rolniczo-Pedagogiczna, ul. Nowotki 12, 08-110 Siedlce Poland

PORTER, Timothy, School of Mathematics and Computer Science, University College of North Wales, Bangor, Gwynedd, LL57 2UW, Wales, Great Britain

PUCZYŁOWSKI, Edmund, Institute of Mathematics, University of Warsaw, PKiN, 00-901 Warsaw, Poland

RACINE, Michel, Department of Mathematics, University of Ottawa, 585 King Edward, Ottawa, Ontario, K1N 9B4, Canada

RAZMYSLOV, Juriĭ P., USSR, Moskva 119121, I-iĭ Neopalimovskiĭ pereulok 16/13, kv. 12 a

SANDS, Arthur D., Department of Mathematics, The University, Dundee, DD1 4HN, Scotland, Great Britain

SCHMIDT, E. Tamás, Mathematical Institute, Hungarian Academy of Sciences, H-1053 Budapest, Reáltanoda u. 13-15, Hungary

SLATER, Michael, Department of Mathematics, University of Bristol, University Walk, Bristol, BS8 1TW, England, Great Britain

STEWART, Patrick N., Department of Mathematics, Dalhousie University, Halifax, Nova Scotia, B3H 4H8, Canada

SULIŃSKI, Adam, Institute of Mathematics, Warsaw University, 00-901 Warsaw, Poland

SZÁSZ, Ferenc A., Mathematical Institute, Hungarian Academy of Sciences, H-1053 Budapest, Reáltanoda u. 13-15, Hungary

TISHCHENKO, Alexandr Vladimirovič, Department of Mathematics, All-Union Correspondence Institute of Food Industry, ul. Chkalova 73, 109803 Moscow, USSR

TONOV, Ivan K., Bulgarian Academy of Sciences, Institute of Mathematics, ul. Akad. G. Bončev 8, 1113 Sofia, Bulgaria

TRLIFAJ, Jan, Čimická 15/257, 182 00 - Praha, Czechoslovakia

VARADARAJAN, Kalathoor, Department of Mathematics, University of Calgary, Calgary, Alberta, T2N 1N4, Canada

VOVSI, Samuil Mihaĭlovič, Department of Mathematics, Riga Polytechnic Institute, Riga 226355, USSR

WATTERS, John F., Department of Mathematics, The University, Leicester, LE1 7RH, England, Great Britain

WIEGANDT, Richárd, Mathematical Institute, Hungarian Academy of Sciences, H-1053 Budapest, Reáltanoda u. 13-15, Hungary

YUSUF, S. Mohammad, Faculty of Natural Sciences, Quaid--i-Azam University, Islamabad, Pakistan

ZALESSKIĬ, Aleksandr Efimovič, USSR, 220045 Minsk-45, Širokaĭa 36, kv. 618 "b"

ZEL'MANOV, Efim Isaakovič, Institute of Mathematics, Siberian Branch of the Academy, 630090 Novosibirsk, Universitetskiĭ pr. 4, USSR

SCHEDULE FOR TALKS

Monday, August 2

Morning Session Ch.: A. D. Sands

9^{00} J. Krempa: Radicals and derivations

10^{00} E. Jespers: On radicals of graded rings

10^{20} A. Suliński: Radicals of graded rings

10^{40} N. Divinsky: 2-graded associative rings

11^{20} T. Porter: A relative Jacobson radical with applications

11^{40} P. N. Stewart: Intermediate extensions and radical classes

Afternoon Session Ch.: F. Loonstra

15^{00} B. I. Plotkin - S. M. Vovsi: Radicals in groups and other structures

16^{00} M. Slater: On Amitsur's condition E

16^{20} J. F. Watters: Radicals and ultraproducts

16^{40} A. D. Sands: A class of rings associated with radical theory

17^{20} O. H. Kegel: Equations in nil rings

17^{40} F. Kmet': A note on equalities of radicals in a semigroup

Tuesday, August 3

Morning Session Ch.: S. A. Amitsur

9^{00} J. S. Golan: Topologies defined on the torsion-theoretic spectrum of a ring - some recent results

10^{00} F. Loonstra: Maximal and minimal problems in hereditary torsion theories

10^{20} L. A. Koĭfman: When all radicals in the category of modules are hereditary

10^{40} W. Lex: Lattices of torsion theories for acts

11^{00} B. Osłowski: Purely inseparable algebras

Wednesday, August 4

Morning Session Ch.: N. Jacobson

9^{00} E. I. Zel'manov: Radicals and structure theory of Jordan algebras

10^{00} K. McCrimmon: Invariance of Jordan radicals

10^{20} S. A. Amitsur: Radicals and the Gelfand-Kirillov dimension

10^{40} L. A. Bokut': Some new results in associative and Lie algebras

11^{20} A. E. Zalesskiĭ: Lie solvable associative rings and a radical in varieties of algebras

11^{40} E. N. Kuz'min: Structure theory of Mal'cev algebras

12^{00} Yu. P. Razmyslov: Radicals in algebras satisfying Capelli's identities

Afternoon Session Ch.: L. C. A. van Leeuwen

15^{00} R. Mlitz: Radicals and interpolation in universal algebra

16^{00} G. Betsch: (Non) hereditary radical and semisimple classes of near-rings

16^{20} K. Kaarli: Radicals in finite near-rings

16^{40} H. Kautschitsch: Maximal ideals in polynomial near-rings

17^{00} L. Márki: Radicals in categories

Thursday, August 5

Morning Session Ch.: N. Divinsky

9^{00} B. J. Gardner: Radicals and varieties

10^{00} L. M. Martynov: Semisimple radical and attainable classes of algebras

10^{20} T. L. Jenkins: Essential ideals and radical classes

10^{40} N. V. Loi: Essentially closed radical classes

11^{20} L. C. A. van Leeuwen: Hereditariness of upper radicals

11^{40} F. Szász: Various topics in radical theory

Friday, August 6

Morning Session   Ch.: A. Suliński

9^{00} E. Puczyłowski: Behaviour of properties of rings under some algebraic constructions

10^{00} M. Niemenmaa: On radicals in x-systems with subgroups as ideals

10^{20} T. V. Ivanova: An upper bound for the Amitsur number of tensor products of PI-algebras

10^{40} A. V. Tishchenko: On J-equiradical rings

11^{20} P. Beckmann: Syzygies and prime sequences in the noncommutative case, the transition from left to right quotients

11^{40} J. Trlifaj: On the structure of Ext-rings

Afternoon Session   Ch.: L. A. Bokut'

15^{00} K. I. Beĭdar: Some examples in radical theory

15^{30} N. I. Dubrovin: Examples of simple radical chain rings

16^{20} Problem Session

COLLOQUIA MATHEMATICA SOCIETATIS JÁNOS BOLYAI
38. RADICAL THEORY, EGER (HUNGARY), 1982.

EXAMPLES OF RINGS AND RADICALS*

K. I. BEĬDAR

Except in Section 4, all rings referred to in this paper are assumed to be associative. If A is a ring then the notations $I \triangleleft^{l} A$, $I \triangleleft^{r} A$, $I \triangleleft A$ mean respectively that I is a left, right, two-sided ideal of A. Z is the ring of integers, and Q the field of rationals. Fundamental definitions and facts in the theory of radicals of rings and algebras can be found in the books [2], [5] and [14]. This paper is in four sections, and we here describe what is achieved in each section.

Section 1. We recall a number of definitions. The radical R is called *left strict* if $L \triangleleft^{l} A$ and $R(L)=L$ together imply $L \subseteq R(A)$. It is *left stable* if $L \triangleleft^{l} A$ implies $R(L) \subseteq R(A)$. It is *left hereditary* if $L \triangleleft^{l} A$ and $L \subseteq R(A)$ together imply $R(L)=L$. These three notions are

* This paper was written during the author's stay in Hungary.

discussed in [3, 8, 9, 10, 11]. We construct examples giving positive answers to the following two questions raised by A. D. Sands in [10]: Does there exist a radical which is left hereditary but not right hereditary? Does there exist a radical which is left strict but not right strict?

Section 2. Let β be the Baer lower radical. Recall from [12] that the radical R is said to be *left* β*-regular* if whenever $L \triangleleft^{l} A$ and $R(A)=0$ it follows that $\{AR(L)+R(L)\}/R(L)$ is β-semisimple, and similarly for right β-regularity. R is β*-regular* if it is both left and right β-regular. We construct an example of a radical which is not β-regular, thus answering a question raised in [12]. We also show that for algebras over a field all radicals are β-regular.

Section 3. An example of a simple chain radical domain constructed by Dubrovin in [6] allows us to give a positive answer to a question raised by Andrunakievich and Ryabuhin in [2, p. 15], viz. whether there exists a simple ring without 1 sucht that the lower radical generated by it is an atom in the 'lattice' of all radicals.

Section 4 is concerned with a problem raised in [13], viz. whether for each n there exists a hom. closed class of rings for which the process of constructing the lower radical terminates at exactly the n-th step. This problem received a positive solution in [4]. We show (Theorem 4) that if $\tilde{V}$ is any variety of alternative rings containing Z then there exists a hom. closed class $\tilde{L}^{(n)} \subseteq \tilde{V}$ such that the construction of the lower radical generated by $\tilde{L}^{(n)}$ in $\tilde{V}$ terminates at exactly the n-th step.

I. LEFT STRICT RADICALS, LEFT HEREDITARY RADICALS, AND NORMAL RADICALS

The concepts from ring- and module-theory that we use below can be found in [7].

LEMMA 1. *Suppose* $M \triangleleft^r A$, *and* N *is a right* A*-module such that* $NM=N$. *Then the natural ring embedding* $\mathrm{End}(N_A) \to \mathrm{End}(N_M)$ *is an isomorphism.*

Proof. Suppose $f \in \mathrm{End}(N_M)$; $a \in A$, $x \in N$. The condition yields $x_i \in N$ and $m_i \in M$ such that $x = \sum_1^t x_i m_i$. Then $f(xa) = f((\sum x_i m_i)a) = f\{\sum x_i(m_i a)\} = \sum f(x_i)(m_i a) =$ $= \sum\{f(x_i)\}m_i \cdot a = \{\sum f(x_i)m_i\}a = (\sum f(x_i m_i))a = \{f(\sum x_i m_i)\}a =$ $= f(x)a$. Thus $f \in \mathrm{End}(N_A)$, so that the injection is surjective, thus bijective.

PROPOSITION 1. *Suppose* A *is a simple ring with* 1,

$L \triangleleft^{l} A$, and $M \triangleleft^{r} A$. Suppose L is ring-isomorphic to M. Then M is a finitely generated projective right ideal of A.

Proof. The elements of A act on the left of the module L_L by multiplication, thus giving us a natural ring homomorphism $\phi: A \to \mathrm{End}(L_L) = T$, say. Since $aL=0$ implies $a=0$, ϕ is a monomorphism. We show that $\phi(A) = T$. Given $k \in L$, $a \in A$, $x \in L$, $f \in T$ we have $\{f \cdot \phi(ka)\}(x) = f\{\phi(ka)(x)\} = f(ka \cdot x) = f(k \cdot ax) = f(k)(ax)$ since $ax \in L$, $= \{f(k)a\} \cdot x = \phi\{f(k)a\}(x)$. Thus $f \cdot \phi(ka) =$ $= \phi(f(k)a) \in \phi(A)$. Now since A is simple we have $LA = A$. Thus, given $b \in A$ we have $b = \sum k_i a_i$ for some $k_i \in L$, $a_i \in A$, whence for $f \in T$ we have $f \cdot \phi(b) =$ $= \phi(\sum f(k_i) a_i) \in \phi(A)$. So $\phi(A) \triangleleft^{l} T$. But clearly $\phi(1_A) = 1_T$, so $\phi(A) = T$. Thus T is simple.

Hence L is a generator in T-mod. Since L and M are ring-isomorphic, it follows that M is a generator in $\mathrm{End}(M_M)$-mod. Now $AM = A$ since A is simple, and $MA = M$ since $1 \in A$. So $MM = MAM = MA = M$. Thus Lemma 1 gives us that $\mathrm{End}(M_M) = \mathrm{End}(M_A) = Q$, say, and we have proved that M is a generator in Q-mod. Since $AM = A$, it is also true that M is a generator in mod-A. Hence $\mathrm{End}({}_Q M) = A$, whence M is a finitely generated projective A-module (see [7], Prop. 4.1.3).

As a curious corollary we note the following result.

PROPOSITION 2. *Suppose* A *and* B *are simple associative rings with* 1 *having isomorphic non*-0 *left ideals* L, M *respectively. Then the isomorphism of* L *to* M *extends to an isomorphism of* A *to* B.

Proof. The isomorphism ψ of L to M transfers in an obvious manner to an isomorphism of $\mathrm{End}(L_L)$ with $\mathrm{End}(M_M)$. Then with the notation above we have $A \simeq \phi(A) = \mathrm{End}(L_L) \simeq \mathrm{End}(M_M) = \phi(B) \simeq B$. We may verify directly that this isomorphism θ extends ψ. Explicitly, for $a \in A$, $\theta(a) =$ the unique $b \in B$ such that $\psi(ak) = b\psi(k)$ for all $k \in L$. If $a \in L$ then from $\psi(ak) = \psi(a)\psi(k)$ we deduce that $b = \psi(a)$.

The construction of this section will be concerned with simple domains, i.e. rings without 0-divisors, but not necessarily either commutative or having a 1. We start with an easy lemma.

LEMMA 2. (a) *If* D *is a simple domain with* 1, *and* $L \triangleleft^l D$, *then* L *is simple*.

(b) *If the domain* D *is an essential ideal of the ring* A, *then* A *is also a domain*.

Proof. (a) If $0 \neq I \triangleleft L$ then $0 \neq LIA \triangleleft A$, so $LIA = A$.

Then $I \supseteq LI(AL) = (LIA)L = AL = L$. Thus L is simple.

(b) Given $a,b \in A$ with $ab = 0$, we have $(Da)(bD) = 0$, whence, since D is a domain and $D \triangleleft A$, we have $Da=0$ or $bD=0$. If $Da=0$ then $a \in M = \{m \in A : Dm=0\} \triangleleft A$. Since D is a domain, $M \cap D=0$, and since D is essential in A , it follows that $M=0$. Thus $a=0$. Similarly if $bD=0$ then $b=0$. So A is a domain.

We next give some definitions and results due to Beidar and Salavova [3]. The class $\tilde{P}$ of rings is called a *left-strict special class* if the following conditions hold: (a) Every $A \in \tilde{P}$ is a prime ring; (b) If $L \triangleleft^{l} A \in \tilde{P}$ then $L/r(L) \in \tilde{P}$, where $r(L) = \{k \in L : Lk=0\}$; (c) If A is a prime ring and $0 \neq I \triangleleft A$ for some $I \in \tilde{P}$, then $A \in \tilde{P}$.

If $\tilde{H}$ is a class of rings we write $\tilde{U}\tilde{H}$ for the class of rings without non-0 hom. images in $\tilde{H}$, and $\tilde{I}\tilde{H}$ for the class of rings without non-0 ideals in $\tilde{H}$. If R is a radical we write $\tilde{P}^{*}(R)$ for the set of prime subdirectly-irreducible rings A with $R(A) \neq 0$, and R^{*} for $\tilde{U}\tilde{P}^{*}(R)$. (Compare [3].) Then we have the following results from [3], Theorems 17 and 23.

PROPOSITION 3. (a) *If* $\tilde{P}$ *is a left-strict special class then* $\tilde{U}\tilde{P}$ *is a left-strict special radical.*

(b) *Suppose* R *is a special radical. Then* R *is a*

left-strict radical iff $\tilde{I}\tilde{R}$ *is a left-strict special class.*

(c) *If* R *is an antisimple radical then* $R^{**}=R$.

(d) *Suppose* R *is an antisimple radical. Then* R *is left strict iff* R^{*} *is left hereditary.*

We now have

THEOREM 1. *Suppose* D *is a simple domain with* 1 *but not right noetherian, and* $\tilde{H}$ *is the class of those subdirectly irreducible rings whose hearts are isomorphic to one or another left ideal of* D . *Set* $R=\tilde{U}\tilde{H}$. *Then*

(1) R *is an antisimple left-strict radical*

(2) R *is not right strict*

(3) R^{*} *is an antisimple left-hereditary radical which is not right hereditary.*

Proof. (1) To show that R is left strict it suffices by Prop. 3a to show that $\tilde{H}$ is a left-strict special class. So we verify the defining conditions for this.

(a) Any left ideal of D is a domain, so by Lemma 2b all rings in $\tilde{H}$ are domains, and in particular are prime.

(b) Suppose $0 \neq L \triangleleft^{l} A \in \tilde{H}$. Since A is a domain,

$r(L)=0$, so we must show that $L\in\tilde{H}$. Suppose A has heart I isomorphic to a left ideal K of D . We claim that IL is the heart of L . For if $0\neq M\triangleleft L$, then $I\supseteq LMI\triangleleft A$, giving $LMI=I$. Hence $M\supseteq LM(IL)=(LMI)L=IL$, as required.

Since $IL\triangleleft^{l} I\simeq K$, it follows that $IL\simeq T$ for some $T\triangleleft^{l} K\triangleleft^{l} D$. Since IL is the heart of L we have $(IL)^2=$ $=IL$, so that $T^2=T$. Then $DT=DT^2=(DT)T\subseteq KT\subseteq T$, so that $T\triangleleft^{l} D$. Thus $L\in\tilde{H}$, as required.

(c) If P is a prime ring and $0\neq N\triangleleft P$ with $N\in\tilde{H}$, suppose H is the heart of N (isomorphic to a left ideal of D). Then $H=NHN$ is an ideal of P , hence its heart. Thus $P\in\tilde{H}$, as required.

Finally it is clear that R is antisimple.

(2) Since D is not right noetherian, it has a right ideal M which is not finitely generated. By Lemma 2a M is simple, so that $R(M)=0$ or M . If $R(M)=0$ then $M\notin R$, so M has a non-0 hom. image $M'\in\tilde{H}$. Since M is simple, $M'\simeq M$, so $M\in\tilde{H}$. Thus M , which is its own heart, is isomorphic to some left ideal L of D . So by Prop. 1 M is a finitely-generated right ideal of D . This contradicts our choice of M , and shows that $R(M)=M$. On the other hand $R(D)=0$, since $D\in\tilde{H}$. So R

is not right strict.

(3) This follows from (1) and (2) in view of Prop. 3(d).

Note. We have incidentally shown that $\tilde{H}$ is a special class of domains, closed under the taking of left ideals.

In view of [9], Thm. 7, we deduce the

COROLLARY. *With the notation and hypotheses of Theorem 1, R is a left stable antisimple radical which is not right strict.*

II. AN EXAMPLE OF A NON-β-REGULAR RADICAL

We denote by $\bar{A}$ the ring obtained by the formal adjunction of a 1 to the ring A; by P the set of primes; by F_p the field of p elements (for $p \in P$); by A^+ the additive group of the ring A; by G^0 the zero ring on the abelian group G. We write A^0 in place of $(A^+)^0$.

LEMMA 3. *Suppose G is an abelian group, A is a ring, and $\phi: G \to A^+$ and $\psi: (A^2)^+ \to G$ are group homomorphisms with $\phi\psi$ = the identity on $(A^2)^+$. Define a binary operation $*$ on the group $(G,+)$ by $a*b = \psi(\phi(a)\phi(b))$ for all $a,b \in G$. Then $G^* = (G,+,*)$ is an associative ring.*

Proof. Distributivity of $*$ over $+$ is clear. We check associativity. For clarity write ϕ_a for $\phi(a)$ etc. Then $(a*b)*c = \psi\{\phi_{a*b}\phi_c\} = \psi\{\phi[\psi(\phi_a\phi_b)]\cdot\phi_c\} =$ $= \psi(\phi_a\phi_b\cdot\phi_c)$ (in view of $\phi\psi = \text{id.}$) $= \psi(\phi_a\cdot\phi_b\phi_c) =$ $= a*(b*c)$ similarly.

Now let H be an abelian group isomorphic to the subgroup of Q^+ generated by $\{1/p: p\in P\}$, under an isomorphism taking 1 to $a\in H$, and $1/p$ to $a_p\in H$. Clearly $H/Za \simeq L^+$, where $L = \oplus_{p\in P} F_p$. In the polynomial ring $L[t]$ set $A = tL[t]$, and $G = H \oplus (At)^+$. Write e_p for the 1 of F_p, when regarded as an element of L. Then we define maps $\phi: G \to A^+$ and $\psi: (A^2)^+ \to G$ by

$$(1) \quad \phi(\textstyle\sum n_p a_p, b) = (\sum n_p e_p)t + b, \text{ and } \psi(b) = (0, b).$$

We easily see that ϕ and ψ are well-defined abelian group homomorphisms and that $\phi\psi = \text{id.}$ on $(A^2)^+$. Using Lemma 3 we make G into a ring G^*. Let us regard G as the internal direct sum of subgroups H and $At =$ $= t^2L[t]$, so that each $g\in G$ is uniquely representable in the form $g = h + b$ with $h\in H$, $b\in At$. Then from (1) and the definition of $*$ we find

$$(2) \quad (\textstyle\sum n_p a_p)*b = b*(\sum n_p a_p) = (\sum n_p e_p)tb;$$

(3) $\quad (\sum n_p a_p) * (\sum m_q a_q) = (\sum n_p m_p e_p) t^2$;

(4) $\qquad b * c = bc$;

(5) $\qquad na * (h+b) = 0$

for all n_p , m_q , $n \in Z$; $b, c \in At$, $h \in H$.

THEOREM 2. *Let* R *be the lower radical generated by* H^o . *Then* R *is not a* β*-regular radical.*

Proof. We first show that $\beta(G^*) = aZ$. From (5) we have $aZ \triangleleft G^*$ with $(aZ)^2 = 0$, so that $aZ \subseteq \beta(G^*)$. Consider now $C = G^*/aZ$. Since $aZ \subseteq H$, we have $C^+ \simeq$ $\simeq (H/aZ) \oplus (At)^+ \simeq L^+ \oplus t^2 L[t]^+$. If $*$ is the product in C induced from G^* , then by (2), (3), (4) we have, for $c, d \in L$, $p(t), q(t) \in L[t]$:

$$(c+t^2 p(t)) * (d+t^2 q(t)) = t^2 (c+tp(t))(d+tq(t)) .$$

It is clear from this that C has no nilpotents: indeed $C \simeq A = tL[t]$ under the map $c+t^2 p(t) \mapsto ct+t^2 p(t)$, and $L \simeq \oplus_p F_p$ has no nilpotents. Thus $\beta(C) = 0$, whence $\beta(G^*) \subseteq aZ$, giving equality.

Now in the 2-by-2 matrix ring $M_2(\overline{G^*})$ let D be the subring $e_{12}\overline{G^*} + \sum\{e_{ij} G^*: (ij) \neq (12)\}$ in the usual notation, and let N be the left ideal $N = e_{11} G^* + e_{21} G^*$

of D . Suppose we can show that $R(N) = e_{21}G^*$. Then $N \supseteq DR(N) + R(N) \supseteq e_{12}G^* e_{21}G^* + e_{21}G^* = N$, so that $\{DR(N) + R(N)\}/R(N) = N/R(N) = e_{11}G^* \simeq G^*$, whereas $\beta(G^*) = aZ \neq 0$. Thus to prove that R is not β-regular we must show that $R(N) = e_{21}G^*$ and also (see the definition of β-regularity in the Introduction) that $R(D) = 0$.

First note that $G^0 \simeq e_{21}G^* \triangleleft N$, and $G^0 = H^0 \oplus \sum_2^\infty (Lt^n)^+$ is the direct sum of H^0 and isomorphic copies of $L^+ = H^0/aZ$. Thus $R(G^0) = G^0$, whence $e_{21}G^* \subseteq R(N)$. Also $N/e_{21}G^* \simeq e_{11}G^* \simeq G^*$.

So to complete the proof that $R(N) = e_{21}G^*$ we must show that $R(G^*) = 0$, and to complete the proof of non-β-regularity of R we must also prove that $R(D) = 0$. The arguments for these two facts are very similar.

First note that $\beta(D) = \sum_{(ij)} e_{ij}(aZ) = M$, say. For by (4) we have $M \triangleleft D$ and $M^2 = 0$, so that $M \subseteq \beta(D)$, and $D/M = e_{12}\bar{A} + \sum\{e_{ij}A : (ij) \neq (12)\}$, which is semiprime (since A has no nilpotents). Thus $\beta(D) = M$.

Now to show that $R(D) = 0 = R(G^*)$, note that $R \subseteq \beta$, so that $R(D) \subseteq \beta(D) = M$, and $R(G^*) \subseteq \beta(G^*) = aZ$. Now M and aZ are rings with zero product and additive group isomorphic to Z^n for $n = 4,1$ respectively. So $T = R(D)$ or $R(G^*)$ is also of the form Z^m for some m $(0 \leq m \leq 4)$, as an additive subgroup of a finitely-generated torsion-free abelian group. Suppose $T \neq 0$ (i.e. $m \neq 0$). Then by definition of R there is a non-0 (abelian-group) homomorphism $\alpha\colon H^0 \to T$. But H^0 is torsion-free of rank 1 , so α must be a monomorphism. Thus $T \simeq Z^m$ has a subgroup isomorphic to H^0 . This is false, so $T=0$ and the proof is complete.

We conclude this section by showing that an example like that of Theorem 4 is impossible for algebras over a field.

LEMMA 4. *Suppose* R *is a radical in a universe* $\tilde{U}$ *of algebras over a field* K *, and* $K^0 \in \tilde{U}$ *. Then the following are equivalent:* (1) $R \geq \beta$; (2) $R(A)^2 \neq R(A)$ *for some* $A \in \tilde{U}$; (3) $R(K^0) = K^0$.

Sketch proof. Clearly $1 \to 3 \to 2$. If 2 holds then $R(X) = X$ for $X = R(A)/R(A)^2$, and $0 \neq X = (X^+)^0$. Pick $0 \neq u \in X$, and let Y be a complementary subspace for Ku in X^+ . Then $X/Y^0 \simeq K^0$, giving 3. The same argument shows that if 3 holds then $R(X) = X$ for all $X \in \tilde{U}$ with $X^2 = 0$, whence 1 follows.

PROPOSITION 4. *In a universe of algebras over a field all radicals are* β*-regular.*

Proof. Suppose R is a radical in the universe $\tilde{U}$

of algebras over the field K . Let $L \triangleleft^{l} A \in \tilde{U}$, and set $M = AR(L) + R(L)$. We must show that $\beta(M/R(L) = 0$. Since $R(L) \subseteq M \triangleleft L$, it follows (e.g. [2], Prop. 1 on p. 142) that $R(M/R(L)) = 0$. Suppose first that $R \supseteq \beta$. Then $\beta(M/R(L)) \subseteq R(M/R(L)) = 0$. Suppose then that $R \not\supseteq \beta$. Then by Lemma 4 we have $R(L)^2 = R(L)$. So $R(L) \subseteq M =$ $= \{AR(L)\}R(L) + R(L) \subseteq LR(L) + R(L) = R(L)$, giving equality. So $\beta(M/R(L)) = \beta(0) = 0$. This does it.

Note. The example of R , D and N above allows us to give a positive answer to the following question by A. D. Sands (oral communication). Does there exist a ring A with left ideal L , and a radical R such that $R(L)$ is not a quasi-ideal of A ? (Recall that S is a quasi-ideal of A if $SA \cap AS \subseteq S$.)

Let I be the ideal $(e_{11}a - e_{22}a)Z + e_{12}aZ$ of D , and let f be the natural homomorphism of $D \to D/I$. Set $A = f(D)$; $L = f(N)$. Since $N \cap I = 0$, we have $R(L) =$ $= f(R(N)) \simeq R(N)$. Also $e_{11}a \notin R(N) + I$, whence $f(e_{11}a) \notin$ $\notin R(L)$. But $e_{11}a \in DR(N)$ and $e_{22}a \in R(N)D$. Since $f(e_{11}a) = f(e_{22}a)$, it follows that $f(e_{11}a) \in AR(L) \cap R(L)A$, so that $R(L)$ is not a quasi-ideal of A .

III. AN ATOM IN THE "LATTICE" OF ALL RADICALS

An example is constructed in [6] of a simple Jacobson-radical domain whose left ideals are linearly ordered by inclusion. We have this example in mind when a ring D appears in this section, though our assumptions on D will in general be weaker.

LEMMA 5. *Suppose* D *is a domain which is an essential ideal in the ring* A .

(a) *If the principal left ideals of* D *are linearly ordered by inclusion, and* $A \neq D$, *then there exists* $x \in A \setminus D$ *and* $n \in Z$ *such that* $xa = na$ *for all* $a \in A$.

(b) *If* D *is simple and* $0 \neq x \in D$ *and* $n \in Z$ *are such that* $xd = nd$ *for all* $d \in D$, *then* $D = A$.

Proof. (a) Pick $y \in A \setminus D$ and $0 \neq c \in D$. Then $yc \in D$, so $Dyc + Zyc$ and $Dc + Zc$ are principal left ideals of D . By the hypothesis, one of these is contained in the other. If $Dyc + Zyc \subseteq Dc + Zc$ then $yc = dc + nc$ for some $d \in D$, $n \in Z$, so $xc = nc$ for $x = y - d$. If $Dc + Zc \subseteq$ $\subseteq Dyc + Zyc$, then $c = dyc + myc$ for some $d \in D$, $m \in Z$, so $xc = c$ for $x = dy + my$. Thus in either case we have $x \in A$, $n \in Z$ such that $xc = nc$.

Now given $a \in A$ we have $(ax - na)c = a(xc - nc) = 0$, and since $c \neq 0$, Lemma 2a gives $ax = na$ for all $a \in A$. In the first of the two cases above we had $x=y-d$ with $d\in D$, $y\notin D$, giving $x\notin D$. In the second case we had $n=1$, so that in particular $yx=y\notin D$, whence $x\notin D$ since $D\triangleleft A$. Thus in all cases $x\notin D$.

(b) Since $x\neq 0$ and D is a domain, $DxD\neq 0$. But $DxD\triangleleft D$, so since D is simple we have $DxD=D$. In particular there exist $a_i, b_i\in D$ such that $x=\sum a_i x b_i$. Then $x=\sum a_i(xb_i)=\sum a_i(nb_i)=n\sum a_i b_i=ne$, say, where $e\neq 0$ since $x\neq 0$. Now we have $x=ne=xe=(xe)e$, so that $x(e-e^2)=0$. Since $x\neq 0$, this shows that $e^2=e$. Now given $a\in A$ we have $0=e(a-ea)$, whence $a=ea$ since $e\neq 0$ and A is a domain by Lemma 2a. But $e\in D\triangleleft A$, so $a\in D$. Thus $A\subseteq D$, as required.

LEMMA 6. *Let* L_D *be the lower radical generated by the simple ring* D*, and suppose* A *is a subdirectly irreducible ring with* $L_D(A)=A$*. Then the heart of* A *is isomorphic to* D*.*

Proof. By [2, Prop. 5 on p. 116] A has an accessible subring C isomorphic to D. Let n be minimal such that there exist A_i with $C\triangleleft A_n\triangleleft\ldots\triangleleft A_1=A$. If $n>1$ then from $C\triangleleft A_n\triangleleft A_{n-1}$ and $C^2=C$ we have $CA_{n-1}=$ $=C^2A_{n-1}=C(CA_{n-1})\subseteq CA_n\subseteq C$, and similarly $A_{n-1}C\subseteq C$. Thus $C\triangleleft A_{n-1}\triangleleft\ldots\triangleleft A_1=A$, contradicting the choice of n.

So $n=1$ and $C \triangleleft A$. If H is the heart of A then $0 \neq H \triangleleft C$ with simplicity of C yields $H = C \approx D$, as required.

THEOREM 3.* *Suppose* D *is a simple domain in which the principal left ideals are linearly ordered by inclusion. Then the lower radical generated by* D *is an atom in the "lattice" of all radicals.*

Proof. Suppose R is a radical with $0 \neq R \subseteq L_D$. Clearly there exists a ring $T \neq 0$ with $R(T) = T = L_D(T)$. Let A be a subdirectly irreducible homomorphic image of T , so that $R(A) = A = L_D(A)$. By Lemma 5 the heart H of A is isomorphic to D . If $A=H$ then $R(H) = H$ whence $R(D) = D$, so that $L_D \subseteq R$. Thus $L_D = R$, the required conclusion.

Suppose then that $A \neq H$. By Lemma 5a pick $x \in A \setminus H$ and $n \in Z$ such that $ax = na$ for all $a \in A$. Let V be an ideal of A maximal w.r.t. $x \notin V$, and set $A' = A/V$. Then A' is subdirectly irreducible with heart K containing the image x' of x , where $x' \neq 0$. Since A' is a hom. image of A , we have $L_D(A') = A'$. So by Lemma 6 $K \approx D$. In particular K is a simple domain. Also $a'x' = na'$ for all $a' \in A'$, and in particular $kx' = nk$

---

* A similar example has been obtained independently by H. Korolczuk (oral communication).

for all $k \in K$. Since also $x' \in K$, Lemma 5b shows that $K=A'$. Thus $A' \approx D$. But $R(A') = A'$ since A' is an image of A and $R(A) = A$. So $R(D) = D$. Thus once again we have $L_D \subseteq R$, so $L_D = R$, as required.

We thus have a positive solution to the question raised in [2], p. 15: if D is Dubrovin's ring [6] then D satisfies the hypotheses of Theorem 5, so that L_D is an atom, but D does not have a 1 since it is Jacobson radical.

IV. ON THE CONSTRUCTION OF THE LOWER RADICAL IN VARIETIES OF ALTERNATIVE RINGS

Set $D = Q[i]$, where $i^2=-1$, and let p be a prime of the form $4t+3$. Set $A_0 = Z[i]$; $B_k = Z + A_k$; $A_{k+1} = pB_k$. Thus $B_k = Z + ip^k Z$, and $A_k = pZ + ip^k Z$ for $k \geq 1$.

LEMMA 7. (a) *If* $0 \neq I \triangleleft A_k$, *then* A_k/I *is finite.*

(b) *If* $f\colon A_k \to D$ *is a homomorphism, then* $f(A_k) = 0$ *or* A_k .

(c) $A_k \triangleleft R \subseteq D$ *iff* $R=A_i$ *or* B_i *for* $i=k$ *or (in case* $k>0$ *) for* $i=k-1$.

Proof. Parts (a) and (b) were proved in [4, Lemma 1].

(c) Given $x \in R \subseteq D$, since $p \in A_k \triangleleft R$ we have $px^n \in A_k$ for all n , whence $x \in A_0$. Thus $R \subseteq A_0$. Suppose $R \neq A_k$

or A_{k-1} . Then it was proved in [4, Lemma 1] that $1 \in R$. Thus $Z \subseteq R$. So $R = Z + Ti$, where T is an additive subgroup, hence ideal, of Z . Since $A_k \subseteq R$ we have $p^k Z \subseteq T$, so $T = p^u Z$ for some $u \leq k$. Since $p \in A_k \triangleleft R$ we have $p(p^u i) \in A_k$, whence $u+1 \geq k$. From $u \leq k \leq u+1$ we have $R = B_k$ or B_{k+1} , for both of which A_k is obviously an ideal from the defining formulas.

LEMMA 8. *Suppose* B *is a subring of* $Z[i]$ *containing* $mZ + inZ$ *for some* $m,n \in Z \setminus \{0\}$. *Suppose* W *is an alternative ring, and* W_0 *is a meta-ideal isomorphic to* B . *Set* $H = \{w \in W\colon kw \in W_0$ *for some* $0 \neq k = k(w) \in Z\}$. *Then*

(a) $W_0 \subseteq H \triangleleft W$, *and*

(b) *if* $(W,+)$ *is torsion-free, then the given isomorphism* $f\colon W_0 \to B$ *extends to a monomorphism of* H *into* D .

Proof. (a) Write $[a+ib]$ for $f^{-1}(a+ib)$ $(a+ib \in B$; $a,b \in Z)$. Suppose $W_0 \triangleleft W_1 \triangleleft \ldots \triangleleft W_r = W$, where for convenience we may suppose (by introducing repetitions if necessary) that r is of the form $4s+1$. Then $m^r n^r [a+ib] =$
$= m^r n^r f^{-1}(a+ib) = f^{-1}(m^r n^r a + m^r n^r ib) = f^{-1}(\alpha m^r + \beta (in)^r)$ for $\alpha = n^r a \in Z$ and $\beta = m^r b \in Z$, $= \alpha f^{-1}(m^r) + \beta f^{-1}((in)^r) =$
$= \alpha [m]^r + \beta [in]^r$ since m , $in \in B$.

Now suppose $x \in H$ and $y = y_r \in W = W_r$ are given.

There exists $0 \neq k \in Z$ with $kx \in W_0$, say $kx = [a+ib]$. Then $(m^r n^r k)(xy) = (m^r n^r [a+ib])y = \alpha[m]^r y_r + \beta[in]^r y_r$. Since W is alternative, $[m]^r y_r = [m]^{r-1} y_{r-1}$ for $y_{r-1} = [m]y_r \in W_{r-1}$ since $[m] \in W_0 \subseteq W_{r-1} \triangleleft W_r$. Then $[m]^{r-1} y_{r-1} = [m]^{r-2} y_{r-2}$ for $y_{r-2} = [m]y_{r-1} \in W_{r-2}$ since $[m] \in W_{r-2} \triangleleft W_{r-1}$. Continuing in this way we find that for $i=r, r-1, \ldots$ we have $[m]^r y_r = [m]^i y_i$ with $y_i \in W_i$, and the case $i=0$ gives $\alpha[m]^r y_r \in W_0$. Similarly $\beta[in]^r y_r \in$ $\in W_0$. Thus $(m^r n^r k)(xy) \in W_0$, and since $m^r n^r k \neq 0$ it follows that $xy \in H$. Similarly $yx \in H$. Clearly H is additively closed and contains W_0. Thus $W_0 \subseteq H \triangleleft W$.

(b) Extend f to H as follows. Given $w \in H$, pick $0 \neq k \in Z$ with $kw \in W_0$, so that f is defined at kw, and set $\bar{f}(w) = k^{-1} f(kw) \in D$. Then $\bar{f}$ clearly agrees with f on W_0, and it is easily verified that it is well-defined and a homomorphism. It is injective since H is torsion-free; hence it is an isomorphism of H with a subring of D.

THEOREM 4. *Let* $\tilde{V}$ *be any variety of alternative rings such that* $Z \in \tilde{V}$. *Then for each* n *there exists a hom. closed class* $\tilde{L}^{(n)} \subseteq \tilde{V}$ *such that the Kurosh process for constructing the lower radical generated by* $\tilde{L}^{(n)}$ *in* $\tilde{V}$ *terminates at precisely the* n*-th step.*

Proof. We may take $\tilde{L}^{(0)} = \{0\}$ (or $\tilde{V}$), so we suppose in what follows that n has some fixed value ≥ 1. Since $Z \in \tilde{V}$, it follows that $A_k, B_k \in \tilde{V}$ for all k. Let $\tilde{L}^{(n)}$ consist of all homomorphic images of A_n and all rings $R \in \tilde{V}$ such that $(R,+)$ is a torsion group. Recall that the Kurosh chain in $\tilde{V}$ that starts with the (homomorphically closed) class $\tilde{L}^{(n)} = \tilde{L}_0$ is $\{\tilde{L}_k : k=0, 1,2,\ldots\}$, where $R \in \tilde{L}_{k+1}$ iff $R \in \tilde{V}$ and every non-0 hom. image of R has a non-0 ideal in $\tilde{L}_k$.

By the argument of Lemma 2 in [4] we have $A_{n-k} \in \tilde{L}_k \setminus \tilde{L}_{k-1}$ for $k = 1,2,\ldots,n$. Thus to show that the chain $\{L_k\}$ stabilises exactly at $k=n$ it suffices to show that $\tilde{L}_{n+1} \subseteq \tilde{L}_n$; i.e. that every non-0 hom. image of every $W \in \tilde{L}_{n+1}$ has a non-0 ideal in $\tilde{L}_{n-1}$. Since $\tilde{L}_{n+1}$ is hom. closed, it suffices to show that if $0 \neq W \in \tilde{L}_{n+1}$ then W has a non-0 ideal in $\tilde{L}_{n-1}$.

Let W^* be the set of torsion elements in $(W,+)$. Then $W^* \triangleleft W$ and $W^* \in \tilde{L}_0 \subseteq \tilde{L}_{n-1}$. So we are done unless (as we assume from now on) $(W,+)$ is torsion-free. From the definition of the $\tilde{L}_i$ we can find a chain $W_0 \triangleleft W_1 \triangleleft \ldots \ldots \triangleleft W_{n+1} = W$ with $W_k \in \tilde{L}_k$ and $W_0 \neq 0$. Since $W_0 \in \tilde{L}_0$ is torsion-free, it is a hom. image of A_n, so by Lemma 6a $W_0 \approx A_n$.

Thus the hypotheses of Lemma 8 hold, and we define $H \triangleleft W$ as in Lemma 8. Then $W_0 = W_0 \cap H \triangleleft W_1 \cap H \triangleleft \ldots \triangleleft W_{n+1} \cap H = H \triangleleft W$. Let f be as in Lemma 8, and set $H_i = f(W_i \cap H)$. Then $A_n = H_0 \triangleleft H_1 \triangleleft \ldots \triangleleft H_{n+1} = f(H) \subseteq D$. We will show that $H_{n+1} = A_t$ or B_t for some t.

If for some k we have $H_k = B_t$ for some t, then $1\ (\in D)$ is in H_k, whence $B_t = H_k = H_{k+1} = \ldots = H_{n+1}$, and we are done. If not, then by successive applications of Lemma 7c, starting with $H_0 = A_n$, we have $H_1 = A_n$ or A_{n-1}; $H_2 = A_n$ or A_{n-1} or A_{n-2}; ... ; $H_n = A_n$ or A_{n-1} or ... or A_0; $H_{n+1} = A_n$ or ... or A_0. This does it.

If $H_{n+1} = B_t$ for some t, or $H_{n+1} = A_0\ (=B_0)$, then $pH_{n+1} = pB_t = A_{t+1}$. Thus in all cases W has an ideal isomorphic to A_s for some $s \geq 1$; viz. $I = H = f^{-1}(H_{n+1})$ if $H_{n+1} = A_s$, and $I = pH = f^{-1}(pH_{n+1})$ if $H_{n+1} = B_t$ for $t = s-1$, or $H_{n+1} = A_0\ (=B_t$ for $t=0)$. Then $I \simeq A_s \in \tilde{L}_{n-s} \subseteq \tilde{L}_{n-1}$ completes the proof.

A similar construction involving the A_k and B_k allows us to answer a question raised in [1, Theorem 1, Cor. 1]: viz. does there exist a regular class of semiprime (associative) rings whose essential cover does not coincide with its essential closure.* We start with

---

* *Added in proof*. J. F. Watters has also constructed such a class over rings, see [15].

LEMMA 9. *Let* R *be an alternative ring which is an essential extension of* A_t . *Then the identity map on* A_t *extends to an isomorphism of* R *with* A_t *or* B_t *or (for* $t>0$ *)* A_{t-1} *or* B_{t-1} .

Proof. If P, Q are non-0 ideals of R then $P\cap A_t$ and $Q\cap A_t$ are non-0 ideals of A_t , so $(P\cap A_t)(Q\cap A_t)\neq 0$, so $PQ\neq 0$. Thus R is a prime ring. In an alternative ring the right annihilator of an ideal is an ideal, so if $x\in R$ and $A_t x=0$ then $x=0$.

Write (x,y) for the commutator $xy-yx$, and (x,y,z) for the associator $xy\cdot z - x\cdot yz$. We use the following identity valid in alternative rings: $(xy,z,w) = x(y,z,w) + (x,z,w)y - (x,y,(z,w))$. If $a,b,c\in A_t$ and $r\in R$, then since $A_t \triangleleft R$ is associative and commutative we have $0 = (br,a,c) = b(r,a,c) + (b,a,c)r - (b,r,(a,c)) = b(r,a,c)$. So $A_t(r,a,c) = 0$, whence $(r,a,c) = 0$. Thus $(R,A_t,A_t) = 0$. So for $a,c\in A_t$, $r,s\in R$ we have $0 = (cs,r,a) = c(s,r,a) + (c,r,a)s - (c,s,(r,a)) = c(s,r,a)$. So $A_t(s,r,a) = 0$, whence $(s,r,a) = 0$. So $(R,R,A_t) = 0$, or in other words A_t is contained in the nucleus of R .

Since $A_t \triangleleft R$ is commutative we now have, for $a,b\in A_t$, $r\in R$, $0 = (a,br) = b(a,r) + (a,b)r = b(a,r)$. So $A_t(a,r) = 0$ whence $(a,r) = 0$, so that $(A_t,R) = 0$. Thus

A_t is contained in the center of R. Now given $a \in S = A_t \setminus \{0\}$, set $N_a = \{r \in R: ar=0\}$. Since a is central in R we easily see that $N_a \triangleleft R$, and since $N_a \cap A = 0$ we have $N_a = 0$. So S is a multiplicatively closed subset of non-0-divisors of R contained in the center of R. we may therefore form the central localisation RS^{-1}. On identifying the subring $A_t S^{-1}$ with D, we have an embedding ϕ of $R \to RS^{-1}$ which extends the identity embedding of $A_t \to D$.

Since $0 \neq p \in A_t$, given $r \in R$ we have $\phi(r) = \phi(rp)p^{-1} \in A_t S^{-1} = D$. Thus ϕ is an embedding of R into D. We have $A_t \triangleleft \phi(R) \subseteq D$, so by Lemma 7c $R \simeq \phi(R)$ is of the desired type. □

If $\tilde{M}_0$ is a subclass of a universe $\tilde{U}$ of rings, then the *essential cover* of $\tilde{M}_0$ in $\tilde{U}$ is defined to be $\{R \in \tilde{U}$: R is an essential extension of some $A \in \tilde{M}_0\}$. The *essential closure* is the smallest class $\tilde{N}$ such that $\tilde{M}_0 \subseteq \tilde{N} \subseteq \tilde{U}$ and $\tilde{N}$ is its own essential cover (i.e. $\tilde{N}$ is closed under essential extensions). It is clear that $\tilde{N} = \cup_0^\infty \tilde{M}_i$, where $\tilde{M}_{i+1}$ is the essential cover of $\tilde{M}_i$. It is this process of taking successive essential covers that is referred to in our final theorem.

THEOREM 5. *For each integer* $n \geq 0$ *there exists a*

regular class $\tilde{M}^{(n)}$ *of prime associative rings such that the process for constructing the essential closure of* $\tilde{M}^{(n)}$ *in the universe* $\tilde{U}$ *of all alternative (or all associative) rings terminates at precisely the* n*-th step.*

Proof. Let $\tilde{F}$ be the class of all finite fields, and for the given $n\geq 0$ let $\tilde{M}^{(n)} = \tilde{F}\cup\tilde{S}_0$, where $\tilde{S}_0$ is the closure under isomorphism of $\{A_n, B_n\}$. Since $A_0 = Z[i]$ is a subdirect product of finite fields, so also are all its non-0 subrings, including A_n and B_n . Thus $\tilde{M}^{(n)}$ is a regular class of prime associative rings. Let $\{\tilde{M}_k: k=0,1,...\}$ be the "essential cover chain" in $\tilde{U}$ starting with $\tilde{M}_0 = \tilde{M}^{(n)}$. We show that it stabilises precisely at $k=n$.

Let $\tilde{S}_k$ be the closure under isomorphism of $\{A_i, B_i: n \geq i \geq n-k\}$ for $0\leq k\leq n$, and $\tilde{S}_k=\tilde{S}_n$ for $k>n$. Then the chain $\{\tilde{S}_k\cup\tilde{F}: k=0,1,...\}$ clearly stabilises precisely at $k=n$, so that it suffices to show that $\tilde{M}_k = \tilde{S}_k\cup\tilde{F}$ for all k . Since the $S_k\cup\tilde{F}$ are all classes of associative rings, we will then have shown that the chain $\{\tilde{M}_k\}$ is the same, and hence stabilises at the same point, whether the construction takes place in $\tilde{U}$ or in the smaller universe of all associative rings.

For $k=0$ we have $\tilde{M}_k = \tilde{S}_k\cup\tilde{F}$ by definition. Suppose we have it for k (≥ 0), and $R \in \tilde{M}_{k+1}$. Then R is an essential extension of some $S \in \tilde{M}_k = \tilde{S}_k \cup \tilde{F}$. If S has a 1 then $R = S \in \tilde{S}_k \cup \tilde{F} \subseteq \tilde{S}_{k+1} \cup \tilde{F}$. If not, then $S \simeq A_{n-i}$ for some i with $0 \leq i \leq k$, whence by Lemma 9 $R \simeq A_{n-i}$ or B_{n-i} or (for $i<n$) A_{n-i-1} or B_{n-i-1}. Thus $R \in \tilde{S}_{k+1}$. So we have shown that $\tilde{M}_{k+1} \subseteq \tilde{S}_{k+1} \cup \tilde{F}$. The opposite inclusion is trivial, and completes the induction, and thus the proof.

Note. Arguments like those in Lemma 9 and Lemma 2b easily show that if $\tilde{M}$ is a class of prime rings in a universe $\tilde{U}$ then the essential closure $\tilde{C}$ of $\tilde{M}$ in $\tilde{U}$ is also a class of prime rings. If furthermore $\tilde{U}$ is the universe of all alternative rings, and all the rings in $\tilde{M}$ are of any one of the following types, then all the rings in $\tilde{C}$ are of the same type: (1) associative; (2) associative domain; (3) associative and commutative; (4) associative and commutative domains.

In each such case the "essential cover chain" is the same whether it is constructed in the large universe $\tilde{U}$ or in the smaller universe of all rings of the specified type, or in any intermediate universe.

Acknowledgement. The author expresses his thanks to the referee and translator for several valuable suggestions which simplified some proofs, corrected some errors and improved the presentation of the results, and also for his work in translating the paper.

REFERENCES

[1] T. Anderson, R. Wiegandt, On essentially closed classes of rings, *Annales Univ. Sci. Budapest. Sect. Math.* 24 (1981), 107-111.

[2] V. A. Andrunakievič, Ju. M. Rjabuhin, *Radicals of algebras and structural theory*, Nauka, Moscow, 1979 (Russian).

[3] K. I. Beidar, K. Salavova, On the lattices of N-radicals, left strict radicals, and left hereditary radicals, *Acta Math. Acad. Sci. Hungar.* to appear (Russian).

[4] K. I. Beidar, A chain of Kurosh may have an arbitrary finite length, *Czechoslovak Math. J.* 32 (1982), 418-422.

[5] N. Divinsky, *Rings and radicals*, Allen & Unwin, London, 1965.

[6] N. I. Dubrovin, Chain domains, *Vestnik Moskov. Univ. Ser. Mat. Meh.*, 1980, no. 2, 51-54 (Russian).

[7] C. Faith, *Algebra: Rings, modules and categories I*, Springer, Berlin - New York, 1973.

[8] A. E. Hoffman, W. G. Leavitt, Properties inherited by the lower radical, *Portugal. Math.* 27 (1968), 63-66.

[9] A. Jaegermann, A. D. Sands, On normal radicals, N-radicals and A-radicals, *J. Algebra* 50 (1978), 337-349.

[10] A. D. Sands, On relations among radical properties, *Glasgow Math. J.* 18 (1977), 17-23.

[11] A. D. Sands, On normal radicals, *J. London Math. Soc.* (2) 11 (1975), 361-365.

[12] P. N. Stewart, R. Wiegandt, Quasi-ideals and bi-ideals in radical theory, *Acta Math. Acad. Sci. Hungar.* 39 (1982), 289-294.

[13] A. Suliński, T. Anderson, N. Divinsky, Lower radical properties for associative and alternative rings, *J. London Math. Soc.* 41 (1966), 417-422.

[14] F. Szász, *Radikale der Ringe*, VEB Deutscher Verlag der Wissenschaften, Berlin, 1975.

[15] J. F. Watters, Essential cover and closure, *Annales Univ. Sci. Budapest. Sect. Math.* 25 (1982), 279-280.

K. I. Beĭdar
USSR
Moskva 117192
Vinnickaya 9, kv. 16

COLLOQUIA MATHEMATICA SOCIETATIS JÁNOS BOLYAI
38. RADICAL THEORY, EGER (HUNGARY), 1982.

SUPERNILPOTENT RADICALS AND HEREDITARINESS OF SEMI-SIMPLE CLASSES OF NEAR-RINGS

G. BETSCH and K. KAARLI

The present note continues the investigations in the first author's joint paper with R. Wiegandt [1] . Its purpose is to provide further information on hereditariness of semisimple classes of near-rings. Our main result is that any non-trivial near-ring radical having a hereditary semisimple class is supernilpotent.

The authors are very grateful to T. Anderson and R. Wiegandt for valuable suggestions and remarks.

1. Preliminaries. Notation

We follow the terminology and notation of [1] and [2]. For the sake of convenience we mention the following concepts. All near-rings considered in this paper will be <u>right</u> near-rings:

$N = (N, +, .)$ will be called a *near-ring* iff it satisfies the axioms

(N 1) $N^+ := (N, +)$ is a group (not necessarily commutative) with neutral element 0 ;

(N 2) $a(bc) = (ab)c$

(N 3) $(a + b)c = ac + bc$ for all $a, b, c \in N$.

Throughout this paper N will denote a near-ring. We introduce

$N_o := \{n \in N \mid n0 = 0\}$,

$N_c := \{n \in N \mid n0 = n\}$, and we have

$$N = N_o + N_c = N_c + N_o .$$

N is called *0-symmetric* iff $N = N_o$ (iff $\forall\, n \in N: n0 = 0$),

constant iff $N = N_c$ (iff $\forall\, n \in N: n0 = n$ iff $\forall\, a, b \in N: ab = a$),

a zero near-ring iff $\forall\, a, b \in N: ab = 0$.

On an additive group A, we may build (at least) three trivial near-rings A^o, A^c, and $A^{(c)}$ in the following way:

- A is the common additive group of A^o, A^c, and $A^{(c)}$;
- multiplication in A^o is defined by $\forall\, a, b \in A: ab = 0$;

in A^c is defined by $\forall a,b \in A: ab = a$;

in $A^{(c)}$ is defined by

$$\forall a \in A \; \forall 0 \neq b \in A: ab = a, a0 = 0.$$

Obviously, A^o is a zero near-ring,

A^c is a constant near-ring,

$A^{(c)}$ is a 0-symmetric near-ring, and

$A^{(c)} \cdot A^{(c)} = A^{(c)}$.

Consequently, if $A \neq 0$, then $A^{(c)}$ is not nilpotent. This fact will be used later.

The construction given here applies, in particular, to the additive group N^+ of a near-ring. The resulting near-rings will be denoted by N^o, N^c, and $N^{(c)}$, respectively. Obviously, N is a zero near-ring iff $N = N^o$, and N is constant iff $N = N^c$.

I is called an *ideal* of N ($I \triangleleft N$) iff

(I 1) (I, +) is a normal subgroup of N^+ ;

(I 2) $IN = \{ in \mid i \in I, n \in N \} \subseteq I$;

(I 3) $\forall u_1, u_2, v \in N$: $u_1 \equiv u_2$ mod I implies $vu_1 \equiv vu_2$ mod I .

A class $\mathbb{X}$ of near-rings is called *hereditary* iff it is hereditary with respect to ideals (i.e. $I \triangleleft N \in \mathbb{X}$ implies $I \in \mathbb{X}$).

A class $\mathbb{R}$ of near-rings is a *radical (radical class)* (in the sense of Kurosh-Amitsur) iff

(R 1) $\mathbb{R}$ is homomorphically closed;

(R 2) for every near-ring N:

$\mathbb{R}(N) := \sum_{\alpha}\{I_{\alpha} \triangleleft N \mid I_{\alpha} \in \mathbb{R}\} \in \mathbb{R}$;

(R 3) $\mathbb{R}$ is closed under extensions, i.e., if $I \triangleleft N, I \in \mathbb{R}$ and $N/I \in \mathbb{R}$ then $N \in \mathbb{R}$.

$\mathbb{R}$ will always denote a (Kurosh-Amitsur) radical class of near-rings. $\mathbb{R}$ is called *supernilpotent* iff $\mathbb{R}$ contains the class $\mathbb{Z}$ of all zero near-rings. We denote $\mathcal{S}\mathbb{R} := \{N \mid \mathbb{R}(N) = 0\}$, the *semisimple class* of $\mathbb{R}$.

As in the ring case one can prove:

$\mathbb{R}$ is hereditary iff $I \triangleleft N$ implies $\mathbb{R}(I) \supseteq \mathbb{R}(N) \cap I$:

$\mathcal{S}\mathbb{R}$ is hereditary iff $I \triangleleft N$ implies $\mathbb{R}(I) \subseteq \mathbb{R}(N) \cap I$

(iff $I \triangleleft N$ implies $\mathbb{R}(I) \subseteq \mathbb{R}(N)$).

The following remark is often useful. It is a consequence of the "intersection representation" of the radical $\mathbb{R}(N)$ (cf. [5] , Th. 8.5), but we present a simple direct proof.

Remark 1.1 Let $\mathbb{R}$ be a radical. Then $K \triangleleft N$ and

$N/K \in \mathcal{S}\mathbb{R}$ imply $\mathbb{R}(N) \subseteq \mathbb{R}(K)$.

Proof. $\mathbb{R}(N) + K/K \triangleleft N/K \in \mathcal{S}\mathbb{R}$. From (R 1) we infer

$$\mathbb{R}(N) + K/_K \simeq \mathbb{R}(N)/_{\mathbb{R}(N) \cap K} \in \mathbb{R}.$$

Since $^N/K \in \mathcal{S}\mathbb{R}$ the near-ring $^N/K$ has no nonzero ideal in $\mathbb{R}$; hence $\mathbb{R}(N) + K \subseteq K$, that is $\mathbb{R}(N) \triangleleft K$, which implies $\mathbb{R}(N) \triangleleft R(K)$.

We add two more remarks.

Remark 1.2. Tangeman-Kreiling [4] presented a construction for the lower radical $\mathcal{L}\mathbb{C}$ determined by a class $\mathbb{C}$ of not necessarily associative rings. This construction carries over verbatim to near-rings. If $\mathbb{C}$ is hereditary, then so is $\mathcal{L}\mathbb{C}$.

Remark 1.3. Saxena-Bhandari [3] claim that if $\mathbb{Z}$ is the class of all zero near-rings, then $\mathcal{L}\mathbb{Z}$ is the prime radical for near-rings (see [2] for definition), and since $\mathbb{Z}$ is hereditary, the prime radical $\mathcal{L}\mathbb{Z}$ must be hereditary, too. Unfortunately, there are certain gaps in the arguments of Saxena-Bhandari, cf. the review in Zbl. der Mathematik by the first author.

2. Zero near-rings in a semisimple class.

The proof of the following result is somewhat related to the arguments in [1].

THEOREM 2.1. *Let* $\mathbb{R}$ *be a radical class with hereditary semisimple class* $\mathcal{S}\mathbb{R}$ *, and assume that* $\mathcal{S}\mathbb{R}$ *contains a zero near-ring* $N = N^{o}$*, where* $N \neq 0$. *Then* $\mathbb{Z} \subseteq \mathcal{S}\mathbb{R}$ *, i.e.* $\mathcal{S}\mathbb{R}$ *contains* *__all__* *zero near-rings.*

Proof. Let M be an arbitrary additive group.
On $\Sigma^{+} = N^{+} \oplus N^{+} \oplus M$ we define a multiplication by

$$(a_1, b_1, m_1)(a_2, b_2, m_2) := \begin{cases} (b_1, 0, 0), & \text{if } m_2 \neq 0 \\ (0, 0, 0), & \text{if } m_2 = 0. \end{cases}$$

We obtain the structure of a 0-symmetric near-ring Σ with additive group Σ^{+}. Now we denote

$K := \{(a, 0, m) \mid a \in N \text{ and } m \in M\}$,

$L := \{(0, 0, m) \mid m \in M\}$.

Then obviously $K \triangleleft \Sigma$ and $\Sigma/K \simeq N$.
Furthermore, $L \simeq M^{o}$. We prove that $L \in \mathcal{S}\mathbb{R}$, which yields $M^{o} \in \mathcal{S}\mathbb{R}$.
Since $\Sigma/K \simeq N \in \mathcal{S}\mathbb{R}$, we infer from 1.1 that $\mathbb{R}(\Sigma) \subseteq \mathbb{R}(K)$.
Now, $L \triangleleft K$ and $K/L \simeq N \in \mathcal{S}\mathbb{R}$. Again by 1.1, we conclude that $\mathbb{R}(K) \subseteq \mathbb{R}(L)$. Hence $\mathbb{R}(\Sigma) \subseteq \mathbb{R}(L) \subseteq L$.
Suppose that $\mathbb{R}(\Sigma) \neq 0$. Then there exists $0 \neq m \in M$ such that $(0, 0, m) \in \mathbb{R}(\Sigma)$. Let $0 \neq n \in N$. Then $(0, n, 0)\,(0, 0, m) = (n, 0, 0) \notin L$. But this is

impossible, since Σ is 0-symmetric, $\mathbb{R}(\Sigma) \triangleleft \Sigma$,

and $\mathbb{R}(\Sigma) \subseteq L$.

Hence $\mathbb{R}(\Sigma) = 0$, $\Sigma \in \mathcal{S}\mathbb{R}$, and from the hereditariness of $\mathcal{S}\mathbb{R}$ we infer $L \in \mathcal{S}\mathbb{R}$, which completes the proof.

COROLLARY. *Let* $\mathbb{R}$ *be a radical class with hereditary semisimple class* $\mathcal{S}\mathbb{R}$. *Then*

either $\mathbb{Z} \subseteq \mathbb{R}$ *(i.e.* $\mathbb{R}$ *is supernilpotent)*

or $\mathbb{Z} \cap \mathbb{R} = \mathbb{O}$ *= class of all one-element near-rings.*

3. (Non)hereditary semisimple classes of near-rings.

Throughout this section, let N denote an (arbitrary) near-ring, and let N^* be any near-ring having the same additive group N^+ as N. The multiplication in N^* will be denoted by $*$. The case $N = N^*$ (i.e., $*$ is the multiplication in N) is not excluded. In fact, certain results in [1] will follow from our present results by assuming $N = N^*$.

We define a multiplication on $N^+ \oplus N^+ \oplus N^+$ by

(3.1) $(a_1,b_1,c_1)(a_2,b_2,c_2) := (a_1 * a_2, c_1c_2a_2 - c_1 0, 0)$.

It is easy to check that this multiplication is associative. What about right distributivity?

$$(a_1+a_2,b_1+b_2,c_1+c_2)(a_3,b_3,c_3)$$
$$= (a_1*a_3+a_2*a_3,c_1c_3a_3+c_2c_3a_3-c_20-c_10,0) \ ,$$
$$(3.2)\quad (a_1,b_1,c_1)(a_3,b_3,c_3) + (a_2,b_2,c_2)(a_3,b_3,c_3)$$
$$= (a_1*a_3+a_2*a_3,c_1c_3a_3-c_10+c_2c_3a_3-c_20,0).$$

Notation.

$[x,y] := -x-y+x+y$ for $x,y \in N$,

$[X,Y] := \{[x,y] \mid x \in X,\ y \in Y\}$ for $X,Y \subseteq N$.

It is clear from (3.2) that the multiplication (3.1) is right distributive, provided $[N_c,N_c] = [N^3,N_c] = 0$. In particular, this multiplication is right distributive if N is 0-symmetric or abelian.

Finally, we put $K := \{(a,b,0) \mid a,b \in N\}$,

$N_1 := \{(a,0,0) \mid a \in N\}$.

Now we have the following

PROPOSITION 3.1. *Let* N *and* N^* *be near-rings with the same additive group, and assume* $[N_c,N_c] = [N^3,N_c] = 0$. *Then*

a) *The multiplication defined by* (3.1) *yields a near-ring* $\Phi^*(N)$ *with additive group* $N^+ \oplus N^+ \oplus N^+$. *If* N *and* N^* *are 0-symmetric, then so is* $\Phi^*(N)$.

b) $K \lhd \Phi^*(N)$, $K \cong N^* \oplus N^o$, *and* $\Phi^*(N)/K \cong N^o$.

c) $N_1 \lhd K$ *and* $N_1 \cong N^*$.

The proof follows by easy computation.

PROPOSITION 3.2. *Let* $\mathbb{R}$ *be a radical with hereditary semisimple class* $\mathcal{S}\mathbb{R}$, *and let* N *be a near-ring satisfying* $[N_c,N_c] = [N^3,N_c] = 0$. *If* $N^* \in \mathbb{R}$ and $N^o \in \mathcal{S}\mathbb{R}$, *then* $R(\Phi^*(N)) = N_1$, $xyz = xy0$ *for all* $x,y \in N$, *and* $N^3 = N^2 0 = N_c$.

Proof. Firstly, from $K \simeq N^* \oplus N^o$ and our assumptions we infer $\mathbb{R}(K) = N_1$. Furthermore, $K \triangleleft \Phi^*(N)$, and since $\mathcal{S}\mathbb{R}$ is hereditary, we have $\mathbb{R}(K) \subseteq R(\Phi^*(N))$. On the other hand, $\Phi^*(N)/K \simeq N^o \in \mathcal{S}\mathbb{R}$, yielding $\mathbb{R}(\Phi^*(N)) \subseteq \mathbb{R}(K)$ by 1.1. Hence $N_1 = \mathbb{R}(K) = \mathbb{R}(\Phi^*(N))$, and $N_1 \triangleleft \Phi^*(N)$. Now let $x,y,z \in N$. Then $(z,0,y) \equiv (0,0,y) \bmod N_1$, which implies

$$(0,\ xyz - x0,\ 0) = (0,\ 0,\ x)(z,\ 0,\ y) \equiv (0,0,x)(0,0,y)$$
$$= (0,\ xy0 - x0,\ 0) \bmod N_1.$$

Hence $xyz = xy0$ and $N^3 = N^2 0 \subseteq N_c$. The inclusion $N_c \subseteq N^2 0$ is clear, for $n \in N_c$ implies $n = n0 = n00 \in N^2 0$. The proof is now complete.

COROLLARY. *Let* $\mathbb{R}$ *be a radical with hereditary semisimple class* $\mathcal{S}\mathbb{R}$, *and let* N *be a 0-symmetric near-ring. If* $N^* \in \mathbb{R}$ *and* $N^o \in \mathcal{S}\mathbb{R}$, *then* $N^3 = 0$.

We are now in a position to prove our main result.

THEOREM 3.3. *Let* $\mathbb{R}$ *be a radical for near-rings*

with hereditary semisimple class $\mathcal{S}\mathbb{R}$.
Then either $\mathbb{R}$ *is supernilpotent* ($\mathbb{Z} \subseteq \mathbb{R}$) *or*
$\mathbb{R} = \mathbb{O}$ = *class of all one-element near-rings.*

Proof. Suppose that $\mathbb{R}$ contains a near-ring $N^* \neq 0$ and assume that $\mathbb{R}$ is not supernilpotent. Then there is a zero near-ring $M \neq 0$ such that $M \notin \mathbb{R}$. Now $0 \neq {}^M/\mathbb{R}(M)$ is a zero near-ring in $\mathcal{S}\mathbb{R}$, hence $\mathbb{Z} \subseteq \mathcal{S}\mathbb{R}$ by Theorem 2.1; in particular $N^o \in \mathcal{S}\mathbb{R}$. Finally, take $N = N^{(c)}$ and apply the Corollary to Proposition 3.2. We get $(N^{(c)})^3 = 0$, which contradicts the fact mentioned in section 1. The theorem is proven.
The following result is a simple but useful restatement of Theorem 3.3.
COROLLARY. *If a radical* $\mathbb{R}$ *is not supernilpotent although it contains a near-ring* $N \neq 0$, *then* $\mathcal{S}\mathbb{R}$ *is not hereditary.*

In [1] a radical $\mathbb{R}$ was called *subidempotent* iff $\mathbb{R} \subseteq \mathcal{U}\mathbb{Z}$:= { N | N has no nonzero homomorphic image in the class $\mathbb{Z}$ of all zero near-rings}.
Obviously, subidempotent radicals cannot be supernilpotent. Hence, as an important consequence of Theorem 3.3 we have the following result, which

improves the Corollary to Theorem 2 in [1].

THEOREM 3.4. *A subidempotent radical with hereditary semisimple class is precisely the class* $\mathbb{O}$ *of one-element near-rings.*

We conclude with the remark that the Tangeman-Kreiling construction [4], which we mentioned in section 1 , provides a tool to construct an abundancy of subidempotent radicals, hence also of near-ring radicals with non-hereditary semisimple classes.

Let $\mathbb{B}$ be a homomorphically closed class of near-rings which satisfies the conditions (i) $\mathbb{B} \neq \mathbb{O}$,

(ii) $\mathbb{B} \cap \mathbb{Z} = \mathbb{O}$.

Then the lower radical $\mathcal{L}\mathbb{B}$ of $\mathbb{B}$ must be subidempotent, hence $\mathcal{S}\mathcal{L}\mathbb{B}$ is not hereditary in view of Theorem 3.4. -

References

[1] G. Betsch and R. Wiegandt, *Non-hereditary Semisimple Classes of near-rings*, Studia Sci. Math. Hungar. (to appear).

[2] G. Pilz, *Near-rings*, North-Holland, 1977.

[3] P. K. Saxena and M. C. Bhandari, *General radical theory for near-rings*, Tamkang J. Math. 12 (1981), 91-97.

[4] R. Tangeman and D. Kreiling, *Lower radicals in nonassociative rings*. J. Austral. Math. Soc. 24 (1972), 419-423.

[5] R. Wiegandt, *Radical and semisimple classes of rings*, Queen's papers in pure and appl. math., No. 37, Kingston, Ontario, 1974.

G. Betsch
Mathematisches Institut der Universität
Auf der Morgenstelle 10
D 7400 Tübingen 1
Federal Republic of Germany

K. Kaarli
Department of Algebra and Geometry
Tartu State University
Tartu, Estonia, USSR

COLLOQUIA MATHEMATICA SOCIETATIS JÁNOS BOLYAI

38. RADICAL THEORY, EGER (HUNGARY), 1982.

SIMPLE GRINGS AND INVARIANT RADICALS

N. DIVINSKY, A. SULIŃSKI and T. ANDERSON

Let $\bar{G}$ be an abelian group and suppose it is a direct sum of subgroups G_0 and G_1 so that $\bar{G} = G_0 + G_1$. Let G be the set of elements of $\bar{G}$ which belong to either G_0 or G_1. Then G does not contain any sum of the form $x + y$ where x is in G_0 and y is in G_1 and where $x \neq 0$ and $y \neq 0$. Thus G is not a group because it is not closed under addition. We call the elements of G homogeneous; we denote G as (G_0, G_1); and we say that G is a 2-graded abelian group.

The nonzero elements of G_0 are said to be homogeneous of degree 0 and the nonzero elements of G_1 are homogeneous of degree 1. We shall say that the 0 element of G has no degree defined for it.

Suppose that an associative multiplication is

is defined in G such that $G_0G_0 \subseteq G_0$; $G_1G_1 \subseteq G_0$; $G_0G_1 \subseteq G_1$ and $G_1G_0 \subseteq G_1$. And suppose further that this multiplication is both right and left distributive with respect to the addition of elements of the same degree. Then G is a 2-graded ring or just graded ring. We shall call such a structure an associative gring, or just a gring.

Thus if (R_0, R_1) is a gring then R_0 is an ordinary associative ring and R_1 is an R_0 bimodule with the property that $R_1R_1 \subseteq R_0$. Of course in general, (R_0, R_1) itself is not a ring because it is not closed under addition. The notion of grings generalizes the notion of rings because every ring R_0 is also a gring, namely $(R_0, 0)$.

The motivation for studying graded structures came originally from physics when Corwin, Ne'eman and Sternberg studied Graded Lie Algebras in 1975 (see [1]). Kaplansky classified simple graded Lie algebras in 1977 and Djokovic also studied Lie algebras in 1976.

Examples:

1. Let R_0 = real numbers, let R_1 = pure complex numbers = $\{b\sqrt{-1}\}$ where b is real. Then $(R_0, R_1) \subsetneq$ complex numbers, and it is a gring.

2. Let R_0 = integers, let $R_1 = \{n\sqrt{-1}\}$ where n is an integer. Then (R_0, R_1) is a gring.

3. Let R_0 be a finite field $= \{\bar{0}, \bar{1}, \bar{2}, \cdots, \overline{p-1}\}$ where p is a prime. Let R_1 be the integers. Define $x_1 \cdot y_1$ as a product modulo p and then it can be thought of as being in R_0 , where x_1 and y_1 are in R_1 . If x_0 is in R_0 define $x_0 x_1$ as an ordinary integer product and thus it is in R_1 . However this is not associative because if $p = 5$ then take 2, 3 and 5 all in R_1 . Then $(2.3)\,5 = \bar{1}.5 = 5$ in R_1 whereas $2(3.5) = 2.\bar{0}$ $= 0$ in R_1 . Therefore (R_0, R_1) is not a gring.

4. Let $R_0 = \left\{\begin{pmatrix} 0 & 0 & a \\ 0 & 0 & 0 \\ 0 & 0 & 0 \end{pmatrix}\right\}$ where a is real, and let

$R_1 = \left\{\begin{pmatrix} 0 & b & 0 \\ 0 & 0 & c \\ 0 & 0 & 0 \end{pmatrix}\right\}$ where b and c are real. Then R_0 is a zero ring and (R_0, R_1) is a gring.

5. Let $R_0 = \{\sum c_\alpha x_\alpha\}$, $R_1 = \{\sum k_\gamma y_\gamma\}$ where the c_α's and the k_γ's are real and where $x_\alpha x_\beta = x_{\alpha+\beta}$ or 0 if $\alpha + \beta \geqq 1$, where $0 < \alpha < 1$, $0 < \beta < 1$ and $0 < \gamma < 1$.

Also $x_\alpha y_\gamma = y_{\alpha+\gamma}$ or 0 if $\alpha + \gamma \geqq 1$ and $y_\alpha y_\beta = x_{\alpha+\beta}$ or 0 if $\alpha + \beta \geqq 1$. Then (R_0, R_1) is a gring.

6. Let $R_0 = \left\{\begin{pmatrix} a & b & 0 \\ c & d & 0 \\ 0 & 0 & e \end{pmatrix}\right\}$ and let $R_1 = \left\{\begin{pmatrix} 0 & 0 & f \\ 0 & 0 & g \\ h & i & 0 \end{pmatrix}\right\}$

where a to i are real. Again (R_0,R_1) is a gring.

7. Let R_0 be any ring and think of it as being painted red. Let $R_1 = R_0$ only think of it as being painted blue. Define red·red = red, red·blue = blue, blue·red = blue and blue·blue = red. Then if x_r and y_r are in R_0 then $x_r y_r = (xy)_r$ in R_0. Now x_b and y_b are the same two elements, only in R_1. Then $x_b y_b = (xy)_r = x_r y_r$. And $x_r y_b = (xy)_b$ in R_1. Then (R_0,R_1) is a gring. This example is an extension of example 5 where $R_0 = \{\sum c_\alpha x_\alpha\} \cong R_1$.

A subset of (A,B) of a gring (R_0,R_1) is said to be a *gring ideal* of (R_0,R_1) if A is an additive subgroup of R_0 and B is an additive subgroup of R_1 and:

$$AR_0 \subseteq A,\ AR_1 \subseteq B,\ BR_0 \subseteq B,\ BR_1 \subseteq A \text{ and}$$

$$R_0A \subseteq A,\ R_1A \subseteq B,\ R_0B \subseteq B,\ R_1B \subseteq A .$$

This essentially means that R_0 leaves both A and B alone, in that R_0 sends A into A and B into B. On the other hand R_1 interchanges them.

We note that it then follows that $R_1AR_1 \subseteq A$, and that $R_1BR_1 \subseteq B$.

It is clear that if (A,B) is a gring ideal of (R_0,R_1) then A is an ideal of the ring R_0. On the other hand if we begin with an ideal I_0 of a ring R_0 and assume that R_0 is part of a gring (R_0,R_1) then it is not clear whether there is a gring ideal with I_0 in the first component i.e. (I_0,I_1).

We shall call a nonzero ideal I_0 of a ring R_0 in a gring (R_0,R_1), *special*, if $R_1I_0R_1 \subseteq I_0$. In a commutative gring, every nonzero ideal of R_0 is special.

LEMMA 1. If I_0 is a special ideal of R_0 in (R_0,R_1), let $I_1 .\equiv. I_0R_1 + R_1I_0$. Then (I_0,I_1) is a nonzero gring ideal of (R_0,R_1).

Proof: Since I_0 is an ideal of R_0, both R_0I_0 and $I_0R_0 \subseteq I_0$. By definition both I_0R_1 and $R_1I_0 \subseteq I_1$. Then $I_1R_0 = I_0R_1R_0 + R_1I_0R_0 \subseteq I_0R_1 + R_1I_0 \subseteq I_1$. Similarly $R_0I_1 \subseteq I_1$. Finally $I_1R_1 = I_0R_1R_1 + R_1I_0R_1 \subseteq I_0R_0 + R_1I_0R_1 \subseteq I_0$ since I_0 is special. And similarly $R_1I_1 \subseteq I_0$.

Remarks: For any special ideal I_0, $I_1 = I_0R_1 + R_1I_0$ is the smallest partner of I_0 so that (I_0,I_1) is a gring ideal.

If we begin with a nonzero ideal I_0 of R_0

which is not special, then we can take $\overline{I_0}$.≡. I_0 + $R_1I_0R_1$ and it is easy to see that $\overline{I_0}$ *is* a special ideal of R_0 . It is the special ideal generated by I_0 .

The ring R_0 might be simple in which case it has no proper special ideals. However it is possible for R_0 to have proper nonzero ideals and yet have no proper special ideals. When R_0 has no proper, special ideals we will say that R_0 is *minisimple*. This depends on the gring (R_0,R_1) of course and R_0 may be minisimple in one gring and not in another.

For any ideal I_0 , $I_0R_1 + R_1I_0$ is the same as $\overline{I_0}R_1 + R_1\overline{I_0} = (I_0 + R_1\ {}_0R_1)R_1 + R_1(I_0 + R_1I_0R_1) = I_0R_1 +$ $R_1I_0 +$ parts contained in I_0R_1 and in R_1I_0 and this is then equal to $I_0R_1 + R_1I_0$.

Later on we shall consider ideals I_0 of R_0 which are special in any gring (R_0,R_1) which has R_0 in the first component. Such an ideal will be called *invariant*. In other words I_0 is an invariant ideal of R_0 if $R_1I_0R_1 \subseteq I_0$ for any R_1 such that (R_0,R_1) is a gring.

We shall call a nonzero subgroup I_1 of R_1 *special* if $I_1R_0 + R_0I_1 + R_1I_1R_1 \subseteq I_1$. This definition depends on the gring (R_0, R_1). We can call I_1 *invariant* if it is special for every gring which has R_1 in the second component. We shall say that R_1 is *minisimple* if it has no proper special subgroups.

LEMMA 2. If I_1 is a special subgroup of R_1, we define $I_0 \;.\!\equiv\!.\; I_1R_1 + R_1I_1$. Then I_0 is special and (I_0, I_1) is a nonzero gring ideal of (R_0, R_1).

Proof: $I_0R_0 = I_1R_1R_0 + R_1I_1R_0 \subseteq I_1R_1 + R_1I_1 \subseteq I_0$ and $R_0I_0 = R_0I_1R_1 + R_0R_1I_1 \subseteq I_1R_1 + R_1I_1 \subseteq I_0$ and $R_0I_0 = R_0I_1R_1 + R_0R_1I_1 \subseteq I_1R_1 + R_1I_1 \subseteq I_0$. Then $R_1I_0R_1$ $= R_1I_1R_1R_1 + R_1R_1I_1R_1 \subseteq R_1I_1R_0 + R_0I_1R_1 \subseteq R_1I_1 + I_1R_1 \subseteq I_0$. Therefore I_0 is a special ideal of R_0. By definition I_1R_1 and $R_1I_1 \subseteq I_0$. Also I_1R_0 and $R_0I_1 \subseteq I_1$. Finally $I_0R_1 = I_1R_1R_1 + R_1I_1R_1 \subseteq I_1R_0 + I_1 \subseteq I_1$ and similarly $R_1I_0 \subseteq I_1$.

Remark: $I_0R_1 + R_1I_0 \subseteq I_1$. Thus though I_0 is the minimal partner of I_1, it is not clear whether I_1 must be the minimal partner of I_0.

LEMMA 3. If (I_0, I_1) is a gring ideal of (R_0, R_1) with both $I_0 \neq 0$ and $I_1 \neq 0$ then both I_0 and I_1

are special. (This is now obvious).

It may be that $R_1^2 = 0$ but if $R_1^2 \neq 0$ then R_1^2 is a special ideal of R_0. To see this we note that $R_1^2R_0 \subseteq R_1.R_1R_0 \subseteq R_1R_1$ and similarly $R_0R_1^2 \subseteq R_1^2$. Finally $R_1.R_1^2.R_1 \subseteq R_1^3R_1 \subseteq R_1R_1$.

Now if R_0 is minisimple then either $R_1^2 = 0$ or $R_1^2 = R_0$.

We also note that R_1^3 is either 0 or is a special subgroup of R_1 because $R_1^3R_0 + R_0R_1^3 + R_1R_1^3R_1 \subseteq R_1^3$.

Thus if R_1 is minisimple then either $R_1^3 = 0$ or $R_1^3 = R_1$.

We would like to know what simple grings are like, where a simple gring is one that has no nonzero proper gring ideals. If I_0 is a proper special ideal of R_0 then Lemma 1 tells us that $(I_0, I_0R_1 + R_1I_0)$ is a nonzero, proper gring ideal of (R_0, R_1), and thus (R_0, R_1) is not simple. Thus if (R_0, R_1) is simple then R_0 must be minisimple and similarly by Lemma 2, R_1 must be minisimple.

If (R_0, R_1) is simple with $R_0 \neq 0 \neq R_1$, then neither $(R_0, 0)$ nor $(0, R_1)$ can be gring ideals. Therefore $R_1R_1 \neq 0$ and since R_0 is minisimple we must have $R_1R_1 = R_0$. Furthermore $R_0R_1 \neq 0$ and thus $R_1^3 = 0$, and since R_1 is minisimple, $R_1^3 = R_1$. Thus $R_0R_0 = R_1^4 = R_1^2 = R_0$. Also $R_1^3 = R_1R_0 = R_0R_1 = R_1$.

In the case when $R_1 = 0$, all ideals of R_0 are special and therefore $(R_0,0)$ is simple if and only if R_0 is simple. Similarly in the case when $R_0 = 0$, all subgroups of R_1 are special and therefore $(0,R_1)$ is simple if and only if R_1 has no proper nonzero subgroups. Putting these ideas together we have:

THEOREM 1. (R_0,R_1) is a simple gring if and only if

a) $R_1 = 0$ and R_0 is a simple ring; or

b) $R_0 = 0$ and R_1 is a simple group (abelian of course) i.e. R_1 is a simple zero ring; or

c) $R_1 \neq 0$, $R_0 \neq 0$, R_0 is minisimple and $R_0^2 = R_0 = R_1^2$ and R_1 is minisimple and $R_0R_1 = R_1R_0 = R_1 = R_1^3$.

Proof: If (R_0,R_1) is a simple gring we have already established a), b) and c). Conversely a) and b) both yield simplicity. In the case when $R_0 \neq 0 \neq R_1$ there are nine kinds of possible gring ideals, namely:

$$\begin{array}{ccc} (R_0,0) & (R_0,I_1) & (R_0,R_1) \\ (I_0,0) & (I_0,I_1) & (I_0,R_1) \\ (0,0) & (0,I_1) & (0,R_1) \end{array}$$

where I_0 is a proper nonzero ideal of R_0 and where I_1 is a proper nonzero subgroup of R_1.

If we know that R_0 is minisimple then the middle three $(I_0,0)$, (I_0,I_1) and (I_0,R_1) cannot occur as gring ideals. If we know that R_1 is minisimple then (R_0,I_1) and $(0,I_1)$ cannot occur. If $R_1R_0 \neq 0$ then $(R_0,0)$ is not a gring ideal and finally if $R_1^2 = R_0 \neq 0$ then $(0,R_1)$ is not a gring ideal. Thus these seven cases are out and that leaves only the zero ideal $(0,0)$ and (R_0,R_1) itself, so (R_0,R_1) is a simple gring.

A slightly different version is:

THEOREM 2. If (R_0,R_1) has $R_0 \neq 0$ and $R_1 \neq 0$ then it is a simple gring if and only if R_0 is minisimple, $R_1^2 = R_0 = R_0^2$, $R_1 = R_1^3 = R_0R_1 = R_1R_0$ and R_1 has no total annihilators (i.e. x in R_1 with $xR_1 = R_1x = 0$ implies $x = 0$).

Proof: If (R_0,R_1) is a simple gring then we know that R_0 is minisimple, $R_1^2 = R_0 = R_0^2$, $R_1^3 = R_1 = R_0R_1 = R_1R_0$. We also know that R_1 is minisimple and thus has no proper special subgroups. If there is a nonzero x in

R_1 such that $xR_1 = R_1x = 0$, let I_1 be the subgroup generated by x. Then $I_1 = \{nx\}$. Then I_1 is a special subgroup of R_1 because $I_1R_0 + R_0I_1 + R_1I_1R_1 = 0 \subseteq I_1$ since $R_0 = R_1^2$. Now $I_1 \subsetneq R_1$ since $R_1R_1 \neq 0$. Thus R_1 would not be minisimple. Thus if (R_0,R_1) is simple, no such total annihilator can exist.

Conversely we must only show that R_1 is minisimple to be able to use Theorem 1 and obtain that (R_0,R_1) is simple. Suppose then that R_1 has a proper special subgroup I_1. Then consider $I_1R_1 + R_1I_1 = I_0$.

By Lemmas 2 and 3, I_0 is a special ideal of R_0, or is 0. It cannot be 0 for then I_1 would consist of total annihilators of R_1, and we are assuming there are none. Since R_0 is minisimple, I_0 must be all of R_0, i.e. $I_1R_1 + R_1I_1 = R_0$. Multiplying by R_1 we have $R_1I_1R_1 + R_1R_1I_1 = R_1R_0$. Now $R_1R_0 = R_1^3 = R_1$ but $R_1I_1R_1 \subseteq I_1$ and $R_1R_1I_1 = R_0I_1 \subseteq I_1$. Thus $I_1 = R_1$ and R_1 is minisimple.

Example 1 is a simple gring. R_0 is a simple ring with unity, $R_0^2 = R_0 = R_1^2$ and $R_1 = R_1^3 = R_1R_0 = R_0R_1$. There are no total annihilators in R_1 and by Theorem 2, (R_0,R_1) is simple.

Example 2 has $R_0^2 = R_0 = R_1^2$ and $R_1 = R_1^3 = R_1R_0$

$= R_0R_1$ but it is not a simple gring because R_0 is not minisimple. For example $(\{2m\},\{2n\sqrt{-1}\})$ is a gring ideal. Note that $(0,\{2n\sqrt{-1}\})$ is not a gring ideal.

Example 3 is of course not an associative gring.

Example 4 has $R_1^2 = R_0$ but $R_0^2 = R_1^3 = R_0R_1 = R_1R_0 = 0$. Thus this is not a simple gring. For example $(R_0,0)$ is a gring ideal.

Example 5 has $R_0^2 = R_0 = R_1^2$ and $R_1 = R_1^3 = R_1R_0 = R_0R_1$ but it is also not simple because R_0 is not minisimple. For example, let $I_0 = \{x_\alpha : \alpha > \frac{1}{2}\}$ and let $I_1 = \{y_\alpha : \alpha > \frac{1}{2}\}$. Then $I_0R_1 = I_1$, $I_1R_0 = I_1$, $I_1R_1 = I_0R_0 = I_0$ and (I_0,I_1) is a gring ideal.

Example 6 is more interesting than the previous examples. It has $R_0^2 = R_0 = R_1^2$ and $R_1 = R_1^3 = R_1R_0 = R_0R_1$. Furthermore R_0 is not simple. In fact R_0 has precisely two nonzero proper ideals, namely

$$I_0 = \left\{\begin{pmatrix} 0 & 0 & 0 \\ 0 & 0 & 0 \\ 0 & 0 & e \end{pmatrix}\right\} \quad \text{and} \quad J_0 = \left\{\begin{pmatrix} a & b & 0 \\ c & d & 0 \\ 0 & 0 & 0 \end{pmatrix}\right\} . \quad \text{And} \quad R_o = I_0 \oplus$$

J_0 . However R_0 is minisimple. To see this we note that $R_1I_0R_1 = J_0$ which is not in I_0 , and $R_1J_0R_1 = I_0$ which is not in J_0 . Since R_1 has no total annihilators, Theorem 2 tells us that this example is a simple gring.

Example 7 is an extension of example 5. If we take R_0 to be any nonzero ring with $R_0^2 = R_0$ then $R_1^2 = R_0$ and $R_1 = R_1^3 = R_1R_0 = R_0R_1$. If R_0 is not a simple ring and has a nonzero proper ideal $I_0 = I_{red} = I_r$ then R_1 has the corresponding set $I_1 = I_{blue} = I_b$. Then (I_0, I_1) is a gring ideal. The point is that I_0 is a special ideal because $R_1I_0R_1$ is the same as $R_0I_0R_0$ which is $\subseteq I_0$. Therefore this example is not a simple gring if R_0 is not a simple ring. On the other hand if R_0 is a simple (nontrivial) ring then it is minisimple. Also R_1 can have no total annihilators and thus by Theorem 2, (R_0, R_1) is a simple gring.

Given a gring (R_0, R_1) we consider the underlying graded abelian group and take $R_0 + R_1$, the set of all $r_0 + r_1$. This is an abelian group and we make it into a ring by defining multiplication as:

$$(r_0 + r_1)(s_0 + s_1) = r_0s_0 + r_1s_1 + r_0s_1 + r_1s_0 .$$

The individual products are given to us from the gring and $r_0s_0 + r_1s_1$ is in R_0 while $r_0s_1 + r_1s_0$ is in R_1 . This multiplication coincides with gring multiplications and extends it to $R_0 + R_1$. It is associative and distributive. The ring $R_0 + R_1$ contains the subring R_0 and the subgroup R_1 , as well as the gring

(R_0, R_1) .

The ring $R_0 + R_1$ is the minimal ring which contains the gring (R_0, R_1) and its simplicity should be intimately related to the simplicity of the gring.

Assume that $R_0 + R_1$ is a simple ring. Then if (I_0, I_1) is a nonzero gring ideal of (R_0, R_1), it is clear that $I_0 + I_1$ is a nonzero ideal of $R_0 + R_1$ and thus is equal to $R_0 + R_1$. So $I_0 = R_0$, $I_1 = R_1$ and $(I_0, I_1) = (R_0, R_1)$. Therefore (R_0, R_1) is a simple gring.

Suppose now that (R_0, R_1) is a simple gring. If $R_0 = 0$ or if $R_1 = 0$ then the overring $R_0 + R_1$ is simple being either R_0 or R_1 . Assume then that $R_0 \neq 0 \neq R_1$. Then R_0 is minisimple, $R_0 = R_0^2 = R_1^2$, $R_1 = R_1^3 = R_0 R_1 = R_1 R_0$ and there are no total annihilators in R_1 . Then $R_0 + R_1$ might be a simple ring. Suppose that it is not simple. Then it has a nonzero proper ideal I .

Let $I_0 \equiv I \cap R_0$. Then I_0 is an ideal of R_0 and it is special (if it is $\neq 0$) because $R_1 I_0 R_1 = R_1 (I \cap R_0) R_1$ is in R_0 and in I and thus in I_0 . Since R_0 is minisimple, $I_0 = 0$ or R_0 . If $I_0 = R_0$ then $R_0 \subseteq I$. Then $R_1 = R_0 R_1 \subseteq I$ and so I

contains everything and $I = R_0 + R_1$. Since I is proper we must have $I \cap R_0 = 0$. Let us next consider $I \cap R_1$. Suppose $I \cap R_1 \neq 0$ and choose $0 \neq x \in I \cap R_1$. Then $x R_1 \subset I \cap R_0 = I_0 = 0$. Similarly, $R_1 x = 0$ as well. Thus x is a total annihilator in R_1 , which, as noted above, is not possible. Thus $x = I \cap R_1 = 0$.

This means that if $r_0 + r_1 \neq 0$ is in I then $r_0 \neq 0$ and $r_1 \neq 0$. Then no other element in I can have the form $r_0 + s_1$ or $s_0 + r_1$, otherwise I would contain $r_1 - s_1$ in R_1 or $r_0 - s_0$ in R_0 . Therefore a given r_0 that appears as the first component in an element of I , only ever appears once. Similarly each r_1 that appears does so only once.

Take $r_0 + r_1 \neq 0$ in I and consider $R_0 r_0 R_0 + R_1 r_0 R_1$. This is in R_0 and is an ideal of R_0 . It contains both $R_0 r_0 R_0$ and $R_1 r_0 R_1$. Assume first that $R_0 r_0 R_0 + R_1 r_0 R_1 = 0$. Then $R_0 r_0 R_0 = 0 = R_1 r_0 R_1$. Then $r_0 R_0$ is an ideal of R_0 . It is special because $R_1 (r_0 R_0) R_1 = R_1 r_0 R_1 = 0$ .It cannot be all of R_0 for then $R_0 \cdot r_0 R_0 = R_0 R_0 = R_0 = 0$. Therefore $r_0 R_0 = 0$. Then $r_0 R_1 = r_0 R_0 R_1 = 0$. Thus $r_0 (R_0 + R_1) = 0$. Let

$Q_0 = \{s_0 \text{ in } R_0 : s_0R_0 = 0\}$. Then Q_0 is an ideal of R_0 and $R_1Q_0R_1 = R_1Q_0.R_0R_1 = 0$. Thus Q_0 is special and if $Q_0 = R_0$ then $R_0R_0 = R_0 = 0$. This is impossible and thus $Q_0 = 0$ and thus if $r_0R_0 = 0$ then $r_0 = 0$. Thus $R_0r_0R_0 + R_1r_0R_1$ cannot be 0 .

Now $R_0r_0R_0 + R_1r_0R_1$ is special since $R_1(R_0r_0R_0 + R_1r_0R_1)R_1 = R_1r_0R_1 + R_0r_0R_0$ and thus it must be all of R_0 . Then for any s_0 ,

$$s_0 = \sum a_{0_i} r_0 b_{0_i} + \sum c_{1_i} r_0 d_{1_i}$$

for some a_{0_i}, b_{0_i} in R_0 and some c_{1_i}, d_{1_i} in R_1 . Then $\sum a_{0_i}(r_0 + r_1)b_{0_i} + \sum c_{1_i}(r_0 + r_1)d_{1_i}$ is in I and equals

$$\sum a_{0_i} r_0 b_{0_i} + \sum c_{1_i} r_0 d_{1_i} + \sum a_{0_i} r_1 b_{0_i} + \sum c_{1_i} r_1 d_{1_i}$$

$= s_0$ + some element in R_1 .

Then every element in R_0 appears as a first component in an element of I . And of course each element appears only once.

Now let s_1 be any nonzero element of R_1 . Then s_1 is in $R_1 = R_0R_1$. Thus $s_1 = \sum a_{0_i} b_{1_i}$. Now each a_{0_i} appears in some $x_i = a_{0_i} + a_{1_i}$ in I .

Then $\sum x_i b_{1_i} = \sum(a_{0_i} + a_{1_i})b_{1_i} = \sum a_{0_i} b_{1_i} + \sum a_{1_i} . b_{1_i}$ $= s_1$ + some element in R_0 . And this is in I . Therefore every s_1 appears as a second component in an element of I , and of course appears only once. Thus every s_0 and every s_1 appear in I but I does not contain all pairs.

The ideal I is then a set of ordered pairs $a_0 + a_1$ or $[a_0, a_1]$, which is precisely a 1-1 correspondence between R_0 and R_1 . This map preserves addition since $a_0 + a_1 + b_0 + b_1 = a_0 + b_0 + a_1 + b_1$ and thus if $a_0 \leftrightarrow a_1$, $b_0 \leftrightarrow b_1$ then $a_0 + b_0 \leftrightarrow a_1 + b_1$. Thus $R_0 \cong R_1$ as abelian groups.

Note also that I is a maximal ideal of $R_0 + R_1$ for if it has one extra element then it must meet R_0 and R_1 and thus be all of $R_0 + R_1$.

Furthermore I is a minimal ideal for if $0 \neq J \subseteq I$ where J is an ideal of $R_0 + R_1$ then the above arguments show that J must contain all the elements of R_0 and R_1 and thus J = I .

If $R_0 + R_1$ has a nonzero proper ideal I , as above, then we can make R_1 into a ring. For any two elements a_1 and b_1 in R_1, we take the pairs a_o +

a_1 and $b_0 + b_1$ in I . Thus we get unique "partners" a_0 and b_0 in R_0 for the given a_1 and b_1 . Then we consider a_1b_1 . This is in R_0 . From $a_0 + a_1$ in I we get $a_0b_1 + a_1b_1$ in I , and from $b_0 + b_1$ in I we get $a_1b_0 + a_1b_1$ in I . Then a_1b_0 must equal a_0b_1 and this is the unique "partner" of a_1b_1 .

Now we can define a multiplication in R_1 as

$$a_1 * b_1 \equiv a_0b_1 = a_1b_0 .$$

This is uniquely defined, closed and associative because $(a_1 * b_1) * c_1 = a_0b_1 . c_0 = a_0 . b_1c_0 = a_1 * (b_1c_0) = a_1 * (b_1 * c_1)$. It is also distributive since $a_1 * (b_1 + c_1) = a_0(b_1 + c_1) = a_0b_1 + a_0c_1 = (a_1 * b_1) + (a_1 * c_1)$. Thus R_1 is a ring.

Now I gives us a ring isomorphism between R_0 and R_1 becuase if $a_0 <-> a_1$ and $b_0 <-> b_1$ then $a_0b_0 <-> a_1 * b_1 = a_1b_0$. And the partner of a_0b_0 *is* a_1b_0 since $a_0 + a_1$ in I means $a_0b_0 + a_1b_0$ is in I. Thus $R_0 \cong R_1$ as rings.

Furthermore if $a_0 + a_1$ is in I , so is $a_0b_1 + a_1b_1$ and so is $a_0b_0 + a_1b_0$. Since $a_1b_0 = a_0b_1$, we must have $a_0b_0 = a_1b_1$.

In this case R_0 is more than minisimple, it must

in fact be simple. To see this let I_0 be any $\neq 0$ ideal of R_0. Then $R_1 I_0 R_1$ is built from elements of the form $a_1 b_0 . c_1 = a_0 . b_0 c_0$ and this is in I_0. Therefore I_0 is special and thus $I_0 = R_0$ and R_0 is a simple (nontrivial) ring.

From $a_0 b_0 = a_1 b_1$ and $a_0 b_1 = a_1 b_0$ we get $a_0(b_1 - b_0) = a_1(b_0 - b_1)$ and $(a_0 + a_1)(b_1 - b_0) = 0$. This is true for every a_0, b_0 and their partners a_1, b_1, in I. Thus $I(b_1 - b_0) = 0$ for every b_0 and its partner b_1.

Thus if I is the set $\{a_0 + a_1\}$, let us call $\{a_0 - a_1\} = I^-$. This is also an ideal of $R_0 + R_1$ and what we have is $II^- = 0$. The ideal $I \cap I^-$ is either 0 or it is equal to both $I = I^-$ and this latter happens only when the characteristic is 2.

The interesting fact is that I and I^- are the only nonzero proper ideals of $R_0 + R_1$, and in the special case when the characteristic is 2 and $I^- = I$, then in fact $R_0 + R_1$ has precisely one nonzero proper ideal, namely I. To prove this we require three observations:

Observation 1. If J is any nonzero proper ideal of $R_0 + R_1$ then either $J = I$ or $J \cap I = 0$.

Proof: $J \cap I$ is an ideal and is $\subseteq I$. Therefore it is 0 or I. If it is I then $I \subseteq J$ and thus $I = J$ since J is proper and I is maximal.

Similarly we have:

Observation 2. If J is any nonzero proper ideal of $R_0 + R_1$ then either $J = I^-$ or $J \cap I^- = 0$.

Observation 3. In any simple nontrivial ring R_0 there exists an element x_0 such that $x_0^2 \neq 0$.

Proof: If $x^2 = 0$ for every x in R_0 then for every x and y in R_0, $(x + y)^2 = 0 = x^2 + y^2 + xy + yx$. Then $xy + yx = 0$ or $xy = - yx$. For any nonzero x in R_0 we have $R_0 = R_0 x R_0 = - x R_0R_0 = - x R_0 = x R_0$. Then there exists a z in R_0 such that $x = xz$ and thus $x = xz = xz^2 = 0$, a contradiction.

We can now prove that I and I^- are the only nonzero proper ideals of $R_0 + R_1$. Let J be a nonzero proper ideal of $R_0 + R_1$ and assume $J \neq I$. Then $J \cap I = 0$ by Observation 1. Take an element x_0 in R_0 such that $x_0^2 \neq 0$ (Observation 3). Then it appears in I as $x_0 + x_1$. It also appears in J, say as $x_0 + y_1$. We must have $x_1 \neq y_1$. Furthermore $(x_0 + x_1)(x_0 + y_1) = 0$ since it is in $I \cap J$. Then $x_0x_0 + x_1y_1 + x_0y_1 + x_1x_0 = 0$ and so $x_0y_1 + x_1x_0 = 0$. Thus $x_0y_1 = - x_1x_0$.

From $x_0 + y_1$ in J we have $x_0x_0 + x_0y_1$ in J or $x_0x_0 - x_1x_0$ in J. On the other hand $x_0 + x_1$ in I means $x_0x_0 + x_1x_0$ in I and thus $x_0x_0 - x_1x_0$ is in I^-. Thus $J \cap I^- \neq 0$ since we are guaranteed that $x_0x_0 \neq 0$. Then $J = I^-$.

Putting all of these things together, we have:

THEOREM 3.

(R_0, R_1) is a simple gring if and only if the overring $R_0 + R_1$ is simple except in one case, which occurs when $R_0 \cong R_1$ as rings, is a simple nontrivial ring and $R_0 + R_1$ has precisely two proper nonzero ideals $I = \{a_0 + a_1\}$ and $I^- = \{a_0 - a_1\}$ when the characteristic is $\neq 2$. If the characteristic is 2 then $I^- = I$. Furthermore the map $a_0 \leftrightarrow a_1$ is a ring isomorphism between R_0 and R_1.

In example 6, (R_0, R_1) and $R_0 + R_1$ are simple, but $R_0 \neq R_1$.

In example 7, if we take R_0 to be simple and nontrivial then (R_0, R_1) is simple. However $R_0 + R_1$ is not simple and $I = \{a_r + a_b\}$ is an ideal of $R_0 + R_1$. The only other proper nonzero ideal is $I^- = \{a_r - a_b\}$.

To define a gring *homomorphism* from (R_0, R_1) onto (S_0, S_1) we take f to be a map which sends

$f : R_0 \to S_0$ and $f : R_1 \to S_1$ [said to be homogeneous of degree 0];

$f(a_0 + b_0) = f(a_0) + f(b_0)$ for all a_0, b_0 in R_0 ;

$f(a_1 + b_1) = f(a_1) + f(b_1)$ for all a_1, b_1 in R_1 ;

$f(xy) = f(x)f(y)$ for all x,y in (R_0,R_1)

or a map g which sends

$g : R_0 \to S_1$ and $f : R_1 \to S_0$ [said to be homogeneous of degree 1];

g preserves addition and multiplication as above.

LEMMA 4. If g is a gring homomorphism of degree 1 from (R_0,R_1) onto (S_0,S_1) then

$$S_0S_0 = S_0S_1 = S_1S_0 = S_1S_1 = 0 .$$

Proof: If x and y are both in R_0 then xy is in R_0 , whereas $g(x)$ and $g(y)$ are both in S_1 and $g(xy)$ is in S_1 . Then $g(x)g(y)$ is in S_0 and it must equal $g(xy)$ in S_1 . Therefore $g(x)g(y) = g(xy) = 0$. Since $g(x)$ is onto S_1 we have $S_1S_1 = 0$. Similarly we obtain the other three all equal to 0 .

To discuss quotient grings, let (A_0,A_1) be a gring ideal of (R_0,R_1) . Then A_0 is an ideal of R_0 and R_0/A_0 is a ring. Since A_1 is a subgroup of the abelian group R_1 , R_1/A_1 is a group. Then we make

$(R_0/A_0, R_1/A_1)$ into a gring by defining:

$$(x_0 + A_0)(y_0 + A_0) = x_0y_0 + A_0 ;$$

$$(x_0 + A_0)(y_1 + A_1) = x_0y_1 + A_1 ;$$

$$(y_1 + A_1)(x_0 + A_0) = y_1x_0 + A_1 ;$$

$$(x_1 + A_1)(y_1 + A_1) = x_1y_1 + A_0 .$$

This gives us a quotient gring which we can denote by $(R_0,R_1)/(A_0,A_1)$.

We defined A_0 to be an invariant ideal of the ring R_0 if for every gring (R_0,R_1), $R_1A_0R_1 \subseteq A_0$ i.e. A_0 is special in every gring having R_0 in the first component.

LEMMA 5. If A_0 is an ideal of a ring R_0 then the following are all equivalent:

(i) A_0 is an invariant ideal of R_0 ;

(ii) For any gring (R_0,R_1), $(A_0,A_0:R_1)$ is a gring ideal of (R_0,R_1) , where $A_0:R_1 .\equiv. \{x \text{ in } R_1 : xR_1 + R_1x \subseteq A_0\}$;

(iii) For any gring (R_0,R_1), there exists a subgroup A_1 of R_1 such that (A_0,A_1) is a gring ideal of (R_0,R_1) ;

(iv) For any gring (R_0,R_1),$(A_0,A_0R_1 + R_1A_0)$ is a gring ideal of (R_0, R_1).

Proof: (i) → (ii).

Let $A_1 = A_0 : R_1$. Then by definition we have $R_1 A_1 \subseteq A_0$ and $A_1 R_1 \subseteq A_0$. Since $R_0(A_0:R_1)R_1 \subseteq R_0 A_0 \subset A_0$ and $R_1 R_0 (A_0:R_1) \subseteq R_1(A_0:R_1) \subseteq A_0$ then $R_0(A_0:R_1) \subseteq A_0 : R_1$. In other words $R_0 A_1 \subseteq A_1$.

Similarly $A_1 R_0 \subseteq A_1$.

Furthermore $R_1 \cdot R_1 A_0 \subseteq R_1^2 A_0 \subseteq R_0 A_0 \subseteq A_0$ and by i, $R_1 A_0 R_1 \subseteq A_0$. Therefore $R_1 A_0 \subseteq A_1$. Similarly $A_0 R_1 \subseteq A_1$. Thus we have (ii).

Clearly (ii) → (iii). From (iii) we know that (A_0, A_1) is a gring ideal and thus $R_1 A_0 R_1 \subseteq A_1 R_1 \subseteq A_0$ which is i . Thus (i), (ii) and (iii) are equivalent.

Clearly (iv) → (iii) and (i) → (iv) is just Lemma 1. This ends the proof.

LEMMA 6. If A_0 is an invariant ideal of R_0 then for any gring ideal (A_0, A_1) of (R_0, R_1) we have $A_0 R_1 + R_1 A_0 \subseteq A_1 \subseteq A_0 : R_1$.

Proof: Since $A_1 R_1 + R_1 A_1 \subseteq A_0$ then $A_1 \subseteq A_0 : R_1$. And $A_0 R_1 + R_1 A_0 \subseteq A_1$.

LEMMA 7. If A_0 is an invariant ideal of R_0 then for any gring (R_0, R_1)

$$0 : {}^{R_1}/_{A_0:R_1} = 0 .$$

Proof: By Lemma 5, $(A_0, A_0:R_1)$ is a gring ideal of (R_0, R_1). Take $x + (A_0:R_1)$ in $0 : {}^{R_1}/_{A_0:R_1}$. Then by definition $(x + (A_0:R_1))(y + (A_0:R_1)) = xy + A_0 = 0$ for any y in R_1. Thus xy is in A_0 i.e. $xR_1 \subseteq A_0$. Similarly $R_1x \subseteq A_0$ and therefore x is in $A_0 : R_1$ and thus $x + A_0:R_1 = 0$.

Gring radical properties are defined in [4].

Let β be a radical property of rings. We can then define a gring property $\bar{\beta}$, namely a gring (R_0, R_1) is a $\bar{\beta}$ gring if $\beta(R_0) = R_0$.

We shall say that a ring radical property β is *invariant* if for every ring R_0, $\beta(R_0)$ is an invariant ideal of R_0.

THEOREM 4. The ring radical property β is invariant and contains all zero rings if and only if the corresponding gring property $\bar{\beta}$ is a gring radical property and for any gring (R_0, R_1) we have $\bar{\beta}(R_0, R_1) = (\beta(R_0), \beta(R_0):R_1)$.

Proof: Assume first that β is a gring radical property and that $\beta(R_0, R_1) = (\beta(R_0), \beta(R_0):R_1)$ for any gring

(R_0, R_1). Then $\beta(R_0)$ is by Lemma 5, an invariant ideal of R_0, for every R_0. Thus β is an invariant radical property.

Next let A be any zero ring i.e. $A^2 = 0$. Then $(0,A)$ is a gring and it is a $\bar{\beta}$ gring. On the other hand $(A,0)$ is a gring which is isomorphic to $(0,A)$ by an isomorphism of degree 1. Then $(A,0)$ is also a $\bar{\beta}$ gring and thus $\beta(A) = A$. Thus β contains all zero rings.

Conversely assume that β is invariant and contains all zero rings. To show that $\bar{\beta}$ is a gring radical property we consider any homomorphic image $f((R_0, R_1))$ of a β gring (R_0, R_1) onto the gring (S_0, S_1). If the degree of f is 0 then $f(R_0) = S_0$ and $f(R_1) = S_1$. Now R_0 is a β ring and therefore S_0 is a β ring and therefore (S_0, S_1) is a $\bar{\beta}$ gring and therefore $\bar{\beta}$ is homomorphically closed for homomorphisms of degree 0.

On the other hand if the degree of f is 1 then $f(R_0) = S_1$ and $f(R_1) = S_0$. Then $S_0S_0 = S_1S_1 = S_0S_1 = S_1S_0 = 0$. Since S_0 is a zero ring it is a β ring and thus again (S_0, S_1) is a $\bar{\beta}$ gring.

Furthermore, since $\beta(R_0)$ is an invariant ideal of R_0, by Lemma 5, $(\beta(R_0), \beta(R_0):R_1)$ is a gring ideal of

(R_0,R_1) . It is itself a $\bar{\beta}$ gring since $\beta(\beta(R_0)) = \beta(R_0)$. We must show that $\beta\left(\beta(R_0),\, \beta(R_0):R_1\right)$ is the largest $\bar{\beta}$ ideal in (R_0,R_1) . So let (I_0,I_1) be any $\bar{\beta}$ gring ideal of (R_0,R_1) . Then $\beta(I_0) = I_0$. Then $I_0 \subseteq \beta(R_0)$ since $\beta(R_0)$ is the β radical of R_0 . Then

$$I_0 : R_1 \subseteq \beta(R_0):R_1 .$$

Also $I_1R_1 \subseteq I_0$ and $R_1I_1 \subseteq I_0$ and therefore $I_1 \subseteq I_0 : R_1$. Therefore $(I_0,I_1) \subseteq (I_0,I_0:R_1) \subseteq (\beta(R_0),\, \beta(R_0):R_1)$. Thus we can call $(\beta(R_0),\, \bar{\beta}(R_0):R_1) = \bar{\beta}(R_0,R_1)$.

Finally we must show that

$$\beta\left({}^{R_0}/_{\beta(R_0)}\,,\ {}^{R_1}/_{\beta(R_0):R_1}\right)$$

$$= \left(\beta({}^{R_0}/_{\beta(R_0)})\,,\ \beta({}^{R_0}/_{\beta(R_0)}) : {}^{R_1}/_{\beta(R_0):R_1}\right)$$

is zero. But $\beta({}^{R_0}/\beta(R_0)) = 0$ because β is a radical property of rings and the second component, which is now $0 : {}^{R_1}/_{\beta(R_0):R_1}$ is 0 by Lemma 7. Therefore $\bar{\beta}$ is a gring radical property and $\bar{\beta}(R_0,R_1) = (\beta(R_0),\beta(R_0):R_1)$.

The Jacobson radical is invariant (see [5]) and the corresponding gring radical can be represented as an intersection of maximum modular one sided gring ideals.

From [3] one can conclude that normal radicals are invariant.

However the Brown-McCoy radical is not invariant. To see this we first need

LEMMA 8. If a ring R is a direct sum of ideals A and B then for any radical β we have $\beta(R) = \beta(A) \oplus \beta(B)$.

Proof: $\beta(A) \oplus \beta(B) \subseteq \beta(R)$ is obvious. For the converse, consider the map f defined by $f(a + b) = [a + \beta(A)] + [b + \beta(B)]$ of R onto $A/\beta(A) \oplus B/\beta(B)$. This is a homomorphism with kernel $\beta(A) + \beta(B)$. If $\beta(R)$ is not in this kernel then $\beta(R)/\beta(R) \cap \text{kernel } f \neq 0$ and $R/\ker f$ contains a nonzero β ideal. But this is impossible since $R/\ker f$ is a direct sum of β-semisimple rings and is thus β-semisimple. Thus $\beta(R) \subseteq \ker f$ and the proof is done.

To see that the Brown McCoy radical is not invariant, let M be the ring of all matrices of some degree over a division ring D i.e. $M = \overset{\infty}{\cup} D_i$.

Let $R_0 = \begin{pmatrix} D & 0 \\ 0 & M \end{pmatrix}$ i.e. all matrices of the form $\begin{pmatrix} d & 0 \\ 0 & m \end{pmatrix}$. Let R_1 be the set of all matrices of the form $\begin{pmatrix} 0 & a \\ b & 0 \end{pmatrix}$, where a is an infinite row (with only a finite number of nonzero elements) and b is an infinite column all with entries in D. Then (R_0, R_1) is a gring and it is simple. To see this, we can use Theorem 1, or observe that if (A_0, A_1) is a nonzero gring ideal of (R_0, R_1) then it must contain some E_{pq} (1 in the pth row q^{th} column and 0's elsewhere) and thus every $E_{ij} = E_{ip}E_{pq}E_{qj}$. Thus $(A_0, A_1) = (R_0, R_1)$.

Now let β be the Brown-McCoy radical (i.e. upper radical determined by all simple rings with unity). Then $\beta(D) = 0$ and $\beta(M) = M$ since M is simple with no unity By Lemma 8 $\beta(R_0) = \begin{pmatrix} \beta(D) & 0 \\ 0 & \beta(M) \end{pmatrix} = \begin{pmatrix} 0 & 0 \\ 0 & M \end{pmatrix}$. This *is* an ideal of R_0 but it is *not* special since $R_1 \beta(R_0) R_1 \subseteq \begin{pmatrix} D & 0 \\ 0 & 0 \end{pmatrix}$. Therefore $\beta(R_0)$ is not invariant and so the Brown McCoy radical is not invariant. Note that R_0 is not simple because it has two $\neq 0$ proper ideals, $\begin{pmatrix} D & 0 \\ 0 & 0 \end{pmatrix}$ and $\begin{pmatrix} 0 & 0 \\ 0 & M \end{pmatrix}$ but neither is special and therefore R_0 is minisimple.

To get some insights into invariant radicals we consider idempotents $e = e^2$ of a ring A. For every idempotent e we get the well known Peirce decomposition

$$A = A_{11} + A_{00} + A_{10} + A_{01}$$

where the sum is direct and

$A_i = \{x \text{ in } A: xe = jx \text{ and } ex = ix\}, \; i,j = 0,1$.

Or $A_{11} = eAe$, $A_{00} = (1 - e)A(1 - e)$; $A_{01} = (1 - e)Ae$ and $A_{10} = eA(1 - e)$. Then

$$A_{ij}A_{k\ell} \begin{array}{l} = 0 \quad \text{if} \quad j \neq k \\ = A_{i\ell} \quad \text{if} \quad j = k \end{array}$$

A_{11} and A_{00} are subrings of A .

We shall say that a ring radical β has the *Exchange Property* if for every ring A and for every indempotent e in A we have

$$A_{01}.\beta(A_{11}).A_{10} \subseteq \beta(A_{00}) .$$

THEOREM 5. β is invariant if and only if β has the exchange property.

Proof: Assume β is an invariant radical and that e is an idempotent of a ring A . Let $R_0 = A_{00} + A_{11}$ and $R_{11} = A_{10} + A_{01}$. Then it is clear that (R_0, R_1) is a gring. By Lemma 8 $\beta(R_0) = \beta(A_{00}) + \beta(A_{11})$. Since β is invariant we have

$$A_{01}\beta(A_{11})A_{10} \subseteq R_1.\beta(R_0).R_1 \subseteq \beta(R_0) = \beta(A_{11}) + \beta(A_{00}) .$$

However $A_{01}.\beta(A_{11})A_{10}$ is clearly $\subseteq A_{00}$ and thus is contained in $\beta(A_{00})$. Thus β has the exchange property. We can obtain another useful relationship:

$$A_{10}\beta(A_{00})A_{01} \subseteq R_1\beta(R_0)R_1 \subseteq \beta(R_0) = \beta(A_{11}) + \beta(A_{00})$$

and since $A_{10}\beta(A_{00})A_{01}$ is clearly in A_{11}, it must be $\subseteq \beta(A_{11})$, i.e.

$$A_{10}.\beta(A_{00}).A_{01} \subseteq \beta(A_{11}) .$$

Conversely assume that β has the exchange property; and let (R_0,R_1) be a gring.

Let R_0^+ be the ring R_0 with a unity element adjoined to it. Then (R_0^+,R_1) is also a gring if we extend the composition $R_0R_1 \to R_1$ as follows:

$$(x + n.1)y = xy + ny$$
$$y(x + n.1) = yx + ny$$

for any x in R_0 and y in R_1 and any integer n. Now let

$A = \begin{pmatrix} R_0^+ & R_1 \\ R_1 & R_0 \end{pmatrix}$. Let $e = \begin{pmatrix} 1 & 0 \\ 0 & 0 \end{pmatrix}$. Then the Pierce decomposition of A relative to e is:

$A_{11} = \begin{pmatrix} R_0^+ & 0 \\ 0 & 0 \end{pmatrix}$, $A_{00} = \begin{pmatrix} 0 & 0 \\ 0 & R_0 \end{pmatrix}$; $A_{10} = \begin{pmatrix} 0 & R_1 \\ 0 & 0 \end{pmatrix}$; $A_{01} = \begin{pmatrix} 0 & 0 \\ R_1 & 0 \end{pmatrix}$.

Now R_0 is an ideal of R_0^+ and thus $\beta(R_0)$ is an ideal of R_0^+ (See [2]). Thus $\beta(R_0) \subseteq \beta(R_0^+)$. Since β has the exchange property, we have

$$\begin{pmatrix} 0 & 0 \\ 0 & R_1\beta(R_0)R_1 \end{pmatrix} = \begin{pmatrix} 0 & 0 \\ R_1 & 0 \end{pmatrix}\begin{pmatrix} \beta(R_0) & 0 \\ 0 & 0 \end{pmatrix}\begin{pmatrix} 0 & R_1 \\ 0 & 0 \end{pmatrix} \subseteq A_{01}\beta(A_{11})A_{10}$$

$$\subseteq \beta(A_{00}) = \begin{pmatrix} 0 & 0 \\ 0 & \beta(R_0) \end{pmatrix}.$$

Thus $R_1\beta(R_0)R_1 \subseteq \beta(R_0)$ and β is an invariant radical.

THEOREM 6. If β is an invariant radical property which contains all zero rings then β is hereditary, i.e. β is then supernilpotent.

Proof: Let I be an ideal of a ring R and let R^+ be R with a unity adjoined. Let $A = \begin{pmatrix} R^+ & I \\ I & I \end{pmatrix}$. Let $e = \begin{pmatrix} 1 & 0 \\ 0 & 0 \end{pmatrix}$ and since β must have the exchange property (Th 5) we have

$$\begin{pmatrix} 0 & 0 \\ I & 0 \end{pmatrix}\begin{pmatrix} \beta(R) & 0 \\ 0 & 0 \end{pmatrix}\begin{pmatrix} 0 & I \\ 0 & 0 \end{pmatrix} \subseteq \begin{pmatrix} 0 & 0 \\ 0 & \beta(I) \end{pmatrix}.$$

Since the left hand side is $\begin{pmatrix} 0 & 0 \\ 0 & I\beta(R)I \end{pmatrix}$ we have $I\beta(R)I \subseteq \beta(I)$.

Now $I.\beta(R)$ is an ideal of R and it is in I.

Then $\frac{I\beta(R) + \beta(I)}{\beta(I)}$ is an ideal of $I/\beta(I)$, a β semisimple ring. However $\left[\frac{I\beta(R) + \beta(I)}{\beta(I)}\right]^2 = 0$ because

$I\beta(R)I\beta(R) \subseteq \beta(I)\beta(R) \subseteq \beta(I)$ and thus the inside is a

β radical ideal in a β semisimple ring. Thus

$\frac{I\beta(R) + \beta(I)}{\beta(I)}$ must be 0 and $I\beta(R) \subseteq \beta(I)$. Then

$\frac{\beta(I) + I^{\cap}\beta(R)}{\beta(I)}$ is an ideal of $\frac{I}{\beta(I)}$ and its square

$\subseteq \frac{\beta(I) + [I^{\cap}\beta(R)]^2}{\beta(I)} \subseteq \frac{\beta(I) + I\beta(R)}{\beta(I)} = 0$. Thus as above

$I^{\cap}\beta(R) \subseteq \beta(I)$. Thus β is hereditary.

Question: We do not know whether this theorem remains true if we drop the assumption that β contains the zero rings, i.e. if I is an ideal of R and $I\beta(R)I \subseteq \beta(I)$ can we conclude that $I^{\cap}\beta(R) \subseteq \beta(I)$?

BIBLIOGRAPHY

(1) L. Corwin, Y. Ne'erman, and S. Sternberg, Graded Lie algebras in mathematics and physics (Bose-Fermi symmetry), *Reviews of Modern Physics* 47 (1975), 573-603.

(2) N. Divinsky, Rings and Radicals, *University of Toronto Press, 1965.*

(3) A. D. Sands, Radicals and Morita Contexts, *J. of Algebra, 24 (1973), pp. 335-345.*

(4) A. Suliński, Radicals of Associative 2-graded rings, *Bull. Acad. Sci. Polon. 29 (1981) pp. 431-434.*

(5) A. Suliński, J. F. Watters, On the Jacobson Radical of Assoc. 2-graded rings. To appear in *Acta Mathematica.*

N. Divinsky and T. Anderson
University of British Columbia
Mathematics Department
Vancouver, B. C., V6T 1Y4
Canada

A. Suliński
Institute of Mathematics
Warsaw University
00-901 Warsaw
Poland

COLLOQUIA MATHEMATICA SOCIETATIS JÁNOS BOLYAI
38. RADICAL THEORY, EGER (HUNGARY), 1982.

RADICALS AND VARIETIES

B. J. GARDNER

While radical theory (in the Kurosh-Amitsur sense) had its origins in ring theory and the quest for structure theory for rings, much of contemporary radical theoretic activity has only the most tenuous connection with these origins. An autonomous branch of mathematics - abstract radical theory - has evolved, in which the objects of study are radical and semi-simple classes of structures as much as the structures themselves, and, of course, many other structures besides rings are involved, not necessarily purely algebraic (e.g. topological groups).

Typical questions considered by radical-theorists for suitable "universal classes" $\mathcal{W}$ are: Are semi-simple classes in $\mathcal{W}$ hereditary?; If a subclass $\mathcal{M}$ of $\mathcal{W}$ has some property, is the same true of its lower radical class?; How many steps are required in the lower radical construction in $\mathcal{W}$?; Does every subclass of $\mathcal{W}$ define an upper radical? These questions have a universal algebraic flavour. (For universal classes of ordered or topological algebraic structures there are other flavours as well.) They also relate to aggregation properties rather than structural properties of algebraic objects.

In view of the central role played in universal algebra by the variety concept, and the fact that, from one point of view, varieties also have to do with aggregation, it was appropriate and inevitable that connections between radical theory and varieties should be investigated.

We shall here survey some of the results of such investigations. In §1 those varieties which are radical or semi-simple classes and those radical and semi-simple classes which are varieties will be characterized, and the known instances of such classes catalogued. Two types of class which are generalizations of varieties - locally equational classes and quasivarieties - will be treated in a similar sort of way in §§2 and 3 respectively.

There is no universally accepted optimum level of generality at which to treat radical theory. In view of this, and of our plan to confine our attention to situations in which all objects are purely algebraic and varieties can be formed, we shall content ourselves with the following rather vague statement of policy: We shall work throughout in a universal class $\mathcal{W}$ which is a variety whose members are rings or groups, possibly with other operations. (Thus we include the class of modules over a ring, the variety of lattice-ordered groups, the class of strongly regular rings, etc.)

When discussing things in full generality we shall indicate the presence of a one-element object by writing $|A| = 1$, rather than $A = 0$ (or 1 or E). The symbol $\triangleleft$ will denote normal subobjects.

1. Varieties

The first substantial radical-theoretic result which involved varieties in a direct way was the characterization by Stewart [71] of the semi-simple radical classes of associative rings as the varieties generated by finite sets of finite fields. It had ealier been proved by Andrunakievich [3] that the class of boolean rings is a radical semi-simple class, but the proof was based on the theory of complementary radicals rather than consideration of varieties. It was observed by Wiegandt [78] that homomorphically closed semi-simple classes of associative rings must be varieties (and hence radical classes). (The analogous result for modules is sufficiently obvious to have gone unremarked, in view of the simple set of closure properties characterizing radical and semi-simple classes of modules; semi-simple radical classes of modules were first discussed by Jans [36].) This result remains true in every universal class (even where semi-simple classes need not be hereditary) as a consequence of Kogalovskii's characterization [39] of varieties as classes closed under subdirect products and homomorphic images.

THEOREM 1.1 *In any universal class $\mathcal{W}$, all homomorphically closed semi-simple classes are varieties.*

Let $\mathcal{C},\mathcal{D}$ be subclasses of a universal class $\mathcal{W}$. We define $\mathcal{C} \circ \mathcal{D}$ to be the class

$$\{A\in\mathcal{W} \mid (\exists I)(I \triangleleft A)(I\in\mathcal{C} \text{ and } A/I\in\mathcal{D})\} .$$

As is well-known, a class $\mathcal{R}$ is a radical class if and only if it is homomorphically closed, closed under unions of chains of normal subobjects and satisfies the condition $\mathcal{R} \circ \mathcal{R} = \mathcal{R}$ (closure under extensions). Since varieties satisfy the first two properties we have

THEOREM 1.2 *(i) A variety $\mathcal{V}$ is a radical class if and only if $\mathcal{V} \circ \mathcal{V} = \mathcal{V}$.*

(ii) A radical class $\mathcal{R}$ is a variety if and only if it is closed under direct products and is hereditary with respect to subobjects.

For a proof of the first assertion (which works for any $\mathcal{W}$) see [16] or [80]; the second assertion is obvious.

Since semi-simple classes are closed under extensions we have

COROLLARY 1.3 *A semi-simple class is a radical class*

if and only if it is homomorphically closed.

Some further terminology is needed before we can characterize those semi-simple classes which are varieties.

Definition 1.4 Let V be a variety. For $A \in W$, let $A(V) = \cap\{I \mid I \triangleleft A \text{ and } A/I \in V\}$. Then V is said to have *attainable identities* if $A(V)(V) = A(V)$ for every A.

Attainability was first discussed by Tamura [76] for semigroups, and (in a formulation slightly different from the one we have given) is meaningful in any equationally defined class of universal algebras. Its importance for our present discussion stems from the following result.

THEOREM 1.5 *(i) A variety V is a semi-simple class if and only if it has attainable identities.*

(ii) The following conditions are equivalent for a semi-simple class S:

(1) S is a variety;

(2) S is homomorphically closed;

(3) S is a radical class.

Proofs of these assertions (for algebras, but applicable elsewhere) are given in [16]. (See also 1.1 and 1.3.)

Using 1.5 and 1.2 (or a direct argument) we get

COROLLARY 1.6 *If a variety V has attainable identities, then $V \circ V = V$.*

The obvious question at this point is when or whether the converse of Corollary 1.6 is valid; the converse *is* true if and only if every extension-closed variety is a semi-simple radical class. The following result is pertinent here.

THEOREM 1.7 *Let* V *be a subvariety of* W *such that* $V \circ V = V$. *Then for every* $A \in W$, $A(V)$ *is the normal subobject of* A *generated by* $A(V)(V)$.

This is proved for algebras in [16]; again the proof works in general. We have an immediate corollary.

COROLLARY 1.8 *If normality of subobjects is transitive in* W, *then every extension-closed subvariety of* W *has attainable identities and is thus a semi-simple radical class.*

Example 1.9 The class Mod(R) of left unital modules over an associative ring R with identity satisfies the conditions of Corollary 1.8. Here the semi-simple radical classes are the classes $V_I = \{M | IM=0\}$ for idempotent ideals I of R (Jans [36]).

Example 1.10 Let F denote the class of strongly regular rings, those associative rings A satisfying the condition

$$(\forall a \in A)(\exists b \in A)(a=a^2b) .$$

If $a \in A \in F$, there is a unique $a' \in A$ such that

$$a = a^2a' = aa'a = a'a^2 \text{ and } a' = (a')^2a = a'aa' = a(a')^2 .$$

Thus A carries an extra, unary, operation $(\ )'$ in addition to the ring operations. When account is taken of the extra operation, F becomes a variety. In F the normal subobjects are just the ring ideals, so since every strongly regular ring is hereditarily idempotent, normality of subobjects is transitive. By Corollary 1.8, all extension-closed varieties in F have attainable identities; in this case all varieties are in fact extension-closed [25].

Transitivity of normality is a rather restrictive condition, but in a number of cases there is a weak version of it which suffices.

Let V be an extension-closed variety. Then for any A, we have $A(V)/A(V)(V) \in V$, with $A(V)$ the normal subobject of A generated by $A(V)(V)$. If we can discover some significant property of objects of the form I^*/I, where $I \triangleleft J \triangleleft A$ and I^* is the normal subobject of A generated by I, then we can show that certain extension-closed varieties - those containing no non-trivial objects with the given property - have attainable identities. Of course, the nature of I^*/I is of considerable interest in other areas of radical theory also.

For a number of kinds of rings and algebras, there are theorems asserting that I^*/I must, if non-zero, contain some sort of nilpotent subobject, Thus, in such contexts, extension-closed varieties with no nilpotent members have attainable identities. How much of a restriction is the requirement of no nilpotent members? In the case of algebras over rings without too complicated ideal structure, the presence of one non-zero nilpotent algebra in an extension-closed variety implies the presence of all of them. If, as frequently happens, free algebras are subdirect products of nilpotent algebras, the presence of nilpotent algebras leads to triviality.

For associative algebras, I^*/I is nilpotent (Andrunakievich [2]); for alternative algebras I^*/I is locally nilpotent (Hentzel and Slater [33]); for Jordan algebras (over a ring containing $\frac{1}{2}$) I^*/I, if non-zero, has a nilpotent ideal (Slin'ko [70]); for (γ,δ)-algebras over a ring containing $\frac{1}{6}$, I^*/I, if non-zero, has an ideal J with $J^2 = 0 \neq J$ (Markovichev [49]). For some other classes of algebras with similar properties, see [1].

These considerations provide us with the following examples.

Examples 1.11 Extension-closed varieties have attainable identities in the classes of

(i) associative rings,

(ii) alternative rings,

(iii) Jordan algebras over a ring containing $\frac{1}{2}$ and

(iv) (γ,δ)-algebras over a ring containing $\frac{1}{6}$.

For the "weak version" of (γ,δ)-algebras, requiring only the identity

$$[x,y,z] + \gamma[y,x,z] + \delta[z,x,y] = 0 ,$$

the same conclusion holds, at least if $(\gamma,\delta) \neq (\pm 1,0),(1,1)$. This is proved, by quite different arguments, in [21], use being made of the structural results of Hentzel and Cattaneo [32]. In [21] the following example is also given.

Example 1.12 Every extension-closed variety of right alternative rings which does not contain rings of characteristic 2 has attainable identities.

As noted above, it was shown by Stewart [71] that the semi-simple radical classes of associative rings are the varieties VAR(K) generated by finite sets K of finite fields. Shevrin and Martynov [64] showed that the VAR(K) are the varieties with attainable identities, while Martynov [51] proved that the VAR(K) are the extension-closed varieties. These varieties were also discussed, from a radical-theoretic point of view, by Gardner and Stewart [28], van Leeuwen and Jenkins [44], Szász [75] and Wiegandt [78], [79]. A quite striking result was recently obtained by Loi [46]: a radical class R of associative rings is closed under essential extensions if and only if R = VAR(K) for some finite set K of finite fields. (Note by way of contrast that essentially-closed radical classes of modules are quite plentiful.)

All the varieties referred to in Examples 1.11 and 1.12 (except in 1.11 (iii) [18]) are the varieties generated by finite sets of finite fields with appropriate characteristics. (There are, therefore, a few open questions floating around for characteristics 2 and 3.)

We can find another condition ensuring that extension-closure implies attainability by looking at what happens in the case of groups. We describe the situation in the following theorem.

THEOREM 1.13 *Suppose that, in* $\mathcal{W}$, *there is a property* (*) *of automorphisms and a property of subobjects, occurrence of which is denoted by* $\prec$, *such that*

(i) $A \prec B \prec C \Rightarrow A \prec C$ *and*

(ii) $A \triangleleft B$ *if and only if* $A \prec B$ *and* $f(A) = A$ *for every automorphism* f *of* B *with property* (*). *Then every extension-closed variety in* $\mathcal{W}$ *has attainable identities.*

Proof. Let $\mathcal{V}$ be a variety with $\mathcal{V} \circ \mathcal{V} = \mathcal{V}$. If $S \triangleleft A \in \mathcal{W}$, then, since

$$S(\mathcal{V}) = \cap\{K \mid S/K \in \mathcal{V}\} ,$$

we have $g(S(\mathcal{V})) = S(\mathcal{V})$ for every automorphism g of S. Let f be an automorphism of A with property (*). Since $S \triangleleft A$, we have $f(S) = S$. Hence f induces an automorphism of S, s $f(S(\mathcal{V})) = S(\mathcal{V})$. Also $S(\mathcal{V}) \triangleleft S \triangleleft A$ implies $S(\mathcal{V}) \prec S \prec A$, so $S(\mathcal{V}) \prec A$. But this means that $S(\mathcal{V}) \triangleleft A$. In particular $A(\mathcal{V})(\mathcal{V}) \triangleleft A$, so by Theorem 1.7 $A(\mathcal{V})(\mathcal{V}) = A(\mathcal{V})$. //

Examples 1.14 The conditions of Theorem 1.13 are met by groups (clearly). They are also met by lattice-ordered groups, as for these, the normal subobjects are characterized by invariance under (group) inner automorphisms, (these being lattice automorphisms) and the transitive property of convexity. Certain varieties of loops also satisfy the conditions of 1.13; see Bruck and Paige [8].

As we shall shortly see, there are no non-trivial extension-closed varieties of groups. In the case of lattice-ordered groups, there is a unique one; this was shown by Glass *et al* [30] (see also Martinez [50]) and is worth a theorem.

THEOREM 1.15 *The unique non-trivial extension-closed variety (equivalently, semi-simple radical class) of lattice-ordered groups is the variety defined by*

$$(x_1 \vee e)(x_2 \vee e) \leq (x_2 \vee e)^2 (x_1 \vee e)^2 .$$

The only examples of which we are aware, in which extension-closure does not imply attainability are the following two.

Example 1.16 [19]. Let $\mathcal{W}$ be the variety of rings defined by the identities

$$2x = 0; \; xx^2 = x^2x; \; (x^2)^m (x^2)^n = (x^2)^{m+n} \forall m,n .$$

Then in $\mathcal{W}$ the variety defined by the identity $x^2 = x$ is extension-closed but does not have attainable identities.

Example 1.17 [23]. Let $\mathcal{W}$ be the class of associative algebras with involution over the ring $\{m/2^n | m,n \in Z\}$. In

$\mathcal{W}$ the extension-closed varieties VAR($x^*=x$) and VAR($x^*=-x$) do not have attainable identities.

Having thus turned to negative results, we now present a criterion for the non-existence of non-trivial semi-simple radical classes.

Definition 1.18 A variety $\mathcal{W}$ is a *Schreier variety* if non-trivial subobjects of free $\mathcal{W}$-objects are free.

THEOREM 1.19 *Let $\mathcal{W}$ be a Schreier variety. Then in $\mathcal{W}$ there are no non-trivial varieties with attainable identities.*

We shall give a proof of this; the proof of the algebra version in [16] is unnecessarily complicated.

Let F be a free object of countably infinite rank in $\mathcal{W}$, $\mathcal{V}$ a proper subvariety of $\mathcal{W}$. Then $F \notin \mathcal{V}$, i.e. $|F(\mathcal{V})| > 1$, so $F(\mathcal{V})$ is free and can therefore be mapped onto a non-trivial object in $\mathcal{V}$. Hence $F(\mathcal{V})(\mathcal{V}) \neq F(\mathcal{V})$. //

Examples 1.20 Relevant examples of Schreier varieties are the varieties of groups, loops (Bruck [7]), non-associative algebras over a field (Kurosh [42]), Lie algebras over a field (Shirshov [66], Witt [81]), commutative and anticommutative algebras over a field (Shirshov [67]); the unital modules over a ring R with identity constitute a Schreier variety if and only if R is a free ideal ring (Cohn [10]). Kaplan [37] showed by a different method that there are no varieties of loops with attainable identities.

Combining Theorems 1.13 and 1.19 for groups, we get

COROLLARY 1.21 *There are no non-trivial extension-closed varieties of groups.*

This also follows from the structure theorem for the ∘-semigroup of group varieties obtained by the Neumanns [53] and Shmel'kin [68] and also from results of Shores [69].

There are a number of situations (see, e.g., [20] and [24]) in which semi-simple classes are virtually never hereditary. In such situations, a variety cannot be a semi-simple class and therefore cannot have attainable identities.

Let $\mathcal{V}$ be a variety with attainable identities in some universal class $\mathcal{W}$. Then there is a radical class $\mathcal{R}$ with $\mathcal{V}$ as semi-simple class. What can be said of $\mathcal{R}$?

PROPOSITION 1.22 *Let $\mathcal{V}$ have attainable identities in $\mathcal{W}$; let $\mathcal{R}$ denote the upper radical class defined by $\mathcal{V}$. Let* F *be a free $\mathcal{W}$-object generated by* $x_1, x_2, x_3, \ldots, I$ *the* T-*ideal (etc.) defining $\mathcal{V}$, and for each* $A \in \mathcal{W}$, *let*

$$I(A) = \{f(a_1, a_2, \ldots, a_n) \mid a_i \in A, f(x_1, x_2, \ldots, x_n) \in I\} .$$

Then $I(A) = \mathcal{R}(A) = A(\mathcal{V})$.

See [16] for a proof for algebras which works generally.

Of course, if $\mathcal{V}$ can be defined by identities in a bounded number of variables, we can use a free object of appropriate finite rank in place of F. This happens with modules: the free module on one generator is used.

If $\mathcal{V}$ and $\mathcal{U}$ are varieties defined by I,J respectively, then (in the notation of Proposition 1.22) $\mathcal{U} \circ \mathcal{V}$ is defined by J(I), the T-ideal (etc.) obtained by substituting elements of I for the variables in the words in J. See [54] for a proof in the group case. Thus $\mathcal{V}$ is extension-closed if and only if I(I) = I. In the case of modules this is the result already mentioned in Example 1.9.

If $\mathcal{V} \circ \mathcal{V} = \mathcal{V}$, i.e. $\mathcal{V}$ (a variety) is a radical class, can the corresponding semi-simple class be a variety? On the other hand, if $\mathcal{U}$ is a semi-simple radical class, can its radical class also be a variety? The answer to both questions is "yes"; the next theorem provides some information.

THEOREM 1.23 *For any* $\mathcal{W}$ *there is a bijection from the radical classes which are varieties and have varieties as semi-simple classes to the ordered pairs* (J,I) *of* T-*ideals (etc.) of a free* $\mathcal{W}$-*object* F *of rank* $\aleph_0$ *which satisfy the following conditions.*

$$J(J) = J,\ I(I) = I,\ |J(I)| = 1,\ I \vee J = F\ .$$

Proof. Let $\mathcal{U},\mathcal{V}$ be varieties such that $\mathcal{U}$ is a radical class and $\mathcal{V} = \{A \mid |\mathcal{U}(A)| = 1\}$. Let J,I be the T-ideals (etc.) defining $\mathcal{U},\mathcal{V}$ respectively. Then $\mathcal{U} \circ \mathcal{U} = \mathcal{U}$ and $\mathcal{V} \circ \mathcal{V} = \mathcal{V}$, so $I(I) = I$ and $J(J) = J$. Since $A/\mathcal{U}(A) \in \mathcal{V}$ for each $A \in \mathcal{W}$, we have $A(\mathcal{V}) \subseteq \mathcal{U}(A)$, and thus $A(\mathcal{V}) \in \mathcal{U}$, for each $A \in \mathcal{W}$. In particular, for a free object F of countably infinite rank, $I = F(\mathcal{V}) \in \mathcal{U}$, so $|J(I)| = 1$. Finally, $F/I \vee J \in \mathcal{V} \cap \mathcal{U}$, so $|F/I\vee J| = 1$, i.e. $F = I \vee J$.

Conversely, if I and J satisfy the stated conditions, let

$$\mathcal{U} = \{A\in\mathcal{W} \mid |J(A)|=1\}, \quad \mathcal{V} = \{A\in\mathcal{W} \mid |I(A)|=1\} .$$

Then $\mathcal{U}$ and $\mathcal{V}$ are both extension-closed, and in particular, $\mathcal{U}$ is a radical class. If $A \in \mathcal{V}$, then $\mathcal{U}(A) \in \mathcal{U} \cap \mathcal{V}$, so $|J(\mathcal{U}(A))| = 1 = |I(\mathcal{U}(A))|$. Hence

$$|\mathcal{U}(A)| = |F(\mathcal{U}(A))| = |(I\vee J)(\mathcal{U}(A))| = |I(\mathcal{U}(A))\vee J(\mathcal{U}(A))| = 1.$$

On the other hand, if $|\mathcal{U}(B)| = 1$, we have $|J(I(B))| = |(J(I))(B)| = 1$, so $I(B) \in \mathcal{U}$ and thus $|I(B)| = 1$, i.e. $B \in \mathcal{V}$. Thus $\mathcal{V}$ is the semi-simple class corresponding to $\mathcal{U}$. //

The above proof is based on one for modules given by Rutter [59]. In the latter, the free module on one generator was used and additional characterizations of the ideals in question were obtained.

A special case occurs when $\mathcal{U}$ is also the semi-simple class corresponding to $\mathcal{V}$. We can describe this situation as follows.

THEOREM 1.24 *Let $\mathcal{U},\mathcal{V}$ be extension-closed varieties in $\mathcal{W}$, defined by* T-*ideals (etc.)* J,I *respectively. The* *following conditions are equivalent.*

(i) $\mathcal{V} = \{A \mid |\mathcal{U}(A)| = 1\}$ *and* $\mathcal{U} = \{A \mid |\mathcal{V}(A)| = 1\}$.

(ii) $A = \mathcal{U}(A) \oplus \mathcal{V}(A)$ *for all* $A \in \mathcal{W}$.

(iii) $F = J \oplus I$, *where* F *is the free object from which* I *and* J *are taken.*

(iv) $\mathcal{W} = \mathcal{U} \vee \mathcal{V}$ *and there is a binary polynomial function* $p(x,y)$ *such that every object in* $\mathcal{U}$ *satisfies* $p(x,y) = x$ *and every object in* $\mathcal{V}$ *satisfies* $p(x,y) = y$.

The equivalence of (i), (ii) and (iii) was established for rings (proof valid generally) in [22]; it was previously proved for modules (with some other characterizations) by Bernhardt [5]. The equivalence of (ii) and (iv) is due to Grätzer *et al* [31].

Example 1.25 For modules, a couple $(\mathcal{U},\mathcal{V})$ satisfying the conditions of Theorem 1.24 is called a *centrally splitting torsion theory*. Such torsion theories correspond to central idempotents: there is such an idempotent e that $\mathcal{U} = \{M \mid eM=0\}$ and $\mathcal{V} = \{M \mid (1-e)M=0\}$. We can take $p(x,y) = ex + (1-e)y$.

Example 1.26 Let $\mathcal{W}$ be the variety of *autodistributed algebras* over Z_2. These are defined by the identities

$$x(yz) = (xy)(xz) \text{ and } (xy)z = (xz)(yz) .$$

These algebras were discussed by Fiedorowicz [12], and in

the associative case by Petrich [57]. Let $\mathcal{U} = \mathrm{VAR}((xy)z = 0 = x(yz))$, $\mathcal{V} = \mathrm{VAR}(x^2=x)$. Then $\mathcal{U},\mathcal{V}$ satisfy the conditions of Theorem 1.24. This was shown by the author [22] and Kepka [38]. Here we can take $p(x,y) = x - (x^3-y^3)^2$.

Definition 1.27 An n-*fold radical theory* T is a finite sequence $(\mathcal{R}_1,\mathcal{R}_2,\ldots,\mathcal{R}_n)$ such that $\mathcal{R}_1,\ldots,\mathcal{R}_{n-1}$ are radical classes and $\mathcal{R}_2,\ldots,\mathcal{R}_n$ respectively are the corresponding semi-simple classes.

This concept was introduced by Kurata [41] for modules. Kurata proved the module version of the following theorem. For a ring version, see [22].

THEOREM 1.28 *The only possible multiple radical theories are those of the following kinds*

(i) $(\mathcal{R},\mathcal{S},\mathcal{T})$ *(non-extendable)*

(ii) $(\mathcal{R},\mathcal{S},\mathcal{T},\mathcal{U})$ *(non-extendable)*

(iii) $(\mathcal{R},\mathcal{S},\mathcal{R},\mathcal{S},\mathcal{R},\mathcal{S},\ldots)$.

Examples of all types are given for varieties of rings in [32]; for modules, see Kurata [41] and Rutter [59].

We have seen when varieties are radical classes and when they are semi-simple classes. Our last results in this section involve, in special cases, the upper and lower radicals defined by varieties in general.

Let $\mathcal{V}$ be a variety in $\mathcal{W}$, $A \in \mathcal{W}$. We define subobjects $A(V^{\alpha})$ as follows:

$$A(V^1) = A(V),\ A(V^2) = A(V)(V),\ A(V^{\alpha+1}) = A(V^{\alpha})(V)\ ,$$

$$A(V^{\beta}) = \bigcap_{\alpha<\beta} A(V^{\alpha}) \text{ if } \beta \text{ is a limit .}$$

THEOREM 1.29 *Let V be a variety in W, R the upper radical class defined by V. The following conditions are equivalent.*

(i) *R is strict (i.e. all R-subobjevts of every object A are contained in R(A)).*

(ii) $R(A) = \bigcap_{\alpha} A(V^{\alpha})$ *for all* $A \in W$.

This is proved for algebras in [16]. Condition (i) holds for all varieties V of associative rings [15] and groups [43]. Condition (ii) for groups was obtained by Shchukin [63].

COROLLARY 1.30 *If the variety V defines an upper radical class which is strict, the smallest semi-simple class containing V is*

$$\{A \mid \exists\alpha \text{ with } |A(V^{\alpha})| = 1\}\ .$$

For varieties V of associative rings, the rings A satisfying the condition

$$A(V^{\alpha}) = 0 \text{ for some } \alpha$$

were discussed by Martynov [51] under the name *V-solvable* rings.

We do not know of a characterization of the semi-simple class defined by a variety more generally; not even for alternative rings.

The following result was obtained by the author and Wiegandt [29].

THEOREM 1.31 *Let V be a variety of associative rings, containing all zerorings. The lower radical class determined by V is $B \circ V$, where B is the Baer lower radical class.*

An analogous result holds for associative algebras over a field. For algebras over a field K of characteristic 0, Ryabukhin and Florea [61] have shown that a variety V defines the same lower radical class as $VAR(M_m(K))$, where $M_m(K)$ is the matrix algebra, and $m = \max\{n | M_n(K) \in V\}$.

We end this section with two remarks on the concept of attainability. As mentioned above, attainability was originally defined for semigroups, and can be defined in areas outside the "traditional" territory of radical theory. For some results relating attainability to "non-standard" versions of radical theory, see the papers of Márki [47], Márki and Wiegandt [48] and the author [27]. There is a generalization of attainability introduced by Shevrin and Martynov [64]: V is *attainable of degree* α if (in the notation of Theorem 1.29) $A(V^{\alpha}) = A(V^{\alpha+1})$ for all A, and α is the smallest such ordinal. For a very comprehensive account of this and other generalizations, see the survey paper by Shevrin and

and Martynov [65].

2. Locally Equational Classes

Nil (associative) rings are those in which every element satisfies an equation $x^n = 0$; locally nilpotent (associative) rings are characterized by the property that every finite subset satisfies an equation $x_1 x_2 \ldots x_n = 0$; torsion groups are those in which every element satisfies an equation $x^n = 1$. Classes such as those described, while not varieties, are characterized by equations, or polynomial identities: the identities are satisfied by subsets of objects rather than whole objects. They provide examples of the following concept, introduced by Hu [35]

Definition 2.1 Let $\mathcal{C}$ be a subclass of a universal class $\mathcal{W}$. Let $E_L(\mathcal{C})$ denote the class of objects $A \in \mathcal{W}$ with the following property. For each finite subset S of A, there is a finite set $\{B_1, \ldots, B_n\} \subseteq \mathcal{C}$ and there are finite subsets $T_i \subseteq B_i$, $i = 1, \ldots, n$, such that every equation satisfied by each T_i is satisfied by S. The class $\mathcal{C}$ is *locally equational*, if $E_L(\mathcal{C}) = \mathcal{C}$.

THEOREM 2.2 [35] *For any $\mathcal{C}$, we have $E_L E_L(\mathcal{C}) = E_L(\mathcal{C})$, and this is the class of directed unions of homomorphic images of subobjects of finite direct products of objects from $\mathcal{C}$.*

There is a connection between locally equational

classes and a certain type of radical class, which we now introduce.

Definition 2.3 (i) A radical class R is *local* if it satisfies the condition

$A \in R \iff$ every finitely generated subobject of A is in R.

(ii) A radical class R is an *n-radical* class ($n=1,2,\ldots$) if it satisfies the condition

$A \in R \iff$ every subobject of A generated by $\leq n$ elements is in R.

Ryabukhin [60] first investigated 1-radical classes of rings (associative and otherwise) under a different name. Stewart [72], [73] studied local and n-radical classes of associative rings (but called the latter n + 1-radical classes).

Let F be a free W-object on $\{x_1,x_2,\ldots\}$. For $f \in F$ we shall write $f = f(x_1,x_2,\ldots)$ (i.e. refer to all variables, though only finitely many are used non-trivially). For $A \in W$, $(a_1,\ldots,a_n) \in A \times \ldots \times A$, let

$$\alpha(a_1,\ldots,a_n) = \{f \in F \mid f(a_1,\ldots,a_n,e,e,e,\ldots)=e\}$$

(where, as a temporary measure, we have used e to denote the zero or group identity, as appropriate). If we define a homomorphism from F to A by

$$x_1 \mapsto a_1;\ \ldots;\ x_n \mapsto a_n;\ x_i \mapsto e \text{ for } i > n\ ,$$

the image is the subobject generated by $\{a_1,\ldots,a_n\}$ and the kernel is $\alpha(a_1,\ldots,a_n)$.

Definition 2.4 Let F be a free $\mathcal{W}$-object on $\{x_1,x_2,\ldots\}$. A set Φ of normal subobjects of F is called a *radical filter* if it satisfies the following conditions.

(i) If $J \in \Phi$, then $x_i \in J$ for almost all i.

(ii) If $J \in \Phi$ and $J \subseteq K \triangleleft F$, then $K \in \Phi$.

(iii) If $J \in \Phi$, then $\alpha(\beta_1+J,\ldots,\beta_n+J)$ (or $\alpha(\beta_1 J,\ldots,\beta_n J)) \in \Phi$ for every finite set $\{\beta_1,\ldots,\beta_n\} \subseteq F$.

(iv) If $K \triangleleft F$, $x_i \in K$ for almost all i and if $\alpha(j_1+K,\ldots,j_m+K)$ (or $\alpha(j_1 k,\ldots,j_m K)) \in \Phi$ for all finite subsets $\{j_1,\ldots,j_m\}$ of some $J \in \Phi$, then $K \in \Phi$.

THEOREM 2.5 *There are bijections between the collections of local radical classes in* $\mathcal{W}$ *and radical filters of normal subobjects of* F, *given by*

$$R \mapsto \{\alpha(a_1,\ldots,a_n) \mid a_1,\ldots,a_n \in A \in R, n=1,2,\ldots\} ;$$

$$\Phi \mapsto \{A \mid \alpha(a_1,\ldots,a_n) \in \Phi \text{ for all } a_1,\ldots,a_n \in A, n=1,2,\ldots\}.$$

This is proved for rings and algebras in [17]. With some minor modifications the proof can be made to work generally. The same goes for the next result.

THEOREM 2.6 *Let* R *be a local radical class in* $\mathcal{W}$. *For every finite subset* S *of an object in* $\mathcal{W}$, *let*

$$\alpha^*(S) = \{f \in F \mid \text{every substitution into } f \text{ from } S \text{ gives } 0 \text{ (or } 1)\} .$$

Then A $\in$ *R if and only if* $F/\alpha^*(S) \in$ *R for all finite* $S \subseteq A$.

It is now straightforward to prove the following (cf. [17]).

THEOREM 2.7 *The following conditions are equivalent for a non-empty class R.*

(i) *R is a local radical class.*

(ii) *R is a locally equational class and* $R \circ R = R$.

Since varieties are special cases of locally equational classes, Theorem 2.7 is a generalization of Theorem 1.2.

For n-radical classes there is an analogue of Theorem 2.5 using radical filters of normal subobjects from a free object on n generators. (Of course, n-radical classes are n+1-radcial classes, and local radical classes.)

In the case of modules, all local radical classes (in fact all hereditary radical classes) are 1-radical classes, and the 1-radical version of Theorem 2.5 is the Gabriel correspondence [14] between torsion classes and idempotent topologizing filters. The 1-radical version of Theorem 2.5 for rings was first obtained by Ryabukhin [60]; this provided the motivation for [17].

Examples 2.8 (i) For groups, the 1-radical classes are the classes R_P defined by sets P of primes as follows.

$$G \in R_P \iff (\forall a \in G)(p \text{ prime and } p \mid 0(a)) \Rightarrow p \in P .$$

(This is because the free group on one generator is abelian, so that 1-radical filters for groups are the same as for abelian groups.) The class of groups without free subgroups of rank ≥ 2 is a 2-radical class [26].

(ii) Local radical classes of associative rings include the classes of locally nilpotent and nil rings, and the class of nil-commutator rings. (The last is a 2-radical class; this example stems from some results of Freidman [13].)

(iii) The locally solvable radical class of Parfenov [56] is a local radical class of Lie algebras.

Results similar to Theorem 2.5, in a very general setting, were obtained by Borceux [6]. (Thanks to Tim Porter for this reference.)

We conclude this section with a generalization of the result of Jans [36] characterizing the semi-simple radical classes of modules. It is proved for rings in [17].

THEOREM 2.9 *Let* R *be a local (or* n-*) radical class,* Φ *the corresponding radical filter of normal subobjects in a free object on* $\aleph_0$ *(or* n*) generators. The following conditions are equivalent;*

(i) Φ *has a smallest element;*

(ii) R *is closed under direct products;*

(iii) R *is an extension-closed variety.*

3. Quasivarieties

Locally equational classes, while providing interesting examples of radical classes, can only be semi-simple classes when they are varieties. Our second generalization of varieties has the opposite behaviour.

A *quasivariety* is defined by implications of the following kind

$$f_1 = g_1 \ \& \ f_2 = g_2 \ \& \ldots \& \ f_n = g_n \Rightarrow f = g$$

where the $f_i = g_i$ and $f = g$ are equations (polynomial identities). Equations are included among the implications: $f = g$ is equivalent to $x = x \Rightarrow f = g$.

Quasivarieties are closed under subobjects, direct products and ultraproducts, and, more generally, filtered products (see, e.g., [11]) and thus in order to be radical classes they must be varieties.

Quasivarities which are semi-simple classes are much more interesting.

Let $T \subseteq W$ be a quasivariety. For $A \in W$, let

$$A(T) = \cap\{I \triangleleft A \mid A/I \in T\} .$$

Exactly as in the case of varieties (Theorem 1.5 (i)) we have

THEOREM 3.1 *A quasivariety* T *is a semi-simple class if and only if* $A(T)(T) = A(T)$ *for every* A.

In cases where semi-simple classes are characterized

by closure under normal subobjects, subdirect products and extensions we can replace the condition $A(T)(T) = A(T)$ for all A by the condition $T \circ T = T$. This is the case for associative and alternative rings (Sands [62], van Leeuwen, Roos and Wiegandt [45]), groups (Chang Wang Hao [9]) and of course modules; see also some examples in [1]. We summarize part of this in the next result.

THEOREM 3.2 *A quasivariety T of associative or alternative rings, groups or modules is a semi-simple class if and only if $T \circ T = T$.*

It is somewhat easier with quasivarieties than with varieties to find examples to show that extension-closure is not sufficient for a semi-simple class:

PROPOSITION 3.3 *Let T be a quasivariety defined by a collection of implications of the form*

$$f(x_1,\dots,x_n) = g(x_1,\dots,x_n) \Rightarrow x_1 = e; \dots ; f(x_1,\dots,x_n) = g(x_1,\dots,x_n) \Rightarrow x_n = e .$$

(We have again used e for the zero or group identity.) *Then $T \circ T = T$.*

Proof. If $I < A$ and if I and A/I are in T, let $a_1,\dots,a_n \in A$ be such that $f(a_1,\dots,a_n) = g(a_1,\dots,a_n)$. Then (denoting cosets modulo I by $(\bar{\ })$) we have $f(\bar{a}_1,\dots,\bar{a}_n) = g(\bar{a}_2,\dots,\bar{a}_n)$, so $\bar{a}_1 = \dots = \bar{a}_n = 0$ or 1 as appropriate. Hence we have $a_1,\dots,a_n \in I \in T$ and

$f(a_1,\ldots,a_n) = g(a_1,\ldots,a_n)$, so $a_1 = \ldots = a_n = 0$ or 1. Thus A is in T. //

When W is the class of all (not necessarily associative) rings, the only hereditary semi-simple classes are those defined in terms of additive structure [20], so we get examples like the following.

Example 3.4 In the class of all rings, let T be defined by

$$x^2 = 0 \Rightarrow x = 0 .$$

The $T \circ T = T$, but T is not a semi-simple class.

It is fairly clear that membership of the quasivariety T in Example 3.4 is not determined by additive structure. In any case, this will become completely obvious after the next result.

THEOREM 3.5 *A non-trivial quasivariety* T *of abelian groups is extension-closed (and thus a semi-simple class) if and only if there is a set* P *of primes such that* T *is defined by the implications*

$$px = 0 \Rightarrow x = 0; \; p \in P .$$

Theorem 3.5 is due to L. M. Martynov and appeared as Proposition 2 in [83].

COROLLARY 3.6 *A quasivariety is a semi-simple class in the class of all rings if and only if it is defined by the implications*

$$px = 0 \Rightarrow x = 0; \; p \in P ,$$

for some set P *of primes.*

Analogous results hold for the universal classes of commutative and anticommutative rings [24]. The same is true for algebras of these types, and in this context there are no non-trivial quasivarieties which are semi-simple classes.

In the class of abelian groups, the semi-simple class of every hereditary radical class is a quasivariety. For modules generally, we have the following result, obtained by Komarnitskii [40] and Prest [58].

THEOREM 3.7 *The semi-simple class corresponding to a hereditary radical class* R *of modules is a quasivariety if and only if the idempotent topologizing filter associated with* R *has a cofinal subset of finitely generated left ideals.*

For the remainder of the paper we shall work entirely in the class of associative rings. When a non-trivial variety of associative rings is a semi-simple class, the corresponding radical class is supernilpotent. The corresponding statement is false for quasivarieties: Proposition 3.3 provides easy counterexamples; e.g. consider the class defined by the condition

$$x^2 = x \Rightarrow x = 0 .$$

In fact it seems likely from Proposition 3.3 (and Theorem 3.2) that the problem of classifying semi-simple quasivarieties will be much more difficult than that for semi-simple varieties.

In the sequel we shall only consider supernilpotent (not necessarily hereditary) radical classes with quasivarieties for semi-simple classes. This means that we shall be concerned with radical classes containing the generalized nil radical class N_g, as subrings of semi-simple rings must be semi-simple and hence semiprime. We shall denote the upper radical class defined by a quasivariety T by $U(T)$.

A standard question asked about hereditary supernilpotent radicals of various types is whether they must be special. We have an affirmative answer in the present case.

THEOREM 3.8 *Let* T *be a quasivariety of semiprime associative rings. Then* $U(T)$ *is hereditary if and only if it is special.*

The proof of Theorem 3.8 follows easily from some results of Stewart [74] in view of the following lemma.

LEMMA 3.9 *Let* T *be a quasivariety,* A *a ring,* $\{I_\lambda | \lambda \in \Lambda\}$ *an ascending chain of ideals of* A *such that* $A/I_\lambda \in T$ *for each* $\lambda \in \Lambda$*. Then* $A/\cup I_\lambda \in T$.

Proof. Let

$$f(x_1,\ldots,x_n) = 0 \Rightarrow g(x_1,\ldots,x_n) = 0$$

be one of a family of implications defining T. If $a_1,\ldots,a_n \in A$ and $f(a_1,\ldots,a_n) \in \cup I_\lambda$, then $f(a_1,\ldots,a_n) \in I_\mu$ for some $\mu \in \Lambda$, so, since $A/I_\mu \in T$, we have $g(a_1,\ldots,a_n) \in I_\mu \subseteq \cup I_\lambda$. //

Turning to the proof of Theorem 3.8, if $U(T)$ is hereditary, then as $N_g \subseteq T$, a combination of Lemma 3.9, and Propositions 1 and 2 of Stewart [74] yields the result that if $A \in T$, then $A/P \in T$ for every minimal prime ideal P of A. Since for rings in T the intersection of all minimal primes is zero, it follows that $U(T)$ is special. The converse is well-known. //

The most difficulty thing to recognize about a quasivariety T as described in Theorem 3.8, once we know that $T \circ T = T$, is the hereditary property of $U(T)$ (or, equivalently, the closure of T under essential extensions (Armendariz [4]). Our final result provides a sufficient condition for $U(T)$ to be hereditary, i.e. special. It bypasses the need to verify extension-closure directly, which can be handy.

THEOREM 3.10 *Let T be a quasivariety of semiprime associative rings defined by the identity*

$$xy = yx$$

and a family of implications

$$f_1 = 0 \ \& \ f_2 = 0 \ \& \ldots \& \ f_t = 0 \Rightarrow g = 0 ,$$

where all monomials in f_i *have the same total degree, for each* i, *and all monomials in* g *have the same total degree. Then* T *is a semi-simple class and* U(T) *is special.*

Proof. Let A be in T and let P be a minimal prime ideal of A. Then A\P is a multiplicative semigroup which is maximal with respect to the exclusion of 0. Suppose $A/P \notin T$.

Then at least one of the defining conditions (not commutativity) is not satisfied by A/P. Let

$$f_1(x_1,\ldots,x_n) = 0 \ \& \ f_2(x_1,\ldots,x_n) = 0 \ \&\ldots\& \ f_t(x_1,\ldots,x_n) = 0 \Rightarrow g(x_1,\ldots,x_n) = 0$$

be such a condition. Then there are elements $a_1,\ldots,a_n \in A$ such that P contains $f_1(a_1,\ldots,a_n),\ldots,$ $f_t(a_1,\ldots,a_n)$, but not $g(a_1,\ldots,a_n)$. For $i = 1,\ldots,t$, let $\alpha_i = f_i(a_1,\ldots,a_n)$, and for $a \in A$, let

$$[a] = \{a,a^2,a^3,\ldots\} .$$

We first prove that for some i, the set

$$T_i = \{b\alpha_i^m \mid b\in A\setminus P, m\geq 1\}$$

does not contain 0. Suppose, on the contrary, that $0 \in T_i$ for every i, i.e. for $i = 1,\ldots,t$ there is an element b_i of A\P such that $b_i\alpha_i^{m_i} = 0$ for some m_i. Let $b = b_1b_2\ldots b_t$. Then $b \notin P$ as P is (strongly) prime and $b_1,b_2,\ldots,b_k \notin P$. Also $b\alpha_i^{m_i} = 0$ for each i, so $(b\alpha_i)^{m_i} = b^{m_i}\alpha_i^{m_i} = 0$, whence

$b\alpha_i = 0$. Let k_i be the total degree of f_i. Then we have

$$f_i(ba_1,\dots,ba_n) = b^{k_i} f_i(a_1,\dots,a_n) = b^{k_i}\alpha_i = 0$$

for each i. Hence $g(ba_1,\dots,ba_n) = 0$. Let ℓ be the total degree of g. Then

$$b^{\ell} g(a_1,\dots,a_n) = g(ba_1,\dots,ba_n) = 0 \in P ,$$

while $b \notin P$. This means that $g(a_1,\dots,a_n) \in P$ - a contradiction.

Hence $0 \notin T_i$ for some i. Hence also $0 \notin [\alpha_i]$. But

$$(A \backslash P) \cup [\alpha_i] \cup T_i$$

is a multiplicative semigroup excluding 0 and containing the maximal such semigroup $A\backslash P$, which means that

$$\alpha_i = f_i(a_1,\dots,a_n) \in A\backslash P .$$

This is contradictory to our supposition. Hence $A/P \in \mathcal{T}$ for every minimal prime ideal P of A. Since A is semiprime, we conclude that every ring in $\mathcal{T}$ is a subdirec product of prime rings in $\mathcal{T}$.

Now let R be a prime ring in $\mathcal{T}$ and let V be an essential extension of R (i.e. $R \triangleleft V$ and $R \cap I \neq 0$ whenever $0 \neq I \triangleleft R$). As R is prime and commutative, so i V (see, e.g. Osborn [55], p.309). Let

$$f_1(x_1,\dots,x_n) = 0 \;\&\; f_2(x_1,\dots,x_n) = 0 \;\&\dots\&\; f_t(x_1,\dots,x_n) = 0 \Rightarrow g(x_1,\dots,x_n) = 0$$

be one of the other defining conditions for $\mathcal{T}$. Let

$v_1, \ldots, v_n$ in V be such that $f_i(v_1, \ldots, v_n) = 0$. If $v_j \neq 0$, then $Vv_j \cap R \neq 0$. Thus for each j there is a non-zero element s_j of V such that $s_j v_j \in R$. Let $s = s_1 \ldots s_n$. Then $sv_j \in R$ for each j. Let k_i be the total degree of f_i. Then as V is commutative, we have

$$f_i(sv_1, \ldots, sv_n) = s^{k_i} f_i(v_1, \ldots, v_n) = 0$$

for each i, so, $sv_1, \ldots, sv_n$ being in R, $g(sv_1, \ldots, sv_n) = 0$. Let ℓ be the total degree of g. Then

$$s_1^{\ell} \ldots s_n^{\ell} g(v_1, \ldots, v_n) = s^{\ell} g(v_1, \ldots, v_n) = g(sv_1, \ldots, sv_n) = 0,$$

with $s_1, \ldots, s_n$, $g(v_1, \ldots, v_n)$ in the commutative prime ring V, so $g(v_1, \ldots, v_n) = 0$. Thus V satisfies all the implications and so $V \in \mathcal{T}$. Let $\mathcal{P}$ be the class of prime rings in $\mathcal{T}$. We have shown that $\mathcal{P}$ is closed under essential extensions. It is clearly hereditary also, so by a theorem of Heyman and Roos [34] is a special class. As shown above, every ring in $\mathcal{T}$ is a subdirect product of rings in $\mathcal{P}$. Since $\mathcal{T}$ is closed under subrings and direct products and $\mathcal{P} \subseteq \mathcal{T}$, the converse is also true. Hence $\mathcal{T}$ is the semi-simple class generated by $\mathcal{P}$ and, of course, $U(\mathcal{T}) = U(\mathcal{P})$ is special. //

The first part of the proof of Theorem 3.10 is a generalization of an argument used by Thierrin [77]. Thierrin's paper is the source of the idea behind the first of the following examples.

Examples 3.11 The quasivarieties of associative

rings defined by the following sets of conditions are semi-simple classes with special upper radicals.

(i) $x^n + y^n + z^n = 0 \Rightarrow x + y + z = 0;\ xy - yx = 0.$

(ii) $x^n + y^n = 0 \Rightarrow x = 0,\ y = 0;\ xy - yx = 0.$

Note that the implication $x^2 = 0 \Rightarrow x = 0$ is a consequence of the first condition in each case. Note also that $xy = yx$, in the form $x = x \Rightarrow xy - yx = 0$, satisfies the "homogeneity" conditions of Theorem 3.10. It played a special role in the proof and so was stated separately. Mason [52] has considered commutative rings satisfying the conditions of 3.11 (ii) with n = 2 (and als other properties encompassed by Theorem 3.10.)

Theorem 3.10 has a lot of hypotheses and is thus a bit unsatisfactory; in particular, one would like to get rid of commutativity. It's worth observing, however, that even with commutativity, extension-closure and semiprimeness do not suffice to yield a hereditary radica class.

Example 3.12 Let T be the quasivariety of associative rings defined by

$$x^3 = x^2 \Rightarrow x = 0;\ xy - yx = 0 .$$

Then all rings in T are commutative and semi-prime. But $2Z \in T$ while Z has no non-zero homomorphic image in T, so T has a non-hereditary upper radical class.

References

[1] T. Anderson and B.J. Gardner, Semi-simple classes in a variety satisfying an Adrunakievich lemma, *Bull. Austral. Math. Soc.* 18 (1978), 187-200.

[2] V.A. Andrunakievič, Radicals of associative rings I, *Amer. Math. Soc. Transl.* (2) 52 (1966), 95-128. Russian original: *Mat. Sb.* 44 (1958), 179-212.

[3] V.A. Andrunakievič, Radicals of associative rings II, *Amer. Math. Soc. Transl.* (2) 52 (1966), 129-149. Russian original: *Mat. Sb.* 55 (1961), 329-346.

[4] E.P. Armendariz, Closure properties in radical theory *Pacific. J. Math.* 26 (1968), 1-7.

[5] R.L. Bernhardt, Splitting hereditary torsion theories over semiperfect rings, *Proc. Amer. Math. Soc.* 22 (1969), 681-687.

[6] F. Borceux, Algebraic localizations and elementary toposes, *Cahiers Topologie Geom. Diff.* 21 (1980), 393-401.

[7] R.H. Bruck, *A Survey of Binary Systems*, Springer, Berlin-New York, 1971.

[8] R.H. Bruck and L.J. Paige, Loops whose inner mappings are automorphisms, *Ann. of Math.* (2) 63 (1956), 308-323.

[9] Chang Wang Hao, On semi-simple classes of groups, *Sibirskii Mat. Zhurnal* 3 (1962), 943-949 (in Russian).

[10] P.M. Cohn, *Free Rings and Their Relations*, Academic Press, London-New York, 1971.

[11] P.M. Cohn, *Universal Algebra*, D. Reidel Publishing Company, Dordrecht-Boston-London, 1981.

[12] Z. Fiedorowicz, The structure of autodistributive algebras, *J. Algebra*, 31 (1974), 427-436.

[13] P.A. Freidman, On the theory of the radical of an associative ring, *Izv. Vyshch. Ucheb. Zaved. Mat.* No.3 (4) (1958), 225-232 (in Russian).

[14] P. Gabriel, Des catégories abéliennes, *Bull. Soc. Math. France* 90 (1962), 323-448.

[15] B.J. Gardner, Some radical constructions for associative rings, *J. Austral. Math. Soc.* 18 (1974), 442-446.

[16] B.J. Gardner, Semi-simple radical classes of algebras and attainability of identities, *Pacific J. Math.* 61 (1975), 401-416.

[17] B.J. Gardner, Radical properties defined locally by polynomial identities, I, *J. Austral. Math. Soc. Ser. A* 27 (1979), 257-273.

[18] B.J. Gardner, Radical properties defined locally by polynomial identities, II, *J. Austral. Math. Soc. Ser. A* 27 (1979), 274-283.

[19] B.J. Gardner, Extension-closed varieties of rings need not have attainable identities, *Bull. Malaysian Math. Soc.* (2) 2 (1979), 37-39.

[20] B.J. Gardner, Some degeneracy and pathology in non-associative radical theory, *Annales Univ. Sci. Budapest. Sect. Math.* 22-23 (1979-80), 65-74.

[21] B.J. Gardner, Radicals related to the Brown-McCoy radical in some varieties of algebras, *J. Austral. Math. Soc. Ser. A* 28 (1979), 283-294.

[22] B.J. Gardner, Multiple radical theories, *Colloq. Math* 41 (1979), 345-351.

[23] B.J. Gardner, Extension-closure and attainability for varieties of algebras with involution, *Comment. Math. Univ. Carolinae* 21 (1980), 285-292.

[24] B.J. Gardner, Some degeneracy and pathology in non-associative radical theory II, *Bull. Austral. Math. Soc* 23 (1981), 423-428.

[25] B.J. Gardner, Radical classes of regular rings with artinian primitive images, *Pacific J. Math.* 99 (1982) 337-349.

[26] B.J. Gardner, Radical properties defined by the absence of free subobjects, *Annales Univ. Sci. Budapest. Sect. Math.* 25 (1982), 53-60.

[27] B.J. Gardner, Radical decompositions of idempotent algebras, *J. Austral. Math. Soc. Ser. A* (to appear).

[28] B.J. Gardner and P.N. Stewart, On semi-simple radical classes, *Bull. Austral. Math. Soc.* 13 (1975), 349-353.

[29] B.J. Gardner and R. Wiegandt, Characterizing and constructing special radicals, *Acta Math. Acad. Sci. Hungar.* 40 (1982), 73-83.

[30] A.M.W. Glass, W.C. Holland and S.H. McCleary, The structure of ℓ-group varieties, *Algebra Universalis* 10 (1980), 1-20.

[31] G. Grätzer, H. Lakser and J. Plonka, Joins and direct products of equational classes, *Canad. Math. Bull.* 12 (1969), 741-744.

[32] I.R Hentzel and G.M.P. Cattaneo, Simple (γ,δ) algebras are associative, *J. Algebra* 47 (1977), 52-76.

[33] I.R. Hentzel and M. Slater, On the Andrunakievich lemma for alternative rings, *J. Algebra* 27 (1973), 243-256.

[34] G.A.P. Heyman and C. Roos, Essential extensions in radical theory, *J. Austral. Math. Soc. Ser. A* 23 (1977), 340-347.

[35] T.K. Hu, Locally equational classes of universal algebras, *Chinese J. Math.* 1 (1973), 143-165.

[36] J.P. Jans, Some aspects of torsion, *Pacific J. Math.* 15 (1965), 1249-1259.

[37] V.M. Kaplan, On attainable classes of groupoids, quasigroups and loops, *Sibirsk. Mat. Zhurnal* 19 (1978), 604-616 (in Russian).

[38] T. Kepka, On a class of non-associative rings, *Comment. Math. Univ. Carolinae* 18 (1977), 531-540.

[39] S.R. Kogalovskii, Structural characteristics of universal classes, *Sibirskii Mat. Zhurnal* 4 (1963), 97-119 (in Russian).

[40] N. Ya. Komarnitskii, On the axiomatizability of certain classes of modules, connected with torsion, *Algebraicheskie Struktury (Mat. Issled. vyp. 56)*, Kishinev (1980), 92-109 (in Russian).

[41] Y. Kurata, On an n-fold torsion theory in the category ${}_R M$, *J. Algebra* 22 (1972), 559-572.

[42] A.G. Kurosh, Non-associative free algebras and free products of algebras, *Mat. Sb. (Receuil Mathématique)* 20 (1947), 239-262 (in Russian).

[43] A.G. Kuroš, Radicals in the theory of groups, *Rings, Modules and Radicals (Proc. Conf. Keszthely, 1971)*, pp. 271-296. Colloq. Math. Soc. J. Bolyai, 6, North-Holland, Amsterdam, 1973.

[44] L.C.A. van Leeuwen and T.L. Jenkins, A note on radical semisimple classes, *Publ. Math. Debrecen* 21 (1974), 179-184.

[45] L.C.A. van Leeuwen, C. Roos and R. Wiegandt, Characterizations of semisimple classes, *J. Austral. Math. Soc. Ser. A* 23 (1977), 172-182.

[46] N.V. Loi, Essentially closed radical classes, *J. Austral. Math. Soc. Ser. A* 35 (1983), 132-142.

[47] L. Márki, Radical semisimple classes and varieties of semigroups with zero, *Algebraic Theory of Semigroups (Proc. Conf. Szeged, 1976)* pp. 357-369 Colloq. Math. Soc. J. Bolyai 20, North-Holland, Amsterdam, 1979.

[48] L. Márki and R. Wiegandt, On semisimple classes of semigroups with zero, *Semigroups* (Oberwolfach, 1978), pp. 55-72. Lecture Notes in Math. Vol. 855, Springer, Berlin-New York, 1981.

[49] A.S. Markovichev, On hereditary radicals of rings of type (γ,δ), *Algebra i Logika* 17 (1978), 33-55 (in Russian).

[50] J. Martinez, Varieties of lattice-ordered groups, *Math. Z.* 137 (1974), 265-284.

[51] L.M. Martynov, On solvable rings, *Ural. Gos. Univ. Mat. Zap.* 8, *Tetrad'* 3 (1972), 82-93 (in Russian).

[52] G. Mason, z-ideals and prime ideals, *J. Algebra* 26 (1973), 280-297.

[53] B.H. Neumann, H. Neumann and P.M. Neumann, Wreath products and varieties of groups, *Math. Z.* 80 (1962), 44-62.

[54] H. Neumann, *Varieties of Groups*, Springer, New York, Berlin, 1967.

[55] J.M. Osborn, Varieties of algebras, *Advances in Math.* 8 (1972), 163-369.

[56] V.A. Parfenov, On the weakly solvable radical of Lie algebras, *Sibirsk. Mat. Zhurnal* 12 (1971), 171-176 (in Russian).

[57] M. Petrich, Structure des demi-groupes et anneaux distributifs, *C.R. Acad. Sci. Paris Sér A-B* 268 (1969), A849-A852.

[58] M. Prest, Torsion and universal Horn classes of modules, *J. London Math. Soc.* (2) 19 (1979), 411-416.

[59] E.A. Rutter, Jr., Four-fold torsion theories, *Bull. Austral. Math. Soc.* 10 (1974), 1-8.

[60] Yu. M. Ryabukhin, Semi-strictly hereditary radicals in primitive classes of rings, *Issled. po Obshch. Algebre*, Kishinev (1965), 112-122 (in Russian).

[61] Yu. M. Ryabukhin and R.S. Floria, Radicals and the classification of varieties I, *Algebraicheskie Struktury (Mat. Issled. vyp.* 56*)*, Kishinev (1980), 123-131.

[62] A.D. Sands, Strong upper radicals, *Quart. J. Math. Oxford* 27 (1976), 21-24.

[63] K.K. Shchukin, On verbal radicals of groups, *Kishinevskii Gos. Univ. Uchenye Zapiski*, 82 (1965), 97-99 (in Russian).

[64] L.N. Shevrin and L.M. Martynov, On attainable classes of algebras, *Sibirskii Mat. Zhurnal* 12 (1971), 1363-1381.

[65] L.N. Shevrin and L.M. Martynov, Attainability and solvability for classes of algebras, *Semigroups (Proc. Conf. Szeged, 1981)* Colloq. Math. Soc. J. Bolyai (to appear).

[66] A.I. Shirshov, Subalgebras of free Lie algebras, *Mat. Sb.* 33 (1953), 441-452 (in Russian).

[67] A.I. Shirshov, Subalgebras of free commutative and free anticommutative algebras, *Mat. Sb.* 34 (1954), 81-88 (in Russian).

[68] A.L. Shmel'kin, The semigroup of varieties of groups, *Dokl. Akad. Nauk. SSSR* 149 (1963), 543-545 (in Russian).

[69] T.S. Shores, A note on products of normal subgroups, *Canad. Math. Bull.* 12 (1969), 21-23.

[70] A.M. Slin'ko, On radicals of Jordan rings, *Algebra in Logika* 11 No.2 (1972), 206-215 (in Russian).

[71] P.N. Stewart, Semi-simple radical classes, *Pacific J. Math.* 32 (1970), 249-254.

[72] P.N. Stewart, Strongly hereditary radical classes, *J. London Math. Soc.* (2) 4 (1972), 499-509.

[73] P.N. Stewart, On the locally antisimple radical, *Glasgow Math. J.* 13 (1972), 42-46.

[74] P.N. Stewart, Radicals and functional representations, *Acta Math. Acad. Sci. Hungar.* 27 (1976), 319-321.

[75] F. Szász, Beiträge zur Radikaltheorie der Ringe, *Publ. Math. Debrecen* 17 (1970), 267-272.

[76] T. Tamura, Attainability of systems of identities on semigroups, *J. Algebra* 3 (1966), 261-276.

[77] G. Thierrin, Sur les idéaux fermatiens d'un anneau commutatif, *Comment. Math. Helv.* 32 (1957-58), 241-247.

[78] R. Wiegandt, Homomorphically closed semisimple classes, *Stud. Univ. Babes-Bolyai, Ser. Math.-Mech.* 17 (1972), 17-20.

[79] R. Wiegandt, Radical-semisimple classes, *Period. Math. Math. Hungar.* 3 (1973), 243-245.

[80] R. Wiegandt, *Radical and Semisimple Classes of Rings*, Queen's University, Kingston, Ontario, 1974.

[81] E. Witt, Die Unterringe der freien Lieschen Ringen, *Math. Z.* 64 (1956), 195-216.

[82] E. N. Zakharova, On radical-semisimple classes of rings and algebras, *Algebraicheskie Struktury (Mat. Issled. vyp. 56)*, Kishinev (1980), 51-61 (in Russian).

[83] L. M. Martynov, On the attainable classes of groups and semigroups, *Mat. Sb.* 90 (1973), 235-245 (in Russian).

B. J. Gardner
Mathematics Department
University of Tasmania
G.P.O. Box 252C
Hobart Tas. 7001
AUSTRALIA

COLLOQUIA MATHEMATICA SOCIETATIS JÁNOS BOLYAI
38. RADICAL THEORY, EGER (HUNGARY), 1982.

TOPOLOGIES DEFINED ON THE TORSION-THEORETIC SPECTRUM OF A NONCOMMUTATIVE RING: SOME RECENT RESULTS AND OPEN PROBLEMS

J. S. GOLAN

The organizers of this conference have asked me to present a survey of current research on some aspect of torsion theories on module categories and I have chosen to concentrate on a problem which, in my opinion, is far from being resolved--namely that of the "proper" definition of a topology on the torsion-theoretic spectrum of a noncommutative ring. This problem has interested me for the past ten years or so and, since the publication of [17], has attracted a considerable amount of interest and work by various other mathematicians around the world as well. Yet, it is safe to say, there is no consensus as yet as to which are the "best" topologies on the spectrum. (Note that in the commutative case there is a definitely-agreed-upon "best" topology on the ideal-theoretic spectrum of a ring, namely the Zariski topology.) Indeed, there is not even an agreed-upon criterion by which proposed topologies can be judged. Several possible criteria suggest themselves; among them:

(I) Generalization of the commutative case. A "good" topology on the torsion-theoretic spectrum of a ring should certainly be homeomorphic to the Zariski topology on the ideal-theoretic spectrum of the ring when the ring is commutative and noetherian.

(II) Lattice-theoretic considerations. One would expect that a "good" topology on the torsion-theoretic spectrum of a ring arises in some natural manner from the structure of the complete brouwerian lattice of all torsion theories on modules over that ring.

(III) Amenability for representations. A "good" topology on the torsion-theoretic spectrum should allow for a representation of the ring and modules over it naturally and easily in terms of sections of structure sheaves defined over it.

In this talk I will present a small selection of some of the topologies which have been studied, each of which is acceptable by at least one of the above criteria. I would like to make it clear, however, that the survey I am about to present is representative rather than comprehensive. It is impossible for me to talk about all--or even most--of the topologies currently under consideration or to go into detail about any of them. The best I can do is present some of them which, in my opinion, show promise as avenues of future work.

0. Background information, notation, and terminology. In all of the following, R will denote an associative ring with unit element 1 and R-mod will denote the category of all (unitary) left R-modules. The injective hull of a left R-module M will be denoted by E(M). Morphisms in R-mod will be written as acting on the right.

The complete brouwerian lattice of all (hereditary) torsion theories on R-mod will be denoted by R-tors. Notation and terminology concerning torsion theories will always follow [18]. In particular, if $\tau \in$ R-tors then $T_\tau(\_)$ will denote the τ-torsion endofunctor of R-mod and $Q_\tau(\_)$ will denote the τ-localization endofunctor of R-mod. The endomorphism ring of the left R-module $Q_\tau(R)$ will be denoted by R_τ. For any left R-module M, $Q_\tau(M)$ is, in a natural way, also a left R_τ-module.

A torsion theory τ in R-tors is said to be *jansian* if and only if the class of all τ-torsion left R-modules is closed under taking arbitrary direct products; it is called *stable* if and only if the class of all τ-torsion left R-modules is closed under taking injective hulls. The ring R is said to be *left stable* if and only if every element of R-tors is stable.

If M is a left R-module then we will denote the meet of all elements of R-tors relative to which M is torsion by $\xi(M)$ and the join of all elements of R-tors relative to which M is torsionfree by $\chi(M)$. Then $\xi = \xi(0)$ is the unique minimal element of R-tors and $\chi = \chi(0)$ is the unique maximal element of R-tors. A torsion theory of the form $\xi(R/I)$, for some left ideal I of R, is said to be *basic*.

If $\tau \in$ R-tors then a nonzero left R-module M is said to be τ-*cocritical* if and only if it is τ-torsionfree while every proper homomorphic image of it is τ-torsion. Cocritical left R-modules are always uniform. If there exists a τ-cocritical left R-module M satisfying $\tau = \chi(M)$ then the torsion theory τ is said to be *prime*. Prime torsion theories were first introduced by Goldman in [25]. The set of all prime torsion theories in R-tors is called the *left spectrum* of R and is denoted by R-sp. A torsion theory in R-tors which is the meet of some nonempty subset of R-sp is said to be *semiprime*.

Since the lattice R-tors is complete brouwerian, we know that any element τ of R-tors has a meet pseudocomplement, which we will denote by $\tau^{\perp}$. Tebyrcè [48] has given a simple characterization of this meet pseudocomplement, namely

$$\tau^{\perp} = \wedge\{\chi(M) \mid M \text{ is a } \tau\text{-torsion simple left R-module}\}.$$

In particular, this implies that the meet pseudocomplement of any element of R-tors is semiprime.

If we endow the set R-sp with any topology whatsoever and if we select a left R-module M then, following [18], we can construct a separated presheaf $\bar{Q}(\_,M)$, taking values in R-mod, as follows:

(1) If U is an open subset of R-sp then $\bar{Q}(U,M) = Q_{\wedge U}(M)$;

(2) If $\iota_{U,V}:U \to V$ is the inclusion map between open subsets of R-sp then $\bar{Q}(\iota_{U,V},M):\bar{Q}(V,M) \to \bar{Q}(U,M)$ is defined to be the unique R-homomorphism α making the diagram

$$\begin{array}{ccc} M & \longrightarrow & \bar{Q}(V,M) \\ {\scriptstyle =}\downarrow & & \downarrow{\scriptstyle \alpha} \\ M & \longrightarrow & \bar{Q}(U,M) \end{array}$$

commute.

For further details of this construction, refer also to [22].

In general, the separated presheaf $\bar{Q}(\_,M)$ thus constructed is not a sheaf. While we can always pass to the associated sheaf $\tilde{Q}(\_,M)$ if necessary, it is of definite interest to know when $\bar{Q}(\_,M)$ is itself a sheaf. This problem can be approached in three manners:

(1) Given a method of topologizing R-sp, we can search for those rings R such that $\bar{Q}(\_,M)$ is a sheaf for all M.

(2) Given a method of topologizing R-sp, and given the ring R, we can try and find the largest class of modules M such that $\bar{Q}(\_,M)$ is a sheaf.

(3) We can try and select a topology on R-sp so that, given R, $\bar{Q}(\_,M)$ is a sheaf for as large a class of modules M as possible.

It is this third approach which I would like to pursue here. In particular, I want to consider a number of possible candidate topologies on R-sp. The relations between them are summarized by the following diagram (refinement increases as one ascends):

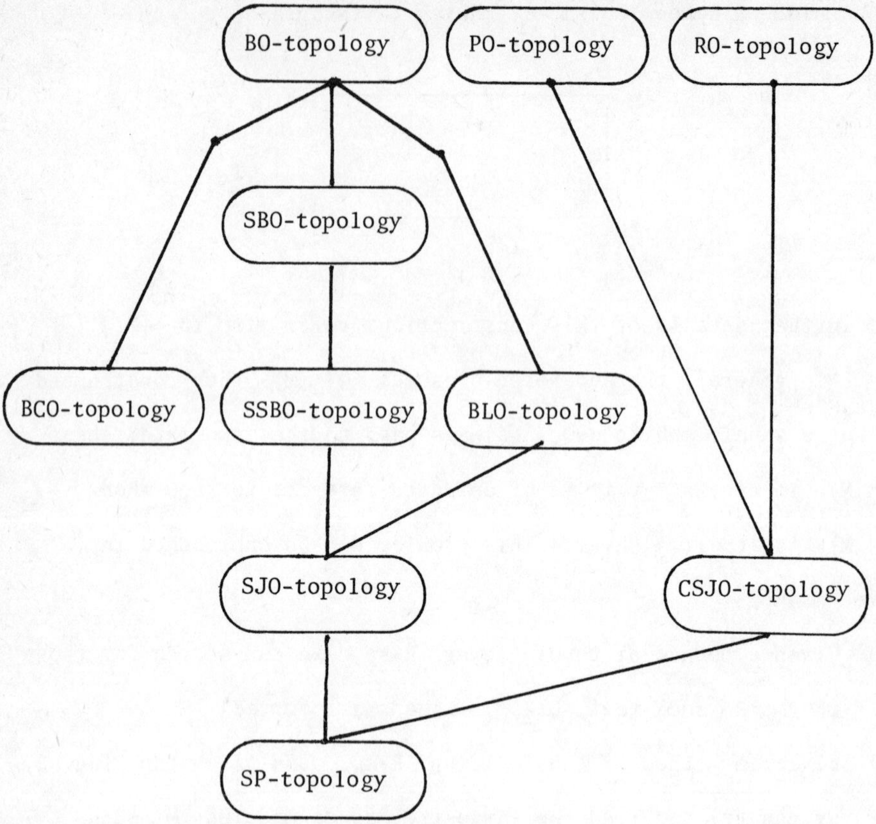

<u>1. *The basic order topology*</u>. The first topology we consider is the *basic order topology (BO-topology)*. This topology is defined by taking as a base of open sets the family of subsets of R-sp of the form $\mathbb{P}(\tau) = \{\pi \in \text{R-sp} \mid \pi \geq \tau\}$, where τ is a basic torsion

theory in R-tors. This topology was introduced in [17] under the name "finitary open topology" and it has proven, in the long run, to be the most convenient topology on R-sp defined so far and the one most used by researchers in the area. It has been extensively discussed in [18, 22] and so I won't go into details about it here, except to cite the following results:

(1.1) PROPOSITION: *If R is a commutative noetherian ring then R-sp, endowed with the basic order topology, is homeomorphic to the ideal-theoretic spectrum of R endowed with the Zariski topology.*

Proof: See [17, Proposition 4.4]. □

Indeed, in the above proposition one need not assume that the ring R is commutative. For example, it is enough to assume that R is a fully left bounded classical left noetherian ring. See [51] for details.

(1.2) PROPOSITION: *Let R be a left stable left noetherian ring and left R-sp be endowed with the basic order topology. If M is a left R-module having finite uniform dimension then $\bar{Q}(\_,M)$ is a sheaf.*

Proof: See [39]. □

Note that if R is a left stable left noetherian ring then R-sp, endowed with the basic order topology, is homeomorphic to the topological space considered by Goldston and Mewborn in [26] and so Proposition 1.2 is just another version of their result.

2. The Stone-Pierce topology. We are now going to shift our attention to a topology much coarser than the basic order topology. For any central idempotent element e of R, the torsion theory $\tau_e = \xi(R/Re)$ is a basic element of R-tors. Torsion theories of this form are called *centrally splitting*. They have been studied by Bernhardt [3,4,5], Golan [18], Ikeyama [31], Jans [33], and others.

(2.1) EXAMPLE: If R is a quasi-frobenius ring then a torsion theory in R-tors is stable if and only if it is centrally splitting. See [3].

(2.2) EXAMPLE: If R is a semiprime right noetherian ring then every stable torsion theory on R-mod is centrally splitting. See [42].

(2.3) EXAMPLE: If R is a commutative noetherian ring then every jansian torsion theory on R-mod is centrally splitting. See [35].

(2.4) PROPOSITION: *For any ring* R *the set of all centrally splitting torsion theories is a complemented sublattice of* R-*tors*.

Proof: If e and f are central idempotent elements of R then so are ef and $e+f-ef$. It is easy to check that

$$\tau_e \wedge \tau_f = \tau_{e+f-ef}$$

and that

$$\tau_e \vee \tau_f = \tau_{ef}.$$

This proves that the set of all centrally splitting torsion theories is a sublattice of R-tors. Moreover, if e is a central idempotent element of R then τ_{1-e} is the complement of τ_e in R-tors, and τ_{1-e} is itself centrally splitting. □

In particular, we note that every centrally splitting torsion theory in R-tors is the complement of some other such theory in R-tors and so is semiprime.

If e and f are central idempotent elements of R then we have

$$\mathbb{P}(\tau_e \vee \tau_f) = \mathbb{P}(\tau_e) \cap \mathbb{P}(\tau_f)$$

and

$$\mathbb{P}(\tau_e \wedge \tau_f) = \mathbb{P}(\tau_e) \cup \mathbb{P}(\tau_f).$$

In particular, the family of all subsets of R-sp of the form $\mathbb{P}(\tau_e)$, where τ_e is a centrally splitting torsion theory in

R-tors, is the base for a topology on R-sp, which we will call the *Stone-Pierce topology (SP-topology)*. This topology is coarser than the BO-topology on R-sp since every centrally splitting torsion theory in R-tors is, as previously observed, basic.

(2.5) PROPOSITION: *If* τ_e *is a centrally splitting torsion theory in R-tors then the localization endofunctor* $Q_{\tau_e}(\_)$ *of R-mod is naturally equivalent to the endofunctor defined by multiplication on the left by the ideal eR.*

Proof: If M is a left R-module then $T_{\tau_e}(M) = M/eM = (1-e)M$ so eM is τ_e-torsionfree. By Proposition 22.14 of [18] we see that E(eM) is also τ_e-torsionfree and so by Propositions 22.14 and 5.5 of [18] we see that E(eM)/eM is τ_e-torsionfree. The result now follows from Proposition 6.3 of [18]. □

As a consequence of Proposition 2.5 we see that for any central idempotent element e of R and for any left R-module M we have $\bar{Q}(\mathbb{P}(\tau_e),M) \cong eM$ canonically. As a result of this, the following proposition is now straightforward.

(2.6) PROPOSITION: *If R-sp is endowed with the Stone-Pierce topology then, for any left R-module M,* $\bar{Q}(\_,M)$ *is a sheaf.*

Note that in the special case of $M = R$ the sheaf $\bar{Q}(\_,R)$ yields the same representation of the ring as the sheaf first studied for commutative rings by Pierce [40] and for noncommutative rings by Burgess and Stephenson [8,9]. The relation between these two constructions is explained in detail in [21, 22]. If the ring R is biregular (i.e., if every principal ideal of R is generated by a central idempotent element) then this sheaf representation is just that given by Dauns and Hofmann in [12].

3. The stable jansian order topology. In the recent work of Borceux, Simmons, and Van den Bossche [6, 7, 47] we are introduced to another topology on R-sp. This topology, coarser than the BO-topology but finer than the SP-topology, still has the property that the presheaf $\bar{Q}(\_,M)$ is a sheaf for any left R-module M. In order to define this topology, we need some more background information.

We have already mentioned that a torsion theory τ in R-tors is jansian if and only if the class of all τ-torsion left R-modules is closed under taking direct products. These torsion theories were first studied by Jans [33] under the name of TTF-theories and have since been extensively investigated. The basic properties of jansian torsion theories are given in Section 22 of [18]. In particular, we note that there is a bijective correspondence between the set of jansian elements of R-tors and the set of idempotent two-sided ideals of R, which assigns to each jansian torsion theory τ the ideal $L(\tau)$ which is the intersection of all left ideals I of R satisfying the condition that R/I is τ-torsion. In particular, $\tau = \xi(R/L(\tau))$ and a left R-module M is τ-torsion if and only if each of its elements is annihilated by $L(\tau)$. The set of all jansian elements of R-tors is closed under taking arbitrary meets. Indeed, if U is a nonempty set of jansian torsion theories then $L(\wedge U) = \Sigma\{L(\tau) \mid \tau \in U\}$.

We are particularly interested in the set of all those elements of R-tors which are both jansian and stable and will denote this set by R-sjans. The elements of R-sjans have been characterized in many different ways. Several such characterizations are given in Proposition 22.10 of [18]. Refer also to results of Azumaya [1], Baccella [2], Colavita, Raggi, and Rios [11], Goel [16], Hacque [28], Miller [37], and Ramamurthi [42]. (See also the paper [10] of Call and Shores, where such torsion theories are refered to as torsion theories "of type I".) To cite but one of these characterizations: a jansian torsion theory τ is stable if and only if $R/L(\tau)$ is flat as a right R-module. In general, not every jansian torsion theory in R-tors is stable.

(3.1) EXAMPLE: If $\{F_i \mid i \in \Omega\}$ is an infinite set of fields and if $R = \Pi_{i\in\Omega} F_i$ then there exists a stable jansian torsion theory τ in R-tors satisfying the condition that

$L(\tau) = \oplus_{\tau \in \Omega} F_i$. See [28].

(3.2) EXAMPLE: The Goldie torsion theory τ_G belongs to R-sjans if and only if nonzero nonsingular left ideals of R have nonsingular socles. See [49]. If R satisfies the stronger condition that every nonzero left ideal has a nonzero socle then $L(\tau_G) = \mathrm{soc}(R)^2$. Moreover, R itself is left nonsingular if and only if $\mathrm{soc}(R) = \mathrm{soc}(R)^2$ so that, in this case, $L(\tau_G) = \mathrm{soc}(R)$. See [36].

(3.3) EXAMPLE: It is easy to see that $\mathbb{Z}$-sjans = $\{\xi, \chi\}$.

(3.4) EXAMPLE: If the ring R is left perfect then every jansian torsion theory in R-tors is stable if and only if R is isomorphic to a finite direct product of left perfect right local rings. See Proposition 1.5 of [19].

(3.5) EXAMPLE: A ring R having the property that R/I is a flat right R-module for any two-sided ideal I is said to be *left weakly regular* or *fully left idempotent.* For such rings it is clearly true that every jansian torsion theory is stable. Left weakly regular rings are studied by Baccella [2], Fisher [13], Hansen [29], Hirano [30], and Ramamurthi [41]. In particular, Fisher has shown that left V-rings are left weakly regular.

The following result was established by Borceux, Simmons, and Van den Bossche [6]:

(3.6) PROPOSITION: *For any ring R, the set R-sjans is a sublattice of R-tors closed under taking arbitrary meets.*

Proof: We have already remarked that the set of all jansian elements of R-tors is closed under taking arbitrary meets. This is true also of the set of all stable elements of R-tors. (See Proposition 11.6 of [18].) Therefore R-sjans is closed under taking arbitrary meets and all we are left to show is that it is also closed under taking joins.

Let τ_1 and τ_2 be elements of R-sjans and let $K = L(\tau_1) \cap L(\tau_2)$. If $a \in K$ then by Proposition 22.10(5) of [18] there exist elements b of $L(\tau_1)$ and c of $L(\tau_2)$, together with elements d and d' of (0:a), satisfying $d + b = 1 = d' + c$. This implies that $a = ba = ca$ and so $a = bca$, where bc is an element of K. Thus we see that for every element a of K there exists an element u of K satisfying $a = ua$. In particular, $K = K^2$ and so $K = L(\tau)$ for some jansian torsion theory τ in R-tors. Indeed, by Proposition 22.10 of [18] we see that $\tau \in$ R-sjans. Moreover, by Proposition 5.9 of [22] we have $\tau = \xi(R/K) = \xi(R/L(\tau_1)) \vee \xi(R/L(\tau_2)) = \tau_1 \vee \tau_2$. □

With a jansian torsion theory τ on R-mod we can associate another torsion theory τ^* defined by the condition that a left R-module M is τ^*-torsion if and only if $m \in L(\tau)m$ for any element m of M. See Proposition 22.8 of [18]. If $\tau \in$ R-sjans then a left R-module M is τ^*-torsion if and only if $L(\tau)M = M$ and M is τ^*-torsionfree if and only if $L(\tau)M = 0$, i.e. if and only if M is τ-torsion. Moreover, in this case $\tau^* = \tau^{\perp}$ and this is the (unique) complement of τ in the lattice R-tors. Thus we see that $\tau = \tau^{\perp\perp}$ and so, in particular, every element of R-sjans is semiprime.

Note that, in general, $\tau^{\perp}$ need not be jansian for every stable jansian torsion theory τ. Indeed, in such a situation $\tau^{\perp}$ is jansian if and only if there exists an idempotent element e of R satisfying the condition that $L(\tau) = eR$. See [43].

If U is a nonempty subset of R-tors then, as an immediate consequence of the definitions, we see that

$$\mathbb{P}(\vee U) = \cap\{\mathbb{P}(\tau) \mid \tau \in U\}.$$

Moreover, if U is finite then

$$\mathbb{P}(\wedge U) = \cup\{\mathbb{P}(\tau) \mid \tau \in U\}$$

as well. For stable jansian torsion theories we have a stronger result.

(3.7) PROPOSITION: *If U is a nonempty subset of R-sjans then* $\mathbb{P}(\wedge U) = \cup\{\mathbb{P}(\tau) \mid \tau \in U\}$.

Proof: If $\tau \in U$ then $\wedge U \leq \tau$ and so $\mathbb{P}(\wedge U) \supseteq \mathbb{P}(\tau)$. This implies that $\mathbb{P}(\wedge U) \supseteq \cup\{\mathbb{P}(\tau) \mid \tau \in U\}$. Assume that this inclusion is proper. Then there exists a prime torsion theory π belonging to $\mathbb{P}(\wedge U)$ but not belonging to $\mathbb{P}(\tau)$ for any element τ of U. In particular, if M is a π-cocritical left R-module then M is not τ-torsionfree. Since τ is stable and M is uniform, this implies that M is τ-torsion and so $L(\tau)m = 0$ for any element m of M. Moreover, since this is true for any element τ of U, we have $Km = 0$ for all such m, where $K = \Sigma\{L(\tau) \mid \tau \in U\}$. But we have already seen that K is just $L(\wedge U)$ and so M is $(\wedge U)$-torsion, contradicting the assumption that $\pi \geq \wedge U$. Thus we must have the equality we seek. □

As a consequence of Proposition 3.7 we see that the family of subsets of R-sp of the form $\mathbb{P}(\tau)$, where τ is a stable jansian element of R-tors, is the family of open sets for a topology on R-sp, which we will call the *stable jansian order topology (SJO-topology)*. Since every stable jansian torsion theory is basic, this topology is coarser than the basic order topology. On the other hand, since τ_e is stable and jansian for any central idempotent element e of R, this topology is finer than the Stone-Pierce topology on R-sp. By Example 2.2 we see that if R is a semiprime right noetherian ring then the SJO-topology and the SP-topology on R-sp coincide.

As previously noted, every element of R-sjans is semiprime. This implies that if τ and τ' are elements of R-sjans satisfying $\mathbb{P}(\tau) = \mathbb{P}(\tau')$ then $\tau = \wedge\mathbb{P}(\tau) = \wedge\mathbb{P}(\tau') = \tau'$. Thus the map $\tau \mapsto \mathbb{P}(\tau)$ is a bijection between R-sjans and the family of all subsets of R-sp open in the SJO-topology.

If $\tau \in$ R-sjans then Hacque [28], Kato [34], and Ohtake [38] have all observed that the localization endofunctor $Q_\tau(\_)$ of R-mod is naturally equivalent to $\mathrm{Hom}_R(L(\tau),\_)$. Therefore we see that for any $\tau \in$ R-sjans we have

$$\bar{Q}(\mathbb{P}(\tau),M) \cong \mathrm{Hom}_R(L(\tau),M)$$

canonically. Hence we might as well assume that we have equality (rather than isomorphism) here.

We now come to the major observation of Borceux, Simmons, and Van den Bossche.

(3.8) PROPOSITION: *If R-sp is endowed with the stable jansian order topology then, for any left R-module M, $\bar{Q}(\_,M)$ is a sheaf.*

Proof: Let $\tau \in$ R-sjans and assume that $\mathbb{P}(\tau) = \cup\{\mathbb{P}(\sigma) \mid \sigma \in U\}$ for some nonempty subset U of R-sjans. Moreover, assume that for each $\sigma \in U$ there exists an R-homomorphism α_σ in $\mathrm{Hom}_R(L(\sigma),M)$ such that α_σ and $\alpha_{\sigma'}$ agree on $L(\sigma) \cap L(\sigma')$ for all $\sigma, \sigma' \in U$. We must show that there exists a unique R-homomorphism $\beta: L(\tau) \to M$ which extends each of the α_σ.

By what we have already seen, $\tau = \wedge U$ and so $L(\tau) = \Sigma\{L(\sigma) \mid \sigma \in U\}$. Thus we see that for each element a of $L(\tau)$ there exists a finite subset U_a of U having the property that $a \in \Sigma\{L(\sigma) \mid \sigma \in U_a\}$. It therefore suffices to show that for any finite subset U' of U there exists a unique R-homomorphism from $\Sigma\{L(\sigma) \mid \sigma \in U'\}$ to M which extends α_σ for each element σ of U'. Indeed, since R-sjans is closed under taking meets, it suffices to consider the case of $U' = \{\sigma_1, \sigma_2\}$ and then proceed by a straightforward induction.

Therefore let us consider two elements σ_1 and σ_2 of R-sjans and R-homomorphisms $\alpha_i: L(\sigma_i) \to M$ which agree on the intersection of their domains. If $d \in L(\sigma_1) + L(\sigma_2)$ then we can write $d = a_1 + a_2$ which $a_i \in L(\sigma_i)$ for $i = 1,2$. If we also have $d = b_1 + b_2$ with $b_i \in L(\sigma_i)$ for $i = 1,2$ then

$$a_1 - b_1 = b_2 - a_2 \in L(\sigma_1) \cap L(\sigma_2)$$

and so we have $a_1\alpha_1 - b_1\alpha_1 = (a_1 - b_1)\alpha_1 = (b_2 - a_2)\alpha_2 = b_2\alpha_2 - a_2\alpha_2$. This implies that $a_1\alpha_1 + a_2\alpha_2 = b_1\alpha_1 + b_2\alpha_2$ and so we have a well-defined R-homomorphism from $L(\sigma_1) + L(\sigma_2)$ to M given by $(a_1 + a_2) \mapsto a_1\alpha_1 + a_2\alpha_2$ which extends both α_1 and α_2. It is clearly unique. □

4. The complement stable jansian order topology. If τ_1 and τ_2 are stable jansian torsion theories in R-tors then, by

De Morgan's identity, we know that $\tau_1 \vee \tau_2 = (\tau_1 \wedge \tau_2)^{\perp}$ and so the set of all complements of stable jansian torsion theories in R-tors is closed under taking finite joins. This means that the family of all subsets of R-sp of the form $\mathbb{P}(\tau^{\perp})$, where τ is a stable jansian torsion theory, is the base for a topology on R-sp which we will call the *complement stable jansian order topology (CSJO-topology)*. This topology is finer than the SP-topology on R-sp but is not in general coarser than the BO-topology since complements of stable jansian torsion theories need not be basic. Indeed, to see where this topology comes from, we note that there is another standard method of defining a topology on R-sp: that of taking as open sets all subsets of R-sp of the form R-sp $\setminus \mathbb{P}(\tau)$, where τ ranges over all elements of R-tors. This topology has been variously called the *hull-kernel topology*, the *Stone topology*, or the *reverse order topology (RO-topology)* and has been studied by Gierz et. al [15], Raynaud [44], and Golan [17, 18]. Note that subsets of R-sp open with respect to the basic order topology are closed with respect to the reverse order topology. However, if we restrict our attention to subsets of R-sp open with respect to the SJO-topology we obtain the following interesting, though elementary, result:

(4.1) PROPOSITION: *If* $\tau \in$ *R-sjans then* $\mathbb{P}(\tau^{\perp}) =$ *R-sp* $\setminus \mathbb{P}(\tau)$.

Proof: Let $\pi \in$ R-sp. If $\pi \notin \mathbb{P}(\tau)$ then $\xi = \tau \wedge \tau^{\perp} \leq \pi$ and $\tau \not\leq \pi$ so, by primeness, we must have $\tau^{\perp} \leq \pi$. Therefore $\pi \in \mathbb{P}(\tau^{\perp})$. Conversely, if $\pi \in \mathbb{P}(\tau^{\perp}) \cap \mathbb{P}(\tau)$ then $\chi = \tau \vee \tau^{\perp} \leq \pi$, which is impossible. Therefore $\mathbb{P}(\tau^{\perp}) \cap \mathbb{P}(\tau) = \emptyset$, which proves the equality we seek. □

Thus we see that the CSJO-topology on R-sp is in fact coarser than the RQ-topology.

If M is a left R-module and if τ is a stable jansian torsion theory in R-tors then $\bar{Q}(\mathbb{P}(\tau^{\perp}), M) = Q_{\tau^{\perp}}(M) \cong M/L(\tau)M$. See [43] for details. Thus the values of $\bar{Q}(\_, M)$ on basic open subsets of R-sp are easy to calculate. Indeed, $Q_{\tau^{\perp}}(R)$ is just isomorphic to $R/L(\tau)$. This ring can also be characterized in another way, namely as $R[S^{-1}]$, where S is the left Öre set consisting of all elements of R of the form 1-a for some element a of $L(\tau)$. See [28, 32] for details.

The CSJO-topology on R-sp has not as yet been properly investigated, though it does seem to hold definite promise, especially for those rings R for which the lattice R-tors is continuous in the sense of [15].

5. The symmetric basic order topology and the symmetric stable basic order topology. A torsion theory $\tau \in$ R-tors is said to be *symmetric* if and only if the following conditions on a left ideal I are equivalent:

(1) R/I is τ-torsion;

(2) I contains a two-sided ideal H of R such that R/H is τ-torsion.

Symmetric torsion theories are examined in detail in [50] and [22]. If the ring R is left noetherian then a torsion theory τ in R-tors is symmetric if and only if there exists a family $\mathcal{A}$ of two-sided ideals of R such that a left R-module M is τ-torsion when and only when each of its elements is annihilated by a product of finitely-many elements of $\mathcal{A}$.

A torsion theory in R-tors of the form ξ(R/I), where I is a two-sided ideal of R, is said to be *symmetric basic*. The family of all symmetric basic torsion theories in R-tors is closed under taking finite meets and joins. Indeed, if I and H are two-sided ideals of a ring R then

$$\xi(R/I) \wedge \xi(R/H) = \xi(R/[I + H])$$

and

$$\xi(R/I) \vee \xi(R/H) = \xi(R/IH) = \xi(R/[I \cap H]).$$

(5.1) PROPOSITION: *If R is a left noetherian ring and if* $\{I_j \mid j \in \Omega\}$ *is a set of two-sided ideals of R, then* $\xi(R/[\Sigma_{j\in\Omega} I_j]) = \wedge_{j\in\Omega}\, \xi(R/I_j)$.

Proof: Set $I = \Sigma_{j\in\Omega} I_j$. Then R/I is a homomorphic image of R/I_j for each j and so $\xi(R/I) \leq \xi(R/I_j)$ for each j. Therefore $\xi(R/I) \leq \wedge_{j\in\Omega} \xi(R/I_j)$. Conversely, since R is left noetherian we know that I is finitely-generated and so there exists a finite subset Λ of Ω satisfying $I = \Sigma_{j\in\Lambda} I_j$. Thus, by the above remarks, we have $\xi(R/I) = \wedge_{j\in\Lambda} \xi(R/I_j) \geq \wedge_{j\in\Omega} \xi(R/I_j)$, proving equality. □

Thus we see that if R is left noetherian then the family of all symmetric basic torsion theories is closed under taking infinite meets as well.

The family of all subsets of R-sp of the form $\mathbb{P}(\tau)$, where τ is a symmetric basic torsion theory in R-tors, forms the base of a topology on R-sp called the *symmetric basic order topology (SBO-topology)*. This topology is clearly coarser than the BO-topology. If R is a fully left bounded left noetherian ring then every element of R-tors is symmetric [46] and so the basic order topology and the symmetric basic order topology on R-sp coincide.

(5.2) PROPOSITION: *If R is a left noetherian ring then the family of all stable jansian torsion theories in R-tors is closed under taking arbitrary meets and joins.*

Proof: We have already remarked in the proof of Proposition 3.6 that the family of all stable torsion theories in R-tors is always closed under taking arbitrary meets. Now let U be a set of stable torsion theories in R-tors and let $\tau = \vee U$. Let M be a τ-torsion left R-module. Since R is left noetherian, we can write $E(M) = \oplus_{h \in \Omega} E_h$, where the E_h are indecomposable injective left R-modules. Since M is large in $E(M)$, it has a nonzero intersection with each of the E_h and so none of the E_h is τ-torsionfree. In other words, for each $h \in \Omega$ there exists an element τ_h of U such that E_h is not τ_h-torsionfree. Since E_h is uniform and τ_h is stable, this implies that E_h is τ_h-torsion and hence τ-torsion. Thus $E(M)$ is τ-torsion, proving the stability of τ. □

Let us denote by R-ssb the family of all stable symmetric basic torsion theories in R-tors. If the ring R is left noetherian, the members of this family are easy to describe. Indeed, a symmetric basic torsion theory $\xi(R/I)$ is stable if and only if I has the Artin-Rees property with respect to every finitely-generated left R-module. That is to say, if M is a finitely-generated left R-module and if N is a submodule of M then for every natural number n there exists a natural number $h(n)$ such that $I^{h(n)}M \cap N \subseteq I^n N$. See Corollary 5.12 of [22] for details.

By Propositions 5.1 and 5.2 we see that if R is a left noetherian ring then R-ssb is closed under taking finite joins and arbitrary meets.

(5.3) PROPOSITION: *If* U *is a nonempty subset of* R-ssb *then* $\mathbb{P}(\wedge U) = \cup\{\mathbb{P}(\tau) \mid \tau \in U\}$.

Proof: If $\tau \in U$ then surely $\mathbb{P}(\tau) \subseteq \mathbb{P}(\wedge U)$ and so $\cup\{\mathbb{P}(\tau) \mid \tau \in U\} \subseteq \mathbb{P}(\wedge U)$. Conversely, assume that $\pi \in \mathbb{P}(\wedge U)$ and let M be a π-cocritical left R-module. Then M is $(\wedge U)$-torsionfree. Therefore there exists an element τ of U such that M is not τ-torsion. Since τ is stable and M is uniform, this implies that M is τ-torsionfree and so $\pi \geq \tau$. Therefore we have $\pi \in \mathbb{P}(\tau)$. □

Combining all of the above results, we see that if R is a left noetherian ring then the family of all subsets of R-sp of the form $\mathbb{P}(\tau)$, where τ is a stable symmetric basic torsion theory in R-tors, is the family of open sets for a topology on R-sp, which we will call the *stable symmetric basic order topology (SSBO-topology)*. Since we have already seen that R-sjans $\subseteq$ R-ssb, we see that this topology is finer than the SJO-topology on R-sp. On the other hand, it is clearly coarser than the SBO-topology on R-sp.

(5.4) PROPOSITION: *Let* R *be a left noetherian ring and* *let R-sp be endowed with the SSBO-topology. If M is a finitely-generated left R-module then $\bar{Q}(\_,M)$ is a sheaf.*

Proof: The proof of this result can be obtained, with slight modifications, from the proof of the result of Papp cited as Proposition 1.2 above. See also the proof of Proposition 10.8 in [22]. □

6. The basic linear order topology and the basic compact order topology. I now want to briefly define two additional topologies on R-sp, both of them coarser than the basic order topology, which look very promising but about which little research has been done as yet.

Following the terminology of [14], we say that a torsion theory τ in R-tors is *linear* if and only if

$$\tau \vee (\wedge U) = \wedge\{\tau \vee \tau' \mid \tau' \in U\}$$

for any nonempty subset U of R-tors. The family of all linear torsion theories in R-tors is clearly closed under taking finite joins.

(6.1) EXAMPLE: A ring R is said to be *left convenient* if and only if every torsion theory in R-tors other than χ is semiprime and $\mathbb{P}(\wedge U) = \cup\{\mathbb{P}(\tau) \mid \tau \in U\}$ for any nonempty subset U of R-tors. These rings are studied in detail in [44] (under another name) and in Chapter 3 of [22]. Several examples of left convenient rings are given there, among them left stable left noetherian rings. By [22, Proposition 3.10] we see that if every element of R-tors other than χ is semiprime then the ring R is left convenient if and only if every element of R-tors is linear.

(6.2) PROPOSITION: *If* $\tau \in$ *R-tors has a complement in R-tors then* τ *is linear.*

Proof: See page 317 of [14]. □

Now define a topology on R-sp by taking as a base of open sets the family of subsets of R-sp of the form $\mathbb{P}(\tau)$, where τ is a basic linear torsion theory in R-tors. We call this the *basic linear order topology (BLO-topology)* on R-sp. This topology is coarser than the BO-topology on R-sp and, by Proposition 6.2, it is finer than the SJO-topology. If R is left convenient then the basic linear order topology and the basic order topology on R-sp coincide.

A torsion theory τ in R-tors is said to be *compact* if and only if $\tau \leq \vee U$ for some nonempty subset U of R-tors implies that there exists a finite subset U' of U satisfying $\tau \leq \vee U'$. Clearly the family of all compact elements of R-tors is closed under taking finite joins. Compact torsion theories have been studied in [20, 22]. See also the work of Gorbunov [27].

(6.3) EXAMPLE: By Propositions 3.2 and 4.1 of [22] we see that if R is a left convenient ring then every compact torsion theory in R-tors is basic. Indeed, by Proposition 4.12 of [22] we see that if R is left stable and left noetherian then the basic torsion theories and the compact torsion theories in R-tors coincide.

Since the set of all basic compact torsion theories in R-tors is closed under taking finite joins, the family of all subsets of R-sp of the form $\mathbb{P}(\tau)$, where τ is a basic compact torsion theory in R-tors, forms the base of a topology on R-sp, which we will call the *basic compact order topology (BCO-topology)*. This topology is, again, coarser than the basic order topology on R-sp. If the ring R is left stable and left noetherian then the basic compact order topology and the basic order topology on R-sp coincide.

7. *The perfect order topology*. Finally, I wish to look at one more topology which is not coarser than the basic order topology on R-sp. Recall that a torsion theory τ in R-tors is said to be *perfect* if and only if the localization endofunctor $Q_\tau(\_)$ is naturally equivalent to $R_\tau \otimes_R \_$. Such torsion theories

are among the oldest investigated and they have been studied very thoroughly by a large number of researchers. A summary of several of their properties can be found in Chapter 17 of [18]. In particular, we note that by [18, Proposition 17.4] the set of all perfect torsion theories in R-tors is closed under taking arbitrary joins.

(7.1) EXAMPLE: If $\tau \in$ R-sjans then we have already noted that for any left R-module M we have $Q_{\tau^\perp}(M) = M/L(\tau)M$. This is naturally isomorphic to $[R/L(\tau)] \otimes_R M = R_{\tau^\perp} \otimes_R M$ and so we see that $\tau^\perp$ is perfect.

(7.2) EXAMPLE: If R is a left hereditary left noetherian ring then every element of R-tors is perfect. See Corollary 6.4 and Proposition 17.1 of [18].

We can define a topology on R-sp by taking as a base of open sets all those subsets of R-sp of the form $\mathbb{P}(\tau)$, where τ is a perfect torsion theory in R-tors. Call this the *perfect order topology (PO-topology)* on R-sp. By Example 7.1, we see that the PO-topology on R-sp is a refinement of the CSJO-topology on R-sp. On the other hand, since perfect torsion theories are not in general basic, it is not as a rule coarser than the basic order topology. Van Oystaeyen [51] has shown that if R is a left noetherian

Azumaya algebra then the perfect order topology on R-sp is a refinement of the basic order topology. In this connection, see also [24].

The following result is essentially due to Shapiro [45]:

(8.3) PROPOSITION: *Every perfect torsion theory in R-tors is semiprime.*

Proof: Let $\tau \in$ R-tors be perfect. If N is a simple left R_τ-module, then N is τ-torsionfree as a left R-module by [18, Proposition 17.1]. In fact, it is easy to check that N is co-critical as a left R-module. Therefore,

$$\tau' = \wedge\{\chi(N) \mid N \text{ is a simple left } R_\tau\text{-module}\}$$

is a semiprime torsion theory in R-tors satisfying $\tau' \geq \tau$. For each simple left R_τ-module N, let $E'(N)$ be its injective hull in R_τ-mod. By Corollary 6.8 of [18], this is also the injective hull of N in R-mod. If $E = \Pi\{E'(N) \mid N \text{ is a simple left } R_\tau\text{-module}\}$ then $\tau' = \chi(E)$ and E is a cogenerator of R_τ-mod. If M is a τ-torsionfree left R-module then M embeds in $Q_\tau(M)$ which, as a left R_τ-module, in turn embeds in a direct product of copies of E. Therefore M is τ'-torsionfree and so $\tau' \leq \tau$, proving equality. Thus τ is semiprime. □

In particular, we see that if M is a left R-module and if

$U = \mathbb{P}(\tau)$ is a basic open subset of R-sp, endowed with the perfect order topology, then $\bar{Q}(U,M)$ is canonically isomorphic to $\bar{Q}(U,R) \otimes_R M$. The general behavior of the presheaf $\bar{Q}(\_,M)$ on R-sp, endowed with the perfect order topology, awaits further investigation.

REFERENCES

[1] G. Azumaya, Some properties of TTF-classes, *Proceedings of the Conference on Orders, Group Rings, and Related Topics (Ohio State University, Columbus, Ohio, 1972)*, Lecture Notes in Mathematics #353, Springer, Berlin, 1973.

[2] G. Baccella, On flat factor rings and fully right idempotent rings, preprint (1981).

[3] R. L. Bernhardt, Splitting hereditary torsion theories over semiperfect rings, *Proc. Amer. Math. Soc.* 22 (1969), 681-687.

[4] R. L. Bernhardt, On splitting in hereditary torsion theories, *Pacific J. Math.* 39 (1971), 31-38.

[5] R. L. Bernhardt, On centrally splitting, *Duke Math. J.* 40 (1973), 903-905.

[6] F. Borceux, H. Simmons, and G. Van den Bossche, A sheaf representation for modules with applications to Gelfand rings, preprint (1982).

[7] F. Borceux and G. Van den Bossche, Algebra in a localic topos with applications to ring theory, preprint (1982).

[8] W. Burgess and W. Stephenson, Pierce sheaves of non-commutative rings, *Comm. Algebra* 4 (1976), 51-75.

[9] W. Burgess and W. Stephenson, An analogue of the Pierce

sheaf for non-commutative rings, *Comm. Algebra* 6 (1978), 863-886.

[10] F. W. Call and T. S. Shores, The splitting of bounded torsion submodules, *Comm. Algebra* 9 (1981), 1161-1214.

[11] L. R. Colavita, F. Raggi Cárdenas, and J. Rios Montes, Sobre filtros estables de Gabriel. V, *An. Inst. Mat. Univ. Nac. Autónoma México* 17 (1977), 1-16. [Spanish]

[12] J. Dauns and K. H. Hofmann, The representation of biregular rings by sheaves, *Math. Z.* 91 (1966), 103-123.

[13] J. Fisher, Von Neumann regular rings versus V-rings, *Ring Theory (Proc. Conf., Univ. Oklahoma, Norman, Okla., 1973)*, Lecture Notes in Pure and Appl. Mathematics #7, Marcel Dekker, New York, 1974.

[14] M. P. Fourman and D. Scott, Sheaves and logic, *Applications of sheaves (Proc., Durham 1977)*, Lecture Notes in Mathematics #753, Springer, Berlin, 1979.

[15] G. Gierz et. al., *A Compendium of Continuous Lattices*, Springer, Berlin, 1980.

[16] V. K. Goel, A study in cotorsion theory, *Ring Theory (Proc. Ohio State Conference)*, Lecture Notes in Pure and Appl. Mathematics #25, Marcel Dekker, New York, 1977.

[17] J. S. Golan, Topologies on the torsion-theoretic spectrum of a noncommutative ring, *Pacific J. Math.* 51 (1974), 439-450.

[18] J. S. Golan, *Localization of Noncommutative Rings*, Marcel Dekker, New York, 1975.

[19] J. S. Golan, Colocalization at idempotent ideals, *Ring Theory (Prof. 1978 Antwerp Conference)*, Lecture Notes in Pure and Appl. Mathematics #51, Marcel Dekker, New York, 1979.

[20] J. S. Golan, Some torsion theories are compact, *Math. Japon.* 24 (1980), 635-639.

[21] J. S. Golan, Two sheaf constructions for noncommutative rings, *Houston J. Math.* 6 (1980), 59-66.

[22] J. S. Golan, *Structure Sheaves over a Noncommutative Ring*, Lecture Notes in Pure and Appl. Mathematics #56, Marcel Dekker, New York, 1980.

[23] J. S. Golan and R. W. Miller, Cotorsionfree modules and colocalizations, *Comm. Algebra* 6 (1978), 1217-1230.

[24] J. S. Golan, J. Raynaud, and F. Van Oystaeyen, Sheaves over the spectra of certain noncommutative rings, *Comm. Algebra* 4 (1976), 491-502.

[25] O. Goldman, Rings and modules of quotients, *J. Algebra* 13 (1969), 10-47.

[26] B. Goldston and A. Mewborn, A structure sheaf for a non-commutative noetherian ring, *J. Algebra* 47 (1977), 18-28.

[27] V. A. Gorbunov, Canonical decompositions in complete lattices, *Algebra i Logika* 17 (1978), 495-511. [Russian]

[28] M. Hacque, Mono sous catégories d'une catégorie de modules, *Publ. Dép. Math. Lyon* 6 (1969), 13-48.

[29] F. Hansen, On one-sided prime ideals, *Pacific J. Math.* 58 (1975), 79-85.

[30] Y. Hirano, On fully right idempotent rings and direct sums of simple rings, *Math. J. Okayama Univ.* 22 (1980), 43-49.

[31] T. Ikeyama, Splitting torsion theories, *Comm. Algebra* 8 (1980), 1267-1282.

[32] T. Ikeyama, Four-fold torsion theories and rings of fractions, *Comm. Algebra* 9 (1981), 1027-1037.

[33] J. P. Jans, Some aspects of torsion, *Pacific J. Math.* 15 (1965), 1249-1259.

[34] T. Kato, Duality between colocalization and localization, *J. Algebra* 55 (1978), 351-374.

[35] Y. Kurata, On an n-fold torsion theory in the category ${}_RM$, *J. Algebra* 22 (1972), 559-572.

[36] J. N. Manocha, On rings with essential socle, *Comm. Algebra* 4 (1976), 1077-1086.

[37] R. W. Miller, TTF classes and quasi-generators, *Pacific J. Math.* 51 (1974), 499-507.

[38] K. Ohtake, Colocalization and localization, *J. Pure Appl. Algebra* 11 (1977), 217-241.

[39] Z. Papp, Spectrum, topologies and sheaves for left noetherian rings, *Module Theory, Papers and Problems (Proc. Seattle 1977)*, Lecture Notes in Mathematics #700, Springer, Berlin, 1979.

[40] R. S. Pierce, *Modules over Commutative Regular Rings*, AMS Memoir #70, Amer. Math. Soc., Providence, 1967.

[41] V. S. Ramamurthi, Weakly regular rings, *Canad. Math. Bull.* 16 (1973), 317-321.

[42] V. S. Ramamurthi, On splitting cotorsion radicals, *Proc. Amer. Math. Soc.* 39 (1973), 457-461.

[43] V. S. Ramamurthi and E. A. Rutter, On cotorsion radicals, *Pacific J. Math.* 62 (1976), 163-172.

[44] J. Raynaud, Localisations premieres et copremieres. Localisations stables par enveloppes injectives, *Ring Theory (Proc. 1977 Antwerp Conference)*, Lecture Notes in Pure and Appl. Mathematics #40, Marcel Dekker, New York, 1978.

[45] J. Shapiro, T-torsion theories and central localizations, *J. Algebra* 48 (1977), 17-29.

[46] S. K. Sim, On ring of which every prime kernel functor is symmetric, *Nanta Math.* 7 (1974), 69-70.

[47] H. Simmons, *The Frame of Localizations of a Ring*, preliminary version (1981).

[48] E. I. Tebyrcè, The Boolean nature of the lattice of torsions in modules, *Mat. Issled.* 8 (1973), 92-105. [Russian]

[49] M. Teply, Some aspects of Goldie's torsion theory, *Pacific J. Math.* 29 (1969), 447-459.

[50] F. Van Oystaeyen, *Prime Spectra in Non-commutative Algebra*, Lecture Notes in Mathematics #444, Springer, Berlin, 1975.

[51] F. Van Oystaeyen, A note on left balanced rings, *Nederl. Akad. Wetensch. Proc. Ser. A.* 79 (1976), 213-216.

J. S. Golan
Department of Mathematics
University of Haifa
Haifa
Israel

COLLOQUIA MATHEMATICA SOCIETATIS JÁNOS BOLYAI
38. RADICAL THEORY, EGER (HUNGARY), 1982.

ESSENTIAL IDEALS AND RADICAL CLASSES

T. L. JENKINS

In [10], Olson and Jenkins defined $E(M)$ to be the class of all rings each nonzero homomorphic image of which contains either a nonzero M-ideal or an essential ideal where M is any class or rings. $E(M)$ was proven to be a radical class and various classes M were considered. In this article two subclasses of $E(M)$ are considered; H the class of all rings each nonzero homomorphic image of which has a proper essential ideal and the class USM of all rings each nonzero homomorphic image of which contains an M-ideal. It is shown that H is a radical class and under certain conditions USM is also a radical class. Various properties placed on M yield several known radical classes and an infinite number of supernilpotent nonspecial radical classes is constructed. All rings considered will be associative and simple rings will be prime. The major knowledge of radical theory required for our purposes can be found in [3] and [13].

Currently there are two ways of interpreting essential ideals; one where the ring is an essential ideal of itself and another where only proper ideals may be essential. For our purposes, if I is an essential ideal in R then I meets all nonzero ideals of R and thus R is essential in itself. When I must be a proper essential ideal it will be stated. In section 1, we consider only proper essential ideals, whereas in section 2 we consider essential ideals.

SECTION 1. The H radical class.

Let M be any class of rings. In [10], Olson and Jenkins showed that

$E(M) = \{R|$ every nonzero homomorphic image has either a nonzero M-ideal or an essential ideal$\}$

is a radical class. We now consider the subet H of $E(M)$.

Definition 1. Let $H = \{R|$ every nonzero homomorphic image of which has an essential ideal.$\}$

From the definition it is clear that simple rings, as well as rings with maximal ideals, cannot belong to H. Let $S = \{R|$ R is either simple or a prime order zero ring$\}$ and $T = \{$all direct sums of members of $S\}$. From [10], we need the following result:

THEOREM 1. The following are equivalent for any ring R.

(1) R has no essential ideals.

(2) Each ideal of R is a direct summand of R.

(3) R is a member of T.

With S as definited above, let US be the upper radical determined by S. We now show that H is an upper radical class.

THEOREM 2. H is a radical class and $H = US$.

Proof: From Theorem 1 and the definition of H it follows that $H = UT$ so since T is a hereditary class $H = UT$ is a radical class. Also $S \subseteq T$ so $UT \subseteq US$. Conversely, if $R \in US$ then R, having no image in S, cannot have an image in T. Thus $R \in UT$ and so $H = UT = US$.

EXAMPLE. Consider the sequence $\aleph_0, \aleph_1, \aleph_2, \cdots$. Then $\aleph_\omega = U\aleph_n > \aleph_n$ for all positive integers n. Suppose V is a vector space over a division ring and that the dimension of V is $\aleph_\omega$. Let R be the ring of all linear transformations of V of rank $< \aleph_\omega$. Then R is an ideal of the ring L of all linear transformations of V. Now if an ideal of R contains a linear transformation of a certain rank, then it also contains all linear transformations of smaller rank [6]. Therefore, it follows that the only ideals of R are of the type:

I_n = {all linear transformations of rank $< \aleph_n$, n a nonnegative integer}.

This implies that R has no maximal ideals and hence $R \in H$. Furthermore, R contains as an ideal the simple ring $I_0 \in SH$ so H is not hereditary.

If β_s denotes the upper radical class determined by the class of simple prime rings it is clear that $H \subsetneq \beta_s$. A partial solution as to the position of H in the diagram of well-known radical classes is given in the following proposition and discussion.

PROPOSITION 1. $H \subsetneq \beta_s$ and $H \cap P \neq 0$ for any supernilpotent radical class P.

Proof. The ring $W = \{\frac{2x}{2y+1} \mid (2x, 2y+1) = 1,\ x, y \in Z\}$ [3, Example 10] belongs to β_s. All the ideals of W are of the form $W = (2) \supset (2)^2 \supset \cdots$. Since W can be mapped onto a nonzero simple ring, it follows that $W \notin H$. Furthermore, the ring p^∞ [3, Example 1] is a zero ring and, therefore, belongs to every supernilpotent radical class P. Since p^∞ has no maximal ideals we have $p^\infty \in H$. Hence $H \cap P \neq 0$.

In [8] van Leeuwen gave the following diagram relating the radical classes. We have only extended it to include the Behrens radical J_B. We follow the standard notation where the first row is the lower Baer, Levitzki, nil, Jacobson, Behrens and Brown-McCoy radicals respectively. The second row consists of the upper radicals determined by subdirectly irreducible rings with idempotent hearts, Levitzki semisimple hearts, nil semisimple hearts, Jacobson semisimple hearts, idempotent hearts with idempotent elments and G respectively. The last row gives the upper radicals determined by all simple prime, simple prime N-semisimple, simple primitive and simple J_B-semisimple rings.

$$\begin{array}{ccccccccccccc}
 & & \beta & \subset & L & \subset & N & \subset & J & \subset & J_B & \subset & G \\
 & & \cap & & \cap & & \cap & & \cap & & \| & & \| \\
 & & \beta_\phi & = & L_\phi & \subseteq & N_\phi & \subset & J_\phi & \subset & (J_B)_\phi & \subset & G_\phi \\
 & & \cap & & \cap & & \cap & & \cap & & \| & & \| \\
H & \subset & \beta_s & = & L_s & \subseteq & N_s & \subset & J_s & \subset & (J_B)_s & \subset & G_s
\end{array}$$

The strictness of the inclusions was shown in [8]. The example following Theorem 2 is $(J_B)_\phi$-semisimple yet $(J_B)_s$ -radical. In Example 3 [4] all the proper one-sided ideals of the ring X are J_s-semisimple, but $(J_B)_s$ -radical. The radical H is independent with all the radicals of the first two rows of the diagram except for G. In order to see this, any simple zero ring is β-radical but H-semisimple. The ring T in the example following Theorem 2 is H-radical, but J_B -semisimple.

We note that if M is a class of rings having no simple rings then

PROPOSITION 2. If M is any class of rings containing no simple rings, then $E(M) = H$.

Proof. From Definition 1 and Theorem 2 we have $US = H \subseteq E(M)$. Now let $A \in E(M)$ and suppose

$A \notin H = US$. This implies that A has a nonzero image, say A/I, where A/I is a simple ring. Since A/I has no essential ideals and no M-ideals, we have $A/I \notin E(M)$. This contradicts $A \in E(M)$ and $E(M)$ is a radical class. Thus $H = E(M)$.

SECTION 2. The class USM

Definition 2. Let M be any class of rings. Then $USM = \{R|$ each nonzero homomorphic image of which contains an M-ideal.$\}$

THEOREM 3. If M is any hereditary class of rings containing no or all nilpotent rings then SM is a hereditary class and so USM is radical.

Proof. Suppose it were possible for $0 \neq J \triangleleft I \triangleleft R \in SM$ with $J \in M$. If J' is the ideal generated by J in R then $J' \neq J$ for $R \in SM$ and so has no M-ideals and since $(J')^3 \subseteq J$ it follows from the hereditary property of M that $R \in SM$ is contradicted unless $(J')^3 = 0$. Now $J \subset J'$ so M could not be a class with no nilpotent rings. But if M contains all nilpotent rings then $J' \in M$ again contradicting $R \in SM$. Hence SM is hereditary and it follows that USM is radical.

In order to show that the radical class of Theorem 3 is hereditary we need the following lemma. It is for this reason that in this section we allow a ring to be essential in itself.

LEMMA [5]. Let I be any nonzero ideal of a ring R. Then there exists a homomorphic image of R containing as isomorphic copy I' of I such that I' is essential in this image.

Lemma 2. Let M be a hereditary class of rings and $0 \neq K \triangleleft I \triangleleft R$ with $I/K \in SM$. If $K \triangleleft R$ then $R \notin USM$.

Proof. Since $K \triangleleft R$ we have $I/K \triangleleft R/K$. By Lemma 1 there exists a nonzero homomorphic image of R/K, say R/P, containing an isomorphic copy of I'/P of I/K as an essential ideal. If $R \in USM$ then R/P has a nonzero M-ideal, say K/P, by definition. But then since I'/P is essential in R/P and I'/P is isomorphic to $I/K \in SM$ we have $0 \neq I'/P \cap K/P$ is an ideal of $I'/P \in SM$ and is an ideal of $K/P \in M$, a contradiction. Hence, $R \notin USM$.

Recall, a radical class R is called *supernilpotent* if R is hereditary and contains all nilpotent rings [3] and is called *subidempotent* if R is hereditary and all rings are idempotent [1].

THEOREM 4. Let M be a hereditary class of rings. If M contains all nilpotent rings USM is a supernilpotent radical and if M contains no nilpotent rings USM is a subidempotent radical.

Proof. Let M be hereditary and contain all nilpotent rings. Let $R \in USM$ and suppose $0 \neq I \triangleleft R$ and

$I \notin \mathcal{USM}$. Then there exists a nonzero homomorphic image I/K of I with no nonzero $\mathcal{M}$-ideals. Since $\mathcal{M}$ contains all nilpotent rings we have that I/K is a semiprime ring. This implies that K is an ideal of R. But then by Lemma 2, $R \notin \mathcal{USM}$ a contradiction. Thus $I \in \mathcal{USM}$ and $\mathcal{USM}$ is supernilpotent.

Now suppose $\mathcal{M}$ contains no nilpotent rings. First we show that $\mathcal{USM}$ in this case is a class of hereditarily idempotent rings. Suppose $R \in \mathcal{USM}$ has a nonzero ideal I with $I \neq I^2$. According to [1], R can be mapped onto a nonzero subdirectly irreducible ring R/J with nilpotent heart H/J. Since $R/J \in \mathcal{USM}$, R/J has a nonzero $\mathcal{M}$-ideal T/J. But $0 \neq H/J \triangleleft T/J \in \mathcal{M}$, a contradiction to $\mathcal{M}$ being hereditary and containing no nilpotent rings. Therefore, $R \in \mathcal{USM}$ implies that R is hereditarily idempotent. Now let $R \in \mathcal{USM}$ and suppose $0 \neq I \triangleleft R$ and $I \notin \mathcal{USM}$. Then there exists a nonzero homomorphic image I/K of I with no ideals. Since, in this case, R is hereditarily idempotent we have $0 \neq K \triangleleft R$. Again by Lemma 2, $R \notin \mathcal{USM}$, a contradiction so $I \in \mathcal{USM}$ and $\mathcal{USM}$ is subidempotent.

We remark here that if $\mathcal{M}$ contains all nilpotent rings then $\beta \subseteq \mathcal{M}$ implies $\mathcal{SM} \subseteq \mathcal{S}\beta$ so $\mathcal{SM}$ is a class of semiprime rings. On the otherhand, if $\mathcal{M}$ contains no nilpotent rings then $\mathcal{SM}$ contains all nilpotent rings. Hence $\mathcal{SM}$ is a hereditary class of rings containing no or all nilpotent rings. Now if we let

$(SM)_k$ be the class of all essential extensions of rings belonging to SM then $U(SM)_k$ is hereditary because USM is hereditary and $USM = U(SM)_k$.

If M is any homomorphically closed class of rings and the lower radical class determined by M is $LM = \{R|$ every nonzero homomorphic image of which has an accessible subring in $M\}$ we have as corollary to THEOREM 4.

COROLLARY. (a) If M is homomorphically closed, $M \subseteq USM \subseteq LM$

(b) If M is hereditary and homomorphically closed and in addition contains no or all nonzero nilpotent rings then $USM = LM$.

Proof. (a) Clear

(b) From Theorem 4 we have that USM is a radical class. Since LM is the smallest radical class containing M we conclude from (a) that $USM = LM$.

Recall that β is the lower Baer radical, that is the lower radical determined by the class of all nilpotent rings. This can also be thought of as the class of all rings each nonzero homomorphic image of which has a nonzero nilpotent ideal. We also let F denote the Blair radical [2] which is the class of all hereditarily idempotent rings.

THEOREM 5. (a) If M is the class of all nilpotent rings, then $USM = \beta$.

(b) If M is the class of all semiprime rings, then $USM = F$.

(c) If M is any hereditary class of rings containing no nilpotent rings, then $UCM \subseteq F$.

Proof. (a) Clear

(c) Any hereditarily idempotent ring is in F

(b) If M is the class of all semiprime rings, we have that $USM \subseteq F$ from (c). Since F is a hereditary radical class of semiprime rings we have that every nonzero image of a ring from F has a nonzero ideal, considered as a ring, which is semiprime. Hence $F \subseteq USM$.

Now let Y denote the upper radical class determined by all fields. We know of many supernilpotent nonspecial radical classes. Those in [11] and [12] contain Y properly. Those of [9] are contained in Y properly whereas those of [7] are independent of Y. We propose to construct an infinite number of supernilpotent nonspecial radical classes of the latter type. For this purpose consider the ring A constructed in [9]. Three of the properties of the ring A are:

(1) Every ideal of A contains zero divisors.

(ii) The only prime image of A is Z_2.

(iii) A is a Boolean ring.

Let P be any hereditary class of prime rings containing Z_2. Consider C = {all nilpotent rings, P}. Then

THEOREM 6. USC is a supernilpotent nonspecial radical class.

Proof. From Theorem 4 it follows that USC is a supernilpotent radical class. Let X be the class of all prime USC semisimple rings. UX is then the smallest special radical class containing USC. The only prime image of the ring A is Z_2 and hence $A \in UX$. A has no nilpotent ideals since it is Boolean. Since every ideal of A has zero divisors, we conclude that A cannot have prime rings as ideals and hence $A \notin USC$. So USC is nonspecial.

In order to construct supernilpotent radical classes independent of Y consider the following: Let E be any prime ring that cannot be mapped onto a field, for example any simple ring which is not a field will do as the quaternions, and β any hereditary class of prime rings containing Z_2 but not E nor any ideals of E. Consider now K = {all nilpotent rings, β}. Then

COROLLARY 2. USK is a supernilpotent nonspecial radical class independent of $\mathcal{Y}$.

Proof. From Theorem 6 we conclude that USK is supernilpotent and nonspecial. Furthermore $Z_2 \in USK$ but $Z^2 \notin \mathcal{Y}$. The ring E above is in $\mathcal{Y}$ but $E \notin USK$.

REFERENCES

[1] V. A. Andrunakievic, "Radicals of Associative Rings I," *Mat. Sbornik,* 44 (1958),179-211 (Russian).

[2] R. L. Blair, "A Note of f-regularity in Rings," *Proc. Amer. Math. Soc.,* 6 (1955), 511-515.

[3] N. J. Divinsky, *Rings and Radicals*, Univ. of Toronto Press, (1965).

[4] N. Divinsky, J. Krempa, A. Sulinski, "Strong Radical Properties of Alternative and Associative Rings," *J. of Alg.,* 17, (1971), 369-388.

[5] G. A. P. Heyman, H. J. le Roux, "On Upper and Complementary Radicals," *Math. Japonicae,* to appear).

[6] N. Jacobson, *Structure of Rings,* Amer. Math. Soc., Colloq. Pub., vol. 37, Providence, (1956).

[7] W. G. Leavitt, "A Minimally Embeddable Rings," submitted.

[8] L.C.A. van Leeuwen, "On a Generalization of Jenkins Radical," *Arch. Der. Mat.,* Fasc. 2, vol. 22, (1971), 155-160.

[9] L.C.A. van Leeuwen and T. L. Jenkins,"A Supernilpotent Non-special Radical Class," *Bull. Austral. Math. Soc.,* vol. 9, no. 3, (1973), 343-348.

[10] D. M. Olson and T. L. Jenkins, "Upper Radicals and Essential Ideals." submitted, *J. of Austral. Math Soc.*

[11] Ju. M. Rjabuhin, "On Hypernilpotent and Special Radicals," *Studies in Algebra and Math. Analysis, Izdat. "Kartja Moldovenjaske,"* Kishinev 1965.

[12] Robert L. Snider, "Lattices of Radicals," *Pac. J. Math.* 40 (1972), 207-220.

[13] R. Wiegandt, *Radical and Semisimple Classes of Rings,* Queen's papers in pure and applied math., no. 37, Kingston, Ontario, (1974).

T. L. Jenkins
Mathematics Department
University of Wyoming
220 Ross Hall
Laramie, WY 82070
USA

COLLOQUIA MATHEMATICA SOCIETATIS JÁNOS BOLYAI

38. RADICAL THEORY, EGER (HUNGARY), 1982.

MAXIMAL IDEALS IN THE NEARRING OF POLYNOMIALS

H. KAUTSCHITSCH

Abstract: Still up today maximal ideals in the nearring of polynomials (R[x],+,) over a commutative ring R with 1 are known only in the case that R is a field. In this paper all maximal ideals over a large class of commutative R with 1, which contains all fields with more then two elements, are determined. One get a connection between the radical Rad R[x], defined as the intersection of all maximal ideals and the Jacobson-radical of R, which differs from that one in the ring-case.

1. Introduction: Throughout this paper let R denote a commutative ring with identity 1 and R[x] the set of all polynomials in one indeterminate x with coefficients of R. By $R_o[x]$ we denote the set consisting of all polynomials in one indeterminate x without constant term and the zeropolynomial 0. Beside the two ringoperations addition (+) and multiplication (.) a very often used operation with polynomials is the composition ($\circ$): $p \circ q := p(q(x))$. In polynomial structures with composition we consider the radical Rad(R[x]), defined as the intersection of all maximal ideals of R[x] ([5]). If (R[x],+,.) is the usual polynomial-ring, this radical coincides with the Jacobson-radical J(R[x]). In the composition-ring (R[x],+,.,$\circ$) the radical Rad(R[x]) differs from the Jacobson-radical ([5]). In the right nearring (R[x],+,$\circ$) we get a similar result, but the proofs are quite different, because we have no multiplication. Still up today maximal ideals are known only for the case, that R is a field ([2],[3],[7]). In this paper all maximal ideals for a large class of rings, which contains all fields $\neq \mathbb{Z}_2$ are determined. For $\mathbb{Z}_2$ all maximal ideals have been determined in ([2]). The results also show that the maximal ideals of the polynomial nearring are related very closely to the maximal ideals of R and therefore also to the radicals. Since in commutative rings with unit our radical Rad(R) coincides with the Jacobson-radical J(R) of R, we get a connection of the radical Rad(R[x],+,$\circ$) and the Jacobson-radical of the coefficient-ring R, but this connection differs from that one in the polynomialring-case ([1]). The connection differs also for the zero-symmetric nearring (R_0[x],+,$\circ$) and for the whole nearring (R[x],+,$\circ$).

We remember that a subset $I \subseteq (R[x],+,\circ)$ is a nearring-ideal iff:

(i) I is a subgroup of (R[x],+)

(ii) if $f \in I$, then $f \circ r \in I$ for all $r \in R[x]$

(iii) if $f \in I$, then $p \circ (q+f) - p \circ q \in I$ for all $p,q \in R[x]$

We often will use the fact that $p \circ (q+f) - p \circ q = f.s$ for a suitable $s \in R[x]$.

2. *Maximal ideals in the zero-symmetric nearring* $(R_o[x],+,\circ)$*:* To determine all maximal ideals, we proceed as follows:

(i) If $J \trianglelefteq R$, we set $\{J\} := \{f \in R_o[x] \mid f'(0) \in J\}$, where f' denotes the formal derivation of f. $\{J\} \trianglelefteq (R_o[x],+,\circ)$, because for $f,g \in \{J\}$ and $r,p,q,s \in R[x]$ we get:

$(f-g)'(0)=f'(0)-g'(0) \in J$

$(f\circ r)'(0)=f'(r\circ 0).r'(0)=0 \in J$

$(p\circ(q+f)-p\circ q)'(0)=(f.s)'(0)=$

$=f'(0).s(0)+f(0).s'(0)=f'(0).s(0) \in J.$

(ii) If $I \trianglelefteq (R_o[x],+,\circ)$, we set $I':=\{f'(0) \mid f \in I\}$. $I' \trianglelefteq R$, called the *derivation-ideal* of I:

If $a,b \in I'$, then

$a-b=f'(0)-g'(0)=(f-g)'(0) \in I'$.

If $r \in R$, then $ra=(rx\circ f)'(0)=r.f'(0) \in I'$, because $rx\circ f=rx\circ(0+f)-rx\circ 0 \in I$ for $f \in I$.

(iii) If $I \trianglelefteq (R_o[x],+,\circ)$, we set $I^o := \{c \in R \mid cx \in I\}$. $I^o \trianglelefteq R$: If $a,b \in I^o$, $r \in R$, then

$(a-b)x=ax-bx \in I$, hence $a-b \in I^o$ and

$(ra)x=(rx)\circ ax \in I$, hence $ra \in I^o$.

(iv) If $J \trianglelefteq R$, we denote by $\langle J \rangle$ the sum of such ideals I of $(R_o[x],+,\circ)$ with $ax \in I$ iff $a \in J$. For example $\{J\}$ are such ideals. But $\langle J \rangle$ may also contain a polynomial of the form $x+a_2x^2+\ldots+a_rx$.
Clearly: $\langle J \rangle^o = J$.

THEOREM 1: *All maximal ideals* **M** *of* $(R_o[x],+,\circ)$ *are given by the ideals* $M=\{J\}$ *and* $M=\langle J \rangle$ *, where J is a maximal ideal of R.*

Proof: We introduce the map $\varphi: R_o[x] \to R$, $\varphi(f)=f'(0)$ for $f \in R_o[x]$. Since we restrict ourselves to the zerosymmetric nearring of polynomials, φ is a homorphism of $(R_o[x],+,\circ)$ on $(R,+,.)$.

By the above remarks (i) and (ii) we see that for ideals $I \trianglelefteq (R_o[x],+,\circ)$ and $J \trianglelefteq R$ the sets $\varphi(I)=I'$ and $\varphi^{-1}(J)=\{J\}$ are ideals too. If now M is a maximal ideal of $(R_o[x],+,\circ)$ such that $M' \neq R$, then also M' is maximal, otherwise there is an ideal $J \neq R$ with $J \supset M'$. But then $\varphi^{-1}(J)=\{J\} \supset M$. Hence $\{J\}=R_o[x]$ and $1=\varphi(x) \in J$, which is a contradiction.

If J is maximal in R, then $\varphi^{-1}(J) = \{J\}$ is maximal in $(R_0[x],+,\circ)$: Let $B \supset J$, then $B' \supset J$, hence $B'=R$ and B contains at least one polynomial f of the form $f=x+b_2x^2+\dots b_rx^r$. But $x^i \in \{J\} \subset B$ for all $i \geq 2$, therefore $b_ix^i=b_ix\circ(0+x^i)-b_ix\circ 0 \in B$, hence $x \in B$ and $B=R_0[x]$.

Now we consider the case $M'=R$. (This case is possible, the ideal $\langle J\rangle$ for example may contain a polynomial f of the form $f=1.x+a_2x^2+\dots+a_rx^r$).

If now J is maximal in R, then also $\langle J\rangle$ is maximal in $(R_0[x],+,\circ)$: Let $B \supset \langle J\rangle$, then there is a polynomial $bx \in B$ with $bx \notin \langle J\rangle$, hence $B^0 \supset J$ and so $B^0=R$ and $x \in B$ or $B=R_0[x]$.

If $\langle J\rangle$ is maximal in $(R_0[x],+,\circ)$, then J is maximal in R: Let $B \supset J$, and $b \notin J$, then $\langle B\rangle \supset \langle J\rangle$, hence $\langle B\rangle = R_0[x]$ or $x \in \langle B\rangle$. Then $1 \in \langle B\rangle^0=B$.

COROLLARY 1: Rad $(R_0[x],+,\circ) = \{J(R)\}$

Proof: $\mathrm{Rad}(R_0[x],+,\circ) = (\bigcap \langle J\rangle) \cap (\bigcap\{J\}) =$
$= \langle \bigcap J\rangle \cap \{\bigcap J\} = \{\bigcap J\} = \{J(R)\}$, (J maximal in R).

3. Maximal ideals in the nearring $(R[x],+,\circ)$: Now the map $\varphi: R[x] \longrightarrow R$ of Theorem 1 is not a homomorphism and this is also valid for the map $r_0+r_1x+r_2x^2+\ldots+r_nx^n \rightarrow r_0$. To solve the problem, we need other ideals. Beside the derivation-ideal we construct for each ideal $I \trianglelefteq (R[x],+,\circ)$ the ideal $I_c := I \cap R \trianglelefteq R$ and for each ideal $J \trianglelefteq R$ the nearringideal $[J] := \{f \in R[x] \mid f \circ r_0 \in J \text{ for all } r_0 \in R\} \trianglelefteq (R[x],+,\circ)$. We get a complete answere for the following class of rings:

THEOREM 2: *Let R be a commutative ring with 1, which contains a unit $u \neq 1$, such that $u-1$ (or $u+1$) is again a unit. All maximal ideals M of $(R[x],+,\circ)$ are given by the ideals $M=[J]$, where J is maximal in R.*

Proof: Let M be a maximal ideal of R[x]. First we conclude that $M_c=R$ is impossible, otherwise $1 \in M$ and in all the following cases we would get $x \in M$, which inplies the contradiction $M=R[x]$:

First case: $char(R) \neq 2$ and $char(R) \neq 3$

$x^2 \circ (x+1) - x^2 \circ x = 2x+1 \in M$, hence $2x \in M$ and

$ux \circ (0+2x) - ux \circ 0 = 2ux \in M$ and hence $2x^i \in M$, $2u^j x^i \in M$. Therefore $x^3 \circ (x+u) - x^3 \circ x = 3ux^2 + 3u^2x + u^3 \in M$ and (since $M_C = R$) $ux^2 + u^2x \in M$, so $u^{-1}x \circ (ux^2 + u^2x) = x^2 + ux \in M$. Also $x^3 \circ (x+1) - x^3 \circ x = 3x^2 + 3x + 1 \in M$ or $x + x^2 \in M$.

If we add or subtract the last two polynomials, we get $2x^2 + (u+1)x \in M$ or $(u-1)x \in M$ and since $u \pm 1$ is a unit and $2x^2 \in M$, we get $x \in M$.

Second case: char(R)=2

$x^3 \circ (x+1) - x^3 \circ x = x + x^2 + 1 \in M$ and $x + x^2 \in M$.

$x^3 \circ (x+u) - x^3 \circ x = 3ux^2 + 3u^2x + u^3 \in M$ and $x^2 + ux \in M$.

By addition or subtraction we get again

$2x^2 + (u+1)x \in M$ or $(u-1)x \in M$.

Third case: char(R)=3

$x^2 \circ (x+1) - x^2 \circ x = 2x + 1 \in M$ or $2x \in M$ and therefore $2x^i = 2x \circ x^i \in M$ for all $i \geq 1$ and $2r_0 x^i =$

$= r_0 x \, (0 + 2x^i) - r_0 x \circ 0 \in M$ for all $r_0 \in R$.

$x^4 \circ (x+1) - x^4 \circ x = x + x^3 + 1 \in M$ or $x + x^3 \in M$.

$x^5 \circ (x+1) - x^5 \circ x = 2x^4 + x^3 + x^2 + 2x + 1 \in M$ or $x^2 + x^3 \in M$,

together we get $x + x^2 \in M$.

$x^4 \circ (x+u) - x^4 \circ x = ux^3 + u^3x + u^4 \in M$ or $x^3 + u^2x \in M$,

$x^5 \circ (x+u) - x^5 \circ x = 2ux^4 + u^2x^3 + u^3x^2 + 2u^4x + u^5 \in M$ or

$u^2x^3 + u^3x^2 \in M$ or (because u is a unit): $x^3 + ux^2 \in M$.

Adding $x^3+u^2x \in M$ we get $2x^3+ux^2+u^2x \in M$ or $ux+x^2 \in M$.

Again by adding and subtraction we get

$(u+1)x+2x^2 \in M$ or $(u-1)x \in M$ and therefore $x \in M$.

Now we conclude that M_c must be maximal:

If $J \supset M_c$, then $M \subseteq [M_c] \subset [J]$, since the constant polynomial $f(x)=j \in J$ with $j \notin M_c$ is an elemnt of $[J]$ but not of $[M_c]$. But then $[J]=R[x]$ and $1 \in J$.

If on the other side J is a maximal ideal of R, then $[J]$ is maximal in $R[x]$: Assume $B \supset [J]$, then $B_c \supset J$, because for $f \in B$ with $f \notin [J]$ we get $f\circ r_o \in B \cap R = B_c$ and $f \circ r_o \notin J$ for a suitable $r_o \in R$. Hence $B_c=R$ and again we conclude that $B=R[x]$.

COROLLARY 2: *If R is a ring like in Theorem 2, then* $Rad(R[x],+,\circ)=(J[R])$

Proof: $Rad(R[x],+,\circ)=\bigcap[J]$ (J maximal in R) $=[\bigcap J]=[J(R)]$.

Remarks: Examples for such rings are all fields with more then two elements, so that all the known results are included (see Corollary 3).

If $R=\mathbb{Z}_2$, then all maximal ideals are determined in ([2]). But also many Z_m, $m \neq 2$ are examples for such rings.

COROLLARY 3: *If F is a finite field* $F \neq Z_2$, *then* $M = \{f \in F[x] \mid f \circ r_o = 0 \text{ for all } r_o \in F\}$ *is the unique maximal ideal, if F is infinite, then (0) is the only maximal ideal of* $(F[x], +, \circ)$.

COROLLARY 4: *If R is a Jacobson-radical-ring, then* $Rad(R[x], +, \circ) = R[x]$.

So the nearring-radical of R[x] differs substantially from the ring-radicals $Rad(R[x],+,.) = J(R[x],+,.) = N[x]$, where N denotes the maximal Nilideal of R. But it coincides with the radical of the composition-ring $(R[x],+,.,\circ)$, see ([5]).

References:

[1] S.A. Amitur, Radicals of polynomial rings, *Canadian J. Math.* 8 (1965), 355-361.

[2] J.L. Brenner, Maximal ideals in the nearring of polynomials mod 2, *Pacific J. Math.* 52 (1974), 595-600.

[3] J. Clay and D. Doi, Maximal ideals in the nearring of polynomials over a field, *Rings, Modules and Radicals (Proc. Conf. Keszthely,*

1974), pp. 117-133. Colloqu. Math. Soc. J. Bolyai, 6, North-Holland, Amsterdam, 1973.

[4] H. Kautschitsch, Maximal ideals in the near-ring of formal power series, *Near-rings and Near-fields (Proc. Conf. San Benedetto del Tronto, 1981).*

[5] R. Mlitz, Ein Radikal für universale Algebra und seine Anwendung für Polynomringe mit Komposition, *Monatsh. Math.* 75 (1971), 144-152.

[6] W. Nöbauer, Über die Operation des Einsetzens in Polynomringen, *Math. Ann.* 134 (1958), 248-259.

[7] G. Pilz, *Near-rings*, North-Holland/American Elsevier, Amsterdam, 1977.

H. Kautschitsch
Institut für Mathematik
UBW Klagenfurt
Universitätsstraße 65-67
A-9010 Klagenfurt
Austria

COLLOQUIA MATHEMATICA SOCIETATIS JÁNOS BOLYAI

38. RADICAL THEORY, EGER (HUNGARY), 1982.

RADICALS AND DERIVATIONS OF ALGEBRAS

J. KREMPA

§ 1. MOTIVATIONS

In this paper K will always denote a commutative ring with identity element and by an algebra we shall always mean a K-algebra. By a commutative ring we shall always mean a commutative and associative ring.

One of the most important results of the theory of radicals is the following theorem of Anderson, Divinsky and Suliński:

THEOREM 1.1 [3]. *If* S *is any radical property in the class of associative algebras, then for any associative algebra* R *and any ideal* I *of* R, $S(I)$ *is an ideal of* R.

This theorem has motivated the following:

Definition. Let $\mathfrak{M}$ be a class of algebras and let S be a radical in $\mathfrak{M}$. We shall say that S is an ADS-radical in $\mathfrak{M}$ if for any algebra $R\in\mathfrak{M}$ and any ideal I of R, $S(I)$ is an ideal of R.

Besides Theorem 1.1, it is also proven in [3] that in the class of alternative rings every radical is an ADS-radical. Much later Slin'ko [21] proved that in the class of Jordan rings every supernilpotent radical is an ADS-radical. Afterwards Nikitin [15] extended Slin'ko's results to all radicals of Jordan rings.

In what follows we shall use the following:

Definition [1]. An ideal I of an algebra A is said to be characteristic if $d(I)\subseteq I$ for any K-linear derivation $d:A\to A$. A radical S in the class $\mathfrak{M}$ of algebras is said to be characteristic if for any algebra $A\in\mathfrak{M}$, $S(A)$ is a characteristic ideal.

Obviously trivial radicals are characteristic. Moreover, by the definition of derivation it follows that an idempotent ideal of any algebra is characteristic. In particular, every subidempotent radical is characteristic.

The theory of radicals has also been intensively developed in the variety of Lie algebras. An important result in this case is

PROPOSITION 1.2 [2]. *Let S be a radical in the class of all Lie algebras. Then S is an ADS-radical* *iff it is a characteristic radical.*

Proof. Let S be an ADS-radical and let $d:I \to I$ be a derivation of a Lie algebra I. Let $I[d] = I \oplus Kd$ as a K-module, be an algebra in which multiplication is determined by the multiplication in I and the conditions: $d \cdot x = d(x) = -x \cdot d$ (for $x \in I$) and $d \cdot d = 0$. It is known that I is an ideal of $I[d]$ and $I[d]$ is a Lie algebra. Hence, by the assumption, it follows that $S(I)$ is an ideal of $I[d]$, i.e. $d(S(I)) \subseteq S(I)$.

Assume now that S is not an ADS-radical. Then there exists a Lie algebra R and an ideal I of R such that $S(I)$ is not an ideal of R. Thus there exists $r \in R$ such that $rS(I) \nsubseteq S(I)$. Putting d to be the left multiplication by r, we get $d(S(I)) \nsubseteq S(I)$.

Various important consequences of Theorem 1.1 and Proposition 1.2 have aroused the interest of many authors in the interdependence of radicals and derivations. This interdependence has appeared to be interesting not only in the case of Lie algebras.

Our aim is precisely to present some known and some new results on this subject. We shall restrict ourselves to the case when the base ring K is a field. This restriction will enable us to avoid problems connected with the additive torsion and it will also enable us to apply linear algebra and tensor products. A good deal of the methods presented here goes back to associative or even commutative algebras.

Having adopted the assumption that K is a field, we obtain immediately the following.

PROPOSITION 1.3. *Let* $\mathfrak{M}$ *be a class of algebras and let* S *be a radical in* $\mathfrak{M}$ *. Then* S *is either supernilpotent or subidempotent.*

All the necessary results on radicals and on varieties of algebras can be found in the books cited in the references.

§ 2. LATTICES OF RADICALS

It is known [22] that if $\mathfrak{M}$ is any class of algebras then all radicals in $\mathfrak{M}$ form a complete lattice.

THEOREM 2.1. *Let* $\mathfrak{M}$ *be a class of algebras. Then the ADS-radicals in* $\mathfrak{M}$ *form a complete sublattice of the lattice of all radicals in* $\mathfrak{M}$.

Proof. Let $\{S_i\}$ be a family of ADS-radicals and let $S=\cap S_i$. We shall show that S is an ADS-radical. Let I be an ideal of an algebra A and put $J=S(I)$. Consider the ideal $\bar{J}$ generated by J in A . Then

$J = S(I) \subseteq S(\bar{J}) \subseteq S_i(\bar{J})$ for any index i. Since S_i is an ADS-radical, $S_i(\bar{J})$ is an ideal of A and by the definition of $\bar{J}$ we get that $S_i(\bar{J}) = \bar{J}$. Since i is arbitrary, we have $S(\bar{J}) = \bar{J}$, hence $S(I) = \bar{J}$ is an ideal of A. Thus the class of ADS-radicals is closed with respect to the intersection. Immediately, by the definition, it also follows that this class contains the join of any chain of radicals. Therefore it is enough to prove that the join of two ADS-radicals is again an ADS-radical. Let S_1, S_2 be ADS-radicals and let $S = S_1 \vee S_2$. Assume that I is an ideal of an algebra A and $J = S(I)$, moreover let $\underline{J}$ be the sum of all ideals of A contained in J. Putting $A_1 = A/\underline{J}$, $I_1 = I/\underline{J}$ and $J_1 = J/\underline{J}$, we get that $S(I_1) = J_1$ and J_1 contains no nonzero ideals of A_1. Suppose that $J_1 \neq 0$. Then, by the definition of S, it follows that J_1 contains a non-zero, e.g. S_1-radical accessible subalgebra B_0. Thus there exists a chain $B_0 \triangleleft B_1 \triangleleft \ldots \triangleleft B_n = J_1 \triangleleft I_1$. Since S_1 is an ADS-radical and $S_1(B_0) = B_0 \neq 0$, one can show by induction that $S_1(I_1) \neq 0$, i.e., $S_1(I_1)$ is a nonzero ideal of A_1. Obviously $S_1(I_1) \subseteq S(I_1) = J_1$, which is impossible as J_1 contains no nonzero ideals of A_1. Hence $J_1 = 0$, which implies that $J = \underline{J}$ is an ideal of A.

For the characteristic radicals we shall prove a slightly weaker result.

THEOREM 2.2. *Let* $\mathfrak{M}$ *be a class of algebras. Then the characteristic radicals in* $\mathfrak{M}$ *form a complete lattice in which the join of any chain and the meet of any family of elements are the same as in the lattice of all radicals in* $\mathfrak{M}$.

Proof. Suppose that I is any algebra, J is an ideal of I and $d: I \to I$ is a derivation. Let J^d be the smallest d-invariant subspace of I containing J, and let J_d be the largest d-invariant subspace of I contained in J. The existence of these subspaces follows by the fact that the class of all d-invariant subspaces is closed with respect to intersection and algebraic sums. Now the proof is analogous to that of Theorem 2.1, one can obtain it by replacing $\bar{J}$ by J^d and $\underline{J}$ by J_d and by taking into consideration the following.

LEMMA 2.3. *In the above notation*

(a) J_d *is an ideal of* I ,

(b) J^d *is an ideal of* I , *moreover,* $J^d =$ $= J + d(J) + d^2(J) + \ldots$.

Proof of the Lemma. (a) follows immediately from

the fact that if $V \subseteq I$ is a d-invariant subspace then VI and IV are also d-invariant subspaces. To prove (b) it is enough to observe that $J^d = J + d(J) + d^2(J) + \dots$ and that if V is an ideal of I then $V + d(V)$ is also.

The lattices considered above do not coincide in general.

Example 2.4. Let $\mathfrak{M}$ be the class of all algebras. It is known [4] that in $\mathfrak{M}$ there are no ADS-radicals except the trivial radicals. However, there are non-trivial subidempotent radicals which, as we have noticed, are characteristic.

Example 2.5. Let $\mathfrak{M}$ be the class of all associative algebras. By Theorem 1.1 we know that the Jacobson radical J is an ADS-radical. But it is not characteristic. Indeed, let $C = K[[t]]$ be a formal power series algebra. Then $J(C) = tC$ is not d-invariant, where d is a standard derivation.

Considering Theorems 2.1, 2.2 and the above examples we arrive at the following natural question:

- do characteristic radicals form a sublattice of the lattice of all radicals?

or in a different manner:

- is a join of characteristic radicals again a characteristic radical?

Now we shall show that in some cases the answer is positive.

LEMMA 2.6. *Let* S_1 *and* S_2 *be characteristic ADS-radicals in the class* $\mathfrak{M}$. *Then* $S_1 \vee S_2$ *is also.*

Proof. By Theorem 2.1 it follows that $S = S_1 \vee S_2$ is an ADS-radical. Let $I \in \mathfrak{M}$ and $S(I) = J$. Dividing, if necessary, by J_d we may assume that J contains no nonzero d-invariant subspaces. In this case it is enough to show that $J=0$. Were $J \neq 0$ then similarly to the end of the proof of Theorem 2.1 we would get that e.g. $S_1(I) \neq 0$. Since S_1 is a characteristic radical, $S_1(I)$ is then a d-invariant subspace. Obviously $S_1(I) \subseteq J$, which is impossible.

COROLLARY 2.7. *Let* $\mathfrak{M}$ *be a class of algebras. Then the characteristic radicals in* $\mathfrak{M}$ *form a complete sublattice in the lattice of all radicals in any of the following cases:*

1) $\mathfrak{M}$ *is a class of alternative algebras,*

2) $\mathfrak{M}$ *is a class of Jordan algebras,*

3) $\mathfrak{M}$ *is a class of Lie algebras.*

Proof follows immediately from Lemma 2.6 and the results in paragraph 1.

§ 3. THE POSITIVE CHARACTERISTIC CASE

From the theory of Lie algebras it follows that the property of being a characteristic ideal or a characteristic radical may depend on the characteristic of the field K. And in fact this is so. In this paragraph we shall assume that K is a field of characteristic $p>0$.

Let $\mathfrak{M}$ be a variety of algebras closed with respect to tensor multiplying by commutative algebras. For any such variety the following holds.

PROPOSITION 3.1. *Let* S *be a nontrivial radical in* $\mathfrak{M}$. *Then* S *is characteristic iff it is subidempotent.*

Proof. Let S be a nontrivial radical in $\mathfrak{M}$ which is not subidempotent and let A be a nonzero S-semisimple algebra. Consider the algebra C_p with basis $1, c, \ldots, c^{p-1}$ and with multiplication induced by the condition $c^p=0$. Now consider the ideal $A \otimes cC_p$ of the algebra $B = A \otimes_K C_p$. From Proposition 1.3 it follows that S is a supernilpotent radical and hence $I \subseteq S(B)$. On the other hand, the algebras B/I and A are isomorphic, hence $S(B) = I$. Now let d be the derivation of C_p induced by the condition $d(c)=1$. Since K is a field of characteristic p, d is well defined. Thus $1 \otimes d$ is a derivation of B for which the ideal $I = S(B)$ is not invariant. Hence the radical S is not characteristic.

The other implication is obvious.

COROLLARY 3.2. *Characteristic radicals in* $\mathfrak{M}$ *form a complete sublattice of the lattice of all radicals.*

Proof follows easily from the fact that the join of subidempotent radicals is also a subidempotent radical.

The author does not know whether the condition imposed on $\mathfrak{M}$ in the above Proposition and Corollary is essential.

From Proposition 3.1 and Proposition 1.2 it follows, in particular, that in the class of Lie algebras a non-trivial radical is an ADS-radical iff it is subidempotent.

Obviously, ADS-radicals are semisimply hereditary on ideals. From the proof of Proposition 1.2 it follows that in any class of Lie algebras characteristic radicals are ADS-radicals.

Example 3.3. Let $\mathfrak{M}$ be the class of all Lie algebras. By $\mathfrak{M}_n$ we shall denote the subclass of $\mathfrak{M}$ consisting of all at most n-dimensional algebras. Moreover, let S be the upper radical in $\mathfrak{M}$ determined by a simple finite-dimensional algebra $R\in\mathfrak{M}$. We shall show that there exists an $n>1$ such that

1) S is a characteristic ADS-radical in $\mathfrak{M}_k$ for $k<n$,

2) S is an ADS-radical in $\mathfrak{M}_n$ and is not a characteristic radical in this class,

3) S is not an ADS-radical in $\mathfrak{M}_{n+1}$, but it is semisimply hereditary on ideals in this class.

Consider the algebra $R \otimes C_p$, where C_p is the algebra defined in the proof of Proposition 3.1. Then $S(R \otimes C_p)$ is not a characteristic ideal and clearly $R \otimes C_p$ is a finite dimensional algebra. Now let n be the smallest positive integer such that there exists an algebra $A \in \mathfrak{M}_n$ for which $S(A)$ is not a characteristic ideal of A. From the choice of n it follows that S is a characteristic radical in $\mathfrak{M}_k$ for $k<n$ and it is not characteristic in $\mathfrak{M}_n$. Now let $B \in \mathfrak{M}_n$ and let I be an ideal of B. If $I=B$ then $S(I)$ is an ideal of B. On the other hand, if $I \neq B$ then $I \in \mathfrak{M}_{n-1}$. Thus $S(I)$ is a characteristic ideal of I. Since B is a Lie algebra, also in this case we get that $S(I)$ is an ideal of B. Thus S is an ADS-radical in $\mathfrak{M}_n$.

Next we shall show that S is not an ADS-radical in $\mathfrak{M}_{n+1}$. Consider an n-dimensional algebra A such that $S(A)$ is not a characteristic ideal, i.e., there exists a derivation $d : A \to A$ such that $d(S(A)) \not\subseteq S(A)$. Now the algebra $A[d]$ defined in the same way as in the proof of Proposition 1.2 lies in $\mathfrak{M}_{n+1}$, A is an ideal of $A[d]$ but $S(A)$ is not an ideal of $A[d]$.

At last we shall show that the radical S is semisimply hereditary on ideals in the class $\mathfrak{M}_{n+1}$. Let $0 \neq B \in \mathfrak{M}_{n+1}$ be an S-semisimple algebra and let I be an ideal of B . By the definition of the radical S it follows that there exists a homomorphism ϕ of B onto R . Let $J = \ker \phi$. Since $S(I)$ is an accessible subalgebra of B and R is a simple algebra, $\phi(S(I)) = 0$, i.e., $S(I) \subseteq J$. Replacing, in case of need, I with $I \cap J$ we may assume that $I \subseteq J$. Since R is a Lie algebra, we have $\dim_K R \geq 3$ [10] . Hence $\dim_K I \leq n-2$. Thus, as we have already shown, $S(I)$ is a characteristic ideal of I . Taking into consideration this and the fact that B is a Lie algebra, we get that $S(I)$ is an ideal of B . As B is an S-semisimple algebra, $S(I) = 0$.

The above example and the remark preceding it suggest the following question:

- Does there exist, in the class of all Lie algebras, a nontrivial supernilpotent radical which is semisimply hereditary on ideals?

§ 4. SOME EFFECTIVE CONSTRUCTIONS

Henceforth K will denote a field of characteristic 0 and throughout this paragraph $\mathfrak{M}$ will denote a variety of algebras.

From Proposition 3.1 we know that in the case of algebras over a field of positive characteristic there are hardly any supernilpotent characteristic radicals. In the present case of algebras over a field of characteristic 0 there exists an effective method of construction of supernilpotent characteristic radicals [16].

The same construction has been in fact used in [20], [13]. Now we shall describe this construction.

Let $K[x_1, x_2, \ldots] = F$ be the free algebra in $\mathfrak{M}$ with $x_1, x_2, \ldots$ as the set of free generators. For $p(x_1, x_2, \ldots, x_n)$, $q(x_1, x_2, \ldots, x_r) \in F$ we put

$$p \circ q = p(q(x_1, \ldots, x_r), q(x_{r+1}, \ldots, x_{2r}), \ldots, q(x_{n+1\ r+1}, \ldots, x_{nr})) .$$

If $p \in F$ then $\bar{p}$ will denote the smallest subset of F containing p and closed under the operation $\circ$.

Since K is infinite, F is indeed a graded algebra, with gradation given by the total degree of polynomials. This enables us to introduce the following:

Definition. Let $p \in F$ be a homogeneous element of

degree greater than 1. An algebra A is called an S_p-algebra if for any finite dimensional subspace $V \subseteq A$ there exists a $q \in \bar{p}$ which is the identity on V.

The following theorem was, in fact, proved in [16].

THEOREM 4.1. *S_p is a characteristic radical in* $\mathfrak{M}$.

The best-known S_p radical is the weakly solvable radical given by the condition $p = x_1 x_2$ defined in [16].

It is clear that the radicals considered in Theorem 4.1 are supernilpotent, i.e. $S_p \geq L_0$, where L_0 denotes the lower radical determined in $\mathfrak{M}$ by the algebras with the zero multiplication. The following question seems to be interesting:

Suppose that the radical L_0 is nontrivial. Does there exist a homogeneous $p \in F$ such that the radical S_p is nontrivial?

It is known [16] that in the variety of all Lie algebras the weakly solvable radical is nontrivial. It seems that the following fact may be useful in discussing the above question.

Remark. If the radical S_p is trivial in $\mathfrak{M}$ then for any $n \geq 1$ there exists a polynomial $q_n \in \bar{p}$ which is the identity on any n-dimension subspace of any algebra in $\mathfrak{M}$.

The following question was put in [8] (Problem II.26):

Is every radical in the class of Lie algebras characteristic?

It is obvious that this question can be considered in any variety. In fact in [13] it is shown that if $\mathfrak{M}$ contains a division algebra D then the lower radical determined by $tD[[t]]$ is not characteristic. We shall show that the existence of a simple algebra in $\mathfrak{M}$ is sufficient for the construction of such a radical in $\mathfrak{M}$. For this we need the following results (for the necessary terminology and results see [10]):

LEMMA 4.2. [10] *Let* L *be a field extension of* K , R *be a central simple* L*-algebra, and* C *be a commutative* L*-algebra with identity. Then every ideal of* $R\otimes_L C$ *is of the form* $R\otimes_L I$ *where* I *is an ideal of* C .

LEMMA 4.3. *Suppose that* R *is an* L*-algebra and* J *is an ideal of a commutative* L*-algebra* C . *Let* $\phi : R\otimes_L J \to S$ *be a homomorphism onto a prime ring* S . *Then* ϕ *can be extended to a homomorphism of the ring* $R\otimes_L C$ *onto a prime ring* S_1 *containing* S .

Proof. Let D be the centroid of the ring S . Since $R\otimes_L J$ is a C-algebra and $\ker\phi$ is a prime ideal, ϕ induces a homomorphism $\psi : C \to D$ given by $\psi(c)\phi(r\otimes i) = \phi(r\otimes ic)$ for $r\in R$, $i\in I$, $c\in C$. Assume that $\psi(J)=0$. Then $0 = \psi(J)\phi(R\otimes J) = \phi(R\otimes J^2) = \phi(R\otimes J)^2 = S^2$, which is impossible. Pick an element $x\in J$ such that $\psi(x) \neq 0$. Since D is a domain, $\psi(x)$ is not nilpotent. Let S_1 be the localization of S at the multiplicative set composed of all powers of $\psi(x)$. Since S is prime, S_1 is also and $S\subset S_1$ in a natural way. Now let $\bar{\phi}(r\otimes c) = \psi(x)^{-1}\phi(r\otimes cx)$ for $r\in R$, $c\in C$. It is easy to check that $\bar{\phi}$ is a homomorphism of $R\otimes C$ into S_1 and an extension of ϕ.

COROLLARY 4.4. *Let R be a central simple L-algebra and J be a commutative L-algebra. Then for any prime ideal I of $R\otimes_L J$ there exists a prime ideal H of J such that $I = R\otimes_L H$.*

Proof. If J has an identity element then Lemma 4.2 yields the result. Suppose now that J is an algebra without identity. Let C be the standard extension of J to an L-algebra with identity element. Then, by Lemma 4.3, we get that there exists a prime ideal $\bar{I}$ of $R\otimes C$ such that $\bar{I} \cap (R\otimes J) = I$. From the first part of the proof it follows that there exists an ideal $\bar{H}$ of C such that $\bar{I} = R\otimes\bar{H}$. Thus it suffices to put $H = \bar{H}\cap J$.

COROLLARY 4.5. *Let R be a central simple L-algebra and J be a Jacobson radical commutative L-algebra. Then there is no homomorphism of $R\otimes_L J$ onto any simple ring.*

Proof. Suppose that $\phi : R\otimes_L J \to S$ is a homomorphism onto a simple ring S. Applying Corollary 4.4 we get that there exists an ideal $H\subseteq J$ such that $\ker \phi = R\otimes H$, thus $S \simeq R\otimes_L (J/H)$. Since S is a simple ring, J/H is a field, which is impossible.

THEOREM 4.6. *Let $\mathcal{U}$ be a non-empty class of simple algebras from $\mathfrak{M}$. Let $\mathcal{S}$ be the upper radical determined by $\mathcal{U}$. Then $\mathcal{S}$ is not characteristic.*

Proof. It is known that the radical $\mathcal{S}$ exists. Let $R\subseteq \mathcal{U}$ and let L denote the centroid of R. If $C=L[[t]]$, $J=tC$, then the algebra $R\otimes_L C$ belongs to $\mathfrak{M}$ as a homomorphic image of the algebra $R\otimes_K C$. Corollary 4.5 easily implies that $\mathcal{S}(R\otimes_L C) = R\otimes J$. This ideal is not invariant under the derivation induced by the standard derivation of C.

In view of Theorem 4.6 and Proposition 1.2 it is interesting to ask whether in the class of Lie algebras every radical is semisimply hereditary on ideals. We show that the answer is in the negative.

Example 4.7. Let R be a simple Lie algebra with centroid L. Let $C = S^{-1}L[t]$, where $S = \{f \in L[t] : f(0) \neq 0\}$ and $J = J(C) = tC$. Put $I = R \otimes_L J$, $A = R \otimes_L C$, $B = A[d]$, where d is the L-linear derivation of A induced by the standard derivation of C. Let S be the lower radical determined by I. Clearly $I \subseteq S(A)$. If $S(A) \neq I$ then $S(A) = A$ since $A/I \simeq R$ is a simple algebra. Then $S(R) = R$, whence there exists a homomorphism of I onto R. By Corollary 4.5 this is impossible, so $0 \neq I = S(A)$.

Since the characteristic of K is 0, I contains no d-invariant ideals, so one can easily check that A is the unique proper ideal in B. Suppose $S(B) = B$. Then there exists an ordinal number α such that $B \in S_\alpha(I)$, where $S_\alpha(I)$ stands for the successive classes in the Kurosh construction of the lower radical determined by I. If $\alpha > 0$ then $A \in S_\beta(I)$ for some $\beta < \alpha$. This means $S(A) = A$, a contradiction. Thus $\alpha = 0$, and hence there exists a homomorphism ϕ of I onto B. Since B is a prime algebra, by Corollary 4.4 there exists a prime ideal $J \neq H \subseteq C$ such that $\ker\phi = R \otimes_L H$. As J contains no prime ideal except 0, ϕ is an injection and hence an isomorphism. But this is impossible since $B^2 = A$ is an idempotent algebra whereas $I^2 = (R \otimes J)^2$ is not an idempotent algebra. Thus we have shown that $S(B)\neq B$. Since $S(A)\neq A$, $S(B)=0$ but $S(A)=I\neq 0$ and S is not semisimply hereditary on ideals.

One can construct an analogous example by putting $C=L[[t]]$. If we assume in addition that R is a division Lie algebra then, in fact, we get the radical constructed in [13]. The fact that the latter radical is not semisimply hereditary on ideals was checked in a different way in [5].

Theorem 4.1, Proposition 1.2 and Example 4.7 suggest the question

- whether in a variety of Lie algebras every radical which is semisimply hereditary on ideals, is characteristic.

§ 5. SEMIPRIME IDEALS

Since subidempotent radicals are characteristic, it suffices to investigate whether supernilpotent radicals are also. A supernilpotent radical of any algebra is a semiprime ideal. Thus in this paragraph we shall examine the question under what conditions a semiprime ideal is characteristic. In this paragraph A shall denote a K-algebra, I an ideal of A , and d a derivation of A .

For our investigations we use the following construction of powers of I, depending on the algebra A.

Definition. For any $n \geq 1$ let $(I,A)^n$ be the K-subspace generated by all products of elements of A containing at least n factors from I.

As a straightforward consequence of this definition we obtain

LEMMA 5.1. *For any* $n, p, q \geq 1$ *the following hold.*

1) $(I,A)^n$ *is an ideal of* A,

2) $(I,A)^{n+1} \subseteq (I,A)^n$,

3) $(I,A)^p (I,A)^q \subseteq (I,A)^{p+q}$,

4) $d(I,A)^{n+1} \subseteq (I,A)^n$.

Let I_d be the largest d-invariant subspace contained in I. By Lemma 2.3 it follows that I_d is an ideal of A and by Lemma 5.1 $\bigcap_{n=1}^{\infty} (I,A)^n \subseteq I_d$.

LEMMA 5.2. (a) *If* I *is a semiprime ideal then* I_d *is also,*

(b) *If* I *is a prime ideal then* I_d *is also.*

Proof. Let B be an ideal of A such that $B^2 \subseteq I_d$. Put $B_0 = B$, $B_n = B + dB + \ldots + d^n B$ for $n>0$, $B^d = \bigcup_{n=0}^{\infty} B_n$. We shall prove that $B^d \subseteq I$. It is enough to show that $B_n \subseteq I$ for any $n \geq 0$. If $n=0$ then $B_n^2 = B^2 \subseteq I_d \subseteq I$. Thus applying semprimeness of I we get that $B_n \subseteq I$. Let $n>0$ and $x, y \in B_n$. Then $x = x_1 + d^n x_2$, $y = y_1 + d^n y_2$, where $x_1, y_1 \in B_{n-1}, x_2, y_2 \in B$. Then, as char $K=0$,

$$d^n x_2 \cdot d^n y_2 = \binom{2n}{n}^{-1} \left[d^{2n}(x_2 y_2) - \sum_{\substack{i=0 \\ i \neq n}}^{2n} \binom{2n}{i} d^i x_2 d^{2n-i} y_2 \right] .$$

Since $x_2 y_2 \in I_d$, we have $d^{2n}(x_2 y_2) \in I_d \subseteq I$. Moreover, by the induction hypothesis it follows that if $0 \leq i < n$ then $d^i x_2 \in I$ and if $n < i \leq 2n$ then $d^{2n-i} y_2 \in I$. Thus in both cases $\binom{2n}{i} d^i x_2 d^{2n-i} y_2 \in I$. Hence $d^n x_2 d^n y_2 \in I$. From this it follows easily by the induction hypothesis that $xy \in I$, i.e. $B^2 \subseteq I$. Therefore $B_n \subseteq I$, as B_n is an ideal of A and I is a semiprime ideal of A . As we have already observed in § 2, B^d is a d-invariant subspace of A and therefore $B^d \subseteq I$ implies that $B^d \subseteq I_d$, i.e., $B \subseteq B^d \subseteq I_d$.

(b) Let I be a prime ideal and B and C be ideals of A such that $BC \subseteq I_d$. We may assume that $I_d \subseteq B \cap C$. Let B' be the largest ideal of A such that $B \subseteq B'$ and $B'C \subseteq I_d$. We shall show that B' is d-invariant. Let $x \in B'$, $y \in C$. Then $dx \cdot y = d(xy) - xdy$. Hence $dx \cdot y \in B' \cap C$, as $xy \in I_d$, i.e. $d(xy) \in I_d \subseteq B'$ and $x \cdot dy \in B'$. Now $(B' \cap C)^2 \subseteq I_d$, therefore applying (a) we get that $B' \cap C \subseteq I_d$. Thus $dx \cdot y \in I_d$ and this means that $(B' + dB')C \subseteq I_d$. Therefore $B' = B' + dB'$, i.e. $dB' \subseteq B'$. Similarly as above one can verify that the largest ideal

C' of A such that $C \subseteq C'$ and $B'C' \subseteq I_d$ is also d-invariant. Now $B'C' \subseteq I_d \subseteq I$ implies that $B' \subseteq I$ or $C' \subseteq I$. Since B' and C' are d-invariant ideals, $B' \subseteq I_d$ or $C' \subseteq I_d$. This means that I_d is a prime ideal.

COROLLARY 5.3. *Let* I *be a semiprime ideal of* A. *If there exists an* n *such that* $(I,A)^n \subseteq I_d$ *then* I *is* d*-invariant.*

Proof. Let $r \geq 1$ be the smallest integer such that $(I,A)^r \subseteq I_d$. If $r>1$ then by Lemma 5.1 we get that $((I,A)^{r-1})^2 \subseteq (I,A)^{2r-2} \subseteq (I,A)^r \subseteq I_d$. From Lemma 5.2 $(I,A)^{r-1} \subseteq I_d$. Thus $r=1$.

THEOREM 5.4. *If* I *is a minimal semiprime or minimal prime ideal of* A *then* I *is characteristic.*

Proof. In each of the considered cases Lemma 5.2 implies that $I_d = I$.

THEOREM 5.5 (see also [2]). *Let* A *be an algebra with* **dcc** *on ideals. Then any semiprime ideal of* A *is characteristic.*

Proof. By Lemma 5.1 it follows the existence of an n such that $(I,A)^n = (I,A)^{n+1}$. Thus $(I,A)^n$ is a characteristic ideal and in fact $(I,A)^n \subseteq I_d$. This ends the proof by Corollary 5.3.

Similarly one can prove the following.

PROPOSITION 5.6. *If* I *is a semiprime and finite dimensional ideal of* A *then* I *is characteristic.*

From Theorem 5.5 or Proposition 5.6 we immediately get the following.

COROLLARY 5.7 (cf. Example 3.3). *Let* A *be a finite-dimensional algebra. Then any semiprime ideal of* A *is characteristic.*

We shall show that Theorem 5.5 does not hold for algebras with acc on ideals and Corollary 5.7 cannot be generalized to countable-dimensional algebras.

Example 5.8. Assume that the field K is countable and R is an at most countable-dimensional central simple K-algebra. Let $C = S^{-1}K[t]$ where $S = \{f \in K[t] : f(0) \neq 0\}$, and $J = J(C) = tC$. Put $A = R \otimes_K C$, $I = R \otimes_K J$, and let d be the derivation of A induced by the standard derivation of C. It is easy to see that A is a countable-dimensional algebra and I is not a d-invariant ideal of A. Moreover, by Lemma 4.2 it follows that A satisfies acc on ideals and I is a prime ideal of A. Using the methods of the previous paragraph one can show that I is the upper radical of A determined by R as well as the lower radical of A determined by I.

Let us recall the following

Definition ([18]). A derivation d of an algebra A is said to be algebraic if for any $a \in A$ there exists an $n \geq 1$ such that $d^n a$ belongs to the subalgebra $\langle a, da, d^2 a, \ldots, d^{n-1} a \rangle$ generated by the elements $a, da, \ldots, d^{n-1} a$.

Algebraic derivations of bounded index, i.e., algebraic derivations for which n may be chosen independently of a, were considered in [13].

THEOREM 5.9 ([13]). *If I is a semiprime ideal of A and d is an algebraic derivation of bounded index then I is d-invariant.*

Proof. Choose n such that for any $a \in A$ $d^n a \in \langle a, da, \ldots, d^{n-1} a \rangle$. Let $x \in (I,A)^n$. Thus by Lemma 5.1 $x, dx, \ldots, d^{n-1} x \in I$. Therefore $\langle x, dx, \ldots, d^{n-1} x \rangle$ is a d-invariant subspace contained in I. Hence $x \in I_d$ and consequently, $(I,A)^n \subseteq I_d$. Now the result follows from Corollary 5.3.

§ 6. LOCALLY FINITE DERIVATIONS

For the further investigations of some classes of algebraic derivations we need some results on graded algebras.

Let G be a group. An algebra A is said to be G-graded if $A = \bigoplus_{g \in G} A_g$, where the A_g $(g \in G)$ are subspaces of A such that $A_g A_h \subseteq A_{gh}$. Then the elements of the subspaces A_g $(g\in G)$ are called homogeneous elements and if $a = \sum_{g\subset G} a_g$, where $a_g \in A_g$, then the elements a_g are called the homogeneous components of a. It is clear that the elements a_g are almost all 0. A subspace $I\subseteq A$ is called homogeneous if $I = \bigoplus_{g\in G} I\cap A_g$.

Let K^* and K^+ denote the multiplicative and the additive group of the field K, respectively. From the definition of a G-graded algebra we immediately get the following.

LEMMA 6.1. *Let* A *be a* G*-graded algebra,* $\alpha : G \to K$ *any mapping and let* $\phi_\alpha : A \to A$ *be a linear mapping such that* $\phi_\alpha(a) = \alpha(g)a$, *for* $a\in A_g$, $g\in G$. *Then*

(1) *(cf. [14], [17]) if* α *is a homomorphism of* G *into* K^* *then* ϕ_α *is an automorphism of the algebra* A;

(2) *if* α *is a homomorphism of* G *into* K^+ *then* ϕ_α *is a derivation of the algebra* A.

Let us recall that a group G is called K-complete [17] if for any $g,h \in G$, $g\neq h$ there exists a homomorphism $\beta : G \to K^+$ such that $\beta(g) \neq \beta(h)$.

LEMMA 6.2. *Let* A *be a* G*-graded algebra and let* G *be a* K*-complete group. Then any subspace* $I \subseteq A$ *invariant under the automorphisms of* A *is homogeneous.*

This lemma is in fact proved in [14].

Now we are going to consider some special algebraic derivations.

A derivation d of an algebra A is called locally finite [1], if for any $a \in A$, the subspace generated by $a, da, d^2a, \ldots$ is finite-dimensional.

We shall consider A as a $K[t]$-module in which $ta = da$. Under this interpretation d is a locally finite derivation iff A is a torsion $K[t]$-module. In this case it is well known [6] that $A = \bigoplus_{f \in F} A_f$, where F is the set of all monic irreducible polynomials in $K[t]$ and $A_f = \{a \in A : f^r a = 0 \text{ for some } r \geq 1\}$. It is clear that some components A_f may be zero, thus e.g. [1]: d is a nil-derivation iff $A_f = 0$ for $f \neq t$.

Definition. A derivation d of an algebra A is said to be a split derivation if it is locally finite and $A_f = 0$ for any polynomial f of degree greater than 1.

Thus any nil-derivation is a split derivation. Moreover, if K is algebraically closed then any locally finite derivation is split. Lemma 6.1 allows us to construct other examples of split derivations.

Henceforth we shall assume that d is a split derivation. Under this assumption $A = \bigoplus_{f \in G} A_f$, where G is the set of all monic polynomials of degree 1 . In the set G we shall define an operation $\circ$ by the formula: $(t-k)\circ(t-l) = t-k-l$. Then G is a group under $\circ$. Let G_d be the subgroup of G generated by the elements g such that $A_g \neq 0$.

From the definition of the operation $\circ$ in G and hence in G_d it follows that the mapping $\alpha : G_d \to K^+$ given by $\alpha(t-k) = k$ is an embedding.

LEMMA 6.3 (cf. [12]). *The algebra* A *is graded by* G_d .

Proof. Obviously $A = \bigoplus_{g\in G_d} A_g$. Now let $g = t-k$, $h = t-l \in G_d$, $x \in A_g$, $y \in A_h$. Then $(g\circ h)(xy) = (t-k-l)(xy) = d(xy)-kxy-lxy = (dx)y + xdy - kxy - lxy = [(t-k)x]y + x[(t-l)y] = (gx)y + x(hy)$. It is easy to prove by induction that for any $n\geq 1$ we have $(g\circ h)^n(xy) = \sum_{i=0}^{n} \binom{n}{i}(g^i x)(h^{n-i}y)$. From the assumption it follows that there exists an m such that $g^m x = 0$, $h^m y = 0$. Putting $n=2m$ we get that $(g\circ h)^n(xy) = 0$, i.e., $xy \in A_{g\circ h}$, which ends the proof.

Henceforth we shall consider A as a G_d-graded algebra.

LEMMA 6.4. *Let* $I\subseteq A$ *be a subspace invariant under the automorphisms of the algebra* A . *Then* I *is* d*-invariant iff it is homogeneous.*

Proof. It is clear that $\bigoplus_{g\in G_d}(I\cap A_g)\subseteq I$. By Lemma 6.1 we get that the natural embedding $\alpha : G_d \to K^+$ induces the derivation ϕ_α . Let $\delta = d-\phi_\alpha$. It is clear that δ is also a derivation of A . This δ is a nil derivation. In fact, if $g=t-k\in G_d$ and $x\in A_g$ then $\delta(x) = dx-\phi_\alpha x = dx-kx = (t-k)x$. Hence $\delta^n x = (t-k)^n x$ for any $n\geq 1$ and thus $\delta^n=0$ for sufficiently large n . Since I is a subspace invariant under automorphisms, it is [1] a δ-invariant subspace. It is clear that if I is a homogeneous subspace then it is also a ϕ_α-invariant subspace. Then I is also d-invariant, as $d=$ $=\delta+\phi_\alpha$. On the other hand, if I is a d-invariant subspace then it is also a ϕ_α-invariant subspace, as $\phi_\alpha = d-\delta$. Now let $x\in I$, $x = x_1+\ldots+x_r$, where $x_i \in A_{g_i}$ for $g_i = t-k_i$, $i=1,2,\ldots,r$. Then $\phi_\alpha{}^s(x) =$ $\sum_{i=1}^{r}(k_i)^s x_i \in I$ for $s = 1,2,\ldots$. Using the standard argument of linear algebra with the Vandermonde determinant we get that $x_i\in I$, $i = 1,2,\ldots,r$, i.e., I is homogeneous.

THEOREM 6.5. *Let* d *be a split derivation of an algebra* A *and let* $I\subseteq A$ *be a subspace invariant under the automorphisms of* A . *If the group* G_d *is* K-*complete then* I *is a* d-*invariant subspace.*

Proof. By Lemma 6.2 it follows that I is a homogeneous subspace and using Lemma 6.4 we get that I is d-invariant.

COROLLARY 6.6 [12]. *If a field K is algebraically closed then any subspace of A which is invariant under automorphisms, is invariant under locally finite derivations.*

Proof. Since char $K = 0$, any abelian group is K-complete in our case [17] and every locally finite derivation is a split derivation.

COROLLARY 6.7. *If A is a finitely generated algebra over K and $I \subseteq A$ is a subspace invariant under automorphisms then I is invariant under any split derivation of A.*

Proof. Let d be a split derivation of A and let G_d be the corresponding group. We may assume that the generators $a_1,\dots,a_r$ of A are homogeneous. Let $a_i \in A_{g_i}$ $(i = 1,2,\dots,r)$, where $g_i \in G_d$. Since A is graded by G_d, it is easy to see that G_d is generated by $g_1,\dots,g_r$. Since G_d is a torsionfree group, there exists an embedding of G_d into $Q^* \subseteq K^*$, where Q denotes the field of rational numbers. Hence G_d is K-complete and by Theorem 6.5 it follows that I is a d-invariant subspace.

The author does not know whether this corollary holds for all locally finite derivations.

In the case of some radicals one can replace a field K with its algebraic closure $\bar{K}$ (see [19]) since the latter is a Galois extension of K. This change of the base field allows us to apply Corollary 6.6. Using this method one can prove [12] that the Jacobson radical of associative algebras is invariant under all locally finite derivations though it is not characteristic (cf. Example 2.5).

At last we shall give an example showing that the assumptions made in this paragraph and in Theorem 5.9 are essential.

Example 6.8. Let T be a set of indeterminates and $K = Q(T)$ be the field of rational functions in the indeterminates ranging over T. Let R be a central simple K-algebra and let G be a torsionfree divisible abelian group of a rank 1 [9]. Consider the algebra $A = R[G] \simeq R \otimes_K K[G]$. Let J be the ideal of $K[G]$ generated by $\{1-g : g \in G\}$ and let $I = R \otimes J$. If S is the upper radical determined by R then $S(A) = I$. Indeed, since $K[G]/J \simeq K$, we have $A/I \simeq R$ and thus $S(A) \subseteq I$. Suppose now that $\phi : I \to R$ is a K-homomorphism of I onto R.

Since R is a central simple K-algebra, similarly to the proof of Lemma 4.3 we have an induced homomorphism $\psi : K[G] \to K$, hence also from G into K. Were $\psi(G) \neq 1$, then $\psi(G)$ would be a divisible subgroup of K^*, but by the definition of K it follows that K contains no nontrivial divisible subgroups. Hence $\psi(G) = 1$ and thus $J \subseteq \ker \psi$, i.e., $\phi(I) = \psi(R \otimes J) = 0$, which is impossible. Hence $I \subseteq S(A)$, i.e., $I = S(A)$.

Since K^+ is a torsionfree divisible group, there exists an embedding α of G into K^+. Let $d = \phi_\alpha$. By Lemma 6.1 it follows that d is a derivation of A. It is clear that A has a base composed of eigenvectors. Moreover, the corresponding group G_d is isomorphic with G and in fact $A = \bigoplus_{g \in G} R \otimes g$ is a decomposition inducing the gradation by G_d. Now it is clear that I is not a d-invariant ideal of A though it is invariant under the automorphisms of A.

Taking in the above construction a sufficiently large set of indeterminates T and a countable dimensional K-algebra R we obtain an example of a countable dimensional algebra A whose radical I is not invariant under split derivations, hence also under algebraic derivations, while the field K may be of arbitrarily large cardinality.

Many other results on characteristic ideals of Lie algebras can be found in [1].

REFERENCES

[1] R. K. Amayo and I. Stewart, *Infinite-dimensional Lie algebras*, Noordhoff International Publishing, Leyden, 1974.

[2] T. Anderson, Hereditary radicals and derivations of algebras, *Canad. J. Math.* 21 (1969), 372-377.

[3] T. Anderson, N. Divinsky, A. Suliński, Hereditary radicals in associative and alternative rings, *Canad. J. Math.* 17 (1965), 594-603.

[4] V. A. Andrunakievič and Ju. M. Rjabuhin, *Radicals of algebras and structure theory* (Russian), Nauka, Moscow, 1979.

[5] E. Behr, *On semisimple heredity of radicals in classes of nonassociative rings*, M. A. thesis, Warsaw, 1981.

[6] N. Bourbaki, *Algèbre, chap. VII*, Hermann et Cie, Paris, 1958.

[7] N. Divinsky, *Rings and radicals*, Allen and Unwin, London, 1965.

[8] *The Dniester notebook: Unsolved problems in the theory of rings and modules* (Russian), Akad. Nauk. Moldav. SSR, Kishinev, 1969.

[9] L. Fuchs, *Infinite Abelian groups, Vol. I*, Academic Press, New York and London, 1970.

[10] N. Jacobson, *Lie Algebras*, Interscience, New York, 1962.

[11] N. Jacobson, *PI-algebras, an Introduction, Lecture Notes in Mathematics*, 441, Springer, Berlin - New York, 1975.

[12] J. Krempa, On invariant subspaces of locally finite derivations, *Bull. London Math. Soc.* 12 (1980), 374-376.

[13] A. J. Kuczyński, Radicals and derivations of Lie algebras, *Bull. Acad. Polon. Sci. Sér. Sci. Math. Astronom. Phys.* 27 (1979), 649-655.

[14] A. J. Kuczyński and E. R. Puczylowski, On semisimplicity of group rings, *Bull. Acad. Polon. Sci. Sér. Sci. Math. Astronom. Phys.* 22 (1974), 1103-1106.

[15] A. A. Nikitin, Heredity of radicals of rings (Russian), *Algebra i Logika* 17 (1978), 303-315.

[16] V. A. Parfenov, On a weakly solvable radical of Lie algebras, *Sibirsk. Mat. Ž.* 12 (1971), 171-176.

[17] D. S. Passman, *The algebraic structure of group rings*, Wiley Interscience Publication, New York, 1977.

[18] B. I. Plotkin, Radical and semisimple groups and Lie algebras, *Proceedings of the third all-Union mathematical conference*, Moscow, June-July 1956.

[19] E. R. Puczylowski, On radicals of tensor products, *Bull. Acad. Polon. Sci. Sér. Sci. Math.* 28 (1980), 243-247.

[20] A. M. Slin'ko, A remark on radicals and derivations of rings (Russian), *Sibirsk. Mat. Ž.* 13 (1972), 1395-1397, 1422.

[21] A. M. Slin'ko, The radicals of Jordan rings, *Algebra i Logika* 11 (1972), 206-215, 239.

[22] R. L. Snider, Lattices of radicals, *Pacific J. Math.* 40 (1972), 207-220.

[23] F. Szász, *Radicals of rings*, Akadémiai Kiadó, Budapest, 1981.

[24] R. Wiegandt, *Radical and semisimple classes of rings. Queen's papers in pure and applied mathematics* 37, Queen's University, Kingston, Ontario, 1974.

[25] K. A. Ževlakov, A. M. Slin'ko, I. P. Šestakov, A. I. Širšov, *Rings that are nearly associative*, Nauka, Moscow, 1978.

Jan Krempa
Institute of Mathematics
University of Warsaw
00-901 Warsaw
POLAND

COLLOQUIA MATHEMATICA SOCIETATIS JÁNOS BOLYAI
38. RADICAL THEORY, EGER (HUNGARY), 1982.

STRUCTURE THEORY OF MAL'CEV ALGEBRAS

E. N. KUZ'MIN

Mal'cev algebras arose for two reasons: the first is related to the generalization of classical Lie group theory to the case of analytic Moufang loops, and the second - to alternative algebras. In 1955 A. I. Mal'cev [17] observed that the tangent algebra can be easily defined for every (locally) analytic loop. He showed then that the tangent algebra of a (locally) analytic Moufang loop, which is characterized by the identity

$$(x \cdot yz)x = xy \cdot zx \quad ,$$

satisfies the identities

$$(1) \qquad x^2 = 0 \ , \quad J(x,y,xz) = J(x,y,z)x$$

where, by definition, $J(x,y,z) = xy \cdot z + yz \cdot x + zx \cdot y$. It is precisely the algebras with identities (1) that were later called Mal'cev algebras (for example in an article [18] by A. A. Sagle, 1961). Since Lie algebras are characterized by the identities $x^2 = 0$, $J(x,y,z) = 0$ (in the notations above) the class of Lie algebras is contained (properly) in the class of Mal'cev algebras.

There is a second well-known way of obtaining Lie algebras: start with an associative algebra, and replace the product by commutation. If we start with an alternative algebra instead of an associative algebra, then its commutator algebra again satisfies the identities (1), thus it proves to be a Mal'cev algebra in the modern terminology. This was shown in the same article by A. I. Mal'cev [17].

While it is well known that from a given Lie algebra one can always recover the corresponding (local) Lie group, the corresponding question for Mal'cev algebras remained an open problem for a long time. Only in 1970 was an affirmative answer to this problem obtained by the present author [14, 15]: every finite-dimensional real Mal'cev algebra is the tangent algebra of some local analytic Moufang loop. Meanwhile the structure theory of finite-dimensional Mal'cev algebras over a field of characteristic 0 was being developed. The methods of representation theory allowed us not only to extend almost all the basic results of the classical theory of Lie algebras to the more general case of Mal'cev algebras, but also to classify all non-Lie simple Mal'cev algebras [13, 19]. Moreover, non-Lie simple Mal'cev algebras could be characterized not only in characteristic 0 but in prime characteristic $p \neq 2,3$, whereas the description of simple finite-dimensional Lie algebras of prime characteristic is still a very difficult and very interesting open problem.

The situation here is the same as in the case of alternative algebras. Every simple alternative algebra is either associative or a Cayley-Dickson algebra of dimension 8 over its center. Similarly every simple Mal'cev algebra is either a Lie algebra or a 7-dimensional algebra over its centroid and in the last case our simple Mal'cev algebra can be obtained in a definite way from a simple Cayley-Dickson algebra over the same field. One of the main tools in proving this result was an analogue of Engel's theorem, which is true not only for Mal'cev algebras, but for binary Lie algebras of arbitrary characteristic and for many other varieties of anticommutative algebras too [10].

After our work on the existence of a local analytic loop with a given tangent Mal'cev algebra, the problem remained open whether an arbitrary local analytic Moufang loop can be embedded in a global (simple-connected) Moufang loop. This problem was successfully solved by F. S. Kerdman [9]. His result turned out to be completely analogous to the corresponding ones in the theory of Lie groups.

A new step in the development of the theory of Mal'cev algebras is connected with the name of V. T. Filippov. Before him infinite-dimensional Mal'cev algebras were studied only occasionally. For example, we proved the existence of the locally nilpotent and locally finite radicals in arbitrary Mal'cev algebras [11, 12] (in the case of Lie algebras this result is due to B. I. Plotkin) The main subject of Filippov's is precisely the infinite-dimensional Mal'cev algebras, which are studied mainly via the study of free Mal'cev algebras.

A beautiful result by Filippov is the theorem that Mal'cev algebras with the n th Engel condition are locally nilpotent [2, 3], this is an exact analogue of the famous theorem of A. I. Kostrikin for Lie algebras. Further Filippov obtained the description of simple and even prime Mal'cev algebras without restriction on dimension [4, 5] (finiteness of dimension is not assumed). From this it follows that every semiprime Mal'cev algebra is special in the sense that such an algebra can be embedded into the commutator algebra of some alternative algebra. Note, however, that in the general case the problem of speciality (analogue of the Birkhoff-Witt theorem) remains open. We also do not know whether the analogue of the Ado-Iwasawa theorem holds. This would say that every finite-dimensional Mal'cev algebra of characteristic not 2 or 3 can be embedded as a subalgebra in the commutator algebra of some finite-dimensional alternative algebra. In the case of Lie algebras there is another formulation of Ado's theorem: the existence of a faithful finite-dimensional representation. Filippov showed that in this formulation the analogue of Ado's theorem is false: he constructed a counter-example [6]. There is a very interesting result of Filippov, which states that the chain of varieties M_n generated by the free Mal'cev algebra with n free generators is strongly ascending at each step, except possibly for coincidence of M_3 and M_4 [6].

But even in the theory of finite-dimensional Mal'cev algebras not all is done yet, and work continues. Recently, for example, three authors (R. Carlsson, A. N. Griškov, E. N. Kuz'min) independently and simultaneously proved that every finite-dimensional Mal'cev algebra over a field of characteristic 0 splits into the semi-direct sum of the solvable radical and a semisimple subalgebra [1, 8, 16]. This is the analogue of Levi's theorem for Lie algebras. Conjugacy of semisimple subalgebras - the analogue of the Mal'cev-Harish-Chandra theorem - was proved earlier by R. Carlsson. We would like to conclude with one example of a radical (in the Amitsur-Kurosh sense) in the class of all Mal'cev algebras, which was constructed by V. T. Filippov [7]. The semisimple algebras corresponding to this radical are exactly the algebras with no nil-elements of index two.

REFERENCES

[1] Carlsson, R., On the exceptional central simple non-Lie Malcev algebras, *Trans. Amer. Math. Soc.* 244 (1978), 173-184.

[2] Filippov, V. T., On Mal'cev algebras satisfying the Engel condition (Russian), *Algebra i Logika* 14 (1975) 441-455.

[3] Filippov, V. T., On Mal'cev algebras with Engel condition (Russian), *Algebra i Logika* 15 (1976), 89-109.

[4] Filippov, V. T., Central simple Mal'cev algebras (Russian), *Algebra i Logika* 15 (1976), 235-242.

[5] Filippov, V. T., On the theory of Mal'cev algebras (Russian), *Algebra i Logika* 16 (1977), 101-108.

[6] Filippov, V. T., On chains of varieties generated by free Mal'cev and alternative algebras (Russian), *Doklady Akad. Nauk SSSR* 260 (1981), 1082-1085.

[7] Filippov, V. T., On nil-elements of index two in Mal'cev algebras (Russian), *Sibirsk. Mat. Ž.* 22 (1981), 232.

[8] Griškov, A. N., An analogue of Levi's theorem for Mal'cev algebras (Russian), *Algebra i Logika* 16 (1977), 389-396.

[9] Kerdman, F. S., On analytic Moufang loops in the large (Russian), *Doklady Akad. Nauk SSSR* 249 (1979), 533-536.

[10] Kuz'min, E. N., On anticommutative algebras satisfying the Engel condition (Russian), *Sibirsk. Mat. Ž.* 8 (1967), 1026-1034.

[11] Kuz'min, E. N., The locally nilpotent radical of Mal'cev algebras satisfying the nth Engel condition (Russian), *Doklady Akad. Nauk SSSR* 177 (1967), 508-510.

[12] Kuz'min, E. N., Algebraic sets in Mal'cev algebras (Russian), *Algebra i Logika* 7 (1968), no. 2, 42-47.

[13] Kuz'min, E. N., Mal'cev algebras and their representations (Russian), *Algebra i Logika* 7 (1968), no. 4, 48-69.

[14] Kuzmine, E. N., La relation entre les boucles de Moufang analytiques, *C. R. Acad. Sci. Paris* 271 (1970), 1152-1155.

[15] Kuz'min, E. N., On a connection between Mal'cev algebras and analytic Moufang loops (Russian), *Algebra i Logika* 10 (1971), 3-22.

[16] Kuz'min, E. N., Levi's theorem for Mal'cev algebras (Russian), *Algebra i Logika* 16 (1977), 424-431.

[17] Mal'cev, A. I., Analytic loops (Russian), *Mat. Sb.* 36 (1955), 569-576.

[18] Sagle, A. A., Malcev algebras, *Trans. Amer. Math. Soc.* 101 (1961), 426-458.

[19] Sagle, A. A., Simple Malcev algebras over fields of characteristic zero, *Pacific J. Math.* 12 (1962), 1057-1078.

E. N. Kuz'min
Institute of Mathematics, Siberian Branch of the Academy
630090 Novosibirsk
Universitetskiĭ pr. 4, USSR

COLLOQUIA MATHEMATICA SOCIETATIS JÁNOS BOLYAI
38. RADICAL THEORY, EGER (HUNGARY), 1982.

HEREDITARINESS OF UPPER RADICALS

L. C. A. van LEEUWEN

Introduction

In this paper only associative and alternative rings are considered. If a result is specified to the class of associative rings, this will be stated explicitly.

Let $\mathbb{M}$ be a class of rings such that the upper radical $U\,\mathbb{M}$ exists. The class $U\,\mathbb{M}$ is a *hereditary class* if $U\,\mathbb{M}$ satisfies the condition B ideal of A, $A \in U\,\mathbb{M} \Rightarrow B \in U\,\mathbb{M}$.

A class $\mathbb{M}$ of rings is said to be a *regular class* if every non-zero ideal of a ring in $\mathbb{M}$ has a non-zero homomorphic image in $\mathbb{M}$.

A ring A will be called an *essential extension* of the ring B, if B is an essential ideal of A. For an arbitrary class $\mathbb{M}$ of rings the class $\mathbb{M}_k$, consisting of all essential extensions of rings belonging to M, will be called the *essential cover* of the class $\mathbb{M}$.

The *upper radical operator* U and the *semisimple operator*

S acting on a class $\mathbb{M}$ are defined as

$$\mathcal{U}\,\mathbb{M} = \{A \mid A \text{ has no non-zero homomorphic image in } \mathbb{M}\}$$

and

$$\mathcal{S}\,\mathbb{M} = \{A \mid A \text{ has no non-zero ideal in } \mathbb{M}\}.$$

A radical $\mathbb{R}$ has the *intersection property* relative to the class $\mathbb{M}$, if for every ring A its radical $\mathbb{R}(A)$ equals to

$$\mathbb{R}(A) = \cap \{I \text{ ideal in } A \mid A/I \in \mathbb{M}\}.$$

We will use the following well-known result: If B is a non-zero ideal in R and B is not essential in R, then there exists an ideal C of R such that R/C contains an essential ideal isomorphic to B (See Prop. 1, [2]).

1. Let $\mathbb{M}$ be a hereditary class of associative (semi)-prime rings. A number of equivalent conditions in order that $\mathcal{U}\,\mathbb{M}$ be hereditary has been determined in [3] (Theorem 7) in terms of the essential cover $\mathbb{M}_k$ of $\mathbb{M}$. A generalization of this result was given in [2], where it is assumed that $\mathbb{M}$ is a regular class of semiprime rings and the essential cover $\mathbb{M}_k$ is replaced by $\mathbf{M}^c$, the essential closure of $\mathbb{M}$, i.e. the smallest essentially closed class containing $\mathbb{M}$. Then again a generalization of this last result was obtained by le Roux, Heyman and Jenkins in [5]. This is theorem 7 in [5], which reads:

Let $\mathbb{M}$ be a regular class of associative rings containing no or all zero rings. The following statements are equivalent:

(i) $U\,\mathbb{M}$ *is hereditary*

(ii) $\mathbb{M}_k \subseteq S\,U\,\mathbb{M}$

(iii) $U\,\mathbb{M} = U\,\mathbb{M}_k$

(iv) $U\,\mathbb{M}$ *has the intersection property relative to* $\mathbb{M}_k$

(v) $U\,\mathbb{M} \cap \mathbb{M}_k = 0$.

A class $\mathbb{M}$ of semiprime rings satisfies the condition

(F) C ideal in B, B ideal in A and $B/C \in \mathbb{M}$ imply C ideal in A. In this paper we show that the above 5 conditions are equivalent if we assume that $\mathbb{M}$ is a regular class of rings satisfying (F).

It might be conjectured that any class $\mathbb{M}$ satisfying (F) consists of semiprime rings, and during the conference I believed that this was true. In a letter R. Wiegandt kindly informed me that he and K. Kaarli have found that the above conjecture is not true. The following 3 propositions I owe to them and I would like to thank them also for the cordiality of correcting my original (false) proof.

PROPOSITION 1

If C is an ideal in B and B is an ideal in A and if B/C has a unit element, then C is an ideal of B.

Proof. Let $e + C$, $(e \in B)$, be the unit element of B/C. For any element $a \in A$ we have $ca \in B$, if $c \in C$, and so $(ca + C)(e + C) = (ca)e + C = ca + C$ or $ca = (ca)e + t$ with a suitable $t \in C$. Since $(ca)e = c(ae) \in C(AB) \subseteq CB \subseteq C$, it follows $ca = (ca)e + t \in C$. A similar reasoning shows $ac \in C$.

An obvious consequence is

PROPOSITION 2

Let $\mathbb{M}$ be any class of rings with 1. If C is an ideal in B, B an ideal in A and $B/C \in \mathbb{M}$, then C is an ideal of A.

Remark. Let $\mathbb{M}$ be a class of rings with 1 but not every ring in $\mathbb{M}$ is semiprime, e.g. the Dorroh extension of a zero-ring is in $\mathbb{M}$. By proposition 2, $\mathbb{M}$ satisfies (F) but $\mathbb{M}$ does not consist of semiprime rings. On the other hand, any class $\mathbb{M}$ of semiprime rings satisfies (F). As not every semiprime ring has a unit element, e.g. a simple prime ring without 1, the condition that any ring in $\mathbb{M}$ should have a unit element is not necessary in order that $\mathbb{M}$ should satisfy (F).

One can also prove

PROPOSITION 3

Let $\mathbb{M}$ be the class of all rings with 1 and $\mathbb{Q}$ that of all semiprime rings. If C is an ideal in B, B an ideal of A and $B/C = I/C \oplus K/C$ such that $I/C \in \mathbb{M}$ and $K/C \in \mathbb{Q}$, then

C is an ideal of A.

Proof. We have that K is an ideal in B, B an ideal in A. Since $B/K \cong \frac{B/C}{K/C} \cong I/C \in \mathbb{M}$, proposition 2 implies: K is an ideal in A. Then C an ideal in K, K an ideal in A and K/C is semiprime yields that C is an ideal in A.

PROPOSITION 4

Let $\mathbb{M}$ *be a regular class of rings satisfying* (F). *Then* $\mathbb{M}_k$ *is also a regular class.*

Proof. Since the proof is completely analogous to the corresponding proof for $\mathbb{M}$ a regular class of semiprime rings ([2], theorem 1), we omit it.

Remark. I do not know whether $\mathbb{M}_k$ satisfies (F). It is known, however, that the essential cover $\mathbb{M}_k$ of a class $\mathbb{M}$ of semiprime rings consists of semiprime rings.

2. THEOREM 1

Let $\mathbb{M}$ *be a regular class of rings. Then the following statements are equivalent:*

(i) $U\,\mathbb{M}$ *is a hereditary radical;*

(ii) $\mathbb{M}_k \subseteq S\,U\,\mathbb{M}$ *and* $U\,\mathbb{M}_k$ *is hereditary;*

(iii) $U\,\mathbb{M} = U\,\mathbb{M}_k$ *and* $U\,\mathbb{M}_k$ *is hereditary;*

(iv) $U\,\mathbb{M} \cap \mathbb{M}_k = 0$ *and* $U\,\mathbb{M}_k$ *is hereditary.*

Proof (i) ⇒ (ii). Let $R \in \mathbb{M}_k$ and suppose $R \notin S\,U\,\mathbb{M}$. Then there exists a nonzero ideal I in R such that $I \in U\,\mathbb{M}$. On the other hand there exists an essential ideal J in R with $J \in \mathbb{M}$. Hence $I \cap J \neq 0$. Since $(I \cap J)$ is an ideal in I and $I \in U\,\mathbb{M}$, $U\,\mathbb{M}$ is hereditary implies $(I \cap J) \in U\,\mathbb{M}$. Also $(I \cap J)$ is an ideal of J and $J \in \mathbb{M}$, $\mathbb{M}$ is regular, imply that $(I \cap J)$ has a non-zero homomorphic image in $\mathbb{M}$. But this is impossible, since $(I \cap J) \in U\,\mathbb{M}$. Hence $\mathbb{M}_k \subseteq S\,U\,\mathbb{M}$. Using the upper radical operator we get $U(S\,U\,\mathbb{M}) \subseteq U\,\mathbb{M}_k$ or $U\,\mathbb{M} \subseteq U\,\mathbb{M}_k$. Also $U\,\mathbb{M}_k \subseteq U\,\mathbb{M}$, so $U\,\mathbb{M} = U\,\mathbb{M}_k$ and $U\,\mathbb{M}$ is hereditary implies $U\,\mathbb{M}_k$ is hereditary.

(ii) $\Rightarrow$ (iii). Clear from (i) $\Rightarrow$ (ii) (last part of the proof).

(iii) $\Rightarrow$ (iv). Suppose $R \in U\,\mathbb{M} \cap \mathbb{M}_k$. Then $R \in U\,\mathbb{M}_k \cap \mathbb{M}_k$, as $U\,\mathbb{M} = U\,\mathbb{M}_k$. Hence $R = 0$ and $U\,\mathbb{M} \cap \mathbb{M}_k = 0$.

(iv) $\Rightarrow$ (i). If $U\,\mathbb{M}$ is not hereditary, then there exists a ring $R \in U\,\mathbb{M}$ having an ideal $A \notin U\,\mathbb{M}$. Suppose that there exists a non-zero homomorphic image of R, say $R/I \in \mathbb{M}_k$. Then by (iv) $R/I \notin U\,\mathbb{M}$. But $R \in U\,\mathbb{M}$ implies $R/I \in U\,\mathbb{M}$, hence R has no non-zero homomorphic image in $\mathbb{M}_k$, i.e. $R \in U\,\mathbb{M}_k$. As $U\,\mathbb{M}_k$ is hereditary, it follows that $A \in U\,\mathbb{M}_k$. But $U\,\mathbb{M}_k \subseteq U\,\mathbb{M}$, so $A \in U\,\mathbb{M}$, which is again a contradiction. Therefore $U\,\mathbb{M}$ is hereditary.

THEOREM 2

Let $\mathbb{M}$ be a class of rings such that $U\,\mathbb{M}$ exists. Then $U\,\mathbb{M}$ is hereditary if and only if $U\,\mathbb{M} = U\,\mathbb{M}_k$ and $U\,\mathbb{M}_k$ is hereditary.

Proof. Suppose that $U\mathbb{M}$ is hereditary. Let $R \in U\mathbb{M}$ and suppose that $R \notin U\mathbb{M}_k$. Then there exists a non-zero image $R/I \in \mathbb{M}_k$. So there is an essential ideal A/I in R/I with $A(I \in \mathbb{M}$. But $R \in U\mathbb{M}$ implies that $R/I \in U\mathbb{M}$ and $U\mathbb{M}$ is hereditary yields $0 \neq A/I \in U\mathbb{M} \cap \mathbb{M}$, which is a contradiction. So $U\mathbb{M} \subseteq U\mathbb{M}_k$ and as the reverse inclusion is always true we get $U\mathbb{M} = U\mathbb{M}_k$. Hence $U\mathbb{M}_k$ is hereditary. The converse is trivial.

Note that the regularity of $\mathbb{M}$ is not needed in the proof of theorem 2.

THEOREM 3.

Let $\mathbb{M}$ *be a class of rings such that* $U\mathbb{M}$ *exists. Then* $U\mathbb{M}$ *is hereditary if and only if* $U\mathbb{M} \subseteq S(S\,U\mathbb{M})$.

Proof. Let $U\mathbb{M}$ be hereditary and suppose $A \in U\mathbb{M}$. Let I be any ideal in A, then $I \in U\mathbb{M}$. So, if $I \neq 0$, then $I \notin S\,U\mathbb{M}$, i.e. A has no non-zero ideals in $S\,U\mathbb{M}$. Hence $A \in S(S\,U\mathbb{M})$ which implies $U\mathbb{M} \subseteq S(S\,U\mathbb{M})$.

Conversely, let $U\mathbb{M} \subseteq S(S\,U\mathbb{M})$ and let $A \in U\mathbb{M}$ with a non-zero ideal I. Now $U\mathbb{M}(I)$ is an ideal in A(A-D-S-property). Hence $I/U\mathbb{M}(I)$ is an ideal in $A/U\mathbb{M}(I) \in U\mathbb{M}$. But $U\mathbb{M} \subseteq S(S\,U\mathbb{M})$, so $A/U\mathbb{M}(I) \in S(S\,U\mathbb{M})$, hence $A/U\mathbb{M}(I)$ has no non-zero ideal in $S\,U\mathbb{M}$. On the other hand, $I/U\mathbb{M}(I) \in S\,U\mathbb{M}$, so $I/U\mathbb{M}(I) = 0$ or $I = U\mathbb{M}(I)$. Hence $I \in U\mathbb{M}$ and $U\mathbb{M}$ is hereditary.

We extend the equivalent conditions of theorem 7 [5] in the following

THEOREM 4

$\mathbb{M}$ is a regular class of rings satisfying (F). *The following statements are equivalent:*

(i) $U\,\mathbb{M}$ *is hereditary*

(ii) $U\,\mathbb{M} \subseteq S\,\mathbb{M}$

(iii) $U\,\mathbb{M} \subseteq S(S\,U\,\mathbb{M})$.

Proof. (i) ⇒ (ii). Let A be a ring in $U\,\mathbb{M}$ and suppose B is a non-zero ideal in A. Then $B \in U\,\mathbb{M}$, so $B \neq 0$ implies $B \notin \mathbb{M}$. Hence $A \in S\,\mathbb{M}$ and $U\,\mathbb{M} \subseteq S\,\mathbb{M}$.

(ii) ⇒ (iii). Let $A \in U\,\mathbb{M}$ and suppose A has a non-zero ideal $I \in S\,U\,\mathbb{M}$. Then $I \notin U\,\mathbb{M}$ so I has a non-zero image $I/J \in \mathbb{M}$. As $\mathbb{M}$ satisfies (F) we get that J is an ideal in A Now $A/J \in U\,\mathbb{M}$ has a non-zero ideal $I/J \in \mathbb{M}$. But $U\,\mathbb{M} \subseteq S\,\mathbb{M}$, so $A/J \in S\,\mathbb{M}$, which leads to a contradiction. Hence A has no non-zero ideal in $S\,U\,\mathbb{M}$ or $A \in S(S\,U\,\mathbb{M})$. It follows that $U\,\mathbb{M} \subseteq S(S\,\ddot{U}\,\mathbb{M})$.

(iii) ⇒ (i) (see theorem 3).

LEMMA 1

$\mathbb{M}$ is a regular class of rings satisfying (F).

Then $A \in U\,\mathbb{M}_k$ *if and only if any ideal of* A *is in* $U\,\mathbb{M}$.

Proof. Suppose that $A \in U\,\mathbb{M}_k$ and let B be any ideal in A. If $B \notin U\,\mathbb{M}$ then some non-zero homomorphic image $0 \neq B/I$

$\in \mathbb{M}$. Since $\mathbb{M}$ satisfies (F), I is an ideal in A. Hence, if B/I is essential in A/I, we get $A/I \in \mathbb{M}_k$ which is a contradiction, as $A \in U\,\mathbb{M}_k$. If B/I is not essential in A/I, then there exists an ideal C/I of A/I such that $\frac{A/I}{C/I} \cong A/C$ contains an essential ideal isomorphic to B/I ([2], Prop. 1). Hence $A/C \in \mathbb{M}_k$, which is again a contradiction. We conclude that $B \in U\,\mathbb{M}$.

Conversely, suppose that any ideal of A is in $U\,\mathbb{M}$. If $A \notin U\,\mathbb{M}_k$ then some non-zero image $A/I \in \mathbb{M}_k$. So there exists a non-zero essential ideal B/I in A/I with $B/I \in \mathbb{M}$. Hence $B \notin U\,\mathbb{M}$ which is a contradiction.

COROLLARY 2

$\mathbb{M}$ is a regular class of rings satisfying (F). *Then $U\,\mathbb{M}$ is hereditary if and only if $U\,\mathbb{M} = U\,\mathbb{M}_k$ (cf.* [5], *theorem 5).*

Proof. Since the proof is an easy consequence of lemma 1 we omit it.

THEOREM 5

$\mathbb{M}$ is a regular class of semiprime rings. Then $U\,\mathbb{M}_k$ is the maximal hereditary radical contained in $U\,\mathbb{M}$.

Proof. It is known that the maximal hereditary radical $\mathbb{H}$ in $U\,\mathbb{M}$ consists of all rings each non-zero accessible subring of which has no non-zero homomorphic image in $S\,U\,\mathbb{M}$. Since $U\,\mathbb{M}_k \subseteq U\,\mathbb{M}$ and $U\,\mathbb{M}_k$ is hereditary, it follows that $U\,\mathbb{M}_k \subseteq \mathbb{H}$. On the other hand, if $A \in \mathbb{H}$, then any non-zero ideal of A has no non-zero image in $S\,U\,\mathbb{M}$, so $0 \neq A/I \in \mathbb{M}$ is impossible since $\mathbb{M} \subseteq S\,U\,\mathbb{M}$ ($\mathbb{M}$ is regular). Then any non-zero ideal of A is in $U\,\mathbb{M}$, hence $A \in U\,\mathbb{M}_k$ by lemma 1.

Let $\mathbb{R}$ be an arbitrary radical class and let $\mathbb{H}$ be the maximal hereditary radical contained in $\mathbb{R}$. Let $\mathbb{C} = \{A|$ every non-zero accessible subring of A belongs to $\mathbb{R}\}$. Then $\mathbb{C} = \mathbb{H}$.

Indeed, if $A \in \mathbb{C}$ and A_n is a non-zero accessible subring of A, then $A_n \in \mathbb{R}$ implies that any $0 \neq A_n/I \notin S\,\mathbb{R}$, so $A \in \mathbb{H}$. Conversely, let $A \in \mathbb{H}$ and A_n be a non-zero accessible subring of A. If $A_n \notin \mathbb{R}$, then $0 \neq A_n/\mathbb{R}(A_n) \in S\,\mathbb{R}$, which contradicts $A \in \mathbb{H}$. Hence $A_n \in \mathbb{R}$ and $A \in \mathbb{C}$.

Now let $\mathbb{M}$ be a regular class of semiprime rings. Combining the different characterizations of $U\,\mathbb{M}_k$ we have

$U\,\mathbb{M}_k = \{A|$any non-zero ideal of A is in $U\,\mathbb{M}\}$ or

$U\,M_k = \{A|$any non-zero accessible subring of A is in $U\,\mathbb{M}\}$ or

$U\,\mathbb{M}_k$ = maximal hereditary radical contained in $U\,\mathbb{M}$.

Example. Let $\mathbb{M}$ be a class of associative simple prime rings. Then $\mathbb{M}$ is the class of simple prime $U\,\mathbb{M}$-semisimple rings. In general, $U\,\mathbb{M}$ is not hereditary.

Let $\mathbb{C}$ be the class of subdirectly irreducible rings with $U\,\mathbb{M}$-semisimple hearts, i.e. the class of subdirectly irreducible rings with idempotent hearts H, $H \in \mathbb{M}$.

Then $\mathcal{C} = M_k$ and M_k is a class of prime rings, in fact, M_k is a special class of rings.

The special radical $U\,\mathbb{M}_k$ is the maximal hereditary radical contained in $U\,\mathbb{M}$. Moreover, M_k is the smallest special class containing $\mathbb{M}$ ([3], theorem 6).

3. The next theorem was inspired by Prop. 3, [2].

THEOREM 6

Let $\mathbb{M}$ *be a regular class of rings satisfying* (F).

Then $U\,\mathbb{M}$ *is hereditary if and only if the following condition holds: if* $B \in \mathbb{M}$ *is a non-zero ideal in A, then* $B^2 = (0)$ *or ann* $B = (0)$ *(in A) implies* $A \notin U\,\mathbb{M}$.

Proof. If $U\,\mathbb{M}$ is hereditary and $B \in \mathbb{M}$ is a non-zero ideal in A with $B \in \mathbb{M}$, then $A \in U\,\mathbb{M}$ would imply $B \in U\,\mathbb{M}$, so $0 \neq B \in \mathbb{M} \cap U\,\mathbb{M}$, which is a contradiction.

Conversely assume that the condition is satisfied. Suppose that $U\,\mathbb{M}$ is not hereditary, so there exists a ring $A \in U\,\mathbb{M}$ containing a non-zero ideal $B \notin U\,\mathbb{M}$. Then B has some non-zero image $B/C \in \mathbb{M}$ and by (F) we get that C is an ideal in A. This implies that $D = \{x \in A \mid xB + Bx \subseteq C\}$ is an ideal in A, so $B \cap D$ is an ideal in A. As $C \subseteq D$, it follows that C is an ideal in $B \cap D$, hence $(B \cap D)/C$ is an ideal in $B/C \in \mathbb{M}$. Now $\mathbb{M}$ is regular implies that, in case $B \cap D \neq C$, there exists a non-zero image $(B \cap D)/E$ of $(B \cap D)/C$ with $(B \cap D)/E \in \mathbb{M}$. Again by (F), it follows that E is an ideal in A. Hence $(B \cap D)/E$ is a non-zero ideal in A/E; $(B \cap D)^2 \subseteq C \subseteq E$, so $[(B \cap D)/E]^2 = (0)$ and $A/E \notin U\,\mathbb{M}$, which is a contradiction. Thus $B \cap D = C$. The annihilator of $(B + D)/D$ in A/D is

$$G/D = \{x + D \in A/D \mid (x + D)B + B(x + D) \subseteq D\} =$$
$$= \{x + D \in A/D \mid xB + Bx \subseteq D\}.$$

Now $xB + Bx \subseteq B$ implies $xB + Bx \subseteq B \cap D = C$, so $x \in D$ which implies $G = D$. Also $0 \neq B/C \cong B/(B \cap D) \cong (B + D)/D$, so $(B + D)/D$ is a non-zero ideal in A/D with $(B + D)/D \in \mathbb{M}$. As ann $(B + D)/D = (0)$ it follows that $A/D \notin U\,\mathbb{M}$, which is again a contradiction. Consequently $U\,\mathbb{M}$ is hereditary.

One may ask what the contents are of theorem 6, in case we specialize $\mathbb{M}$ requiring that $\mathbb{M}$ should be a regular class of semiprime rings. The next result is contained in proposition 10 [4], theorem 2 [2] and in theorem 7 [5] for associative rings.

COROLLARY 3

Let $\mathbb{M}$ be a regular class of semiprime rings. Then $U\,\mathbb{M}$ is a hereditary radical if and only if $U\,\mathbb{M} \cap \mathbb{M}_k = (0)$.

Proof. It can easily be seen that if $\mathbb{M}$ is a class of semiprime rings the statement that $0 \neq B \in \mathbb{M}$ is an ideal in A with ann $B = (0)$ is equivalent to the statement: $B \in \mathbb{M}$ is an essential ideal in A. So by theorem 6 we get: $U\,\mathbb{M}$ is hereditary if and only if $(0 \neq)B \in \mathbb{M}$ is an essential ideal in A implies $A \notin U\mathbb{M}$. This amounts to the same as: $U\mathbb{M}$ is hereditary if and only if $A \in \mathbb{M}_k$ implies $A \notin U\mathbb{M}$ i.e. $U\mathbb{M} \cap \mathbb{M}_k = (0)$.

With the aid of theorem 6 we can get a new proof of the characterization of regular classes determining supernilpotent radicals, due to Rjabuhin and Wiegandt ([4], theorem 2).

THEOREM 7

Let $\mathbb{M}$ be a regular class of rings. The radical class $U\mathbb{M}$ is supernilpotent if and only if $\mathbb{M}$ consists of semiprime rings and satisfies:

(E) *if $B \neq 0$ is an essential ideal in A and $B \in \mathbb{M}$ then $A \notin U\mathbb{M}$.*

Proof. If $U\mathbb{M}$ is supernilpotent then $\mathbb{M}$ consists of semiprime rings. Also $U\mathbb{M}$ is hereditary. Let $B \in \mathbb{M}$ be a non-zero ideal in A, then $B^2 = (0)$ is impossible. Now $\text{ann } B \cap B$ is an ideal in B, so if $\text{ann } B \cap B \neq 0$, then $B \in \mathbb{M}$ and $\mathbb{M}$ is regular imply that some non-zero image of $\text{ann } B \cap B$ is in $\mathbb{M}$. But clearly $[\text{ann } B \cap B]^2 = 0$ and any non-zero image of $\text{ann } B \cap B$ is also a zero-ring and cannot be in $\mathbb{M}$. Therefore $\text{ann } B \cap B = 0$ and B is essential in A implies $\text{ann } B = 0$. As $\mathbb{M}$ also satisfies (F), by theorem 6 we get: $A \notin U\mathbb{M}$ i.e. (E).

Conversely assume that $\mathbb{M}$ consists of semiprime rings and satisfies (E). Then $\mathbb{M}$ is a regular class of rings satisfying (F). Clearly all nilpotent rings are in $U\,\mathbb{M}$. Suppose that $0 \neq B \in \mathbb{M}$ is an ideal in A. If B is essential in A, then $A \notin U\,\mathbb{M}$ and by theorem 6 we are done. So assume that B is not essential in A. Suppose that $A \in U\,\mathbb{M}$. Then there exists an ideal C of A such that A/C contains an essential ideal isomorphic to B. But then $A/C \notin U\,\mathbb{M}$ by (E), which contradicts $A \in U\,\mathbb{M}$. Hence $A \notin U\,\mathbb{M}$.

An example of a supernilpotent radical class $U\,\mathbb{M}$ is obtained in the following

COROLLARY 4

Let $\mathbb{M}$ be a hereditary class of rings containing all nilpotent rings. Then $S\,\mathbb{M}$ is a weakly special class and $U\,S\,\mathbb{M}$ is a supernilpotent radical.

Proof.

(a) If $A \in S\,\mathbb{M}$ and I is a nilpotent ideal in A, then $I \in \mathbb{M}$ and consequently $I = 0$. A can thus have no non-zero nilpotent ideals and hence is semiprime.

(b) Let $A \in S\,\mathbb{M}$ and I be a non-zero ideal in A. Suppose $I \notin S\,\mathbb{M}$. Then I has a non-zero ideal $J \in \mathbb{M}$. Let J' be the ideal of A generated by J. Recall that $J' \subseteq I$ and $(J')^3 \subseteq J$. Also $(J')^3 \neq 0$ since A is semiprime. Thus, since $J \in \mathbb{M}$, we must have $(J')^3 \in \mathbb{M}$, as $\mathbb{M}$ is hereditary. But this is a non-zero ideal in $A \in S\,\mathbb{M}$, which is impossible. Thus $I \in S\,\mathbb{M}$ and $S\,\mathbb{M}$

is hereditary.

(c) Let $B \in S\mathbb{M}$ with B an ideal of a ring A such that $\text{ann } B = 0$. Suppose $A \notin S\mathbb{M}$. Then A has a non-zero ideal $I \in \mathbb{M}$. Now $I \cap B$ is an ideal of I and, since $\mathbb{M}$ is hereditary, $I \cap B \in \mathbb{M}$. However, $I \cap B$ is also an ideal of B which has no non-zero ideals in $\mathbb{M}$. Thus $I \cap B = 0 \Rightarrow IB = BI = 0$ so that $I \subseteq \text{ann } B = 0$, which is a contradiction. Then $A \in S\mathbb{M}$.

By (a), (b) and (c) we infer that $S\mathbb{M}$ is a weakly special class. As it was proved by Rjabuhin, the upper radical $US\mathbb{M}$ is supernilpotent.

Literature

[1] T. Anderson and R. Wiegandt, Semi-simple classes of alternative rings, *Proc. Edinb.Math.Soc.25* (1982), 21-26

[2] T. Anderson and R. Wiegandt, On essentially closed classes of rings, *Ann. Univ. Sci. Math. Eötvös Sect. Math. 24* (1981), 107-111.

[3] G.A.P. Heyman and C. Roos, Essentially extensions in radical theory for rings, *J. Austr. Math. Soc. Ser. A. 23* (1977), 340-347.

[4] Yu.M. Rjabuhin and R. Wiegandt, On special radicals, supernilpotent radicals and weakly homomorphically closed classes, *J. Austr. Math. Soc. Ser A. 31*

(1981), 151-162.

[5] H.J.Le Roux, G.A.P. Heyman and T.L. Jenkins, Essentially closed classes of rings and upper radicals, *Acta Math. Sci. Hung*, *38* (1981), 63-68.

[6] W.G. Leavitt, Hereditary upper radicals, (preprint).

[7] J.F. Watters, Essential cover and closure, *Ann. Univ. Sci. Budapest. Eötvös Sect. Math.* *25* (1982), 279-280.

L. C. A. van Leeuwen
Rijksuniversiteit te Groningen
Postbus 800
Groningen
The Netherlands

COLLOQUIA MATHEMATICA SOCIETATIS JÁNOS BOLYAI
38. RADICAL THEORY, EGER (HUNGARY), 1982.

SMALL IDEALS AND THE BROWN-McCOY RADICAL

N. V. LOI and R. WIEGANDT

The purpose of this note is to study the Brown-McCoy radical in terms of small ideals; in this way Brown-McCoy radical rings, semisimple rings and the Brown-McCoy radical of a ring will be characterized. Sharpening results of [7] and [14] sufficient condition will be given for a class to have a special upper radical. Also a closure property of varieties will be established which is characteristic to the idempotent varieties of associative or alternative rings. Small ideals in the context of radical theory have been investigated also in the recent paper [3].

Throughout, a ring will mean an associative or alternative one. The class of all (associative or alternative) rings will be denoted by $\mathbb{A}$ and $\mathbb{M}$ will stand for the class of all simple rings with unity. We shall use the operators $\mathcal{U}$ and $\mathcal{S}$ acting on classes $\mathbb{C}$ and defined by

$\mathcal{U}\mathbb{C}=\{A\in\mathbb{A}:$ A has no nonzero homomorphic image in $\mathbb{C}\}$

and

$\mathcal{S}\mathbb{C}=\{A\in\mathbb{A}:$ A has no nonzero ideal in $\mathbb{C}\}$

which are called the *upper radical* and the *semisimple operator*, respectively. As is well-known, the *Brown-McCoy radical class* $\mathbb{G}$ is the upper radical class $\mathcal{U}\mathbb{M}$ of $\mathbb{M}$. For the basic results of radical theory we refer to [2], [13] and [15]. A class $\mathbb{C}$ is *hereditary*, if $I \triangleleft A \in \mathbb{C}$ implies $I\in\mathbb{C}$. An ideal I of a ring A is said to be *small in* A if $I\neq 0$ and for any other ideal K $(\neq A)$ of A, $I+K\neq A$. If $I \triangleleft A$ and A has a unity, then A will be said a *unital extension* of I.

THEOREM 1. *The Brown-McCoy radical class* $\mathbb{G}$ *is the largest radical class disjoint to the class* $\mathbb{E}$ *of all rings with unity.*

Proof. If A is a ring with unity 1, then A has a maximal ideal M excluding 1. Hence $A/M\in\mathbb{M}$ holds, implying $A\notin\mathbb{G}$. Thus $\mathbb{G}\cap\mathbb{E}=\emptyset$ and by the definition of $\mathbb{G}$, it is obvious that $\mathbb{G}$ is maximal with respect to the property $\mathbb{G}\cap\mathbb{E}=\emptyset$. Let us consider any radical class $\mathbb{R}$ such that $\mathbb{R}\cap\mathbb{E}=\emptyset$, and the lower radical $L(\mathbb{R}\cup\mathbb{G})$. In view of Leavitt [6]

$$L(\mathbb{R}\cup\mathbb{G})=\{A: \begin{array}{l}\text{every nonzero homomorphic image of}\\ A \text{ has a nonzero ideal in } \mathbb{R}\cup\mathbb{G}\end{array}\}.$$

(Here the Anderson-Divinsky-Suliński Theorem is used which is valid also for alternative rings.) Again, let M denote a maximal ideal of a ring $A\in\mathbb{E}$ excluding 1 . As $A/M\in\mathbb{M}$, the ring A/M has no nonzero ideal in $\mathbb{R}\cup\mathbb{G}$, and hence $A\notin L(\mathbb{R}\cup\mathbb{G})$. Thus by the maximality of $\mathbb{G}$ it follows $L(\mathbb{R}\cup\mathbb{G})=\mathbb{G}$ and so $\mathbb{R}\subseteq\mathbb{G}$.

The next theorem characterizes the Brown-McCoy radical rings (see also Proposition 1.10 of [3]).

THEOREM 2. *A ring* $A\neq 0$ *is a Brown-McCoy radical ring if and only if* A *is a small ideal in any unital extension of* A .

Proof. Let B be any unital extension of $A\in\mathbb{G}$. Now we have $A\subseteq\mathbb{G}(B)$. Let I be an ideal of B such that $\mathbb{G}(B)+I=B$. The factor ring B/I is either a ring with unity, or $B/I=0$. Since

$$B/I \cong (\mathbb{G}(B)+I)/I \cong \mathbb{G}(B)/(\mathbb{G}(B)\cap I)\in\mathbb{G} ,$$

the first case is not possible. Consequently $B=I$, proving that $\mathbb{G}(B)$ is small in B . Since $A \subseteq \mathbb{G}(B)$, also A is small in B .

Assume that $A \notin \mathbb{G}$ and let B $(\neq A)$ be any extension of A (with or without unity). Since the Brown-McCoy radical is hereditary (also in the alternative case), B is not a Brown-McCoy radical ring. Consequently there are ideals M_α in B such that $B/M_\alpha \in \mathbb{M}$ and $\mathbb{G}(B)=\cap_\alpha M_\alpha$. As $A \notin \mathbb{G}$, also $A \nsubseteq M_\alpha$ for some α . Since M_α is a maximal ideal in B , we have $A+M_\alpha=B$ and hence A is not small in B .

Let us mention that for associative rings Michler [9] has proved in Hilfssatz 3.5 (b) that every small ideal of a ring is contained in the Brown-McCoy radical.

COROLLARY 3. *Let* A *be a ring with unity. The Brown-McCoy radical* $\mathbb{G}(A)$ *of* A *is a small ideal of* A *, and* $\mathbb{G}(A)$ *is the sum of all small ideals of* A .

The assertion is a straightforward consequence of Theorem 2.

PROPOSITION 4. *Let* D_A *denote the Dorroh extension of a ring* $A \neq 0$. *Then* $\mathbb{G}(D_A)=\mathbb{G}(A)$.

Proof. By the Anderson-Divinsky-Suliński Theorem $\mathbb{G}(A) \subseteq \mathbb{G}(D_A)$ holds. Since D_A/A is isomorphic to the ring of integers, and this latter ring is Brown-McCoy semisimple, it follows $\mathbb{G}(D_A) \subseteq \mathbb{G}(A)$.

As an immediate consequence of Corollary 3 and Proposition 4 we get the following representation of the Brown-McCoy radical of a ring.

THEOREM 5. *For any ring* A *its Brown-McCoy radical is*

$$\mathbb{G}(A) = \sum(I \lhd D_A : I \text{ is small in } D_A)$$

where D_A *denotes the Dorroh extension of* A . *In particular, a ring is Brown-McCoy semisimple if and only if its Dorroh extension does not contain small ideals.*

Let $\mathbb{C}$ be an abstract class of rings. We say that $\mathbb{C}$ is *closed under small extensions*, if I is a small ideal in A and $I \in \mathbb{C}$ imply $A \in \mathbb{C}$.

PROPOSITION 6. *Let* $\mathbb{C}$ *be a hereditary class which is closed under small extensions. If* $\mathbb{C} \neq \mathbb{A}$, *then* $\mathbb{C}$ *is contained in the class* $S\mathbb{G}$ *of all Brown-McCoy semisimple rings.*

Proof. Let $A \neq 0$ be a Brown-McCoy radical ring in $\mathbb{C}$ and B an arbitrary ring. Further, let D denote the Dorroh extension of the direct sum $A \oplus B$. It is easy to check that $A \lhd D$ and $B \lhd D$. Since D is a ring with unity and $0 \neq A \in \mathbb{G}$, by Corollary 3 A is small in D . Since $\mathbb{C}$ is closed under small extensions, it follows $D \in \mathbb{C}$. Furthermore, by $B \lhd D$ the hereditariness of $\mathbb{C}$ yields $B \in \mathbb{C}$, contradicting $\mathbb{C} \neq \mathbb{A}$. Thus $\mathbb{C}$ does not contain nonzero Brown-McCoy radical rings. Since $\mathbb{C}$ is hereditary, $\mathbb{C}$ must consist of Brown-McCoy semisimple rings.

The next theorem gives two characterizations of the Brown-McCoy semisimple class $\mathcal{S}\mathbb{G}$.

THEOREM 7. *For a class* $\mathbb{S}$ *the following three conditions are equivalent.*

(i) $\mathbb{S}$ *is the class* $\mathcal{S}\mathbb{G}$ *of all Brown-McCoy semisimple rings;*

(ii) a) $\mathbb{S} \neq \mathbb{A}$,

b) $\mathbb{S}$ *is hereditary,*

c) *if* $I \lhd A$ *and* $I \in \mathbb{S}$, *then* I *is not small in* A , *and* $\mathbb{S}$ *is the largest class with respect to properties* a), b), c);

(iii) $\mathbb{S}$ *satisfies* a), b) *and*

d) $\mathbb{S}$ *is closed under small extensions, and* $\mathbb{S}$ *is the largest class with respect to properties* a), b), d).

Proof. (i) ⇒ (ii). The validity of a) and b) is trivial. Let I be an ideal of a ring A such that

$0 \neq I \neq A$ and $I \in S\mathbb{G}$. Since the Brown-McCoy radical is hereditary, $A \notin \mathbb{G}(A)$ and therefore there exists a maximal ideal M of A such that $A/M \in \mathbb{M}$. Moreover, there is one such maximal ideal which does not contain I , as otherwise I would be contained in $\mathbb{G}(A)$. Hence $I+M=A$ and I is not small in A and c) is satisfied. Next, let $\mathbb{C}$ be a class satisfying a), b) and c). If $\mathbb{C} \nsubseteq S\mathbb{G}$, then there is a ring $B \in \mathbb{C} \setminus S\mathbb{G}$. Now by the hereditariness of $\mathbb{C}$ we have $0 \neq \mathbb{G}(B) \in \mathbb{C}$. By Theorem 2 $\mathbb{G}(B)$ is small in its Dorroh extension and so $\mathbb{C}$ does not satisfy condition c), a contradiction. Thus $\mathbb{C} \subseteq S\mathbb{G}$.

(ii) ⇒ (iii). Straigthforward.

(iii) ⇒ (i). By Proposition 6 we have $\mathbb{S} \subseteq S\mathbb{G}$. Moreover $S\mathbb{G}$ satisfies c) and hence also d).

COROLLARY 8. *A ring* $A \neq 0$ *is Brown-McCoy semisimple, if and only if every ideal* I *of* A *is not small in any extension of* I .

Proof. If $A \in S\mathbb{G}$, then by b) and c) of Theorem 7 the conclusion holds. If $A \notin S\mathbb{G}$, then $\mathbb{G}(A) \neq 0$ and by Theorem 2 $\mathbb{G}(A)$ is small in its Dorroh extension.

THEOREM 9. *Let* $\mathbb{C}$ *be a non-empty subclass of Brown-McCoy semisimple rings satisfying condition*

(1) *if* M *is a maximal ideal of a ring* $A \in \mathbb{C}$ *,*

then also $A/M \in \mathbb{C}$.

Then the class $\mathcal{U}\mathbb{C}$ *is a special radical class.*

Proof. By the assumptions on $\mathbb{C}$ the class $\mathbb{Q}=\mathbb{M}\cap\mathbb{C}$ is not empty. Since $\mathbb{Q}$ is a special class, $\mathcal{U}\mathbb{Q}$ is a special radical. We have obviously $\mathcal{U}\mathbb{C} \subseteq \mathcal{U}\mathbb{Q}$. Let A be a ring taken from $\mathcal{U}\mathbb{Q} \setminus \mathcal{U}\mathbb{C}$. Then A has a nonzero homomorphic image in $\mathbb{C}$ and so by (1) also in $\mathbb{Q}$, contradicting $A \in \mathcal{U}\mathbb{Q}$. Thus also $\mathcal{U}\mathbb{Q} \subseteq \mathcal{U}\mathbb{C}$ holds.

COROLLARY 10. *Let* $\mathbb{C}$ *be a hereditary class of rings satisfying conditions* (1) *and*

(2) *if* $A \in \mathbb{C}$ *and* $A \triangleleft B$ *, then* A *is not small in* B

Then $\mathcal{U}\mathbb{C}$ *is a special radical containing all Brown-McCoy radical rings.*

Proof. Since $\mathbb{C}$ is hereditary, by Corollary 8 condition (2) yields that $\mathbb{C}$ consists of Brown-McCoy semisimple rings. Hence Theorem 9 is applicable.

Remark 1. In Szász-Wiegandt [14] Theorem 5 it has been proved under stronger assumptions that the upper radical $\mathcal{U}\mathbb{C}$ is hereditary. Van Leeuwen - Jenkins have proved in [7] Theorem 7 that $C \subseteq \mathcal{S}\mathbb{G}$ provided that $\mathbb{C}$ is hereditary, homomorphically closed and satisfies condition (2). Thus Theorem 9 and Corollary 10 sharpen the above mentioned results of [7] and [14].

Finally, we are going to determine the varieties of rings which are closed under small extensions.

PROPOSITION 11. *If a variety* $\mathbb{V}$ *of rings does not contain a zeroring* $A\neq 0$ *, then* $\mathbb{V}$ *is a homomorphically closed semisimple class.*

Proof. The assertion strengthens Gardner's [1] Theorem 4.3 (iii) ⇒ (iv) inasmuch as $\mathbb{V}$ need not be idempotent. The proof is a modification of that in [1] and we shall trace back the alternative case to the associative one. By the assumption on $\mathbb{V}$ for any ring $A\in\mathbb{V}$, any subring $[a]$ generated by $a\in A$ must satisfy $[a]^2=[a]$. Moreover, the ring A does not contain nonzero nilpotent elements either, and so by Rjabuhin [10], [12] (or Hentzel [5]) A is a subdirect sum of (alternative) rings without zero-divisors. Hence a result of Stewart [12] (see also [15] Lemma 33.3) along with its proof is applicable and therefore A is a subdirect sum of (alternative) division rings B_α such that each subring $[b]$ $(0\neq b\in B_\alpha)$ is a finite field. Hence each B_α is periodic. Since the Cayley-Dickson algebras are not periodic, each B_α is associative and also commutative. Thus A is associative and $\mathbb{V}$ consists of associative rings. Thus the class $\mathbb{K}$ of all fields in $\mathbb{V}$ is a special class and hence

$$\mathbb{V} = \{A : A \text{ is a subdirect sum of rings in } \mathbb{K}\}$$

is exactly the semisimple class of the special radical $\mathcal{U}\mathbb{K}$. Thus $\mathbb{V}$ is a homomorphically closed semisimple class.

A complete description of all homomorphically closed semisimple classes have been given in [4], they are all contained in $\mathcal{S}\mathbb{G}$ (with the exception of the class $\mathbb{A}$).

COROLLARY 12. *A variety* $\mathbb{V}$ *of rings is closed under small extensions if and only if* $\mathbb{V}$ *is a homomorphically closed semisimple class.*

The assertion follows immediately from Propositions 6 and 11 and Corollary 8.

Remark 2. By Corollary 12 *the varieties closed under small extensions, are precisely the idempotent varieties* ([15] Theorem 34.1 and Gardner [1] Theorem 4.3), or equivalently, the varieties closed under essential extensions ([8] Theorem 1).

REFERENCES

[1] B. J. Gardner, Radical properties defined locally by polynomial identities, II, *J. Austral. Math. Soc.* 27 (1979), 274-283.

[2] B. J. Gardner, Some current issues in radical theory, *Mathematical Chronicle* 8 (1979), 1-23.

[3] B. J. Gardner, Small ideals in radical theory, *Acta Math. Acad. Sci. Hungar.*, to appear.

[4] B. J. Gardner and P. N. Stewart, On semi-simple radical classes, *Bull. Austral. Math. Soc.* 13 (1975), 349-353.

[5] I. R. Hentzel, Alternative rings without nilpotent elements, *Proc. Amer. Math. Soc.* 42 (1974), 373-376.

[6] W. G. Leavitt, Sets of radical classes, *Publ. Math. Debrecen* 14 (1967), 321-324.

[7] L. C. A. van Leeuwen and T. L. Jenkins, Upper radicals and simple rings, *Periodica Math. Hungar.* 6 (1975), 69-74.

[8] N. V. Loi, Essentially closed radical classes, *J. Austral. Math. Soc.* 35 (1983), 132-142.

[9] G. Michler, Kleine Ideale, Radikale und die Eins in Ringen, *Publ. Math. Debrecen* 12 (1965), 231-252.

[10] Ju. M. Rjabuhin, Algebras without nilpotent elements, I, (Russian) *Algebra i Logika* 8 (1969), 181-214.

[11] Ju. M. Rjabuhin, Algebras without nilpotent elements, II, (Russian), *Algebra i Logika* 8 (1969), 215-240.

[12] P. N. Stewart, Semisimple radical classes, *Pac. J. Math.* 32 (1970), 249-254.

[13] F. A. Szász, *Radicals of rings*, Akadémiai Kiadó, 1981.

[14] F. A. Szász and R. Wiegandt, On hereditary radicals, *Periodica Math. Hungar.* 3 (1973), 235-241.

[15] R. Wiegandt, *Radical and semisimple classes of rings*, Queen's papers in pure & appl. math., no. 37, Kingston, Ontario, 1974.

R. Wiegandt and N. V. Loi
Mathematical Institute of the
Hungarian Academy of Sciences
Budapest V., Reáltanoda u. 13-15
H-1053
Hungary

COLLOQUIA MATHEMATICA SOCIETATIS JÁNOS BOLYAI

38. RADICAL THEORY, EGER (HUNGARY), 1982.

ON A CERTAIN CLASS OF MAXIMAL AND MINIMAL SUBMODULES

F. LOONSTRA

1. INTRODUCTION.

In the following only unitary left R-modules over an (associative) ring R are considered. The submodules 0 and M of the R-module M will be called the trivial submodules of M. A submodule $U \leq_R M$ is called a *uniform* submodule of M, if $U \neq 0$, and if any two nonzero submodules U_1 and U_2 of U have a nonzero intersection; in this case U is an essential extension of each of its nonzero submodules U' (i.e. $U' \leq_e U$).

The problem we are discussing can be described by means of the following example.

Let A be a nonzero abelian group; can we find the evt. maximal (resp. minimal) nontrivial subgroups C of A such that A/C is torsionfree ?

In the case of R-modules we can discuss a similar problem using the notion of the *singular* submodule of a given R-module. Then the question will be: find the evt. maximal (resp. minimal) nontrivial submodules N of a given R-module $_RM$ such that the corresponding quotients are nonsingular. A more general discussion needs the notion of a suitable torsion theory.

If $A \neq 0$ is an abelian group with subgroup C such that A/C is torsionfree, then we have $t(A) \subseteq C$, i.e., the torsion subgroup t(A) is the unique minimal subgroup C of A with A/C being torsionfree. If $t(A) = 0$, then we will discuss evt. nontrivial minimal solutions of the problem. Concerning an evt. maximal subgroup C (with A/C torsionfree) we may assume that A is torsionfree; for

$$A/C \cong {}^{A/t(A)}\big/_{C/t(A)} = A'/C',$$

and C is maximal in A (with A/C torsionfree) if and only if C' is maximal in A' (A'/C' being torsionfree). Assuming that A is torsionfree, we know that A/C is torsionfree if and only if C is pure in A.

So the problem is: find a maximal pure subgroup $C \neq A$ of the (torsionfree) group A. A necessary and sufficient condition in order that C be a maximal pure subgroup

($\neq$ A) of A, is the condition that the rank $\tau(A/C) = 1$. Therefore we choose a maximal independent system

$$M = \{a_i\}_{i \varepsilon I}$$

of elements of A and we define $M_i = M \setminus \{a_i\}$, $i \varepsilon I$. Then the smallest pure subgroup $C_i = \langle M_i \rangle_*$, generated by M_i is a maximal pure subgroup of the torsionfree group A($i\varepsilon I$).

Defining $D_i = \langle a_i \rangle_*$, we have $\tau(D_i) = 1$ and D_i is a minimal pure subgroup of A($i\varepsilon I$). However, in general, two different maximal independent systems

$$M = \{a_i\}_{i\varepsilon I}, M' = \{a_i'\}_{i\varepsilon I}$$

give rise to different systems of maximal (resp. minimal) pure subgroups.

The following theorem can be proved (see Loonstra [1] :

Let $M = \{a_i\}_{i\varepsilon I}$ and $M' = \{a_i'\}_{i\varepsilon I}$

be two maximal independent sets of elements of the torsionfree group A; then the sets $\{C_i\}_{i\varepsilon I}$ and $\{C_i'\}_{i\varepsilon I}$ of the corresponding maximal pure subgroups of A coincide, if and only if (after a suitable renumbering of the elements of M') there exist nonzero integers $\{\lambda_i, \lambda_i'\}$, $i\varepsilon I$, such that $\lambda_i a_i = \lambda_i' a_i'$ ($i\varepsilon I$). The same conditions are necessary and sufficient in order that the sets $\{D_i\}$ and $\{D_i'\}$ of the minimal pure subgroups coincide. This problem for abelian groups can be generalized for (left)R-modules $_RM$ if we dispose of a torsion submodule t(M) of $_RM$. Therefore we mention the following equivalences:

a) The ring R has the property that in every R-module $_RM$ the torsion elements form a submodule $t(M) \subseteq M$ such that $\subseteq M/t(M)$ is torsionfree, if and only if R is a left Ore domain.

b) The ring R has the property that for all R-modules M the cardinality of elements of order 0 in maximal independent sets of M only depends on M, if and only if R is a left Ore domain.

If we define in this case purity of the submodule N of M by $\lambda N = N \cap \lambda M (\forall \lambda \varepsilon R)$, then the problem of finding a maximal (resp. minimal) submodule N of M such that M/N is torsionfree can be solved for modules over a left Ore domain just as for abelian groups (Loonstra [1]).

2. GENERALIZATION FOR MODULE OVER AN ARBITRARY RING R.

For a generalization of the problem (discussed in par.1) for modules over any ring R we need the use of a torsion theory. Let σ be any left exact preradical of the category R-Mod, thenthere exists a smallest left exact radical $\bar{\sigma} \geq \sigma$, such that for any R-module M, $\bar{\sigma}(M)$ is the smallest submodule $N \subseteq M$ with the property $\sigma(M/N) = 0$. From the property $\sigma(M) \subseteq_e \bar{\sigma}(M)$ it follows that $\sigma(M) = 0$ if and only if $\bar{\sigma}(M) = 0$. Therefore, if M has the property that $\sigma(M) \neq 0$, then also $\bar{\sigma}(M)$ is a nontrivial submodule of M, and it is the unique "smallest" submodule N of M with $\sigma(M/N) = 0$.

To obtain some information about an evt. (nontrivial) maximal submodule N of M such that M/N is "torsionfree" we need a more elaborate consideration.

We suppose first of all that ρ is a left exact preradical and we call the following properties: if ρ is a left exact preradical, then

a) $\bar{\rho}$ is left exact;

b) $\rho(M) \subseteq_e \bar{\rho}(M)$ for all $M \in$ R-Mod;

c) $M \subseteq_e M'$ implies that $\rho(M) = 0$ if and only if $\rho(M') = 0$.

Let $\mathcal{T}(M)$ be the collection of all left exact preradicals ρ such that $\rho(M) = 0$. Then $M \subseteq_e M'$ implies $\mathcal{T}(M) = \mathcal{T}(M')$.

We choose any fixed injective R-module $E \neq 0$ and we define

$$\tau_E(M) = \bigcap_{\phi} \{\mathrm{Ker}\ \phi \mid \phi \in \mathrm{Hom}_R(M,E)\}.$$

Then

(i) τ_E is a left exact preradical;

(ii) $\rho \in \mathcal{T}(E) \Longleftrightarrow \rho(M) \subseteq \tau_E(M)$ for all $M \in$ R-Mod;

(iii) τ_E is a radical.

If we want to have an evt. "largest" submodule N of M for which M/N is "torsionfree", we restrict ourselves to left exact radicals $\rho\in\mathcal{T}(E)$, i.e. $\rho(E) = 0$, where E is a fixed chosen injective R-module. Then we have $\rho(M)\subseteq\tau_E(M)$ ($\forall M\in$R-Mod). Since $\rho\in\mathcal{T}(E)$, we find

$$\rho\left(M/\tau_E(M)\right)\subseteq\tau_E\left(M/\tau_E(M)\right) = 0 \text{ for all } \rho\in\mathcal{T}(E).$$

That proves: if E is any injective module, ρ a left exact radical, $\rho\in\mathcal{T}(E)$ and M any R-module, then $\tau_E(M)$ is - among the submodules $\rho(M)$ with $\rho\in\mathcal{T}(E)$ - the unique maximal submodule N of M with $\rho(M/N) = 0$ (assuming that $\tau_E(M) \neq M$). We have $\tau_E(M) = M$ if and only if $\mathrm{Hom}_R(M,E)=0$.

3. USE OF THE GOLDIE TORSION THEORY.

It is noteworthy that our original problem for abelian groups leads in a natural way to the use of the Goldie torsion theory. First of all a general remark: the evt. minimal (resp. maximal) submodules N of $_RM$ such that - for a certain torsion theory $(\mathcal{T},\mathcal{F})$ - M/N is $\mathcal{F}$-torsionfree, can be determined supposing that M is $\mathcal{F}$-torsionfree. For let t be the corresponding idempotent radical of the torsion theory, then $M/t(M)\in\mathcal{F}$.

If, conversely, $N\subseteq M$ and $M/N\in\mathcal{F}$, then we must have $t(M)\subseteq N$.

Indeed, since $\mathrm{Hom}_R(t(M), M/N) = 0$, we have $t(M) \subseteq \mathrm{Ker}(\pi)$, where π is the canonical projection $\pi: M \to M/N$. That implies that $t(M)$ is the minimal submodule N of ${}_RM$ with $M/N \in \mathfrak{F}$ (for all R-modules).

To characterize the submodules $N \subseteq M$ with $M/N \in \mathfrak{F}$, we define a submodule N of ${}_RM$ to be $\mathfrak{F}$-closed if $t(M/N)=0$ or $M/N \in \mathfrak{F}$. One can prove that the collection of the $\mathfrak{F}$-closed submodules of M is closed under arbitrary intersections. For the evt. solutions of our problem we need therefore the evt. minimal (resp. maximal) $\mathfrak{F}$-closed submodules N of the $\mathfrak{F}$-torsionfree module M.

We specialize in the following way, using the *singular submodule* $Z(M)$ of M, where

$$Z(M) = \{m \in M \mid Im = 0 \text{ for some essential left ideal } I \text{ of } R\}.$$

M is called a *singular* R-module if $Z(M) = M$, and M is called *nonsingular*, if $Z(M) = 0$. $Z : M \to Z(M)$ is an idempotent preradical of the category R-Mod, but (in general) it is not a radical. The smallest radical $\overline{Z}$, containing Z, is Z_2, determined by

$$Z_2(M) / Z(M) = Z(M / Z(M)).$$

We have therefore:

$$Z(M/Z_2(M)) = 0, \text{ since } Z_2(M/Z_2(M)) = 0.$$

If moreover K is a submodule of M such that $Z(M/K) = 0$ then $Z_2(M) \subseteq K$; so $Z_2(M)$ is the minimal submodule K of M with $Z(M/K) = 0$.

For our problem we may therefore suppose that M is a nonsingular R-module, and we need to determine evt. minimal (resp. maximal) submodules K of M such that $Z(M/K) = 0$.

Let us call a submodule K of M *Z-closed* if $Z(M/K) = 0$. The set $Z^*(M)$ of all Z-closed submodules of M is closed under arbitrary intersections; indeed, if $\{K_i\}_{i\in I}$ is a set of Z-closed submodules of M, then $M/(\bigcap_{i\in I} K_i)$ can be embedded in $\prod_{i\in I} M/K_i$. Since the class of nonsingular modules is closed under direct products and submodules, we see that $M/(\bigcap_{i\in I} K_i)$ is nonsingular. This implies that for any submodule N of M there exists a smallest submodule K of M containing N and such that M/K is nonsingular. Now we need the following property of Z-closed submodules of M:

3.1. *If* M *is nonsingular then a submodule* N *is* Z*-closed if and only if* N *is (essentially) closed.*

Proof: If N is Z-closed, then N cannot have a proper essential extension; for if $N \subseteq_e K \subseteq M$, then K/N is singular; but $K/N \subseteq M/N$ and M/N is nonsingular, so K/N must be also nonsingular, i.e. $K/N = 0$, or $K = N$, and that implies that N is essentially closed. Conversely, suppose that N is an (essentially) closed submodule of M, and M is nonsingular. If $K/N = Z(M/N)$ the $N \subseteq_e K$ since M is nonsingular. This implies that $K = N$, i.e. $Z(M/N) = 0$, and N is Z-closed.

Returning to our original problem in case M is nonsingular we need evt. minimal (resp. maximal) essentially closed submodules N of M. This brings us to the study of closed submodules of a nonsingular R-module M. The description of the closed submodules of a R-module M can be given by means of the following property: a submodule N of M is a closed submodule of M if and only if $N = M \cap N_o$, where N_o is an injective submodule of an injective hull $\hat{M}$ of M. This means that the closed submodules of M can be found as intersections of M with the injective submodules of $\hat{M}$. In our case the injective hull $\hat{M}$ (being an essential extension of the nonsingular module M) is also nonsingular. If I and J are two injective submodules of $\hat{M}$ and $J \subseteq I$, then $J \cap M \subseteq I \cap M$. For a submodule N of M to be maximal (resp. minimal) closed in M, it is therefore necessary and sufficient that $\hat{N}$

is a maximal (resp. minimal) injective submodule of $\hat{M}$. This fact proves already that - in general - in M there do not always exist maximal (resp. minimal) closed proper submodules.

3.2. *Suppose that* N *is a closed submodule of* M *such that* $\widehat{M/N}$ *is isomorphic with an indecomposable injective submodule of* $\hat{M}$; *then* N *is a maximal closed submodule of* M. *Conversely, if* N *is a maximal closed submodule of* M, *then* $\widehat{M/N}$ *is isomorphic with an indecomposable injective submodule of* $\hat{M}$.

Proof: N will be a maximal closed submodule of M if and only if $\hat{N}$ is maximal injective submodule of $\hat{M}$. If N' is a complement of N in M, then $N \oplus N' \subseteq_e M$ implies $\hat{N} \oplus \hat{N}' = \hat{M}$, and therefore $\hat{N}$ is maximal injective in $\hat{M}$ if and only if $\hat{N}'$ is indecomposable injective. Since $N' \cong (N \oplus N')/N \subseteq_e M/N$, it follows that $\hat{N}' \cong \widehat{M/N}$ and that proves our statement.

For the determination of an evt. nontrivial minimal closed submodule N of the (nonsingular) module M we use the following statement:

3.3. *The following properties of a submodule* $N \neq 0$ *of* M *are equivalent:*

(i) N *is a minimal closed submodule of* M;

(ii) N *is a closed and uniform submodule of* M.

Proof: (i) → (ii) If N is a minimal closed submodule of M, N cannot have a proper closed submodule P; otherwise the associative property of being closed implies that P is closed in M. Then $\hat{N}$ must be indecomposable injective and N is a uniform submodule of M.

(ii) → (i) If N is uniform then $\hat{N}$ is an indecomposable injective submodule and N cannot have a proper closed submodule N'.

As a result of 3.2 and 3.3 we see that a maximal closed submodule of M exists if and only if a minimal closed submodule of M exists, and this is equivalent with the fact that $\hat{M}$ has an indecomposable injective direct summand.

To apply the last remark, we discuss a decomposition valid for any R-module M.

Let $\{N_i\}_{i\in I}$ be the family of all uniform submodules of the R-module M and $\mathfrak{F}$ the set of all subsets F of I such that $\sum_{i\in F} N_i$ is direct, then $\mathfrak{F}$ can be ordered by inclusion and $\mathfrak{F}$ is inductive, i.e. there are *maximal direct sums of uniform submodules of* M,

$$N = \bigoplus_{i\in F_o} N_i .$$

If the family of the uniform submodules is empty, we

take $N = 0$. If a submodule K of M contains a uniform submodule N_i, then $N \cap K \neq 0$ for every maximal direct sum N of uniform submodules of M. Let $N = \underset{i \varepsilon F_o}{\oplus} N_i$ be a maximal direct sum of uniform submodules N_i of M; then a new maximal direct sum of uniform submodules can be obtained

a) by replacing each N_i by a (non-zero) submodule $N_i' \subseteq N_i$,

b) by replacing each N_i by an essential extension $\overline{N}_i$ of N_i in M.

One may choose for $\overline{N}_i$ a maximal essential extension of N_i (in M); then the $\overline{N}_i$ are essentially closed uniform submodules of M. If M is injective, $N = \underset{i \varepsilon F_o}{\oplus} N_i$ is a maximal direct sum of uniform submodules, we may even suppose that the $\overline{N}_i$ are indecomposable injective submodules of M.

The set $\mathcal{L}$ of all submodules L_i, of M containing no uniform submodule is ordered by inclusion and $\mathcal{L}$ is inductive; $\{0\}$ is such a submodule. $\mathcal{L}$ is invariant for essential extensions in M effected on its elements L_i, i.e., if $L_i \subseteq_e \overline{L}_i$ ($\subseteq M$), then also $\overline{L}_i \varepsilon \mathcal{L}$.

In particular, every maximal element $L \varepsilon \mathcal{L}$ has no proper essential extension in M, i.e. such a maximal element L is essentially closed in M. If M is injective then the closed submodule L is injective, i.e. L is a direct factor of M, but L cannot contain an indecomposable injective submodule.

Summarizing we find that for every R-module M there exist maximal direct sums N of uniform submodules of M, and maximal submodules L of M containing no uniform submodules. If $A \neq 0$ is a submodule of M, then $L \cap A = 0$ implies $N \cap A \neq 0$; for if $N \cap A = 0$, then A cannot contain any uniform submodule, since N is a maximal direct sum of uniform submodules. For the same reason $N \cap L = 0$, i.e. $N + L = N \oplus L$. Moreover, the first remark implies that $N \oplus L \subseteq_e M$. Indeed, suppose X is a submodule of M, $(N \oplus L) \cap X = 0$, then $N \cap X = 0$ and $L \cap X = 0$; but $L \cap X = 0$ also implies $N \cap X \neq 0$! So $X = 0$, and $N \oplus L \subseteq_e M$.

We have more: suppose S is a submodule of M, then a submodule K of M is a complement of S if and only if

a) K is closed,

b) $K \cap S = 0$ and

c) $K \oplus S \subseteq_e M$.

The necessity of this statement is obvious. Conversely, if K is closed, $K \cap S = 0$ and $K \oplus S \subseteq_e M$, then it follows that K is a complement of S. For if $K' \supseteq K$ is a complement of S, $K \oplus S \subseteq_e M$ implies $K \oplus S \subseteq_e K' + S$, i.e. $K \subseteq_e K'$. But K is closed, so $K' = K$. In the same way S is a complement of K.

Conclusion:

3.4. *For every* R*-module* M *there exist maximal direct sums* N *of uniform submodules of* M, *and maximal submodules* L *of* M *containing no uniform submodules, such that*

a) $N + L = N \oplus L$,

b) $N \oplus L \subseteq_e M$,

c) N *and* L *are (essentially) closed submodules of* $_RM$,

d) N *is a complement of* L *in* M, *and* L *is a complement of* N *in* M.

If M is injective, then M contains an injective hull $\hat{N}$ of N, where $\hat{N}$ is an injective hull of a direct sum of indecomposable injective modules, while L (being a closed submodule of an injective module M) is also injective. In this case $M = \hat{N} \oplus L$, where L has no indecomposable injective submodules.

Concerning the uniqueness we mention that if M is injective, and $M = \hat{N} \oplus L = \hat{N}' \oplus L'$ (i.e. two decompositions of M of the same kind), then there exists an automorphism $\emptyset$ of M, such that $\emptyset(\hat{N}) = \hat{N}'$, and $\emptyset(L) = L'$.

With these results we conclude with

3.5. *Let* M *be any (nonsingular)* R*-module,* $N = \bigoplus_{i \in I} N_i$

a maximal direct sum of uniform submodules N_i *of* M, *and* L *a maximal submodule of* M *containing no uniform submodule of* M, *then*

a) $\hat{M} = \hat{N} \oplus \hat{L}$;

b) *the minimal closed submodules of* M *are of the form* $M \cap \hat{N}_i$, *where* $\hat{N}_i$ *is an indecomposable injective summand of* $\hat{M}$, *hence of* $\hat{N}$;

c) *the maximal closed submodules* M *are of the form* $M \cap (A_i \oplus \hat{L})$ *where* A_i *is a complementary summand of* $\hat{N}_i$ *in* $\hat{N}$.

Important examples of R- modules M for which L = 0 are

a) Noetherian R-modules,

b) R-modules over a Noetherian ring R, over a perfect ring or over a valuation ring.

That not every R-module M has necessarily proper minimal closed submodules is proved by the following example (see Sharp-Vámos [2], p. 56): let R be a ring of all sequences of elements of some field K, addition and multiplication being defined componentwise, and I be the ideal consisting of all sequences with only finitely many nonzer terms. Then $\widehat{R/I}$ has no indecomposable injective submodules, and therefore R/I (if nonsingular) cannot have minimal closed submodules.

If N is a submodule of the finite dimensional R-module M, then also $\hat{N}$ is finite dimensional; if therefore M is nonsingular then also N is nonsingular and the minimal (resp. maximal) closed submodules of N are also known. Dr. P. Vámos paid my attention to the result 3.5. (a) as an exercise (4.5) of his book [2]; in his result we also find that N and L are uniquely determined up to isomorphism. I am indebted to dr. Vámos for his remarks on this subject.

REFERENCES

[1] F. Loonstra, Remark on the maximal torsion subgroup of an abelian group, *Beiträge zur Algebra und Geometrie* 3 (1974), 55-57.

[2] D. W. Sharp, P. Vámos, *Injective modules*, Cambridge University Press, 1972.

F. Loonstra
Haviklaan 25
2566 XB - Den Haag
The Netherlands

COLLOQUIA MATHEMATICA SOCIETATIS JÁNOS BOLYAI

38. RADICAL THEORY, EGER (HUNGARY), 1982.

ON RADICAL, SEMISIMPLE, AND ATTAINABLE CLASSES OF ALGEBRAS

L. M. MARTYNOV

Let R be an arbitrary associative and commutative ring with identity different from zero, and $\mathbb{K}$ be any variety of linear algebras over R (R-algebras). As was pointed out by Gardner [3, Theorem 1.5], being radical and semisimple (in the sense of Kurosh and Amitsur) and being attainable (in the sense of Tamura and Mal'cev) are closely connected properties for classes of R-algebras. Namely, he proved that a proper subclass of $\mathbb{K}$ is radical-semisimple iff it is homomorphically closed and semisimple iff it is an attainable variety. For an account of related results we refer to Gardner's survey [4] in these Proceedings.

This aim of the present paper is to give further characterizations of classes which enjoy the above properties, in the case when R has no non-trivial idempotent ideals and $\mathfrak{K}$ satisfies some natural conditions (these conditions are fulfilled, in particular, by the varieties of all associative, alternative, and Jordan R-algebras, respectively). Many of the previous results in this line can be obtained as corollaries to our main theorem.

The author thanks L. Márki for his help in putting this paper into English.

1. SOME DEFINITIONS AND NOTATIONS

All the classes of R-algebras we shall consider are assumed to be abstract (i.e., closed under isomorphisms), contain the one-element algebra, and are subclasses of our universe which is the variety $\mathfrak{K}$. A class $\mathfrak{X}$ of R-algebras is said to be proper if $\mathfrak{X} \neq \mathfrak{K}$, and non-trivial if, in addition, $\mathfrak{X} \neq \mathfrak{E}$ where $\mathfrak{E}$ is the class of one-element R-algebras. A variety $\mathfrak{M}$ of R-algebras is said to be minimal if $\mathfrak{M} \neq \mathfrak{E}$ and $\mathfrak{E}$ is the only proper subvariety of $\mathfrak{M}$. By a residual class of R-algebras we mean a class which is closed under subdirect products. For further notions (like attainability or product of classes) we refer to the survey [4].

The following notations will be used throughout the paper.

$(M)_A$ - the ideal generated by the subset M in an R-algebra A ;

A^0 - the R-algebra with zero multiplication on the additive group of the R-algebra A ;

A^n $(n>0)$ - all finite sums of products of n elements of the algebra A, with any arrangements of the brackets;

F - the free algebra of countable rank in $\mathfrak{K}$;

$R\langle x\rangle$ - the ring of polynomials in x with coefficients from R and without constant term;

$[\phi]$ - the class of all R-algebras from $\mathfrak{K}$ which satisfy all the formulas from ϕ , where ϕ is a set of formulas;

$\mathfrak{X}(A)$ - the smallest ideal of the algebra A such that the corresponding factor algebra of A lies in the variety $\mathfrak{X}$;

var A - the smallest variety containing the R-algebra A ;

$\mathfrak{A}\vee\mathfrak{B}$ - the smallest variety containing the varieties $\mathfrak{A}$ and $\mathfrak{B}$;

$\mathfrak{A}\mathfrak{B}$ - the $\mathfrak{K}$-product of the subclasses $\mathfrak{A}$ and $\mathfrak{B}$ of K ;

$G(\mathfrak{K})$ - the groupoid of the subvarieties of $\mathfrak{K}$ under $\mathfrak{K}$-product;

Asd(R) - the variety of all power-associative R-algebras;

$\mathrm{Alt}(R)$ - the variety of all alternative R-algebras;

$\mathrm{Ass}(R)$ - the variety of all associative R-algebras;

$\mathrm{Jord}(R)$ - the variety of all Jordan R-algebras;

$\mathbb{Z} = [xy=0]$ - the variety of all R-algebras with zero multiplication;

$\mathbb{B}_I$ - the variety of all R-algebras from $\mathbb{K}$ which are annihilated by the ideal I of R.

A subalgebra B of an R-algebra A is said to be idempotent if $B^2 = B$. Following [6], by an anti-integral element of a power-associative R-algebra A we mean an element which is the root of a polynomial of the form $r_n x^n + \ldots + r_2 x^2 + x$ from $R\langle x\rangle$; the roots of the polynomials $x^n - x$ (for $n>1$) are called the periodic elements. A ring without idempotent ideals is a ring which has no non-trivial idempotent ideals; such is e.g. any field or the ring of integers.

2. THE MAIN RESULT

THEOREM. *Let* R *be a ring without idempotent ideals and* $\mathbb{K}$ *be a variety of power-associative* R*-algebras which satisfies:*

(i) $\mathbb{Z} \subseteq \mathbb{K}$;

(ii) $\bigcap_{n=1}^{\infty} F^n = 0$;

(iii) *if* A, B, C *are algebras from* K *such that* $C \triangleleft B \triangleleft A$, $B \neq C$, *and* $B/C \in [x^2=0 \rightarrow x=0]$, *then* $(C)_A \neq B$;

(iv) *those subdirectly irreducible algebras in* $\mathbb{K}$ *which satisfy* $x^n=x$ *for some* $n>1$, *are simple.*

Then the following properties are equivalent for a proper subclass $\mathbb{X}$ *of* $\mathbb{K}$:

1) $\mathbb{X}$ *is a radical-semisimple class;*

2) $\mathbb{X}$ *is a homomorphically closed semisimple class;*

3) $\mathbb{X}$ *is an attainable variety;*

4) $\mathbb{X}$ *is a residual radical class;*

5) $\mathbb{X}$ *is an idempotent variety;*

6) $\mathbb{X}$ *is a variety of algebras with no non-zero nilpotent elements;*

7) $\mathbb{X}$ *is a variety of idempotent algebras;*

8) $\mathbb{X}$ *is a subvariety of a variety* $[xf(x)=x]$ *for some* $f(x) \in R\langle x \rangle$;

9) $\mathbb{X}$ *is a variety of anti-integral algebras;*

10) $\mathbb{X}$ *is a subvariety of a variety* $[x^n=x]$ *for some* $n>1$;

11) $\mathbb{X}$ *is a variety of periodic algebras.*

This theorem will be proven through a series of lemmas. Some of the properties 1)-11) are equivalent under milder assumptions than those we imposed on R

and $\mathbf{K}$. To make this more transparent, we shall formulate the corresponding statements (some of which are known) as general as we can prove them.

The first four lemmas were proved in [3] (see Theorems 1.5, 1.4, 1.10, and Proposition 1.8).

LEMMA 1. *Properties* 1)-3) *of the Theorem are equivalent for any class* $\mathfrak{X}$ *of* R*-algebras.*

LEMMA 2. *Properties* 4) *and* 5) *of the Theorem are equivalent for any class* $\mathfrak{X}$ *of* R*-algebras.*

LEMMA 3. *If* $\mathfrak{X}$ *is an idempotent in* $G(\mathbf{K})$ *then for any* $A \in \mathbf{K}$ *, the ideal generated by* $\mathfrak{X}(\mathfrak{X}(A))$ *in* A *coincides with* $\mathfrak{X}(A)$.

LEMMA 4. *If* $\mathbf{K}$ *satisfies conditions* (i) *and* (ii), $\mathfrak{X}$ *is an idempotent in* $G(\mathbf{K})$ *and* $\mathfrak{X} \supseteq \mathfrak{Z}$ *, then* $\mathfrak{X} = \mathbf{K}$.

LEMMA 5. *Let* $\mathfrak{X}$ *be a homomorphically closed class of* R*-algebras each of whose subdirectly irreducible algebras is simple, and* $\mathfrak{A}$, $\mathfrak{B}$ *be varieties of* R*-algebras such that* $\mathfrak{A}\mathfrak{B} \subseteq \mathfrak{X}$. *Then* $\mathfrak{A}\mathfrak{B} = \mathfrak{A} \vee \mathfrak{B}$.

Proof. The inclusion $\mathfrak{A} \vee \mathfrak{B} \subseteq \mathfrak{A}\mathfrak{B}$ is obvious. Let $A \in \mathfrak{A}\mathfrak{B} \subseteq \mathfrak{X}$, then A is a subdirect product of simple algebras P_i $(i \in I)$ from $\mathfrak{X}$. Since the class $\mathfrak{A}\mathfrak{B}$ is homomorphically closed, each of the P_i lies in $\mathfrak{A}\mathfrak{B}$, hence $P_i \in A \cup B$ for all $i \in I$. Now it is clear that $A \in \mathfrak{A} \vee \mathfrak{B}$, and we are done.

Remark 1. Lemma 5 holds for arbitrary universal algebras provided that $\mathfrak{A}\mathfrak{B}$ is a variety.

The following statement was formulated in [9] for the variety Ass(R) but, as was observed in [2], Theorem 12, it holds also for the variety Asd(R) .

LEMMA 6. *Let* $\mathfrak{M}$ *be a minimal variety of power-associative* R*-algebras. Then either* $\mathfrak{M} = \mathrm{var}(R/M)^{\circ}$ *for some maximal ideal* M *of* R *, or* $\mathfrak{M} = \mathrm{var}\, R/M$ *for some maximal ideal* M *of finite index. Every variety of these kinds is minimal.*

LEMMA 7. *Let* R *be a ring without idempotent ideals, and* $\mathfrak{K}$ *be a variety of power-associative* R*-algebras satisfying conditions* (i), (ii). *Then every proper idempotent subvariety* $\mathfrak{X}$ *of* $\mathfrak{K}$ *is contained in the quasivariety* $[x^2=0 \to x=0]$.

Proof. Suppose that the conditions of the Lemma are satisfied and that $\mathfrak{X}$ contains an algebra with more than one element and with zero multiplication. By Lemma 6 $\mathfrak{X}$ contains a minimal variety $\mathrm{var}(R/M)^{\circ}$ for some maximal ideal M of R . Let T be the smallest ideal of R such that $(R/T)^{\circ} \in \mathfrak{X}$. Since $(R/M)^{\circ} \in \mathfrak{M} \subseteq \mathfrak{X}$, we have $M \supseteq T$ and therefore $T \neq R$. We are going to show that $T = \{0\}$. Suppose this is not so, then by our condition on R we have $T^2 \neq T$ and by the definition of T $(R/T^2)^\circ \notin \mathfrak{X}$. On the other hand, the annihilator of the R-module T/T^2 contains T, hence the R-algebra T/T^2 is a homomorphic image of a direct sum of algebras isomorphic to $(R/T)^\circ$. This implies $T/T^2 \in \mathfrak{X}$, and by the isomorphism of the R-algebras $(R/T)^\circ$ and $(R/T^2)^\circ/(T/T^2)$ we have $(R/T^2)^\circ \in \mathfrak{X}^2 \setminus \mathfrak{X}$, which contradicts the idempotence of $\mathfrak{X}$. Hence $T = \{0\}$ and so $R^\circ \in \mathfrak{X}$. This implies $\mathfrak{Z} \subseteq \mathfrak{X}$, and now by Lemma 4 $\mathfrak{X} = \mathfrak{K}$, a contradiction. This proves that $\mathfrak{X} \subseteq [x^2=0 \to x=0]$.

Remark 2. In the special case when R is a principal ideal domain, Lemma 7 yields Corollary 1.9 of [3].

Remark 3. The non-existence of non-trivial idempotent ideals in R is an important condition in Lemma 7; namely, if $I \triangleleft R$ is idempotent then the variety $\mathfrak{B}_I$ is idempotent (even attainable) though it may be different from $\mathfrak{K}$ and may contain nilpotent algebras with more than one element.

LEMMA 8. *If* R *is a ring without idempotent ideals and* $\mathfrak{K}$ *is a variety of power-associative* R*-algebras which satisfies conditions* (i)-(iii) *of the Theorem then the properties* 3) *and* 5) *are equivalent for any subclass* $\mathfrak{X}$ *of* $\mathfrak{K}$.

Proof. Attainable varieties are clearly idempotent.

Conversely, if $\mathfrak{X} \in G(\mathbb{K})$ and $\mathfrak{X}^2=\mathfrak{X}$ then Lemma 7 says that $\mathfrak{X} \subseteq [x^2=0 \to x=0]$. Suppose now that $A \in \mathbb{K}$ is such that $\mathfrak{X}(A) \neq \mathfrak{X}(\mathfrak{X}(A))$, then by $\mathfrak{X}(A)/\mathfrak{X}(\mathfrak{X}(A)) \in \mathfrak{X}$ and by condition (iii) we obtain that the ideal C generated by $\mathfrak{X}(\mathfrak{X}(A))$ in A is different from $\mathfrak{X}(A)$, which contradicts Lemma 3. Thus $\mathfrak{X}(A) = \mathfrak{X}(\mathfrak{X}(A))$ for all $A \in \mathbb{K}$, i.e., $\mathfrak{X}$ is an attainable variety.

For proving the equivalence of the properties 6)-11) in the Theorem we shall need the following.

LEMMA 9. *If an integral domain* A *over* R *has more than one element and satisfies the identity*

$$r_n x^n + \ldots + r_2 x^2 + x = 0$$

where $r_2, r_3, \ldots, r_n \in R$, *then* A *is a finite* R*-field.*

Proof. Take an arbitrary element $a \in A$. Putting $e = -(r_n a^{n-1} + \ldots + r_2 a)$, we have $ae = a$, therefore e is a non-zero idempotent hence the identity of A . Then a is a root of the polynomial $f(x) = (r_n e)x^n + \ldots + (r_2 e)x^2 + ex$, whose coefficients lie in A . But in the integral domain A the polynomial $f(x)$ can have at most n roots, and thus A is a finite field.

LEMMA 10. *Properties* 6)-11) *of the Theorem are equivalent for any class* $\mathfrak{X}$ *of power-associative* R*-algebras.*

Proof. The implications 6) ⇒ 7), 8) ⇒ 9), 10) ⇒ 11),

$8)\Rightarrow 6)$, $10)\Rightarrow 7)$ are evident. The validity of $9)\Rightarrow 8)$ and $11)\Rightarrow 10)$ follows from the fact that every relation for the generator of the free monogenic R-algebra in $\mathfrak{X}$ is an identity of $\mathfrak{X}$. Let now $\mathfrak{X}$ be a variety of idempotent R-algebras, and denote by A the free monogenic R-algebra in $\mathfrak{X}$. Then we have $A^2=A$, hence for the free generator a of A an equality of the form $a=af(a)$ holds, where $f(x)\in R\langle x\rangle$. Now it is clear that $\mathfrak{X}\subseteq[xf(x)=x]$. Furthermore, A is associative, commutative, and has no non-zero nilpotent elements, hence it is a subdirect product of integral domains over R (see e.g. [7], Theorem 1.1). Being homomorphic images of A, the latter subdirectly irreducible algebras satisfy the identity $xf(x)-x=0$, hence they are finite R-fields by Lemma 9. This identity yields a common upper bound for the orders of these fields, so there is an $n>1$ such that $a^n=a$, and then the identity $x^n=x$ must hold in $\mathfrak{X}$. All this shows $7)\Rightarrow 8)\Rightarrow 10)$, and Lemma 10 is proven.

LEMMA 11. *Let* R *and* $\mathbb{K}$ *satisfy the conditions of the Theorem. Then the properties* 5) *and* 10) *are equivalent for proper subclasses* $\mathfrak{X}$ *of* $\mathbb{K}$.

Proof. Let $\mathfrak{X}$ be idempotent in $G(\mathbb{K})$. By Lemma 7 $\mathfrak{X}$ is a variety of algebras without nilpotent elements, i.e., it enjoys property 6) of the Theorem, hence also property 10) for the latter two are equivalent by Lemma 10. So we have 5) $\Rightarrow$ 10). Conversely, let $\mathfrak{X}$ be a subvariety of $[x^n=x]$ for some $n>1$. Then $\mathfrak{X}^2$ is a variety of algebras without nilpotent elements, and since properties 6) and 10) are equivalent, we have $\mathfrak{X}^2 \subseteq [x^m=x]$ for some $m>1$. By condition (iv) of the Theorem we can apply now Lemma 5 to the homomorphically closed class $[x^m=x]$ and the varieties $\mathfrak{X}$ and $\mathfrak{X}$, and thus we obtain $\mathfrak{X}^2=\mathfrak{X}$. This proves 10) $\Rightarrow$ 5).

The validity of the Theorem is an immediate consequence of Lemmas 1, 2, 8, 10, 11.

3. COROLLARIES

COROLLARY 1. *Let the ring* R *and the variety* $\mathbb{K}$ *satisfy the conditions of the Theorem. Then* $\mathbb{K}$ *has a non-trivial radical-semisimple subclass if and only if* R *has a maximal ideal of finite index such that* $R/M \in \mathbb{K}$.

Proof. If $M \triangleleft R$ is a maximal ideal of finite index then R/M is a finite R-field and therefore $\mathrm{var}(R/M) \subseteq [x^n=x]$ for some $n>1$. If $R/M \in \mathbb{K}$ then $\mathrm{var}(R/M)$ is a radical semisimple class in $\mathbb{K}$ by the Theorem. Conversely, a non-trivial radical-semisimple class has no non-zero nilpotent algebras by the Theorem, hence by Lemma 6 it contains a minimal variety of the form var(R/M) for some maximal ideal M of finite index in R .

The following two lemmas facilitate applications of our main result to concrete classes of algebras.

LEMMA 12. *If P is a subdirectly irreducible right alternative R-algebra with more than one element and P satisfies $x^n=x$ for some $n>1$ then P is a finite R-field.*

Proof. It is well known (see e.g. [11], Corollary 1, p. 403) that every right alternative R-algebra without nilpotent elements is alternative. Since P satisfies $x^n=x$ for some $n>1$, P is a commutative algebra. As is also well known (see e.g. [11], Lemma 8, p. 171) the identity $(x,y,z)^2=0$ holds in every commutative alternative R-algebra, where (x,y,z) stands for the associator of the elements x,y,z . Since P has no non-zero nilpotent elements, we see that P is an associative algebra. A further well-known result says that every semisimple (in the sense of Jacobson) commutative R-algebra is a subdirect product of fields. Since P is subdirectly irreducible and satisfies $x^n=x$, we infer that it is a finite R-field.

LEMMA 13. *If* $1/2 \in R$ *and* P *is a subdirectly irreducible Jordan* R*-algebra with more than one element and* $x^n=x$ *holds for some* $n>1$ *in* P, *then* P *is simple.*

Proof. Clearly, every element a of P has finite additive order which is a power of a prime number $p\neq 2$. Since $(pa)^m = (p^m a)a^{m-1} = 0$ for some m, by $(pa)^n = pa$ we obtain $pa = 0$. Hence P is a periodic Jordan algebra over the p-element prime field. By [6], Theorem 15.11, P is a subdirect product of simple Jordan rings. Since the maximal ideals of the ring P are ideals of the R-algebra P, P must be a simple Jordan R-algebra.

COROLLARY 2. *If* R *has no non-trivial idempotent ideals then conditions* (i)-(iv) *of the Theorem are satisfied for the varieties* Ass(R), Alt(R) *and* Jord(R) *(*$1/2 \in R$ *is needed in the last case), hence the properties* 1)-11) *of the Theorem are equivalent for each proper subclass* $\mathfrak{X}$ *of any of them.*

Proof. All the varieties listed in our statement are power-associative and contain $\mathfrak{Z}$. Condition (ii) of the Theorem holds in every homogeneous variety, and our varieties are such ([11], p. 24). The validity of (iii) follows from the fact that in each of our cases, if

$C \triangleleft B \triangleleft A$ and $C \neq B$ then the R-algebra $B/(C)_A$ has non-zero nilpotent ideals by results from [1] (Lemma 4), [5], and [8]. Condition (iv) is fulfilled by Lemmas 12 and 13.

Finally, we remark that many previous related results (see [4] and its bibliography) can be obtained as special cases of our theorem by choosing the ring of operators R (in particular, to be the ring of integers or a field) and the variety $\mathbb{K}$ (here our lemmas and corollaries are helpful) in an appropriate way; this is left to the reader. Of course, one has to exclude the results which characterize radical-semisimple classes by identities, if one considers cases when $\mathbb{K}$ is not a variety or is a variety of linear Ω-algebras over R but it does not satisfy the conditions imposed on $\mathbb{K}$ in the Theorem.

4. CONCLUDING REMARKS AND PROBLEMS

From our results it follows that if R has no idempotent ideals then the proper radical-semisimple classes in Ass(R) are just the subvarieties of the varieties $[x^n=x]$, $n = 2,3,\ldots$. By Lemmas 5 and 12 they form a subsemilattice in the groupoid $G(\text{Ass}(R))$. As we noticed in Remark 3, if I is any non-trivial idempotent ideal of

R then the variety $\mathbb{B}_I$ is a radical-semisimple subclass of Ass(R) (the varieties of this kind also form a subsemilattice in $G(\mathrm{Ass}(R))$, namely if $I,J \triangleleft R$ are idempotent then $\mathbb{B}_I \mathbb{B}_J = \mathbb{B}_{IJ}$).

Problem 1. Characterize the radical-semisimple classes of associative algebras over an arbitrary ring R .

If $\mathbb{K}$ is a variety then the condition of being homomorphically closed is rather strong for semisimple classes, and such semisimple classes are rather "small". Therefore it is natural to weaken this condition by requiring only that the semisimple class be a quasi-variety.

Problem 2. Characterize those semisimple classes of rings which are quasi-varieties.

Examples of such semisimple classes are the quasi-varieties of associative rings $[x^2=0 \to x=0]$ and $[px=0 \to x=0 \mid p \in P]$ where P is an arbitrary non-empty set of prime numbers.

Problem 3. Characterize those semisimple classes of associative algebras over a field which are quasi-varieties.

There is also another interesting and related direction: investigations of radical-semisimple subclasses in some important, homomorphically closed and hereditary classes of associative rings which are not varieties. This topic is treated e.g. in [10].

REFERENCES

[1] V. A. Andrunakievič, Radicals of associative rings (Russian), *Mat. Sb.* 44 (1958), 179-212. English translation: *Amer. Math. Soc. Transl.* (2) 52 (1966), 95-128.

[2] V. A. Artamonov, Lattices of varieties of linear algebras (Russian), *Uspehi Mat. Nauk* 33 (1978), no. 2, 135-167.

[3] B. J. Gardner, Semi-simple radical classes of algebras and attainability of identities, *Pacific J. Math.* 61 (1975), 401-416.

[4] B. J. Gardner, Radicals and varieties, *this volume*.

[5] I. R. Hentzel and M. Slater, On the Andrunakievich lemma for alternative rings, *J. Algebra* 27 (1973), 243-256.

[6] J. M. Osborn, Varieties of algebras, *Advances in Math.* 8 (1972), 163-369.

[7] Ju. M. Ryabukhin, Algebras without nilpotent elements II (Russian), *Algebra i Logika* 8 (1969), no. 2, 215-240.

[8] A. M. Slin'ko, On radicals of Jordan rings (Russian), *Algebra i Logika* 11 (1972), no. 2, 206-215.

[9] T. R. Sundararaman, Precomplete varieties of R-algebras, *Algebra Universalis* 5 (1975), 397-405.

[10] E. M. Zacharova, On radical-semisimple classes and strong J-semisimplicity of locally finite PI-algebras (Russian), *Mat. Issled. (Ring Theoretical Constructions)*, 49 (1979), 68-79.

[11] K. A. Ževlakov, A. M. Slin'ko, and I. P. Šestakov, *Rings that are nearly associative* (Russian), Nauka, Moscow, 1978.

L. M. Martynov
USSR 644099 Omsk
Pedagogicheskiĭ Institut, Kafedra Algebry

COLLOQUIA MATHEMATICA SOCIETATIS JÁNOS BOLYAI
38. RADICAL THEORY, EGER (HUNGARY), 1982.

RADICALS AND INTERPOLATION IN UNIVERSAL ALGEBRA

R. MLITZ

The aim of the present paper is to give a survey of results concerning connections between radicals and interpolation properties in a universal algebraic setting. It includes structure theorems for special types of algebras, which are important for the historical development of the area (rings, near-rings) or which illustrate the general theory. Most of the theorems are known and have already been published. Nevertheless some new corollaries and remarks have been added.

A natural connection between radicals and function algebras appears first in the famous

STRUCTURE THEOREM OF WEDDERBURN-ARTIN (1908, 1927) ([31],[2]): *A nil-semisimple ring with d.c.c. on left ideals is isomorphic to a direct sum of finitely many full matrix rings over skew fields.*

The generailsation of this theorem to the chain-condition-free case is given by the

CHEVALLEY-JACOBSON DENSITY THEOREM (1939,1945)

([7],[13],[14],see also [15],[24]), which can be formulated as a structure theorem in the following way:

A Jacobson-semisimple ring is isomorphic to a subdirect product of dense rings of linear transformations of vector spaces over skew-fields .

This theorem shows how a radical (the Jacobson-radical) may be connected with an interpolation property (density in the ring of endomorphisms of a vector space) . Let us first make a

Detailed analysis of the above density theorem :

It naturally decomposes into three main parts :

1. *Definition of the radical by left modules :*

The Jacobson-radical J(R) of an associative ring R can be defined by

$$J(R) = \bigcap_{I(R)} \mathrm{Ann}_R M ,$$

where the intersection runs over the class I(R) of all left modules over R which are irreducible (i.e. do not contain nontrivial submodules) .

2. *Subdirect representation of semisimple rings :*

A ring is Jacobson-semisimple (i.e. satisfies J(R)=0) iff it is a subdirect product of primitive rings (i.e.

rings having a faithful irreducible left module) .

3. *The interpolation property :*

Every (primitive) ring R has an interpolation property on every (faithful) irreducible left module over R . This interpolation property can be formulated in two different ways :

a. *Density in a ring of functions (the bicentraliser) :*

If M is an irreducible left module over R, the ring R is dense in its bicentraliser $R_M^{**} = \mathrm{End}_{\mathrm{End}_R M} M$ over M . Since by Schur's lemma $\mathrm{End}_R M$ is a skew-field if M is irreducible, M may be considered as a vector space over this skew-field and R_M^{**} consists exactly of all linear mappings of this vector space into itself . Thus, the density of R in its bicentraliser R_M^{**} is nothing else than the following interpolation property :
To every finite subset $\{x_1,\ldots,x_s\}$ of M and every linear mapping L in R_M^{**} there is some element r in R (depending upon $\{x_1,\ldots,x_s\}$) which satisfies $r.x_i = L(x_i)$ for $1 \leq i \leq s$.

b. *Interpolation on certain (the linearly independent) subsets :*

To every finite subset $\{x_1,\ldots,x_s\}$ of M which is linearly independent over the skew-field $\mathrm{End}_R M$ and to any choice of elements $y_1,\ldots,y_s$ in M, there is an element r in R such that $r.x_i = y_i$ holds for $1 \leq i \leq s$.

Notice for both formulations that R always acts linearly on M .

Before considering the situation in universal algebra, it is advantageous to have a look at

THE NEAR-RING CASE :

Near-rings are relatively well-behaving since they are close to rings; nevertheless their Jacobson-type density theorems well reflect the general connection "radical - interpolation property" to be discussed later . We follow the notation of Pilz' book [25] which also contains a good survey of the Jacobson-radical-theory for near-rings. A *zero-symmetric near-ring* is an algebra $\langle N,+,.\rangle$ satisfying :

1. $\langle N,+\rangle$ is a group (not necessarily abelian),
2. $\langle N,.\rangle$ is a semigroup ,
3. $(m+n)p = mp + np \qquad \forall\, m,n,p \in N$,
4. $p0 = 0 \qquad \forall\, p \in N$.

Left modules over near-rings (called N-groups in the near-ring literature) are defined similarly to the ring case using the above properties 1.to 4. .
As before, we proceed considering the three main parts of the density theorems :

1. *Definition of Jacobson-type radicals for near-rings* (Betsch 1962,1963 - [4],[5]) :

There are three ways to define "irreducible" left modules over near-rings : call a left module M over a near-ring N

of type 0 if it is *cyclic* (i.e. of the form $M = N.m$ for some $m \in M$ – every such m is called a *generator* of M) and simple ;

of type 1 if it is *strictly cyclic* (i.e. cyclic and containing only generators and elements m which satisfy $N.m = 0$) and simple ;

of type 2 if it does not contain nontrivial submodules .

These three types of left modules over near-rings yield three radicals J_0, J_1, J_2 by :

$$J_i(N) = \bigcap_{M \in I_i(N)} \mathrm{Ann}_N M \qquad i = 0,1,2 \quad ,$$

where $I_i(N)$ denotes the class of all left modules of type i over N .

2. *Subdirect representation of semisimple near-rings* (Betsch 1962,1963 – [4],[5]) :

A near-ring is J_i*-semisimple iff it is a subdirect product of i-primitive near-rings* (i.e. near-rings having a faithful left module of type i) *(i = 0,1,2)* .

3. *The interpolation properties :*

Case i = 0 (Betsch 1973 – [6]) :

Let M *be a (faithful) left module of type 0 over* N ;

then to every finite set $\{x_1,\dots,x_s\}$ *of generators of* M *satisfying* $\mathrm{Ann}_N x_i \neq \mathrm{Ann}_N x_j$ *for* $i \neq j$ *and to every choice of elements* $y_1,\dots,y_s$ *in* M, *there is some* n *in* N *such that* $n.x_i = y_i$ *holds for* $i = 1,\dots,s$.

Notice that the property of M to be faithful - which is used in Betsch's proof - is not really a restriction since M is always faithful over $N/\mathrm{Ann}_N M$ and since $\mathrm{Ann}_N M \neq 0$ only means that there are several elements in N doing the same job on M .

Case $i = 1$ (Ramakotaiah 1969 - [27]) :

Same result as above for $i = 0$, *the generators of* M *being now exactly those* $x \in M$ *which satisfy* $N.x \neq 0$.

Case $i = 2$ (Polin 1971 - [26]) :

Let M *be a (faithful) left module of type 2 over* N ; *then* N *is dense in the sub-near-ring* T *of* $N_M^{**} = \mathrm{End}_{\mathrm{End}_N M} M$ *generated by the* N-*regular functions of the bicentraliser* N_M^{**} . (Here a function f from M into itself is called N-*regular* if for $x,y \in M$ the equality

$$n.x = n.y \quad \forall\, n \in N \quad \text{implies} \quad f(x) = f(y) .)$$

This means that to every finite subset $\{x_1,\dots,x_s\}$ *of* M *and every function* t *in* T , *there is an element* n *in* N *such that* $n.x_i = t(x_i)$ *holds for* $i = 1,\dots,s$.

The above formulation is a simplified version of Polin's density theorem proved originally for zero-symmetric mΩ-near-rings and containing moreover an algebraic description of the m -near-ring T .

Notice that the interpolation properties formulated here are of type a - density in an algebra of functions - for i=2 and of type b - interpolation on certain subsets - for i=0,1 .

The study of these special cases leads to the

General problem :

Are there similar correspondences between radicals, "primitive" algebras and interpolation properties for general algebras ?

We will attack this problem as in the ring and near-ring cases by considering the three main parts of such a correspondence. As a preliminary, we have to introduce radicals and left modules for universal algebras .

THE SITUATION IN UNIVERSAL ALGEBRA :

0. *Definition of general radicals and left modules :*

a. *General radicals* (Hoehnke 1966 - [12]) :

Given a variety V of universal algebras $\langle A,\Omega\rangle$, a *radical* on V is defined by a mapping

$$\rho : A \to \rho A \in \mathrm{Con}\, A$$

satisfying 1. $\varphi(\rho A) \subset \rho(\varphi A)$ for every homomorphism φ defined on A (on ρA, φ has to be taken by components) ;

2. $\rho(A/\rho A) = 0 \in \mathrm{Con}\, A$.

b. *Left modules over universal algebras* :

Left modules are well known over rings, near-rings (see above) or semigroups (acts, polygons in other terminologies); moreover, definitions of left modules over -rings and m -near-rings were given by Skornjakow (1973 - [28]) and Polin (1971 - [26]) . We now give a definition which covers all these special cases .

We start with a variety $V' = V(\Omega',\Lambda')$ of universal algebras (given by a system Ω' of operations and a set Λ' of laws), satisfying

a. the algebras of V' are nonunary, i.e. Ω' contains a $(\mu+1)$-ary operation with $\mu \geq 1$, which we shall denote by "$\circ$" : $(a,b_1,\ldots,b_\mu) \to a\circ(b_1,\ldots,b_\mu)$.

In the above examples we may take for "$\circ$" the product; in this case "$\circ$" has the following properties :

b. "$\circ$" is *hyperassociative* :

$$(a\circ(b_1,\ldots,b_\mu))\circ(c_1,\ldots,c_\mu) = = a\circ(b_1\circ(c_1,\ldots,c_\mu),\ldots,b_\mu\circ(c_1,\ldots,c_\mu)) ;$$

c. "$\circ$" is *left distributive* with respect to all nonnullary operations belonging to $\Omega'\setminus\{\circ\}$:

$$\omega(a_1,\ldots,a_{n(\omega)})\circ(c_1,\ldots,c_\mu) = = \omega(a_1\circ(c_1,\ldots,c_\mu),\ldots,a_{n(\omega)}\circ(c_1,\ldots,c_\mu)).$$

In the above examples, "$\circ$" satisfies in addition

$$\omega\circ(c_1,\ldots,c_\mu) = \omega \qquad \text{for } n(\omega) = 0 ;$$

but if we take $\omega = 1$ in the variety of rings with identit or of monoids, this left distributivity with respect to all nullary operations is not satisfied for "$\circ$" being the product. This is the reason for which in the following "left distributive" will always mean "left distributive with respect to all non-nullary operations" .

The basic constructions (left modules over universal algebras (in this section 0), definition of radicals by left modules (section 1)) and some of the results (subdirect decomposition of semisimple algebras (section 2), first density theorem for the fully retractable radical (section 3a)) require only the presence of a variety $\mathcal{V}'$ satisfying condition a. ; the proofs of the remaining results make heavy use of the laws b. and c. . For the sake of shortness we shall call a variety $\mathcal{V}'$ satisfying the conditions a.,b. and c. (for suitable "$\circ$") a *hyperassociative - left distributive variety* or shortly a *HALD-variety* . Examples of HALD-varieties with a binary operation "$\circ$" (i.e. with $\mu=1$) are provided by all varieties of semigroups, of associative rings, of near-rings, of lattices, of tri-operational algebras in the sense of Menger ([17]) . For $\mu\geq 2$, we have as examples the varieties of superassociative systems (Menger [18]), of simple Menger algebras (Whitlock [33]), of n-semilattices or n-groups (Dicker [9]), of mΩ-near-rings (Polin [26]) . A further example of a HALD-variety with binary "$\circ$" is provided by the conrings introduced recently by Skornjakow

([29],[3o]) .

The following concept of left module over an algebra is a slight modification of the author's definition from 1977 ([19]) :

Let $\langle A,\Omega'\rangle$ be an algebra belonging to a variety $\mathcal{V}' = \mathcal{V}(\Omega',\Lambda')$ satisfying condition a., denote by Σ the system of all nullary operations of Ω' which only appear in laws of Λ' which contain the operation "$\circ$", and denote by the system of operations $\Omega'\setminus(\{\circ\}\cup\Sigma)$. A *left module* over $\langle A,\Omega'\rangle$ will be an algebra $\langle M,\Omega\cup A\rangle$ satisfying :

a. every element a of A is an operator from M^{μ} into M (notation: $a : (m_1,\ldots,m_\mu) \to a\circ(m_1,\ldots,m_\mu) \in M$;

d. $\langle M,\Omega\rangle$ belongs to the variety $\mathcal{V} = \mathcal{V}(\Omega,\Lambda)$, where Λ denotes the set of all those laws of Λ' in which the operation "$\circ$" does not appear ;

and in addition condition b. resp. c. (as above, but unde

$a,a_1,\ldots,a_{n(\omega)},b_1,\ldots,b_\mu \in A$, $c_1,\ldots,c_\mu \in M$, in the case when the operation "$\circ$" in Ω' is hyperassociative (b. resp. left distributive (c.) .

This definition of left modules is in a certain sense the weakest possible, since it only requires that $\langle M,\Omega\rangle$ belongs to the same variety as $\langle A,\Omega\rangle$ and that b. resp. c. (if fulfilled for "$\circ$" in Ω' on the algebras in $\mathcal{V}'$) carry over to the module operation . It is easy to check that for semigroups and near-rings this definition yields exactly the classical left modules; moreover, it coincides with Polin's definition for $m\Omega$-near-rings . In the case of associative rings, it turns out to be weaker than the usual definition, since in particular it does not imply that the elements of A act linearly on the left modules M over A . In order to cover the classical concept of left modules over associative rings and Skornjakow's left modules over Ω-rings, one may add a further condition to the definition :

e. $\langle M,\Omega\cup A\rangle$ satisfies all the identities obtainable from the laws of $\Lambda'\setminus\Lambda$ by replacing the letters appearing in these laws by elements of A and M in a way yielding well-defined words .

The definition of left modules using conditions a.,d.,e. covers not only the above mentioned classical cases but also Skornjakow's left conmodules over conrings (see [29], [3o]); it is relatively strong since it requires that all the identities obtainable from Λ' hold for the module-operation "$\circ$" . In the cases of rings or near-rings with identity resp. of monoids it yields the unitary left modules . Nevertheless, this definition might be too weak as well since it depends upon Λ' : it might happen that many of the laws of $\Lambda'\setminus\Lambda$ cannot be carried over from the operation "$\circ$" in Ω' to the action "$\circ$" of A on M, the cording words being not well-defined for this action . For example, if we describe the variety of Lie algebras over a commutative ring with identity by $\Lambda' \setminus \Lambda$ consisting of the two distributive laws, the law $x^2 = 0$ and the law $x(yz)+y(zx)+z(xy) = 0$, only the two distributive laws carry over to left module laws . In order to get the well-known concept of left module over a Lie algebra (see f.ex. Bahturin's book [3] p.18) from our general definition, we have to add to the above Λ' the law $(xy)z = x(yz)-y(xz)$, which of course is derivable from the above Λ' . The strongest possible definition of left modules would thus be the one using conditions a.,d. and e. , Λ' being taken as the set of all laws holding in $\mathcal{V}'$. For many purposes it might be the best to choose - for the considered variety $\mathcal{V}'$ - a definition between the weakest and the strongest by requiring condition e. only for certain laws valid in $\mathcal{V}'$.

1. *Definition of radicals by left modules :*

The first general result in this direction is the

THEOREM OF ANDRUNAKIEVIČ-RJABUHIN (1964 - [1]) :

Every general class $K = \bigcup_{R \in \mathcal{V}'} K(R)$ *of left modules defines a Kuroš-Amitsur-radical* ρ_K *on the variety* $\mathcal{V}'$ *of associative rings by* $\rho_K(R) = \bigcap_{M \in K(R)} \mathrm{Ann}_R M$.

Moreover, every Kuroš-Amitsur-radical on this variety $\mathcal{V}'$ *can be obtained in this way .*

In this theorem, a *general class of left modules* is defined as follows : suppose that to each ring R in V' we have assigned a (possibly empty) class K(R) of nontrivial left modules over R (i.e. left modules M with $\mathrm{Ann}_R M \neq R$); the class $K = \bigcup_{R \in V'} K(R)$ is then called a general class, if under the definition $r.m = (\varphi r).m$ (φ a homomorphism defined on R)

1. every M belonging to $K(\varphi R)$ becomes an element of K(R) ;
2. every M belonging to K(R) and satisfying $\ker \varphi \subset \mathrm{Ann}_R M$ becomes an element of $K(\varphi R)$;

and if

3. $\rho_K(R) = 0$ is equivalent to : $K(I) \neq \emptyset$ for all ideals $I \neq 0$ of R .

A similar result has been obtained by Skornjakow for Hoehnke-radicals of Ω-rings (1973 - [28]); it will turn out to be a special case of a general theorem . To formulate this theorem, we start with a variety V' of nonunary algebras and an arbitrary but fixed definition of left modules (between the weakest and the strongest one) .
For every left module M over $A \in V'$ and every x in M^μ (μ being the arity of the elements of A as operators on M), we define the A-*annihilator of* x by

$$\mathrm{Ann}_A x = \{(a,b) \in A \times A \;/\; a \circ x = b \circ x\}$$

and the A-*annihilator of* M as the set of all pairs (a,b) in $A \times A$ satisfying $w(a,a_1,\dots,a_n) \circ x = w(b,a_1,\dots,a_n) \circ x$

for all Ω'-words w and all arbitrary choices of the elements $a_1,\ldots,a_n$ in A and x in M^{μ} . The A-annihilator of M will be denoted by $\mathrm{Ann}_A M$; it is always a congruence of $\langle A,\Omega'\rangle$; in the case of a HALD-variety $\mathcal{V}'$, $\mathrm{Ann}_A M$ simplifies to

$$\mathrm{Ann}_A M = \{(a,b)\in A\times A \ / \ a\circ x = b\circ x \quad \forall\ x\in M\}$$

(see [19]) . $\mathrm{Ann}_A x$ is then always a congruence of $\langle A,\Omega\cup A\rangle$, i.e. of A considered as a left module over itself .

THEOREM (the author 1977 - [19]) :

For every variety $\mathcal{V}'$ of nonunary algebras every general class K of left modules over the algebras of $\mathcal{V}'$ defines a Hoehnke-radical ρ_K on $\mathcal{V}'$ by : $\rho_K(A) = \bigcap_{M\in K(A)} \mathrm{Ann}_A M$.

In this theorem, a *general class of left modules* over the algebras of $\mathcal{V}'$ is defined as above for associative rings (ker φ being now the congruence corresponding to φ), but without condition 3. which in the ring case only ensures that the obtained radicals are not only Hoehnke-radicals but satisfy the stronger conditions for Kuroš-Amitsur-radicals .

Notice that the original proof of this theorem was made for a definition of left modules using condition e., but remains valid for any other definition mentioned above . It follows that radicals may be built using different kinds of left modules . For example, the theorem of Andrunakievič-Rjabuhin may be formulated for the variety $\mathcal{V}'$ of all associative rings with identity either with arbitrary or with unitary left modules .

It might be of interest to the reader that in the case when every algebra of V' is embeddable in an algebra of V' having a "right-identity" respective to the operation "$\circ$" (i.e. an element $e \in A^{\mu}$ satisfying $a \circ e = a \quad \forall\, a \in A$), every Hoehnke-radical on V' is of the form ρ_K for a suitable general class K of left modules over the algebras of V' (see [19]) . This is the case for example for the varieties of all rings, of all semigroups, of all associative rings, of all zero-symmetric near-rings, of all lattices,...

2. *Subdirect decomposition of semisimple algebras :*

As we have seen above, for the Jacobson-radicals of rings and near-rings semisimplicity is equivalent to subdirect decomposability into primitive resp. i-primitive (i=0,1,2) (near)-rings . Andrunakievič and Rjabuhin proved a similar theorem for arbitrary Kuroš-Amitsur-radicals of associative rings (1964 - [1]) and Skornjakow showed that the analogous result holds for Hoehnke-radicals of Ω-rings (1973 - [28]) . The formulations of their theorems coincide with the following general result which can be proved similarly to Skornjakow's theorem using general classes of left modules as introduced above .

THEOREM (the author 1977 - [19]) :

Let ρ_K *be a Hoehnke-radical defined by a general class of left modules on a variety* V' *of nonunary algebras . Then an algebra* A *in* V' *is* ρ_K*-semisimple (i.e. fulfils* $\rho_K(A) = 0$*) iff it is isomorphic to a subdirect product of* K*-primitive algebras* . Here, an algebra A is called K-*primitive* if K(A) contains a *faithful* left module M (i.e. a left module M with $Ann_A M = 0 \in Con\ A$) .

3. *The interpolation properties :*

As for the Jacobson-radicals of rings and near-rings we consider interpolation properties formulated in two different ways :

a. *Density in an algebra of functions :*

In the interpolation properties of this type for rings and near-rings the bicentraliser plays an important role . Let us first introduce the bicentraliser of a system of operations in universal algebra (for more details see for example III.3 in Cohn's book [8]) .
Two *operations* ω, π (arities $n(\omega), n(\pi)$) on A are said to *commute on* A, if

case 1 : $n(\omega)$ and $n(\pi) > 0$

for every $n(\omega) \times n(\pi)$-matrix over A application of ω to each column and then of π to the results yields the same as application of π to each row and of ω to the results;

case 2 : $n(\omega) > 0$, $n(\pi) = 0$

$\omega(\pi,\ldots,\pi) = \pi$ in A ;

case 3 : $n(\omega) = n(\pi) = 0$

$\omega = \pi$ in A .

If C is a system of operations on A, the *centraliser* C_A^* *of* C *over* A is the system of all operations on A which commute with the operations of C . The *bicentraliser* C_A^{**} *of* C *over* A is defined by $C_A^{**} = (C_A^*)_A^*$.

Notice that a centraliser is always a *clone* of operations (i.e. a system of operations containing the projections and closed under composition) .

In order to obtain a density property of A over the left modules of K(A), we have to make a suitable choice of the general class K and hence of the corresponding radical ρ_K .

The fully retractable radical :

Let us recall that an algebra $\langle A,\Omega\rangle$ is said to be *fully retractable* if for every natural number n every subalgebra B of the direct power $\langle A,\Omega\rangle^n$ is an epimorphic image of $\langle A,\Omega\rangle^n$

THEOREM (the author 1977 - [2o]) :

Let V' be a variety of nonunary algebras and denote for every $A\in V'$ by F(A) *the class of all left modules* M *over* A *which are fully retractable (as algebras* $\langle M,\Omega\cup A\rangle$ *)* .

Then

1. $F = \bigcup_{A \in V'} F(A)$ *is a general class of left modules over the algebras of* V' *and defines therefore a Hoehnke-radical on* V' *which we call the fully retractable radical ;*

2. $M \in F(A)$ *implies that the clone of operations generated by* $\Omega \cup A$ *on* M *is dense in its bicentraliser* $(\Omega \cup A)_M^{**}$ *; this means that to every function* ϕ *(arity* $n(\phi)$ *) in the bicentraliser* $(\Omega \cup A)_M^{**}$ *and to every finite subset* $\{x_1, \ldots, x_s\}$ *of* $M^{n(\phi)}$ *there is an* $n(\phi)$*-ary function* ψ *, obtainable in a finite number of steps from the operations belonging to* $\Omega \cup A$ *using composition of functions and the projections, which satisfies* $\phi(x_i) = \psi(x_i)$ *for* $i = 1, \ldots, s$. *This applies in particular to the case* $n(\phi) = \mu$ *where* $\mu+1$ *is the arity of the operation " " in the system* Ω' .

3. *The bicentraliser* $(\Omega \cup A)_M^{**}$ *consists of exactly all those operations on* M *which for every natural number* n *- if defined by components on* M^n *- commute with all the endomorphisms of the left module* $\langle M, \Omega \cup A\rangle^n$ *over* A *, a fact which we shortly express by*

$$(\Omega \cup A)_M^{**} = (\bigcup_{n>0} \mathrm{End}_A M^n)_M^{*} .$$

There are of course several problems arising in the application of this theorem :

a. The bicentraliser $(\Omega \cup A)_M^{**}$ is in general smaller than in the classical cases of Jacobson-radicals for rings or near-rings, where it was equal to $(\mathrm{End}_A M)^*_M$; its computation might create some troubles as we can see from the above formula .

b. In general the system A of μ-ary operators acting on M is smaller than the system of all μ-ary operations in the clone generated by $\Omega \cup A$ on M ; hence, density of $\Omega \cup A$ does not in general imply density of A .

c. It might be difficult to determine the fully retractable left modules over an algebra A .

Since density is transitive, problem a. disappears in the case when $(\Omega \cup A)^*_M$ is itself dense in $(\mathrm{End}_A M)^*_M$ over M . A sufficient condition for this density is for example the property that in the variety $V = V(\Omega, \Lambda)$ - to which every left module over A belongs - each congruence of a direct product of two factors is decomposable into a product of two congruences of these factors . Conditions for such varieties were indicated by Fraser and Horn (197o - [1o]) .

To avoid problems b. and c., it is advantageous to use general classes which are contained in F (as in the case of J for rings and J_2 for near-rings) . Since the proofs of the following results make heavy use of conditions b. and c. for the variety V', we restrict ourselves from now on to HALD-varieties .

A radical of Jacobson-type for HALD-varieties :

If we start with a HALD-variety V', then obviously every simple and fully retractable left module M over an algebra A in V' is *cyclic* (i.e. of the form $M = A \circ x$ for some element x of M^μ called then a *generator* of M) . Combining results by the author from 1977 ([21], last theorem) and from 198? ([23], theorem 1 + corollary 2) we obtain the following

THEOREM :

For every $A \in V'$ *(HALD-variety) let* H(A) *denote the class of simple, fully retractable left modules over* A *for which the annihilators of every two generators commute (with respect to the composition of relations) . Then the class* $H = \bigcup_{A \in V'} H(A)$ *is a general class of left modules over the algebras of* V' *and defines therefore a Hoehnke-radical on* V' *. Moreover, if* M *belongs to* H(A)*,* A *is dense in the algebra of the* A*-regular* μ*-ary functions of* $(\Omega \cup A)_M^{**}$*, i.e. to every* A*-regular* μ*-ary function* f *in* $(\Omega \cup A)_M^{**}$ *and every finite subset* $\{x_1, \ldots, x_s\}$ *of* M^μ *there is some* a *in* A *satisfying* $a \circ x_i = f(x_i)$ *for* $i = 1, \ldots, s$.

For this theorem, A-regularity is defined as for $m\Omega$-near-rings by : a μ-ary function f on a left module M over A is called A-*regular* if for $x, y \in M^\mu$ the equality $a \circ x = a \circ y$ for all $a \in A$ implies $f(x) = f(y)$.

Notice that in the case when the left modules over the algebras of $\mathcal{V}'$ are Ω-groups, the annihilators of any two elements of these left modules commute; moreover, in this case every left module without nontrivial submodules is simple and fully retractable . It follows that in this case our theorem is valid especially for all the left modules without nontrivial submodules and we get the corresponding part of Polin's result for J_2-primitive zero-symmetric mΩ-near-rings as a special case .

b. *Interpolation on certain (independent) subsets :*

In the classical ring case (Jacobson-radical, irreducible left modules) we had interpolation on every finite subset of the considered left module which was linearly independent over the centraliser-skew-field . In the case of universal algebras this linear independence will have to be replaced by a more general concept of independence :

We use the definition given by Whitney (1935 - [34]) and extended by Kertész (196o - [16]) : a system S_d of finite subsets of a set P is called the system of *finite independent subsets* for an *abstract dependence relation* d on P, if it satisfies :

1. $\emptyset \in S_d$;
2. $X \in S_d$ and $Y \subset X$ imply $Y \in S_d$;
3. $X, Y \in S_d$ and $|Y| = |X| + 1$ imply the existence of an element y in $Y \setminus X$ such that $X \cup \{y\}$ belongs to S_d (Steinitz' lemma)

An *infinite subset* of P is called *independent* with respect to the relation d if every of its finite subsets belongs to S_d .

(For equivalent definitions see f.ex. Welsh's book [32].)

We are now able to formulate the following

THEOREM (the author 1978 - [22]) :

Let K be a general class of cyclic left modules over the algebras of a HALD-variety V' . Then $M \in K(A)$ *implies the existence of an abstract dependence relation* d *on* M^μ *such that, given a finite subset* $\{x_1,\ldots,x_s\}$ *of* M^μ *, there is to every choice of elements* $y_1,\ldots,y_s$ *in* M *an element* a *in* A *satisfying* $a \circ x_i = y_i$ *for* $i = 1,\ldots,s$ *iff the set* $\{x_1,\ldots,x_s\}$ *is independent with respect to* d . *This means that* A *interpolates every* μ*-ary function on* M *on a given finite subset of* M^μ *iff this subset is independent with respect to a suitable dependence relation* d *defined on* M^μ .

It now only remains to describe the above dependence relation (which coincides for irreducible left modules over associative rings with linear dependence over the centraliser-skew-field) .

THEOREM (the author 198? - [23]) :

Let M *be a cyclic left module over an algebra* A *belonging to a HALD-variety V' . Then a finite subset* $\{x_1,\ldots,x_s\}$

of M^{μ} *is independent with respect to the dependence relation d appearing in the preceding theorem iff it fulfills the following three conditions :*

1. $x_1,\ldots,x_s$ *are generators of* M ;
2. $\bigcap_{i=1}^{j-1} \mathrm{Ann}_A x_i \vee \mathrm{Ann}_A x_j = A\times A$ *in* Con $\langle A,\Omega\cup A\rangle$ *for* $j = 2,\ldots,s$;
3. $\mathrm{Ann}_A x_i$ *and* $\mathrm{Ann}_A x_j$ *commute (with respect to the composition of relations) for all pairs* (i,j) *of numbers in* $\{1,\ldots,s\}$ $(i\neq j)$.

Notice that conditions 2. and 3. are independent and that the above conditions simplify in special cases; for example the following holds :

If M *is a cyclic and simple left module over* A *, in the above theorem condition 2. may be replaced by condition*

2'. $\bigcap_{i=1}^{j-1} \mathrm{Ann}_A x_i \not\subseteq \mathrm{Ann}_A x_j$ *for* $j = 2,\ldots,s$.

But even in the case of cyclic and simple left modules condition 2. does not in general simplify to "$\mathrm{Ann}_A x_i \neq \mathrm{Ann}_A x_j$ for $i\neq j$ in $\{1,\ldots,s\}$" as in the case of zero-symmetric near-rings (see [23]) .

We originally started with one interpolation property of associative rings on their irreducible left modules formulated in two different ways (using that for linear interpolation the images of $x_1,\ldots,x_s$ may be arbitrarily prescribed iff $x_1,\ldots,x_s$ are linearly independent) . Now, for A belonging to a HALD-variety and a left module M in the class H(A) (introduced in the previous section 3.a.) both the last density theorem (section 3.a.) and the above characterisation of independence with respect to d are valid (notice that a finite subset of a left module M belonging to H(A) is independent with respect to d iff it satisfies condition 2'.) . Combining the two results we obtain that if M belongs to H(A), for interpolation by an A-regular μ-ary function in $(\Omega \cup A)_M^{**}$ the images of $x_1,\ldots,x_s \in M^\mu$ may be arbitrarily prescribed iff $\{x_1,\ldots,x_s\}$ is independent with respect to the above dependence relation d . Thus, for the general class H of left modules, the situation is the exact analogue of the classical one .

Another important remark to be made is that the above independence (and hence the interpolation property) depends essentially upon the position of the annihilators $\mathrm{Ann}_A x_i$ $(i=1,\ldots,s)$ in the congruence lattice Con $\langle A, \Omega \cup A\rangle$ of the algebra A considered as a left module over itself . In particular, if n is the maximal length of any chain descending in this lattice from $\mathrm{Ann}_A x$, there is no subset of M^μ consisting of more than n+1 elements and containing the element x which is independent with respect to d . Hence, if Con $\langle A, \Omega \cup A\rangle$ satisfies the descending chain condition, there are no infinite independent subsets with respect to d in the cyclic left modules over A . Specialising this remark to the classical case of irreducible left modules over associative rings, we obtain that every such left module is of finite dimension as a vector space over its centraliser-skew-field - a fact which is already contained in the Wedderburn-Artin theorem .

In the previous sections we have studied the connection "radical-interpolation property" starting from a radical . It seems quite natural to investigate

4. *The converse problem :*

Given an interpolation property, is there a "correspon-ding" radical ?

The natural way to attack this problem is to build

General classes defined by interpolation properties :

a. *The density case :*

It is easy to see that under the equality $a\,x = (\varphi a)\,x$ ($x \in M^{\mu}$, φ a homomorphism defined on A) used in the definition of general classes, density properties of A resp. $\Omega \cup A$ on left modules M over A carry over to the same density properties of φA resp. $\Omega \cup \varphi A$ on M considered as a left module over φA and vice-versa . This observation leads to the general result formulated below .

Let us first introduce some notations :

For an algebra A of the variety $\mathcal{V}'$ (of nonunary algebras as mentioned in section 0.) denote by

D(A) the class of all left modules over A such that $\Omega \cup A$ is dense in the bicentraliser $(\Omega \cup A)^{**}_M$;

D (A) the class of all left modules M over A such that the set of the μ-ary operations of the clone generated by $\Omega \cup A$ is dense in the set of the μ-ary operations of $(\Omega \cup A)^{**}_M$ - where $\mu+1$ is the arity of "$\circ$" in Ω' ;

D'(A) the class of all left modules M over A such that A is dense in the algebra of the A-regular μ-ary functions of $(\Omega \cup A)^{**}_M$.

THEOREM :

The unions $\mathcal{D}$, $\mathcal{D}_\mu$, $\mathcal{D}'$ of the respective classes taken over all the algebras of a variety $\mathcal{V}'$ of nonunary algebras are general classes of left modules over the algebras of $\mathcal{V}'$; moreover, if $\mathcal{K}$ is any of these general classes, every algebra A *of $\mathcal{V}'$ has the density property defining $\mathcal{K}$ on every left module belonging to* K(A) .

It is clear from the construction that $\mathcal{D}$ contains the class $\mathcal{F}$ of all fully retractable left modules and is itself contained in $\mathcal{D}_\mu$; if $\mathcal{V}'$ is a HALD-variety, $\mathcal{D}'$ contains the general class $\mathcal{H}$ introduced in section 3.b. ; for the corresponding radicals the converse inclusions hold . Unfortunately neither the exact position of these general classes $\mathcal{D}, \mathcal{D}_\mu, \mathcal{D}'$ in the lattice of general classes nor good characterisations of the corresponding radicals are known for general universal algebras .

b. *The independence case :*

As in section 3.b. let V' be a HALD-variety and let N be a system of nullary operations which may be defined on M for some left modules M over algebras of V' . For $A \in V'$ and an arbitrary natural number s, let us denote by

$I^{(s)}(A)$ the class of all left modules M over A for which the operations of N are defined, which satisfy $|M^{\mu} \setminus N| \geq s$, and on which A interpolates every μ-ary function defined on M on every (maximally) s-element subset of $M^{\mu} \setminus N$;

$J^{(s)}(A)$ the class of all left modules M over A for which the operations of N are defined and on which A interpolates every μ-ary function defined on M on every maximally s-element subset of $M^{\mu} \setminus N$;

I (A) the class of all left modules M over A for which the operations of N are defined and on which A interpolates every μ-ary function defined on M on every finite subset of $M^{\mu} \setminus N$;

P(A) the class of all simple left modules over A .

THEOREM (the author 198? - [23]) :

Let us denote by $I_N^{(s)}$, $J_N^{(s)}$, I_N , P *the unions of the respective classes from above, taken over all algebras of*

a HALD-variety V' *. Then* $I_N^{(s)}$ *,* $J_N^{(s)}$ *,* I_N *,* P *and arbitrary unions and intersections of these classes (even if taken for different systems* N*) are general classes of left modules over the algebras of* V' *, provided that for every* M *belonging to such a class, the operations from the considered systems* N *are defined on* M^{μ} *in the same way for* M *considered as left module over* A *or over* φA *- where* φ *denotes any homomorphism defined on* A*) . Moreover from our construction follows that, if* K *is any of the general classes obtained by intersection and union from classes of the form* $I_N^{(s)}$ *,* $J_N^{(s)}$ *,* I_N *,* $I_N^{(s)} \cap P$ *,* $J_N^{(s)} \cap P$ *,* $I_N \cap P$*, every algebra* A *of* V' *has the interpolation property corresponding to* K *on every left module* M *belonging to* K(A) *;*

In this context, the interpolation property corresponding to the intersection resp. union of general classes K_x $(x \in X)$ defined by interpolation properties is obtained in the following way :

$M \in \bigcap_{x \in X} K_x(A) \Leftrightarrow$ A has on M simultaneously all the interpolation properties required for $M \in K_x(A)$ $(x \in X)$;

$M \in \bigcup_{x \in X} K_x(A) \Leftrightarrow$ A has on M at least one of the interpolation properties required for $M \in K_x(A)$ $(x \in X)$.

Examples :

I_N is the union of the $J_N^{(s)}$ taken over all natural s ;

$M \in ((I_N^{(s)} \cap P) \cup J_N^{(t)})(A)$ for some pair (s,t) with t<s

the operations of N are defined on M^μ and A interpolates every μ-ary function on M on every maximally u-element subset of $M^\mu \setminus N$, where for u there are two possibilities : u=s if M is simple and satisfies $|M^\mu \setminus N| \geq s$,

u=t otherwise .

Let us notice that for $s \geq 2$, $I_\emptyset^{(s)}$ is always contained in P . As an example to show what may be the role of N let us mention the case of the variety V' of all semigroups ; in this case we may use N to prescribe a system of fixed points under the action of the semigroup A on the left modules belonging to the considered class K(A) .

A problem which arises when a radical is defined by an interpolation property as described above is the "inner" characterisation of this radical, i.e. a characterisation which does not involve left modules over the considered algebra but depends only upon the algebra and its congruences (as algebra or as left module over itself) . To conclude this survey, we indicate some results on

Inner characterisations of semigroup-radicals defined by interpolation properties :

We consider the variety V' of all semigroups and take for a system of at most one nullary operation; furthermore we suppose $s \geq 2$. Under these assumptions, the left modules to be considered are all cyclic and simple (= totally irreducible in Hoehnke's terminology from [11]) . Every such left module M over A is isomorphic to A/λ for some maximal modular left congruences λ of A (more precisely, these left congruences λ are the A-annihilators of the generators of M); we recall that a left congruence λ on a semigroup A is called *modular* if there is an element e_λ in A such that (ae_λ, a) belongs to λ for all a∈A . Since in a cyclic and simple left module over a semigroup every element - with the possible exception of one fixed point - is a generator (for more details see f.ex. [11]), we obtain under our assumptions $|N| \leq 1$ and $s \geq 2$:

$$\rho_K(A) = \cap\ (\lambda\ \ |A/\lambda \in K(A))$$

for every semigroup A and every general class K obtainable from classes $I_N^{(s)}$, $J_N^{(s)}$ and I_N by union and intersection .

In order to write the right hand side of this formula "module-free", let us introduce the following equivalence relation on the set of all maximal modular left congruences of a semigroup A (see [23]) :

λ~μ iff there is some a in A such that

a. $[a]_\lambda$ is not a left ideal in A ;

b. aλ = μ, where aλ is defined by :

$$(c,d) \in a\lambda \Leftrightarrow (ca,da) \in \lambda\ .$$

THEOREM (the author 198? - 23) :

Let K be anyone of the general classes $I_N^{(s)}$, $J_N^{(s)}$, I_N

$(|N|\leq 1, s\geq 2)$ of left modules over semigroups or a general class obtainable by union and intersection of such classes ; then the corresponding radical ρ_K on the variety of all semigroups may be obtained by intersecting all the maximal modular left congruences of the considered semigroup which belong to certain equivalence classes (depending of course upon the considered K) with respect to the above equivalence relation . (For the exact description of these equivalence classes see [23] *.)*

Let us notice that the above radicals ρ_K are in general large since they all contain Hoehnke's radical $\overline{\text{rad}}$ (see [11]) . In the case of commutative semigroups the above radicals are either trivial ($\rho_K(A) = A\times A$ for all commutative semigroups A for $K = I_{\emptyset}^{(s)}$, $J_{\emptyset}^{(s)}$, $I_{\emptyset}$, $I_{\{\nu\}}^{(s)}$ $(s\geq 2)$ or coincide with the upper radical determined by the class of all semilattices ($K = J_{\{\nu\}}^{(s)}$ $(s\geq 2)$, $I_{\{\nu\}}$) .

References

[1] V.A. Andrunakievic and Yu.M.Rjabuhin, Modules and radicals (russian), *Dokl.Akad.Nauk SSSR, Matem.* 156 (1964), 991-994 ; translation: *Soviet Math. Dokl.* 5 (1964), 728-732 .

[2] E. Artin, Zur Theorie der hyperkomplexen Zahlen, *Abh.Math.Sem.Univ.Hamburg* 5 (1927), 251-26o .

[3] J.A. Bahturin, *Lectures on Lie algebras*, Studien zur Algebra und ihre Anwendungen 4, Akademie-Verlag, Berlin, 1978 .

[4] G. Betsch, Ein Radikal für Fastringe, *Math.Z.* 78 (1962), 86-9o .

[5] G. Betsch, *Struktursätze für Fastringe*, Inaugural-Dissertation, Universität Tübingen, 1963 .

[6] G. Betsch, Primitive near-rings, *Math.Z.* 13o (1973), 351-361 .

[7] C. Chevalley, *Communication in the "Math.Club"*, Princeton, 1939 .

[8] P.M. Cohn, *Universal algebra*, Harper & Row, New York-Evanston-London, 1965 .

[9] R.M. Dicker, The substitutive law, *Proc.London Math.Soc.*, 3rd ser., 13 (1963), 493-51o .

[1o] G.A. Fraser and A. Horn, Congruence relations in direct products, *Proc.Amer.Math.Soc.* 26 (197o), 39o-394 .

[11] H.-J. Hoehnke, Structure of semigroups, *Canad.J. Math.* 18 (1966), 449-491 .

[12] H.-J. Hoehnke, Radikale in allgemeinen Algebren, *Math.Nachr.* 32 (1966), 347-383 .

[13] N. Jacobson, The radical and semi-simplicity for arbitrary rings, *Amer.J.Math.* 67 (1945), 3oo-32o .

[14] N. Jacobson, Structure theory of simple rings without finiteness assumptions, *Trans.Amer.Math. Soc.* 57 (1945), 228-245 .

[15] N. Jacobson, *Structure of rings*, Colloquium Publ. 37, Amer.Math.Soc., Providence, 1964 .

[16] A. Kertész, On independent sets of elements in algebra, *Acta Sci.Math.(Szeged)* 21 (196o), 26o-269 .

[17] K. Menger, Tri-operational algebra, *Reports Math. Coll.Notre Dame*, 2nd ser., 5-6 (1945), 3-1o .

[18] K. Menger, Superassociative systems and logical functors, *Math.Ann.* 157 (1964), 278-295 .

[19] R. Mlitz, Modules and radicals of universal algebras (russian), *Izv.Vysš.Učebn.Zaved.,Mat.* 1977, No.6, 77-85 ; translation: *Soviet Math.Izv.VUZ* 21 (1977), No.6, 61-67 .

[2o] R. Mlitz, Jacobson's density theorem in universal algebra, *Contributions to universal algebra (Proc. Conf.Szeged, 1975)*, 331-34o, Colloq.Math.Soc.J. Bolyai 17, North-Holland, Amsterdam, 1977 .

[21] R. Mlitz, A structure theorem in universal algebra, *Anais Acad.Brasil.Cienc.* 49 (1977), 359-363 .

[22] R. Mlitz, Cyclic radicals in universal algebra, *Algebra universalis* 8 (1978), 33-44 .

[23] R. Mlitz, Universal algebras, semigroups and interpolation, *Manuscript - submitted for publication* .

[24] T. Nalayama, Über einfache distributive Systeme unendlicher Ränge I, *Proc.Imp.Acad.Tokyo* 2o (1944), 61-66 .

[25] G. Pilz, *Near-rings*, Mathematics Studies 23, North-Holland Publ.Comp., Amsterdam, 1977 .

[26] S.V. Polin, primitive m-near-rings over multioperator groups (russian), *Mat.Sbornik* 84(126) (1971), 254-272 ; translation: *Math.USSR, Sb.* 13 (1971), 247-265 .

[27] D. Ramakotaiah, Structure of 1-primitive near-rings, *Math.Z.* 11o (1969), 15-26 .

[28] L.A. Skornjakow, On radicals of Ω-rings (russian), *Izbrannye Voprosy Algebry i Logiki* 283-299, Nauka, Novosibirsk, 1973 .

[29] L.A. Skornjakow, Stochastic acts and conmodules, *Semigroup forum* 25 (1982), 269-282 .

[3o] L.A. Skornjakow, Algebra of stochastic distributions (russian), *Izv.Vysš.Učebn.Zaved.,Mat.*, to appear .

[31] J.H.M. Wedderburn, On hypercomplex numbers, *Proc. London Math.Soc.*, 2nd. ser., 6 (19o8), 77-117 .

[32] D.J.A. Welsh, *Matroid theory*, Academic Press, London-New York-San Francisco, 1976 .

[33] H.I. Whitlock, Composition algebras, *Math.Ann.* 157 (1964), 167-178 .

[34] H. Whitney, On the abstract properties of linear dependence, *Amer.J.Math.* 57 (1935), 5o9-533 .

The author wishes to express his thanks to the referee for his valuable remarks which have been the basis for the revised version of the paper .

R. Mlitz
Institut für Angewandte und Numerische Mathematik
TU Wien ,
Gußhausstraße 27-29
A-1o4o W i e n
AUSTRIA .

COLLOQUIA MATHEMATICA SOCIETATIS JÁNOS BOLYAI

38. RADICAL THEORY, EGER (HUNGARY), 1982.

PURELY INSEPARABLE ALGEBRAS AND THEIR GENERALIZATIONS

B. J. OSŁOWSKI

By an algebra we mean an algebra with an identity element over a base field K. All tensor products are taken over K, unless otherwise stated. By A^o we shall denote the algebra opposite to A, by $[A,A]$ the ideal of A generated by all commutators $ab-ba$ $(a,b \in A)$. $Z(A)$ will denote the center of A. If $p: A \to B$ is a morphism of algebras then p^o denotes the corresponding morphism from A^o to B^o. $\gamma_A : A \otimes A^o \to A$, $m_A : A \otimes A \to A$ shall denote the multiplication morphisms, i.e. $\gamma_A\left(\sum a_i \otimes b_i\right) = \sum a_i b_i$, $m_A\left(\sum (a_i \otimes b_i)\right) = \sum a_i b_i$. B will denote the Baer radical, J - the Jacobson radical, K - the Koethe upper nil radical, K_{Ab} - the absolutely nil radical, L - the Levitzki radical and M - the Brown-McCoy radical. Information on radical theory and on the above mentioned radicals in particular can be found e.g. in [13], [1].

Let K be a field. It is known [3] that a field extension $L \supseteq K$ is separable as a field extension iff the algebra $L \otimes_K M$ is J-semisimple for any field extension $M \supseteq K$. This fact motivated the definition of a separable algebra given by Bourbaki. Another very important definition of a separable algebra is connected with Brauer groups of rings. Namely [5], an algebra A is said to be separable if the kernel of the multiplication morphism γ_A is a direct summand of the left $A \otimes A^o$-module $A \otimes A^o$. Taking into account this last notion Sweedler [12] defined, using the Jacobson radical, a certain kind of pure inseparability investigated later in [6].

Our aim is to study an analogous notion induced by other classical radicals. We shall show that the main result of the papers [12], [6] remains valid though we obtain several different kinds of inseparability. Our investigations lead also to some questions concerning radicals of tensor products.

Definition. Let S be a radical. A K-algebra A is called purely S-inseparable if $\ker \gamma_A \subseteq S(A \otimes A^o)$. A is said to be S-radicial if $\ker m_A \subseteq S(A \otimes A)$. The class of all purely S-inseparable algebras will be denoted by I_S .

The notion of purely J-inseparable algebra was introduced by Sweedler in [12] and those of $\mathcal{B}$-radicial and J-radicial algebras by Holleman in [6].

We shall state results only for purely S-inseparable algebras because the corresponding results are also evidently true for S-radicial algebras.

The definition of a separable algebra given in [3] was extended in [9] in the following way: An algebra A is called S-separable if for any field extension $L \supseteq K$ the algebra $A \otimes L$ is S-semisimple. The following proposition shows that our notion of pure S-inseparability is well adjusted to the above S-separability.

PROPOSITION 1. *Let* A *be a* K*-algebra and let* S *be any of the radicals* $\mathcal{B}$, L , K , K_{Ab} , J , M . *Then* A *is* S*-separable and* $A \in I_S$ *iff* $A=K$.

Proof. Since A is S-separable, putting $L=K$ in the definition of an S-separable algebra we get $S(A)=0$. Applying Th. 5.3 of [9] we see that $S(A \otimes A^o) = 0$. Since $A \in I_S$, we have $\ker \gamma_A \subseteq 0$ and thus $A=K$.

The next proposition gives some facts proved in [12].

PROPOSITION 2. (a) *If* $\pi : A \to B$ *is a surjective*

morphism of algebras then $\pi\otimes\pi^o(\ker\gamma_A) = \ker\gamma_B$.

(b) *If* A *is a purely* S*-inseparable* K*-algebra and* L *is a field extension of* K *contained in* $Z(A)$ *then* A *is a purely* S*-inseparable* L*-algebra.*

(c) *If* $\pi : A \to B$ *is a surjective morphism of algebras and* $A \in I_S$ *then* $B \in I_S$.

For the proof of the above proposition see those of Lemma 1.e, Prop. 6.b, 6.e of [12].

LEMMA 3 (cf. [6], [12]). *Let* $\pi : A \to B$ *be a surjective morphism of algebras such that* $\ker\pi\otimes A^o + A\otimes\ker\pi^o \subseteq$ $\subseteq S(A\otimes A^o)$. *Then* $B\in I_S$ *implies* $A\in I_S$.

Proof. Since $B\in I_S$, by Prop. 2.a we get that $\pi\otimes\pi^o(\ker\gamma_A) = \ker\gamma_B \subseteq S(B\otimes B^o)$. Since $\ker(\pi\otimes\pi^o) =$ $= \ker\pi\otimes A^o + A\otimes\ker\pi^o \subseteq S(A\otimes A^o)$ and $\pi\otimes\pi^o$ is surjective, we have $\pi\otimes\pi^o(S(A\otimes A^o)) = S(B\otimes B^o)$. Hence it follows that $\ker\gamma_A \subseteq S(A\otimes A^o) + \ker(\pi\otimes\pi^o) \subseteq S(A\otimes A^o)$, which ends the proof.

The next lemma can be found in [9], [12].

LEMMA 4. *If* A , B *are* K*-algebras and* $x\in A$, $y\in B$ *are such that* $x\otimes 1 + 1\otimes y$ *is invertible in* $A\otimes B$ *then either* x *or* y *is algebraic over* K .

PROPOSITION 5. *If* B *is a subalgebra of an algebra* $A\in I_S$ *and* S *is a radical hereditary on subalgebras then* $B\in I_S$.

Proof. Let $i : B \to A$ denote the inclusion homomorphism. From the commutativity of the diagram

$$\begin{array}{ccc} B\otimes B^{o} & \longrightarrow & B \\ {\scriptstyle i\otimes i^{o}}\downarrow & & \downarrow{\scriptstyle i} \\ A\otimes A^{o} & \longrightarrow & A \end{array}$$

it follows that $\ker \gamma_B = \ker \gamma_A \cap B\otimes B^{o} \subseteq S(A\otimes A^{o}) \cap B\otimes B^{o} \subseteq$ $\subseteq S(B\otimes B^{o})$.

Henceforth we shall assume that S is a supernilpotent, hereditary (on ideals) radical such that $S\leq M$ and that for any K-algebra A , $S(A\otimes L) = A\otimes L$ implies $S(A) = A$, whenever L is a field extension of K which is purely inseparable as a field extension.

PROPOSITION 6. *Let A be a field extension of K . Then the following conditions are equivalent:*

(1) $A \in I_S$.

(2) *A is a purely inseparable algebraic extension of K .*

Proof. Suppose that $A \in I_S$. Since A is commutative, $\gamma_A : A\otimes A^{o} \to A$ is a morphism of algebras. Hence $\ker \gamma_A$ is an ideal of $A\otimes A^{o}$ and by the isomorphism theorem $A\otimes A^{o}/\ker \gamma_A \simeq A$. Since $\ker \gamma_A \subseteq S(A\otimes A^{o})$ and $S\leq M$, $A\otimes A^{o}$ is a local ring with maximal ideal $\ker \gamma_A =$

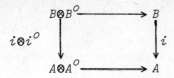

$= S(A\otimes A^o)$. If there were an element $x\in A$ transcendental over K then $x\otimes 1 + 1\otimes(x+1) \notin \ker \gamma_A$ would be, by Lemma 4, a non-invertible element of $A\otimes A^o$. Therefore A is an algebraic extension of K , and thus $A\otimes A$ is also an algebraic algebra as A is commutative. Since $A\otimes A^o$ is commutative, $\ker \gamma_A = S(A\otimes A^o) \subseteq M(A\otimes A^o) = J(A\otimes A^o)$. Hence $\ker \gamma_A$ is a nil ideal of $A\otimes A^o$ and by [7] Th. 21.2 chap. IV we get that A is purely inseparable over K . Conversely, if A is a purely inseparable algebraic extension of K then for any $x\in A$ there exists an n such that $(x\otimes 1 - 1\otimes x)^{p^n} = 0$, where $p = \text{char } K$. Therefore $\ker \gamma_A = B(A\otimes A^o) = S(A\otimes A^o)$.

PROPOSITION 7 (cf. Prop. 2.9 [6]). *Let* $A\in I_S$. *Then*

(a) $[A,A]\otimes A^o + A\otimes[A,A]^o \subseteq S(A\otimes A^o)$,

(b) $[A,A] \subseteq S(A)$,

(c) $S(A)^o = S(A^o)$.

Proof. From the equality $(y\otimes 1)(x\otimes 1 - 1\otimes x) - (x\otimes 1 - 1\otimes x)(y\otimes 1) = (yx-xy)\otimes 1$ it follows that $[A,A]\otimes A^o$ is contained in the ideal generated by $\ker \gamma_A$ and hence it is contained in $S(A\otimes A^o)$. Since S is hereditary on ideals, $[A,A]\otimes A^o$ is S-radical. Now if J is any maximal ideal of A^o containing $[A,A]^o$ then the field $L = A^o/J^o = A/J$ belongs to I_S by Prop. 2c.

From Prop. 6 it follows that L is a purely inseparable algebraic extension of K. Since $[A,A] \otimes_K L$ is an S-radical algebra, from the assumption on S we get that $S([A,A]) = [A,A]$. Similarly we obtain that $A \otimes [A,A]^o \subseteq \subseteq S(A \otimes A^o)$ and $S[A,A]^o = [A,A]^o$. Since $A/[A,A] = = A^o/[A,A]^o$ and $[A,A] \subseteq S(A)$, $[A,A]^o \subseteq S(A^o)$, we have $S(A)/[A,A]) = S(A/[A,A]) = S(A^o)/[A,A]^o$. Therefore $S(A^o) = S(A)^o$.

PROPOSITION 8. *If* $A \in I_S$ *then* $S(A) \otimes A^o + A \otimes S(A)^o \subseteq \subseteq S(A \otimes A^o)$.

Proof. Let $\bar{A} = A/[A,A]$ and $f = \gamma_{\bar{A}}|B$ where $B = S(A)/[A,A] \otimes \bar{A} \triangleleft \bar{A} \otimes \bar{A}^o$. Since $\bar{A}$ is a commutative algebra, f is a morphism of algebras. Now Prop. 2c says that $\bar{A} \in I_S$ and therefore $\ker f = \ker \gamma_{\bar{A}} \cap B$ is an S-radical ideal of B. From this and from the fact that $f(B) = S(A)/[A,A] = S(\bar{A})$ we get $B \in S$. But $B \simeq S(A) \otimes A^o/([A,A] \otimes A^o + S(A) \otimes [A,A]^o)$, hence by Prop. 7a we obtain $S(A) \otimes A^o \subseteq S(A \otimes A^o)$. Analogously one can show that $A \otimes S(A)^o \subseteq S(A \otimes A^o)$.

Now we have the following

PROPOSITION 9. *If* $A \in I_S$ *then* A *has no nontrivial central idempotents and therefore* A *has exactly one maximal ideal.*

Proof. Suppose that $e \in A$ is a central idempotent. Then $e \otimes (1-e) \in \ker \gamma_A \subseteq S(A \otimes A^o)$ is also a central idempotent. Since $S \leq M$, the S-radical rings contain no non-trivial central idempotents and thus $e \otimes (1-e) = 0$ and e is a trivial idempotent. Suppose now that M_1, M_2 are maximal ideals of A, $M_1 \neq M_2$. Then Prop. 2c implies that $A/M_1 \cap M_2$ is a purely S-inseparable algebra and by the Chinese Remainder Theorem $A/M_1 \cap M_2 \simeq A/M_1 \oplus A/M_2$, which contradicts the first part of the proposition.

Using Lemma 4 we get immediately the following result.

PROPOSITION 10 (Th. 10, [12]). *If* $A \in I_S$ *and* $S \leq J$ *then* A *is algebraic over* K.

The next proposition corresponds to Th. 11c of [12] and in fact could have been proved by using that theorem.

PROPOSITION 11. *If* $A \in I_S$ *then* $A/S(A)$ *is a purely inseparable algebraic field extension of* K.

Proof. By Prop. 6 and Prop. 2c it suffices to prove that $\bar{A} = A/S(A)$ is a field. By Prop. 7b, $\bar{A}$ is a commutative algebra, and by Prop. 7 it has exactly one maximal ideal M. Since $\bar{A}$ is commutative, $M = M(\bar{A}) = J(\bar{A})$. By Prop. 2c $\bar{A} \in I_S$, hence $\ker \gamma_{\bar{A}} \subseteq M(\bar{A} \otimes \bar{A}^o) = J(\bar{A} \otimes \bar{A}^o)$. Therefore $1 \otimes 1 + x \otimes 1 - 1 \otimes x = 1 \otimes (1-x) + x \otimes 1$ is invertible in $\bar{A}\otimes\bar{A}^o$, and using Lemma 4 we get that $\bar{A}$ is algebraic over K . Hence $J(\bar{A}) = B(\bar{A}) = 0$ and thus 0 is the maximal ideal of $A/S(A)$, which ends the proof.

PROPOSITION 12. *If* $A \in I_S$ *then* $A\otimes A^o/S(A\otimes A^o) \simeq A/S(A)$ and $\gamma_A^{-1}(S(A)) = S(A\otimes A^o)$.

Proof. Let $\bar{A} = A/S(A)$ and let $p: A \to \bar{A}$ be the natural homomorphism. Since by Prop. 7b $\bar{A}$ is commutative, $\gamma_{\bar{A}} : \bar{A}\otimes\bar{A}^o \to \bar{A}$ is a morphism of algebras. Let $f = \gamma_{\bar{A}} \circ (p\otimes p^o) : A\otimes A^o \to \bar{A}$. Since p is surjective, by Prop. 2a and by Prop. 8 it follows that $\ker f =$ $= \ker \gamma_A + S(A)\otimes A^o + A\otimes S(A)^o \subseteq S(A\otimes A^o)$. Therefore $f(S(A\otimes A^o)) = S(\bar{A}) = 0$, and f induces an isomorphism $\bar{f} : A\otimes A^o/S(A\otimes A^o) \xrightarrow{\sim} \bar{A}$. The last part of the statement follows from the commutativity of the diagram

$$\begin{array}{ccc} A\otimes A^o & \xrightarrow{\gamma_A} & A \\ \downarrow & & \downarrow p \\ A\otimes A^o/S(A\otimes A^o) & \xrightarrow[\bar{f}]{} & \bar{A} \end{array}$$

COROLLARY. $A \in I_S$ *iff* $\gamma_A^{-1}(S(A)) \subseteq S(A\otimes A^o)$.

PROPOSITION 13. *If* $A\otimes A^o/S(A\otimes A^o)$ *is a division algebra then* A *has a non-zero commutative homomorphic image.*

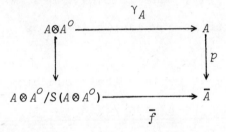

Proof. Let M be a maximal ideal of A. Since $S \leq M$, $S(A \otimes A^o)$ is the unique maximal ideal of $A \otimes A^o$. Hence $M \otimes A^o + A \otimes M^o \subseteq S(A \otimes A^o)$ and thus $\bar{A} \otimes \bar{A}^o / S(\bar{A} \otimes \bar{A}^o)$ is a division algebra, where $\bar{A}$ denotes the algebra A/M. Thus we may assume that A is a simple algebra. Since $A \otimes_{Z(A)} A^o$ is a homomorphic image of $A \otimes_K A^o$, $A \otimes_{Z(A)} A^o / S(A \otimes_{Z(A)} A^o)$ is a division algebra. Since A is a central simple algebra over $Z(A)$, $A \otimes_{Z(A)} A^o$ itself is a division algebra. Using Lemma 4 we get that A is algebraic over $Z(A)$. Suppose now that there exists an element $x \in A \setminus Z(A)$. Then $B =$ $= Z(A)(X) \otimes_{Z(A)} Z(A)(x)$, being a finite dimensional subalgebra of a division algebra, is also a division algebra. Let f be the minimal polynomial of x over $Z(A)$, then $B \simeq Z(A)(x)[t]/f(t)$ has zero divisors, which is a contradiction. Therefore any simple image of A is commutative.

Remark. One cannot replace the assumption on A by the weaker assumption that $A \otimes A^o / S(A \otimes A^o)$ is a simple algebra. In fact, putting $A = M_n(K)$ - the algebra of $n \times n$ matrices over K $(n>1)$ - we get that $A \otimes A^o / S(A \otimes A^o) \simeq$ $\simeq M_{n^2}(K)$ is a simple K-algebra and A has no non-zero commutative homomorphic images.

PROPOSITION 14. *If* $A\otimes A^o/S(A\times A^o)$ *is a simple algebra and* A *has a non-zero commutative homomorphic image then* $A\in I_S$.

Proof. It is clear that under the hypothesis there exists a homomorphism η of A onto a non-zero S-semi-simple commutative algebra B . Since B is commutative, $\eta\circ\gamma_A : A\otimes A^o \to B$ is a surjective morphism of algebras, and $S(B)=0$ implies $S(A\otimes A^o)\subseteq \ker(\eta\circ\gamma_A)$. Hence $\eta\circ\gamma_A$ induces a surjective homomorphism $\bar{\gamma}_A : A\otimes A^o/S(A\otimes A^o) \to B$. Since $B\neq 0$ and $A\otimes A^o/S(A\otimes A^o)$ is simple, $\bar{\gamma}_A$ is an isomorphism. Now from the commutativity of the diagram

$$\begin{array}{ccc} A\otimes A^o & \xrightarrow{\gamma_A} & A \\ \downarrow & & \downarrow \\ A\otimes A^o/S(A\otimes A^o) & \xrightarrow{\sim} & B \end{array}$$

it follows that $S(A\otimes A^o) = \gamma_A^{-1}(\ker \eta) \supseteq \ker \gamma_A$.

THEOREM 15 (cf. [6], [12]). *Let* A *be a* K*-algebra. Then the following conditions are equivalent:*

a) $A\in I_S$.

b) $A\otimes A^o/S(A\otimes A^o)$ *and* $A/S(A)$ *are isomorphic purely inseparable field extensions of* K .

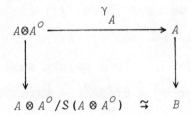

c) $A\otimes A^{o}/S(A\otimes A^{o})$ *is a simple* K*-algebra and* $A/S(A)$ *is a commutative* K*-algebra.*

d) $A\otimes A^{o}/S(A\otimes A^{o})$ *is a simple* K*-algebra and* A *has a non-zero commutative homomorphic image.*

e) $S(A)\otimes A^{o}+A\otimes S(A)^{o}\subseteq S(A\otimes A^{o})$ *and* $A/S(A)$ *is a purely inseparable field extension of* K.

f) $A\otimes A^{o}/S(A\otimes A^{o})$ *is a field extension of* K.

g) $A\otimes A^{o}/S(A\otimes A^{o})$ *is a division algebra.*

Proof. a) ⇒ b) follows from Prop. 11 and 12. The implications b) ⇒ c) ⇒ d) and f) ⇒ g) are obvious. d) ⇒ a) follows by Prop. 14, and a) ⇒ e) follows from Prop. 8 and 11.

e) ⇒ f). Let $\bar{A}=A/S(A)$ and $I=S(A)\otimes A^{o}+A\otimes S(A)^{o}$. Since $I\subseteq S(A\otimes A^{o})$, we have $S(A\otimes A^{o}/I)=S(A\otimes A^{o})/I$. Therefore $A\otimes A^{o}/S(A\otimes A^{o})\simeq(A\otimes A^{o}/I)/S(A\otimes A^{o}/I)\simeq$ $\simeq\bar{A}\otimes\bar{A}^{o}/S(\bar{A}\times\bar{A}^{o})$. Since $\bar{A}$ is a purely inseparable field extension of K, $\ker\gamma_{\bar{A}}=B(\bar{A}\otimes\bar{A}^{o})=M(\bar{A}\otimes\bar{A}^{o})$, and therefore $S(\bar{A}\otimes\bar{A}^{o})=\ker\gamma_{\bar{A}}$. Thus $A\otimes A^{o}/S(A\otimes A^{o})\simeq\bar{A}\otimes\bar{A}^{o}/\ker\gamma_{\bar{A}}\simeq$ $\simeq\bar{A}$ is a purely inseparable field extension of K.

The implication g) ⇒ d) holds by Prop. 13.

COROLLARY 1. *If* $S=B,L$ *then an algebra* A *is purely* S*-inseparable iff* $A/S(A)$ *is a purely inseparable field extension of* K.

Proof. The result follows by Th. 15b.

COROLLARY 2. *If* $S = K_{Ab}, J, M$ *then an algebra* A *is purely* S*-inseparable iff* $A/S(A)$ *is a purely inseparable field extension of* K *and* $S(A)\otimes S(A)^o$ *is an* S*-radical algebra.*

Proof. Suppose that our algebra A is such that $A/S(A)$ is a purely inseparable field extension of K and $S(A)\otimes S(A^o)$ is S-radical. Then $S(A)\otimes A^o/S(A)\otimes S(A^o) \simeq$ $\simeq S(A)\otimes A/S(A)$ is S-radical, in case $S = K_{Ab}$ by the definition of K_{Ab}, in case $S=J$ by [2], in case $S=M$ by [10]. Thus $S(A)\otimes A^o \subseteq S(A\otimes A^o)$ and hence A is purely S-inseparable by Th. 15e.

It is clear that $I_B \subseteq I_L \subseteq I_{K_{Ab}} \subseteq I_K \subseteq I_J \subseteq I_M$.

In [6] it was erronously asserted that $I_B = I_J$. Now we shall show that $I_B \neq I_L$.

Example 1. Let B any non-zero algebra such that $B(B) = 0$ and $L(B) = B$ (see e.g. [13]) and let A be the standard extension of B by K. Then by Cor. 1 of Th. 15 it follows that $A \in I_L \setminus I_B$.

Example 2. Using the methods of [4] chap. 5 it is easy to see that over any field K there exists a central non-algebraic division algebra. Let D be such an algebra. Let S be the algebra of all infinite matrices over D with a finite number of non-zero entries. Then

S is a central simple K-algebra and thus $S \otimes_K S^o$ is a simple K-algebra without identity (since $Z(S \otimes S^o) = Z(S) \otimes Z(S^o) = 0$). Let A be the standard extension of S by K. Then, by Cor. 2 to Th. 15, it follows that $A \in I_M$. On the other hand, A is not algebraic over K and thus one cannot drop the assumption $S \leq J$ in Prop. 10. Moreover, $A \notin I_J$ and thus $I_J \neq I_M$.

Questions:

1) The author does not know whether $I_L \neq I_K$, $I_K \neq I_J$.

2) It is clear that for $S = B, L$ pure S-inseparability coincides with S-radiciality and the question arises whether these two concepts coincide for other radicals.

3) Sweedler has asked whether it is true that if $K(A) = J(A)$ and $A/J(A)$ is a purely inseparable field extension of K then $A \in I_J$. Had it been so, then for any nil algebra B, $B \otimes B^o$ would be a J-radical algebra and there wouldn't exist a simple nil ring. On the other hand, A. Z. Ananin has announced that over a field of zero characteristic there exist nil algebras the tensor product of which is not J-radical. This still unpublished result would imply the negative answer to the above question.

4) Find a necessary and sufficient condition for an algebra A to be such that

a) $J(A\otimes A^{o}) = A\otimes A^{o}$,

b) $K(A\otimes A^{o}) = A\otimes A^{o}$.

In particular: are these conditions equivalent? One may expect one of the conditions $L(A)=A$ or $K_{Ab}(A)=A$ or $K(A)=A$ to be the answer.

5) Sweedler has asked if every subalgebra of a purely J-inseparable algebra is purely J-inseparable. Prop. 5 gives the positive answer to the corresponding question for purely S-inseparable algebras, for a radical hereditary on subalgebras. In [6] it was announced that the answer to Sweedler's question is positive but the proof given there relies on the false equality $I_B = I_J$. Therefore this question is still open. On the other hand, since in PI-algebras M coincides with J and K coincides with B ([7], pp. 17, 55), any PI-algebra belongs to I_J iff it belongs to I_B and any locally PI-algebra belongs to I_J iff it belongs to I_L , hence the answer is positive in these cases.

REFERENCES

[1] S. A. Amitsur, Nil radicals. Historical notes and some new results, *Rings, modules and radicals*, pp. 47-65, *Colloq. Math. Soc. J. Bolyai*, *6*, North-Holland, Amsterdam, 1973.

[2] S. A. Amitsur, The radical of field extensions, *Bull. Res. Council Israel, Sect. 7F* (1957-1958), 1-10.

[3] N. Bourbaki, *Algèbre, Ch. 8*, Hermann, Paris, 1958.

[4] P. M. Cohn, *Skew field constructions*, LMS Lecture Notes 27, London, 1977.

[5] F. DeMeyer and E. Ingraham, *Separable algebras over commutative rings*, Lecture Notes in Math. 81, Springer, Berlin - New York, 1971.

[6] A. Holleman, Inseparable Algebras, *J. of Algebra* 47 (1977), 415-429.

[7] N. Jacobson, *Lectures in abstract algebra, Vol. III*, v. Nostrand, Princeton, 1964.

[8] N. Jacobson, *PI-algebras, an Introduction*, Lecture Notes in Math. 441, Springer, Berlin, 1975.

[9] J. Krempa, On semisimplicity of tensor products, *Ring theory, Proceedings of the 1978 Antwerp Conference*, pp. 105-122.

[10] J. Krempa and J. Okniński, Semilogical, semiperfect and perfect tensor products, *Bull. Polon. Acad. Sci., Ser. Sci. Math.* 28 (1980), 249-256.

[11] E. R. Puczylowski, On radicals of tensor products, *Bull. Acad. Polon. Sci. Ser. Sci. Math.* 28 (1980), pp. 243-247.

[12] M. E. Sweedler, Purely inseparable algebras, *J. of Algebra* 35 (1975), pp. 342-355.

[13] F. A. Szász, *Radicals of rings*, Akadémiai Kiadó, Budapest, 1981.

B. J. Osłowski
Wyższa Szkoła Rolniczo-Pedagogiczna
Zakład Matematyki
08-110 Siedlce
ul. Nowotki 19/21
POLAND

COLLOQUIA MATHEMATICA SOCIETATIS JÁNOS BOLYAI
38. RADICAL THEORY, EGER (HUNGARY), 1982.

RADICALS OF JORDAN ALGEBRAS OF DEGREE 3

H. P. PETERSSON and M. L. RACINE*

1. INTRODUCTION.

Simple exceptional Jordan division algebras are finite dimensional [6,7] and therefore they are generically algebraic of degree 3 over a field Φ. In this paper, results concerning radicals of Jordan algebras of degree 3 over a unital commutative associative ring are obtained. These will be used in subsequent papers to prove some of the results concerning exceptional algebras announced in [5].

First we recall the axiomatization of Jordan algebras of degree 3 given in [3]. Let Φ be a unital commutative associative ring, J a Φ-module endowed with a cubic form N with values in Φ, a quadratic mapping $x \to x^{\#}$ in J and a distinguished element $1 \in J$ satisfying

* The research of the second author is supported in part by an NSERC grant.

(1) $x^{\#\#} = N(x)x$

(2) $N(1) = 1$

(3) $T(x^{\#},y) = \Delta_x^y N$, the directional derivative of N in the direction y, evaluated at x; $T(x,y) = -\Delta_1^x \Delta^y \log N$.

(4) $1^{\#} = 1$

(5) $1 \times y = T(y)1 - y$, where $T(y) = T(y,1)$ and $x \times y = (x+y)^{\#} - x^{\#} - y^{\#}$.

If these hold for every scalar extension of Φ then McCrimmon [3] has shown that

(6) $yU_x = T(x,y)x - x^{\#} \times y$

defines a quadratic Jordan structure (J,U,1) which we denote J(N,#,1). We refer to T as the associated *trace form*, to # as the *adjoint*, to 1 as the *base point* and to N as the norm. Every element of J satisfies

(7) $x^3 - T(x)x^2 + S(x)x - N(x)1 = 0$, where $S(x) = T(x^{\#})$;

(8) $x^{\#} = x^2 - T(x)x + S(x)1$.

The following identities are proved in [3] and will be used in what follows.

(9) $x^{\#} \times (x \times y) = N(x)y + T(x^{\#},y)x$.

(10) $x^{\#} \times y^{\#} + (x \times y)^{\#} = T(x^{\#},y)y + T(y^{\#},x)x$.

(11) $T(x \times y) = T(x)T(y) - T(x,y)$.

(12) $T(x,y \times z) = T(x \times y,z)$.

(13) $T(x,x^{\#}) = 3N(x)$.

(14) $(yU_x)^{\#} = y^{\#}U_{x^{\#}}$.

(15) $N(yU_x)yU_x = N(x)^2N(y)yU_x$.

(16) $x^{\#}U_x = N(x)x$.

(17) $(x^{\#})^2U_x = N(x)^2 1$.

(18) $(x \times y)U_x = T(x^{\#},y)x - N(x)y$.

Before we consider radicals of $J(N,\#,1)$, we must first relate the module and Jordan structures.

2. THE ANNIHILATOR.

Let Φ be a unital commutative associative ring of scalars. We write $\mathrm{Nil}\Phi$ for the nil radical and $\Phi^{\times}$ for the group of units of Φ . We also fix a cubic form $(N,\#,1)$ with adjoint and base point over Φ, put $J = J(N,\#,1)$ and denote by $J^{\times}$ the set of invertible elements in J. For our subsequent investigations it will be important to have some control over the *annihilator* of J, i.e., over the ideal

$$\mathrm{Ann}(J) = \{\alpha \in \Phi;\ \alpha J = 0\}$$

of Φ.

We also need the following result, which is mainly due to McCrimmon [3].

THEOREM 1. a) *An element* $v \in J$ *is invertible in* J *if and only if* $N(v) \in \Phi$ *is invertible in* Φ. *If this is so,* $v^{-1} = N(v)^{-1}v^{\#}$.

b) *For* $v \in J^{\times}$, *the triple* $(N^{(v)}, \#^{(v)}, 1^{(v)})$ *defined by*

$$N^{(v)}(w) = N(v^{-1})^{-1}N(w),\ w^{\#(v)} = N(v)w^{\#}U_v^{-1},\ 1^{(v)} = v^{-1}$$

is a cubic form with adjoint and base point over Φ, *and* $J^{(v)} = J(N^{(v)}, \#^{(v)}, 1^{(v)})$ *is the isotope of* J *relative to* v. For reasons which will soon become apparent we wish to indicate a proof for part a) of Theorem 1.

We shall see later (cf. Theorem 10) that the relation $N(v^{-1}) = N(v)^{-1}$ does not hold in general, so, contrary to what has been stated in [3, Theorem 2], $(N'^{(v)}, \#^{(v)}, 1^{(v)})$, with $N'^{(v)}(w) = N(v)N(w)$ is not in general a cubic form with adjoint and base point since it violates axiom (1.2). First we establish

LEMMA 2. a) *For each* $\alpha \in \Phi$ *satisfying* $\alpha 1 = 0$, *in particular, for each* $\alpha \in \mathrm{Ann}(J)$, *we have* $\alpha^3 = 3\alpha = 0$. *In particular,* $\mathrm{Ann}(J)$ *is a nil ideal in* Φ .

b) *We have* $\mathrm{Ann}(J) = 0$ *provided* Φ *satisfies one of the following conditions*

(i) Φ *is reduced, i.e., contains no nilpotents other than zero.*

(ii) Φ *has no 3-torsion.*

(iii) *The* trace ideal of J, *i.e., the* Φ*-module spanned by* $T(J \times J)$, *contains at least one regular element of* Φ.

Proof. a) $\alpha^3 = N(\alpha 1) = 0$, $3\alpha = T(\alpha 1) = 0$, by (1.2), (1.4) and (1.13).

b) Because of a), the conclusion is clear in cases (i), (ii). Now let γ be a regular element of Φ contained in the trace ideal of J. Then $\gamma = \sum T(v_i, w_i)$, for some $v_i, w_i \in J$, and, given $\alpha \in \mathrm{Ann}(J)$, we obtain
$\gamma\alpha = \sum T(\alpha v_i, w_i) = 0$, hence $\alpha = 0$. □

We are now ready to establish part a) of Theorem 1. Assuming first $N(v) \in \Phi^{\times}$ and arguing as in [3], we put $w = N(v)^{-1} v^{\#}$ and conclude $wU_v = v$, $w^2 U_v = 1$ from (1.16), (1.17). Hence v is invertible in J with inverse w. Conversely, for $v \in J^{\times}$ and $w = v^{-1}$, we again proceed as in [3] and deduce.

$$1 = N(1)1 = N(w^2U_v)w^2U_v$$

$$= N(v)^2N(w^2)w^2U_v = N(v)^2N(w^2)1, \quad \text{by (1.15)}.$$

The original proof of [3] *could* be construed as saying that this implies $N(v)^2N(w^2) = 1$. This is, however, *not* the case as we shall see later. Instead, Lemma 2 a) only yields $N(v)^2N(w^2) \equiv 1 \bmod \mathrm{Nil}\,\Phi$, which suffices for our purpose; we conclude $N(v)^2N(w^2) \in \Phi^\times$, hence $N(v) \in \Phi^\times$, as desired. □

PROPOSITION 3. *Let* $v \in J$ *be invertible. Then*

$$\mathrm{Ann}(J) = \{\alpha \in \Phi : \alpha v = 0\}$$

Proof. For $\alpha \in \Phi$ satisfying $\alpha v = 0$ and $w \in J$, we obtain $2\alpha w = v^{-1}U_{\alpha v, w} = 0$. By Theorem 1 a) and (1.13) we also have $3\alpha = T(\alpha v, v^{-1}) = 0$, hence $\alpha w = 3\alpha w - 2\alpha w = 0$. □

Using Proposition 3, we can now improve (1.15) considerably.

PROPOSITION 4. a) *For* $v, w, x \in J$ *we have*

$$N(v^{\#})w = N(v)^2 w, \quad N(wU_v)x = N(v)^2N(w)x.$$

b) *For* $v \in J$, $w \in J$ *we have* $N(v^{-1})w = N(v)^{-1}w$.

Proof. a) We only treat the first formula since the second will follow analogously. A standard argument, sometimes referred to as "Koecher's Principle" (cf McCrimmon [4], Jacobson [2]), allows us to assume that v is invertible. Then $N(v^{\#})v^{\#} = v^{\#\#\#} = (N(v)v)^{\#} = N(v)^2 v^{\#}$, so $\alpha = N(v^{\#}) - N(v)^2$ annihilates $v^{\#}$. But $v^{\#} = N(v)v^{-1}$ is invertible. Hence, by Proposition 3, $\alpha \in \mathrm{Ann}(J)$, which is precisely what we had to prove.

b) By a) and Theorem 1 a), we obtain

$$N(v^{-1})w = N(N(v)^{-1}v^{\#})w = N(v)^{-3}N(v^{\#})w$$
$$= N(v^{-3})N(v)^2 w = N(v)^{-1}w. \qquad \square$$

A more concise way of expressing Proposition 4 is by means of the annihilator.

COROLLARY 5. a) *For* $v, w \in J$ *we have*

$$N(v^{\#}) \equiv N(v)^2 \bmod \mathrm{Ann}(J), \quad N(wU_v) \equiv N(v)^2 N(w) \bmod \mathrm{Ann}(J).$$

b) *For* $v \in J^{\times}$ *we have* $N(v^{-1}) \equiv N(v)^{-1} \bmod \mathrm{Ann}(J)$. $\square$

The most important special case of this will be stated as follows

COROLLARY 6. *If* $\mathrm{Ann}(J) = 0$, *in particular, if* Φ *is reduced or has no 3-torsion, then*

a) $N(v^{\#}) = N(v)^2$, $N(wU_v) = N(v)^2 N(w)$ *for all* $v, w \in J$

b) $N(v^{-1}) = N(v)^{-1}$ *for all* $v \in J^{\times}$. □

REMARK 7. It will be seen later (cf. Theorem 10) that the formulae of Corollary 6 do not hold in full generality, so the extra hypothesis "$\mathrm{Ann}(J) = 0$" is an essential one. On the other hand, there is of course an easy way to enforce its validity: just pass to $\bar{\Phi} = \Phi/\mathrm{Ann}(J)$, regard J canonically as an algebra $\bar{J}$ over $\bar{\Phi}$ and put $\bar{N} = {}^{-} \circ N$, $^{-}$ being the natural map from Φ to $\bar{\Phi}$. Then $(\bar{N}, \#, 1)$ is a cubic form with adjoint and base point over $\bar{\Phi}$ satisfying $\bar{J} = J(\bar{N}, \#, 1)$, and $\mathrm{Ann}(\bar{J}) = 0$. Thus the formulae of Corollary 6 are forced to hold "by changing scalars", i.e., by passing from Φ to $\bar{\Phi}$. □

COROLLARY 8. *Let* $v, w \in J$. *Then*

a) $N(v^m) \equiv N(v)^m \bmod \mathrm{Ann}(J)$ *for all positive integers* m.

b) $N(v \times w) + N(v)N(w) \equiv T(v^{\#}, w)T(v, w^{\#}) \bmod \mathrm{Ann}(J)$.

Proof. a) follows immediately from Corollary 5 a) by induction. In b), we may assume $\mathrm{Ann}(J) = 0$. Choosing an indeterminate τ, this clearly implies $\mathrm{Ann}(J') = 0$ for $J' = J \otimes \Phi[\tau]$ over $\Phi[\tau]$. Hence Corollary 6 a) gives $N(x^{\#}) = N(x)^2$ for $x = v \otimes 1 + w \otimes \tau$, and comparing coefficients of τ^3 using (1.9) and (1.13), yields b). □

Proposition 4 enables us to complete the proof of Theorem 1 b). We first show that $(N^{(v)}, \#^{(v)}, 1^{(v)})$ is a cubic form with adjoint and base point by noting $v^{-1\#} = \{N(v)^{-1}v^{\#}\}^{\#} = N(v)^{-2}v^{\#\#} = N(v)^{-1}v$, writing $T^{(v)}$ for the associated trace form and checking the axioms. Given $w, x, y \in J$, we compute

$$\begin{aligned} w^{\#(v)\#(v)} &= N(v)\{w^{\#(v)\#}\}U_v^{-1} = \\ &= N(v)\{N(v)w^{\#}U_{v^{-1}}\}^{\#}U_v^{-1} = [(1.14)] \\ &\quad N(v)^3\{w^{\#\#}U_{N(v)^{-1}v}\}U_v^{-1} = N(w)N(v)w \\ &= [\text{Proposition 4 b)}]\ N(v^{-1})^{-1}N(w)w \\ &= N^{(v)}(w)w\ , \end{aligned}$$

which yields (1.1); $N^{(v)}(1^{(v)}) = N(v^{-1})^{-1}N(v^{-1}) = 1$, which yields (1.2);

$$T^{(v)}(x,y) = (\Delta^{x}_{1^{(v)}} N^{(v)})(\Delta^{y}_{1^{(v)}} N^{(v)}) - \Delta^{x}_{1^{(v)}} \Delta^{y} N^{(v)}$$

$$= N(v^{-1})^{-2} T(v^{-1\#},x) T(v^{-1\#},y) -$$

$$N(v^{-1})^{-1} T(v^{-1} \times x,y) = [\text{Proposition 4 b)}]$$

$$N(v^{-1})^{-2} T(N(v^{-1})v,x) T(N(v^{-1})v,y)$$

$$- N(v^{-1})^{-1} T(N(v^{-1})v^{\#} \times x,y)$$

$$= T(v,x)T(v,y) - T(v^{\#} \times x,y)$$

$$= T(T(v,x)v - v^{\#} \times x,y) = T(xU_v,y) ,$$

and then

$$T^{(v)}(x^{\#(v)},y) = T(\{N(v)x^{\#}U_v^{-1}\}U_v,y) = T(N(v)x^{\#},y)$$

$$= [\text{Proposition 4 b)}]\ T(N(v^{-1})^{-1}x^{\#},y)$$

$$= N(v^{-1})^{-1} T(x^{\#},y) = N(v^{-1})^{-1} \Delta^{y}_{x} N$$

$$= \Delta^{y}_{x} N^{(v)} ,$$

which yields (1.3); $1^{(v)\#(v)} = N(v)v^{-1\#}U_v^{-1} = vU_{v^{-1}} = v^{-1} = 1^{(v)}$, which yields (1.4);

$$1^{(v)} \times^{(v)} y = N(v)\{v^{-1} \times y\}U_{v^{-1}} = [(1.18)]N(v)\{T(v^{-1\#},y)v^{-1}$$

$$- N(v^{-1})y\} = [\text{Proposition 4 b)}]\, T(v,y)1^{(v)} - y$$

$$= T(v^{-1}U_v,y)1^{(v)} - y = T^{(v)}(1^{(v)},y)1^{(v)} - y,$$

which yields (1.5). Hence $(N^{(v)},\#^{(v)},1^{(v)})$ is a cubic form with adjoint and base point over Φ. That $J(N^{(v)},\#^{(v)},1^{(v)})$ is the v-isotope of J, may now be proved as in [3, p.501]. □

McCrimmon [3] has raised the question of whether J always satisfies the identities $N(v^{\#}) = N(v)^2$ and $N(wU_v) = N(v)^2 N(w)$. We wish to show that this is not true by giving an example. In view of our previous results, *any* such example must display pathologies of characteristic 3 and a badly behaved trace form. Our approach rests upon an elementary principle for constructing cubic maps with prescribed values at a finite number of points, and prescribed derivative.

LEMMA 9. *Let* V,W *be a vector spaces over a field* Φ_0, $\# : V \to V$ *a quadratic map and* $\tau : V \times V \to W$ *a symmetric bilinear map. Suppose* V *is finite-dimensional, let* $(b_1,\ldots,b_n)$ *be a basis of* V *and* $(w_1,\ldots,w_n)$ *a family of vectors in W. Then the following are equivalent.*

(i) *There exists a cubic map* $\nu : V \to W$ *such that* $\nu(b_i) = w_i$ *for* $1 \leq i \leq n$ *and* $\Delta_v^{v'}\nu = \tau(v^{\#}, v')$ *for all* v, v' *in all scalar extensions of* V.

(ii) *We have* $\tau(b_i^{\#}, b_i) = 3w_i$ *for* $1 \leq i \leq n$, *and the map*

$$V \times V \times V \to W, \ (v_1, v_2, v_3) \to \tau(v_1 \times v_2, v_3),$$

$\times$ *being the bilinearization of* #, *is totally symmetric in all three variables.*

Proof. (i) => (ii) The first set of conditions follows immediately from the Euler Differential Equations, the second one from standard properties of the general differential calculus for rational maps.

(ii) => (i). Given $\beta_1, \ldots, \beta_n$ in a unital commutative associative Φ_0-algebra, we set

$$\nu\Big(\sum_{i=1}^{n} \beta_i b_i\Big) = \sum_{i=1}^{n} \beta_i^3 w_i + \sum_{\substack{i,j=1 \\ i \neq j}}^{n} \beta_i^2 \beta_j \ \tau(b_i^{\#}, b_j) +$$

$$\sum_{\substack{i,j,k=1 \\ i<j<k}}^{n} \beta_i \beta_j \beta_k \ \tau(b_i \times b_j, b_k).$$

It is then straightforward to show that this defines a cubic map $\nu : V \to W$ satisfying the requirements stated in (i). □

THEOREM 10. *The formulae*

$$N(v^{\#}) = N(v)^2 ,$$

$$N(wU_v) = N(v)^2 N(w),$$

$$N(v^{-1}) = N(v)^{-1} \text{ for } v \text{ invertible}$$

do not hold in all Jordan algebras $J(N,\#,1)$, $(N,\#,1)$ *a cubic form with adjoint and base point.*

Proof. The second equation, for v invertible and $w = v^{-1}$, implies the third, the third implies the first for v invertible, and hence in full generality. It therefore suffices to find a counter-example for the first equation. To this end, we start with a field Φ_0 of characteristic 3 and choose a cubic form $(N_0,\#,1)$ in finitely many variables with adjoint and base point over Φ_0 whose associated trace form, T_0, is identically zero. Suppose further that $J_0 = J(N_0,\#,1)$ contains an element a such that $1,a,a^{\#}$ are linearly independent. (A convenient way to meet all these requirements would be to take a finite dimensional purely inseparable field extension E/Φ_0 of exponent 1, other than Φ_0, and to define N_0 as the generic norm, # as the adjoint, 1 as the identity element of E/Φ_0.) For each $v \in J_0$ we have $v^{\#} = v^2$, so J_0 is a linear Jordan algebra with multiplication $vv' = \frac{1}{2} v \times v' = -v \times v'$ since char $\Phi_0 = 3$. Now let W be a vector space over Φ_0 of dimension > 2 and $\lambda : J_0 \to W$ a linear map which satisfies $\lambda(1) = 0$ and is associative relative to the linear structure of J_0. (We allow λ to vanish identically; at the other extreme, if (the linear structure of) J_0 is associative, any linear map from J_0 to W killing 1 will do.) Then the map $\tau : J_0 \times J_0 \to W$ defined by $\tau(v,v') = \lambda(vv')$ is symmetric and bilinear. We claim that τ satisfies the requirements stated in Lemma 9 (ii). Indeed, total symmetry of the expression $\tau(v_1 \times v_2, v_3) = -\lambda((v_1v_2)v_3)$ in $v_1, v_2, v_3 \in J$ being obvious, we compute for $v \in J$,

$$\tau(v^{\#},v) = \lambda(v^{\#},v) = \lambda(T_0(v,v)v - v^{\#} \times v)$$

$$= \lambda(vU_v) = \lambda(v^3) = [(1.7)]\, \lambda(N(v)1) = N(v)\lambda(1) = 0 .$$

Hence, fixing linearly independent vectors $w_1, w_2 \in W$, Lemma 9 produces a cubic map $\nu : J_0 \to W$ such that

$\nu(1) = 0$, $\nu(a) = w_1$, $\nu(a^{\#}) = w_2$ and $\Delta_v^{v'}\nu = \tau(v^{\#},v')$ holds in all scalar extensions of J_0 .

We now pass to the split null extension $\Phi = \Phi_0 + W$ of the Φ_0-module W. By letting W act trivially on J_0, J_0 becomes a unital Φ-module in the natural way. We define $N : J_0 \to \Phi$ by $N(v) = N_0(v) + \nu(v)$ for $v \in J_0$ and claim that $(N,\#,1)$ is a cubic form with adjoint and base point over Φ. By construction, N is cubic over Φ_0 ; for $\alpha \in \Phi_0$, $w \in W$ and $v \in J_0$, we therefore obtain

$$N((\alpha + w)v) = N(\alpha v) = \alpha^3 N(v) = [\text{since } W^2 = 0]$$

$$(\alpha^3 + w^3)N(v) = [\text{since char } \Phi_0 = 3] \ (\alpha + w)^3 N(v),$$

so N is homogeneous of degree 3 over Φ. The remaining assertions follow along similar lines, constantly making use of the fact that W squares to zero and kills J_0. We may thus form $J = J(N,\#,1)$, which, incidentally, is just J_0 regarded as an algebra over Φ. We now compute

$$N(a^{\#}) = N_0(a^{\#}) + \nu(a^{\#}) = N_0(a)^2 + w_2 \text{ , by Corollary 6;}$$

$$N(a)^2 = (N_0(a) + \nu(a))^2 = N_0(a)^2 + 2N_0(a)\nu(a)$$

$$= N_0(a)^2 - N_0(a)w_1 \text{ .}$$

Since w_1, w_2 are linearly independent over Φ_0, this shows $N(a^{\#}) \neq N(a)^2$, as desired. □

It seems worth pointing out that, in the above construction, the trace form of J, though taking values in Ann(J), may be forced to be non-zero, for example when J_0 arises from a purely inseparable field extension of exponent one over Φ_0.

3. RADICALS OF J(N,#,1).

In this section we consider some of the radicals associated with $J = J(N,\#,1)$ over a unital commutative associative ring Φ.

By analogy to the theory of quadratic forms, let

(1) $\quad \mathrm{Rad}\, N = \{z \in J \mid N(z+x) = N(x),\ \forall x \in J\}$.

Letting $x = 0$, $N(z) = 0$ for $z \in \mathrm{Rad}\, N$. By (1.3),

(1') $\quad \mathrm{Rad}\, N = \{z \in J \mid N(z) = 0 = T(z^{\#},x) + T(z,x^{\#}),\ \forall x \in J\}$

Let

(2) $\quad \mathrm{Rad}'N = \{z \in J \mid N(z) = T(z^{\#},x) = T(z,x^{\#}) = 0,\ \forall x \in J\}$

If $T(z,x^{\#}) = 0$ for all x, linearizing, $T(z, x \times y) = 0$ for all x,y, and by (1.5), $0 = T(z, x \times 1) = T(z, T(x)1 - x) = T(x)T(z,1) - T(z,x)$. But $T(z,1) = T(z,1^{\#}) = 0$. So

(2') $\mathrm{Rad}'N = \{z \in J \mid N(z) = T(z^{\#},x) = T(z,x) = 0,\ \forall x \in J\}$.

LEMMA 1. $\mathrm{Rad}'N \subset \mathrm{Rad}\ N$ *and if*

(i) *Φ has no 2 torsion, or*

(ii) *Φ contains a regular element λ with $1-\lambda$ also regular, then equality holds.*

Proof: The inclusion is clear. By (1'), if $z \in \mathrm{Rad}\,N$,

(3) $\lambda T(z^{\#},x) + \lambda^2 T(z,x^{\#}) = 0$ for all $\lambda \in \Phi$.

If (ii) holds then $T(z^{\#},x) + \lambda T(z,x^{\#}) = 0$ for λ regular and by (1') $(1-\lambda)T(z,x^{\#}) = 0$. Therefore $T(z,x^{\#}) = 0 = T(z^{\#},x)$. If (i) holds, letting $\lambda = -1$ in (3) and using (1'), yields $2T(z,x^{\#}) = 0$. Hence $T(z,x^{\#}) = 0 = T(z^{\#},x)$. □

Rad N is closed under addition but in general not under multiplication by scalars. However if Φ is a field then, for $z \in \mathrm{Rad}\ N$ and $\lambda \in \Phi^{\times}$, $N(\lambda z + x) = \lambda^3 N(z + \lambda^{-1}x) = \lambda^3 N(\lambda^{-1}x) = N(x)$. $\mathrm{Rad}'N$ on the other hand is always a sub-module of J.

PROPOSITION 2. *If Φ is a field then* $\mathrm{Rad}'N$ *is a nil ideal containing all absolute zero divisors of* J.

Proof. Let z be an absolute zero divisor. Arguing as in [3 p.507], since $U_z \equiv 0$, $z^3 = z^2 = 0$ so $N(z) = S(z) = T(z) = 0$. Therefore $z^\# = z^2 - T(z)z + S(z)1 = 0$ and $xU_z = T(x,z)z - z^\# \times x = T(x,z)z = 0$. Hence $T(x,z) = 0 = T(x,z^\#)$ and $z \in \mathrm{Rad}'N$.

Since $N(xU_y) = N(x)N(y)^2$ when Φ is a field by Corollary 2.6, $N(zU_y) = 0 = N(yU_z)$ when $N(z) = 0$. If $z \in \mathrm{Rad}'N$, $T(zU_y,x) = T(z,xU_y) = 0$ by (1.6) and (1.12), and $T((zU_y)^\#,x) = T(z^\# U_{y^\#},x) = T(z^\#,xU_{y^\#}) = 0$. Also $yU_z = T(z,y)z - z^\# \times y = -z^\# \times y$, so $T(yU_z,x) = -T(z^\#,y \times x) = 0$. Similarly $T((yU_z)^\#,x) = -N(z)T(z \times y^\#,x) = 0$. Thus zU_y and $yU_z \in \mathrm{Rad}'N$ when $z \in \mathrm{Rad}'N$ and $\mathrm{Rad}'N$ is an ideal of J. □

The above result is also a consequence of the more general results which follow but it seemed worthwhile to consider the radical of the cubic form and proceed directly. Note that for fields of cardinality > 2, $\mathrm{Rad}\, N = \mathrm{Rad}'N$.

We wish to study the nil radical of J.

LEMMA 3. *If J is a unital quadratic Jordan algebra over Φ, let $\Phi[u]$ be the unital subalgebra of J generated by an element $u \in J$, and $\Phi[\lambda]$ the polynomial* *ring in the indeterminate* λ.

(i) *If* $f \in \Phi[\lambda]$ *satisfies* $f(u) = 0$ *then* $(f^m g)(u) = 0$ *for all integers* $m \geq 2$ *and all* $g \in \Phi[\lambda]$.

(ii) *If* u *is nilpotent, then there exists* $r \in \mathbb{N}$ *such that* $u^n = 0$ *for all* $n \geq r$.

(iii) *The nil radical of* $\Phi[u]$ *is given by*

$$\mathrm{Nil}\ \Phi[u] = \{v \in \Phi[u] \mid v \ \text{ nilpotent}\}.$$

Proof.

(i) $(f^m g)(u) = ((f^{m-2}g)(u))U_{f(u)} = 0$

(ii) If $u^k = 0$ for some $k \in \mathbb{N}$, let $r = 2k$ and apply (i) with $f = \lambda^k$, $g = \lambda^{n-r}$, $m = 2$.

(iii) We must show that the set of all nilpotents is an ideal of $\Phi[u]$. Let $v = f(u)$, $w = g(u)$, $f,g \in \Phi[\lambda]$. If $v^k = f^k(u) = 0$, $(wU_v)^k = ((f^2g)(u))^k = (f^{2k}g^k)(u) = 0$, by (i), and $(vU_w)^k = ((g^2f)(u))^k = (g^{2k}f^k)(u) = f^k(u)U_{g^k(u)} = 0$. So wU_v and vU_w are nilpotent.

It remains to show that if v and w are nilpotent so is $v+w$. If $v^k = 0 = w^t$ then

$$(v+w)^{2(k+t)} = \sum_{j=0}^{2(k+t)} \binom{2(k+t)}{j} (f^j g^{2(k+t)-j})(u).$$

By (i), if $2k \le j \le 2(k+t)$ then $(f^j g^{2(k+t)-j})(u) =$ $((f^k)^2 f^{j-2k} g^{2(k+t)-j})(u) = 0$. Similarly $(f^j g^{2(k+t)-j})(u) = 0$ if $2(k+t)-j \ge 2t$. Hence $(v+w)^{2(k+t)} = 0$. □

Let $J = J(N,\#,1)$.

PROPOSITION 4. *For* $u \in J$, *the following are equivalent.*

(i) u *is nilpotent*

(ii) $T(u)$, $T(u^\#)$, $N(u)$ *are nilpotent.*

Proof.

(ii) => (i) : We have $u^3 = T(u)u^2 - T(u^\#)u + N(u)1$ which is nilpotent by Lemma 3 (iii). Hence u is nilpotent.

(i) => (ii) : By Corollary 2.8 a) and Lemma 2.2 a), $N(u)$ is nilpotent. By Lemma 3 (ii), there exists an integer r such that $u^n = 0$ for all $n \ge r$. We prove the nilpotency of $T(u)$ and $T(u^\#)$ by induction on r. If $r = 1$ there is nothing to prove. For $r > 1$ and $n \ge r-1$ we have $(u^2)^n = u^{2n} = 0$, so by the induction hypothesis, $T(u^2)$ and $T(u^{2\#}) = T(u^{\#2})$ are nilpotent. Since

$x^{\#} = x^2 - T(x)x + T(x^{\#})1$, we have

$$T(x)^2 = 2T(x^{\#}) + T(x^2).$$

Specializing x to u and then to $u^{\#}$ yields

$$(4) \quad T(u)^2 = 2T(u^{\#}) + T(u^2),$$

$$T(u^{\#})^2 = 2T(u)N(u) + T(u^{\#2}).$$

Since $N(u)$ and $T(u^{\#2})$ are nilpotent so is $T(u^{\#})$ and by (4) so is $T(u)$. □

By a *#-ideal in* J we mean an ideal I in J satisfying $I^{\#} + J \times I \subset I$.

PROPOSITION 5. *Let* $\mathfrak{a}$ *be an ideal of* Φ. *Then*

(i) $I'_{\mathfrak{a}} = \{v \in J \mid T(v,J) + T(v^{\#},J) \subset \mathfrak{a}\}$ *is a #-ideal in* J, *and, if* $1/2 \in \Phi$, $I'_{\mathfrak{a}} = \{v \in J \mid T(v,J) \subset \mathfrak{a}\}$.

(ii) *If* $\mathfrak{a}$ *contains the annihilator of* J,

$$I_{\mathfrak{a}} = \{v \in I'_{\mathfrak{a}} \mid N(v) \in \mathfrak{a}\}$$

is a #-ideal in J, and

$$I_{\mathfrak{a}} = I'_{\mathfrak{a}} \text{ provided } 1/3 \in \Phi$$

$$I_{\mathfrak{a}} = \{v \in J \mid N(v) \in \mathfrak{a},\ T(v,J) \subset \mathfrak{a}\} \text{ provided}$$

$1/2 \in \Phi$,

$$I_{\mathfrak{a}} = \{v \in J \mid T(v,J) \subset \mathfrak{a}\} \text{ provided } 1/6 \in \Phi .$$

Proof.

(i) Given $u,v \in I'_{\mathfrak{a}}$, $w \in J$, we obtain

$$T((u+v)^{\#},w) = T(u^{\#},w) + T(u \times v,w) + T(v^{\#},w)$$
$$= T(u^{\#},w) + T(u,v \times w) + T(v^{\#},w) \in \mathfrak{a} .$$

So $I'_{\mathfrak{a}}$ is a Φ-submodule of J. Now fix $u \in I'_{\mathfrak{a}}$ and $v,w \in J$. Since $T(u^{\#\#},v) = N(u)T(u,v) \in \mathfrak{a}$, we have $I'^{\#}_{\mathfrak{a}} \subset I'_{\mathfrak{a}}$. Since $T(u \times v,w) = T(u,v \times w) \in \mathfrak{a}$ and $T((u \times v)^{\#},w) = T(u^{\#},v)T(v,w) + T(v^{\#},u)T(u,w) - T(u^{\#},v^{\#} \times w) \in \mathfrak{a}$ by (1.10) and (1.12), we have $J \times I'_{\mathfrak{a}} \subset I'_{\mathfrak{a}}$. It remains to show that $I'_{\mathfrak{a}}$ is an ideal. But $vU_u = T(u,v)u - u^{\#} \times v \in I'_{\mathfrak{a}}$ and

$$T(uU_v,w) = T(u,wU_v) \in \mathfrak{a}$$

$$T((uU_v)^{\#},w) = T(u^{\#}U_{v^{\#}},w) = T(u^{\#},wU_{v^{\#}}) \in \mathfrak{a} .$$

Hence $uU_v \in I'_{\mathfrak{a}}$ and $I'_{\mathfrak{a}}$ is an ideal of J. By (1.11), $T(u,u \times v) = 2T(u^{\#},v)$ and the last statement of (i) holds.

(ii) For $u,v \in I_{\mathfrak{a}}$ we have $N(u+v) = N(u) + T(u^{\#},v) + T(u,v^{\#}) + N(v) \in \mathfrak{a}$. So $I_{\mathfrak{a}}$ is a Φ-submodule of J. Fix $u,v \in J$ and suppose one of them belongs to $I_{\mathfrak{a}}$. Since $\mathrm{Ann}(J) \subset \mathfrak{a}$, Corollary 2.5 a) and Lemma 2.2 a) imply

$$N(uU_v) \equiv N(u)N(v)^2 \equiv 0 \bmod \mathfrak{a} ,$$

whence $I_{\mathfrak{a}}$ is an ideal. Similarly given $u \in I_{\mathfrak{a}}$, $v \in J$, Corollaries 2.5 a), 2.8 b) and Lemma 2.2 a) yield

$$N(u^{\#}) \equiv N(u)^2 \equiv 0 \bmod \mathfrak{a},$$

$$N(u \times v) \equiv T(u^{\#},v)T(u,v^{\#}) - N(u)N(v) \equiv 0 \bmod \mathfrak{a}.$$

Hence $I_{\mathfrak{a}}$ is a #-ideal, clearly contained in $I'_{\mathfrak{a}}$. By (1.13) $T(u,u^{\#}) = 3N(u)$ and this with (i) implies the last three equalities in (ii). □

THEOREM 6. *The nil radical of* J *is given by*

$$\text{Nil } J = \{u \in J \mid N(u),\ T(u,v),\ T(u^{\#},v) \in \text{Nil } \Phi$$

for all $v \in J\}$.

Proof. Setting $\mathfrak{a} = \text{Nil } \Phi$, the right hand side of the above equation is simply $I_{\mathfrak{a}}$. For $u \in I_{\mathfrak{a}}$, $N(u)$, $T(u) = T(u,1)$, $T(u^{\#}) = T(u^{\#},1)$ all belong to Nil Φ. Hence combining Propositions 4 and 5, we see that $I_{\mathfrak{a}}$ is a nil ideal in J and is thus contained in Nil J. Conversely suppose $u \in \text{Nil } J$. Then $N(u) \in \text{Nil } \Phi$ by Proposition 4 and $u^{\#} = u^2 - T(u)u + T(u^{\#})1 \equiv T(u^{\#})1 \bmod \text{Nil } J$. But, by Proposition 4, $T(u^{\#})$ is nilpotent. As $(\text{Nil } \Phi)J$ is obviously a nil ideal in

J and is thus contained in Nil J, we conclude that $T(u^{\#})1 \in \text{Nil } J$. Hence $u^{\#} \in \text{Nil } J$. Furthermore, for every $v \in J$, $vU_u = T(u,v)u - u^{\#} \times v \in \text{Nil } J$ which implies $u^{\#} \times v \in \text{Nil } J$. Hence, by (1.11) and Proposition 4, $T(u^{\#})T(v) - T(u^{\#},v) = T(u^{\#} \times v) \in \text{Nil } \Phi$. Since $T(u^{\#}) \in \text{Nil } \Phi$, we have $T(u^{\#},v) \in \text{Nil } \Phi$. Similarly $w = uU_v = T(v,u)v - v^{\#} \times u \in \text{Nil } J$, whence $T(u,w) = T(u,v)^2 - T(u,u \times v^{\#}) = T(u,v)^2 - 2T(u^{\#},v^{\#})$ is nilpotent which implies $T(u,v) \in \text{Nil } \Phi$ to complete the proof. □

Combining Theorem 6 and Proposition 5 yields.

COROLLARY 7.

(i) $\text{Nil } J = \{v \in J : T(v,J) + T(v^{\#},J) \subset \text{Nil } \Phi\}$ *provided* $1/3 \in \Phi$.

(ii) $\text{Nil } J = \{v \in J | N(v) + T(v,J) \subset \text{Nil } \Phi\}$ *provided* $1/2 \in \Phi$. □

COROLLARY 8. *Let* Φ *be a field. Then*

(i) $\text{Nil } J = \{u \in J | N(u) = T(u,v) = T(u^{\#},v) = 0$, *for all* $v \in J\} = \text{Rad}'N = \text{Rad } J$.

(ii) *Either*

$$\text{Nil } J = \{u \in J \mid T(u,v) = T(u^{\#},v) = 0, \textit{ for all } v \in J\}$$

or T *vanishes identically, in which case* Φ *has characteristic 3 and* $\text{Nil } J = \{u \in J \mid N(u) = 0\}$.

(iii) *If char* $\Phi \neq 2$ *and* T *is not identically* 0,

$$\text{Nil } J = \{v \in J \mid T(v,J) = \{0\}\}.$$

Proof.

(i) Since $\text{Nil } \Phi = \{0\}$ the first two equalities of (i) are immediate. By (1.7) J is an algebraic algebra. Thus, by Theorem 3.3 p. 3.6 of [1], the Jacobson radical Rad J is nil. So we have the last equality.

(ii) Setting $\mathfrak{a} = \{0\}$, we have $\text{Nil } J = I_{\mathfrak{a}} \subset I'_{\mathfrak{a}}$. Therefore if Nil J differs from $I'_{\mathfrak{a}}$, $I'_{\mathfrak{a}}$ contains an element u with $N(u) \neq 0$. This u is invertible and as $I'_{\mathfrak{a}}$ is an ideal this forces $I'_{\mathfrak{a}} = J$, i.e. $T \equiv 0$. The rest is clear.

(iii) Again with $\mathfrak{a} = \{0\}$, char $\Phi \neq 2$ and Proposition 5 (i) yield $I'_{\mathfrak{a}} = \{v \in J \mid T(v,J) = \{0\}\}$. Since T is not identically 0, $\text{Nil } J = I'_{\mathfrak{a}}$, by part (ii). □

We are now ready to give a description of Rad J when Φ is a ring.

THEOREM 9. *If* $J = J(N,\#,1)$ *is a quadratic Jordan algebra over a unital commutative associative ring of scalars* Φ *then the Jacobson radical*

$$\mathrm{Rad}\, J = \{u \in J \mid N(u), T(u,v), T(u^{\#},v) \in \mathrm{Rad}\, \Phi \text{ for all } v \in J\},$$ *where* $\mathrm{Rad}\, \Phi$ *is the Jacobson radical of* Φ .

Proof. Let $\mathfrak{m}$ be a maximal ideal of Φ. Then $J/\mathfrak{m}J \cong J \otimes_{\Phi} \Phi/\mathfrak{m}$ is an algebra over the field $\Phi/\mathfrak{m}$. Now $J/\mathfrak{m}J = J(\bar{N},\bar{\#},\bar{1})$, where $\bar{N},\bar{\#},\bar{1}$ are the natural maps and unit of $J/\mathfrak{m}J$ induced by those of J. By Corollary 8 (i), $\mathrm{Rad}(J/\mathfrak{m}J) = \{u \in J/\mathfrak{m}J \mid \bar{N}(u) = \bar{T}(u,v) = \bar{T}(u^{\bar{\#}},v) = 0 \text{ for all } v \in J/\mathfrak{m}J\}$. But $\Phi/\mathfrak{m} \otimes_{\Phi} \mathrm{Rad}\, J \subset \mathrm{Rad}(\Phi/\mathfrak{m} \otimes_{\Phi} J) = \mathrm{Rad}(J/\mathfrak{m}J)$ since $\Phi/\mathfrak{m} \otimes_{\Phi} \mathrm{Rad}\, J$ is a quasi-invertible ideal in $\Phi/\mathfrak{m} \otimes J$. Indeed if $z_i \in \mathrm{Rad}\, J$, $\alpha_i \in \Phi$, then $z = \sum_{i=1}^{t} \alpha_i z_i \in \mathrm{Rad}\, J$ and is quasi-invertible in J, that is, $1 - z$ is invertible with inverse $1 - w$. Denote $\alpha_i + \mathfrak{m}$ by $\bar{\alpha}_i$; $\sum_{i=1}^{t} = \bar{\alpha}_i \otimes z_i = \sum_{i=1}^{t} \bar{1} \otimes \alpha_i z_i = \bar{1} \otimes z$ is quasi-invertible with quasi-inverse $\bar{1} \otimes w$. Therefore if $u \in \mathrm{Rad}\, J$ and $v \in J$, then $N(u)$, $T(u,v)$ and $T(u^{\#},v) \in \bigcap_{\mathfrak{m}\ \max} \mathfrak{m} = \mathrm{Rad}\, \Phi$ and

$$\mathrm{Rad}\, J \subset \{u \in J \mid N(u),\ T(u,v),\ T(u^{\#},v) \in \mathrm{Rad}\, \Phi,\ \forall v \in J\}.$$

To prove the other inclusion, we consider the

ideal $I = I_{\mathrm{Rad}\ \Phi}$ of J (note that Rad $\Phi \supset$ Nil $\Phi \supset$ Ann (J)). For $u \in I$, we have $N(1-u) = 1 - T(u) + T(u^{\#}) - N(u)$. Since $T(u)$, $T(u^{\#})$, $N(u)$ belong to Rad Φ, $N(1-u) \equiv 1 \bmod \mathrm{Rad}\ \Phi$. Hence $N(1-u) \in \Phi^{\times}$ and $1-u$ is invertible. Thus I is a quasi-invertible ideal in J, and we conclude $I \subset$ Rad J. □

REFERENCES

[1] N. Jacobson, *Lectures on Quadratic Jordan Algebras*, Tata Institute of Fundamental Research Bombay, 1969.

[2] N. Jacobson, *Structure Theory of Jordan Algebras*, 5, The University of Arkansas Lecture Notes in Mathematics, Fayetteville, 1981.

[3] K. McCrimmon, The Freudenthal - Springer - Tits constructions of exceptional Jordan algebras, *Trans. AMS* 139 (1969), 495-510.

[4] K. McCrimmon, Koecher's principle for quadratic Jordan algebras, *Proc. AMS* 28 (1971), 39-43.

[5] H.P. Petersson and M.L. Racine, Exceptional Jordan division algebras, *Contemporary Math.* 13 (1982), 307-315.

[6] E.I. Zelmanov, On prime Jordan rings. *Algebra i Logika*, 18 (1979), 162-175.

[7] E.I. Zelmanov, Jordan division algebras, *Algebra i Logika*, 18 (1979), 286-310.

H. P. Petersson
Fachbereich Mathematik und Informatik
Fernuniversität
Lützowstraße 125
D-5800 Hagen 1
Bundesrepublik Deutschland

M. L. Racine
Department of Mathematics
University of Ottawa
Ottawa, Ontario
K1N 9B4
Canada

COLLOQUIA MATHEMATICA SOCIETATIS JÁNOS BOLYAI

38. RADICAL THEORY, EGER (HUNGARY), 1982.

RADICALS AND VERBALS

B. I. PLOTKIN

The present paper deals essentially with the notion of radical. We shall consider radicals in various structures. They have many features in common but differences also show up. Some problems will be formulated. In the definition of radical it will not be assumed that the corresponding class is closed under extensions. For instance, the locally nilpotent radical is extension-closed in associative rings but is not in Lie rings or in groups. This approach changes quite a lot in radical theory. In particular, it leads to a duality between radicals and verbals.

The author apologizes for the lack of references to many important related papers dealing with extension-closed radicals. This topic will not be treated here.

Thanks are due to L. Márki for his careful reading of the manuscript and for a number of helpful comments on it.

1. RADICALS AND VERBALS AS FUNCTORS

Let K be an arbitrary category. Recall that a one-place covariant functor $\phi: K \to K$ is called a subfunctor of the identity if for every object A in K a monomorphism $f(A): \phi(A) \to A$ is defined in such a way that for any morphism $\alpha: A \to B$ in K, the following diagram is commutative:

$$\begin{array}{ccc} \phi(A) & \xrightarrow{\phi(\alpha)} & \phi(B) \\ {\scriptstyle f(A)}\downarrow & & \downarrow{\scriptstyle f(B)} \\ A & \xrightarrow{\alpha} & B \end{array} .$$

A functor $\phi: K \to K$ is called a factor functor of the identity if to every object A in K an epimorphism $f(A): A \to \phi(A)$ is assigned so that

$$\begin{array}{ccc} \phi(A) & \xleftarrow{\phi(\alpha)} & \phi(B) \\ {\scriptstyle f(A)}\uparrow & & \uparrow{\scriptstyle f(B)} \\ A & \xleftarrow{\alpha} & B \end{array} .$$

is commutative for all morphisms α in K.

A *verbal* in a category K is a factor functor ϕ of the identity such that $\phi^2 = \phi$ and the following condition is satisfied: every commutative diagram of the form

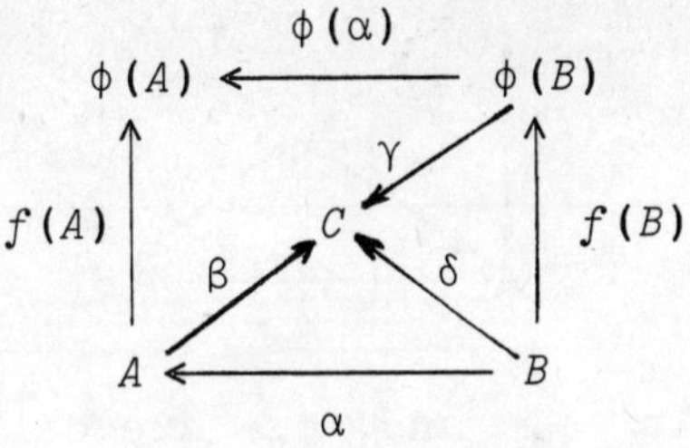

in K in which all morphisms are epimorphisms, admits a (unique) commutative extension

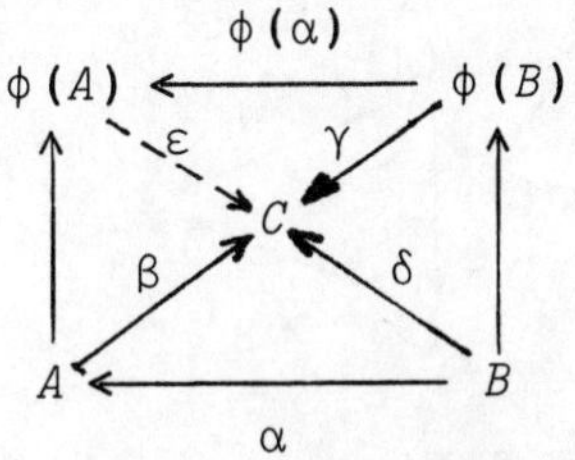

by an epimorphism $\varepsilon: \phi(A) \to C$.

A *radical* in K is defined dually. It is a subfunctor ϕ of the identity such that $\phi^2 = \phi$ and whenever we have a commutative diagram

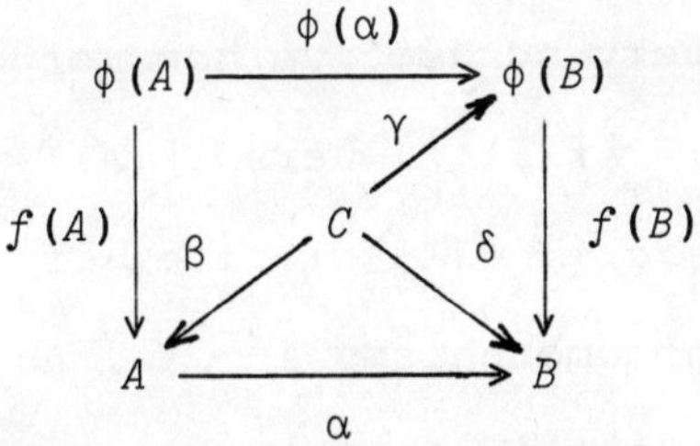

consisting of monomorphisms, we can extend it (uniquely)

by a monomorphism $\varepsilon : C \to \phi(A)$ to a commutative diagram

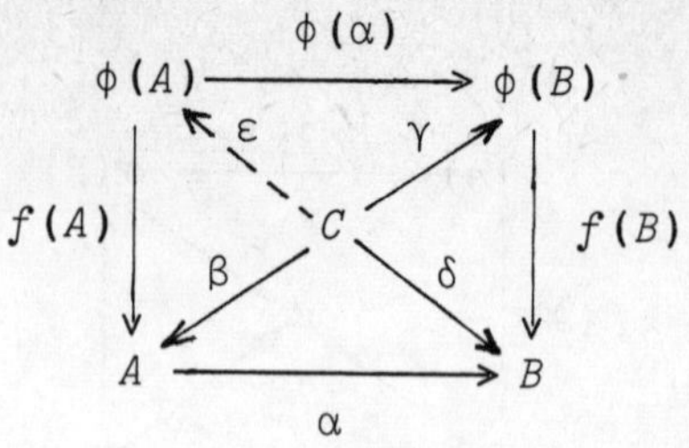

Two radicals ϕ_1 and ϕ_2 are called *equivalent* if for every object A there is an isomorphism $\varphi(A) : \phi_1(A) \to \phi_2(A)$ such that the diagram

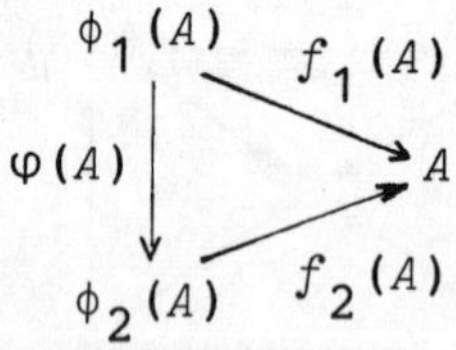

commutes. Dually we define equivalence of verbals.

Assume now that our category K is a variety of universal algebras. We shall consider functions F which assign to each algebra A from K a congruence $F(A)$ on A. Such a function F is called a *verbal function* in K if for every surjective homomorphism $\alpha : A \to B$ in K we have $F(A)^{\alpha} = F(B)$ (where $F(A)^{\alpha} = \{(\alpha a_1, \alpha a_2) : (a_1, a_2) \in F(A)\}$), i.e. if F commutes with the surjective homomorphisms in K. As is easy to see, verbal functions are in a one-to-one correspondence with subvarieties of K : to each F we assign the variety

$\Theta = F^*$ consisting of those $A \in K$ for which $F(A)$ is the equality relation. Then for every $A \in K$, $F(A)$ is the verbal congruence belonging to Θ, i.e. the congruence which yields the greatest homomorphic image of A in Θ.

Let ϕ be a verbal in K, where the latter is a variety. For every algebra A in K, denote by $F(A)$ the kernel equivalence of the epimorphism $f(A)\colon A \to \phi(A)$.

PROPOSITION 1.1. *Let the verbal* ϕ *be such that all the* $f(A)\colon A \to \phi(A)$ *are surjective homomorphisms. Then* F *is a verbal function.*

Proof. Consider a surjective homomorphism $\alpha\colon A \to B$ and let $a_1 F(A) a_2$. Then we have

$$a_1^{f(A)\phi(\alpha)} = a_2^{f(A)\phi(\alpha)}, \text{ hence } a_1^{\alpha f(B)} = a_2^{\alpha f(B)},$$

so $a_1^{\alpha} F(B) a_2^{\alpha}$ and therefore $F(A)^{\alpha} \subseteq F(B)$.

Now put $C = B/F(A)^{\alpha}$ and denote by $\beta\colon B \to C$ the canonical homomorphism. The epimorphism $\alpha\colon A \to B$ induces an epimorphism $\alpha'\colon A/F(A) \to C$. Furthermore, we have a commutative diagram

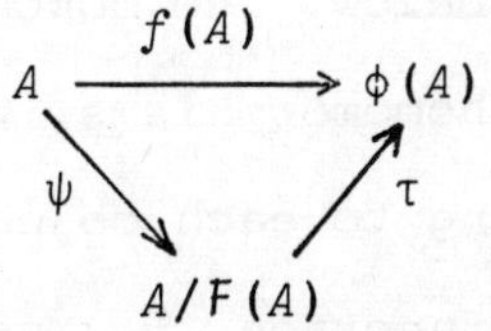

Putting $\delta = \psi\alpha' : A \to C$ and $\gamma = \tau^{-1}\alpha' : \phi(A) \to C$, we arrive at a commutative diagram

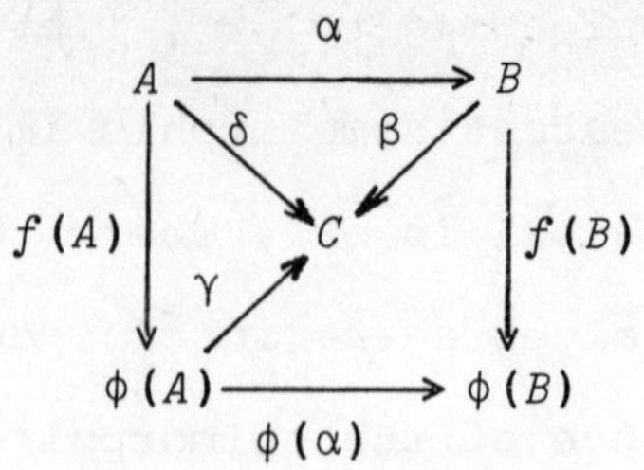

.

By the assumption there is an epimorphism $\varepsilon : \phi(B) \to C$ which extends this diagram to a commutative one. This ε is a surjective homomorphism. Let now b_1 and b_2 be elements of B such that $b_1 \mathcal{F}(B) b_2$. Then $b_1{}^{\beta} = b_1{}^{f(B)\varepsilon} = b_2{}^{f(B)\varepsilon} = b_2{}^{\beta}$, or equivalently, $b_1 \mathcal{F}(A)^{\alpha} b_2$. Thus we have proven that $\mathcal{F}(B) = \mathcal{F}(A)^{\alpha}$.

It is also obvious that the variety $\Theta = \mathcal{F}^*$ coincides with the class of objects $\phi(A)$ for all algebras A in K.

Next we turn to radicals and take for K a variety of multioperator groups. The morphisms in this category will be the accessible homomorphisms, defined after the proof of Theorem 1.3 below. The monomorphisms are precisely the injective homomorphisms. We shall consider functions $\mathcal{F}$ assigning to each object $A \in K$ an ideal $\mathcal{F}(A)$ in A. Such a function $\mathcal{F}$ must be abstract in the sense that whenever $\alpha: A \to B$ is an isomorphism, $F(A)^{\alpha} = F(B)$ must hold. F is called a *radical function* if it satisfies the following two conditions:

1. For every surjective homomorphism $\alpha: A \to B$,

$$F(A)^{\alpha} \subset F(B) .$$

2. If A is an ideal in B then

$$F(A) = A \cap F(B) .$$

Radical functions are in a one-to-one correspondence with radical classes in K. A class Θ of objects in K is called a *radical class* if it is closed under taking ideals and homomorphic images and if every algebra $A \in K$ has an ideal $\Theta'(A)$ which belongs to Θ and contains all ideals of A belonging to Θ. In addition we suppose that whenever A is an ideal in B, $\Theta'(A)$ is also. If Θ is a radical class then $F = \Theta'$ is a radical function. Conversely, if F is a radical function then the class $\Theta = F'$ consisting of all the A for which $F(A) = A$, is a radical class. Here we have $\Theta'' = \Theta$ and $F'' = F$.

Let now ϕ be a radical in our category K and denote by F the function which assigns to each object A the image of the mapping $f(A): \phi(A) \to A$.

PROPOSITION 1.2. *If the radical* ϕ *is such that for every object* A , $F(A)$ *is an ideal in* A , *then* F *is a radical function.*

Proof. Let $\alpha : A \to B$ be a surjective homomorphism and consider the corresponding commutative diagram

$$\begin{array}{ccc} \phi(A) & \xrightarrow{\phi(\alpha)} & \phi(B) \\ {\scriptstyle f(A)}\downarrow & & \downarrow{\scriptstyle f(B)} \\ A & \xrightarrow[\alpha]{} & B \end{array} .$$

Take an $a \in F(A)$ and let $a = a'^{f(A)}$, $a' \in \phi(A)$. Then $a^{\alpha} = a'^{f(A)\alpha} = a'^{\phi(\alpha)f(B)} \in F(B)$, and the first condition of radical functions is verified.

Next, let A be an ideal in B with the canonical injection $\alpha : A \to B$. Taking elements a and a' as above, we have $a = a^{\alpha} \in F(B)$. This yields $F(A) \subset A \cap F(B)$, and it remains to prove the converse inclusion. Put $C = A \cap F(B)$ and consider the canonical embeddings $\beta : C \to A$ and $\delta : C \to B$. By the assumption we have an isomorphism $f(B) : \phi(B) \to F(B)$. Take $f(B)^{-1} : F(B) \to \phi(B)$ and, in view of $C \subset F(B)$, define $\gamma = \delta f(B)^{-1} : C \to \phi(B)$. Then we have the commutative diagram

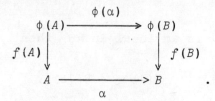

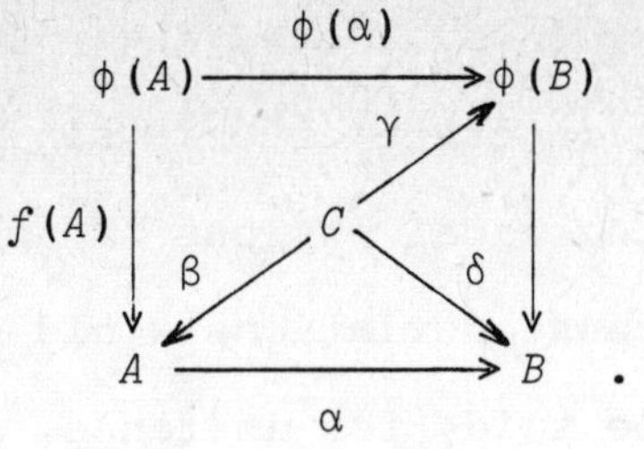

This diagram admits a commutative extension by an $\varepsilon : C \to \phi(A)$. Now we have for every $c \in C$:

$$c = c^{\beta} = c^{\varepsilon f(A)} \in F(A) ,$$

and the proof is complete. Clearly, the radical class corresponding to F consists of the $\phi(A)$, for all objects A in K .

Propositions 1.1 and 1.2 are proven under special assumptions on the verbal and the radical, respectively. For verbals this is due to the fact that epimorphisms in a category of algebras are not always surjective homomorphisms, whereas in the case of radicals it is necessary that the monomorphisms related to radicals define ideals of the corresponding objects. Moreover, one can consider radicals also in categories of multioperator groups which are not varieties, and then it may also happen that the monomorphisms needed are not injective. Therefore, the general categorical definitions of radical and verbal, as presented above, might lead to "deviations from the norm" in concrete cases - similarly to what happens with the categorical notions of epimorphism and monomorphism. Maybe these deviations would merit a special study. What can be said, for instance, about those varieties of algebras which are closed under categorical epimorphisms? *)

In order to avoid anomalies and to develop a natural theory, it is convenient to use the notion of bicategory (these categories are often called categories with a factorization system). Recall the following definition.

Let K be a category, K' and K'' be the subcategories of K consisting of the epimorphisms and the monomorphisms, respectively. We distinguish a subcategory K_1 of K' and a subcategory K_2 of K''. The morphisms of K_1 and K_2 will be called *admissible epimorphisms* and *admissible monomorphisms*, respectively. We assume that the following conditions are satisfied.

1. Each morphism α in K can be decomposed into a product $\alpha = \nu\mu$, $\nu \in K_1$, $\mu \in K_2$.

2. If $\nu\mu = \nu'\mu'$ where $\nu, \nu' \in K_1$, $\mu, \mu' \in K_2$, then

---

*) All varieties of commutative semigroups have this property, see N. M. Khan [2].

there is an isomorphism β such that $\nu' = \nu\beta$.

3. Every isomorphism in K is at the same time an admissible epimorphism and an admissible monomorphism.

The category K together with the distinguished K_1 and K_2 is called a *bicategory*. As a typical example we can take an arbitrary variety of universal algebras with surjective homomorphisms as admissible epimorphisms and injective homomorphisms as admissible monomorphisms.

Radicals and verbals are defined in bicategories in the same way as in arbitrary categories, with the additional requirement that all the $f(A): \phi(A) \to A$ and $f(A): A \to \phi(A)$ as well as all epimorphisms and monomorphisms occurring in the commutative triangles of the definitions be admissible.

Let now K be an arbitrary variety of universal algebras considered as a bicategory as in the example above, and let F be a verbal function in K . Attach a functor ϕ to F by putting for all $A \in K$ $\phi(A) = A/F(A)$, and let $f(A): A \to \phi(A)$ be the corresponding natural homomorphism. As an immediate consequence of the definition of verbal functions we obtain that to each homomorphism $\alpha : A \to B$ a $\phi(\alpha): \phi(A) \to \phi(B)$ is assigned, and that ϕ is an idempotent factor functor of the identity.

THEOREM 1.3. *For every verbal function* F *the corresponding functor* ϕ *is a verbal in* K *. The verbal function belonging to* ϕ *is just the original* F *. If* ϕ *is a verbal in the bicategory* K *and* F *is the corresponding verbal function then the functor attached to* F *is equivalent to the original* ϕ *.*

Proof. Let ϕ be the functor attached to a verbal function F and suppose that an admissible commutative diagram

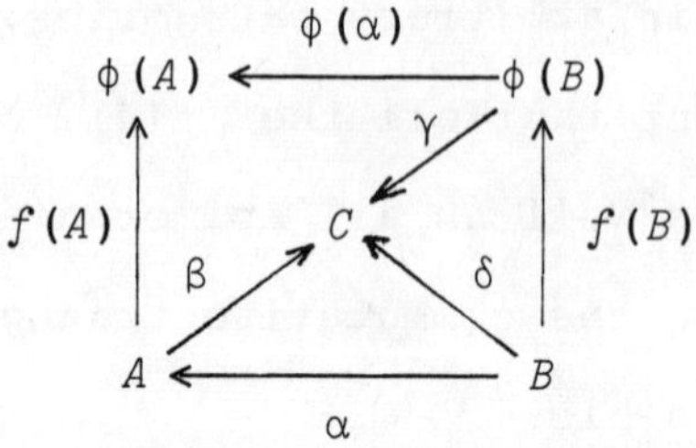

is given. Let Θ be the variety defined by the verbal function F . Since $\phi(B) = B/F(B) \in \Theta$ and $\gamma\colon \phi(B) \to C$ is a surjective homomorphism, we have $C \in \Theta$. Denoting by ρ the kernel equivalence of β , we have therefore $F(A) \subset \rho$. Then there is a natural homomorphism $\varepsilon\colon A/F(A) = \phi(A) \to C$ which makes our diagram commutative.

Thus we have proven that ϕ is a verbal, and it is clear that F is the verbal function belonging to ϕ .

Let now ϕ be an arbitrary verbal, F be the verbal function corresponding to ϕ , and ϕ' be the verbal attached to $\mathcal{F}$. For any object A , $\mathcal{F}(A)$ is the kernel congruence of the surjective homomorphism $f(A):$ $:A \to \phi(A)$. Considering the natural homomorphism $f'(A):$ $:A \to \phi'(A) = A/\mathcal{F}(A)$, we obtain now a commutative diagram

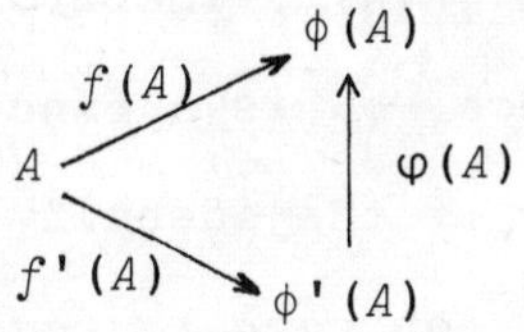

where $\varphi(A)$ is an isomorphism.

Next we turn to the similar theorem on radicals.

Let K be a variety of multioperator groups. If $A \in K$ and B is a subalgebra of A then B is said to be *accessible* in A if there is a chain $B = B_0, B_1, \ldots,$ $B_n = A$ in which each member is an ideal of the following one. A homomorphism $\alpha\colon B \to A$ is called accessible if its image is accessible in A . Clearly, any product of accessible homomorphisms is again accessible. Our category K will now be the variety K together with the accessible homomorphisms as morphisms. This K will be considered as a bicategory in the natural way.*) For each

*) A similar framework (categories with iterated normal morphisms) has been introduced also in papers by B. Terlikowska with the aim of developing a theory of extension-closed radicals in certain categories (see e.g. [5]).

radical function F we define a functor ϕ . Put $\phi(A)= =F(A)$ and let $f(A): \phi(A) \to A$ be the identical embedding. Suppose we have an accessible homomorphism $\alpha: B \to A$ with image B' . Then $F(B)^{\alpha} \subset F(B') = B' \cap F(A)$, and $F(B)^{\alpha}$ is clearly accessible in $F(A)$. Therefore α induces a $\phi(\alpha): \phi(B) \to \phi(A)$ hence ϕ is a functor. This ϕ is an idempotent subfunctor of the identity.

PROPOSITION 1.4. *For each radical function* F *the functor* ϕ *is a radical.*

Proof. Suppose we have an accessible commutative diagram

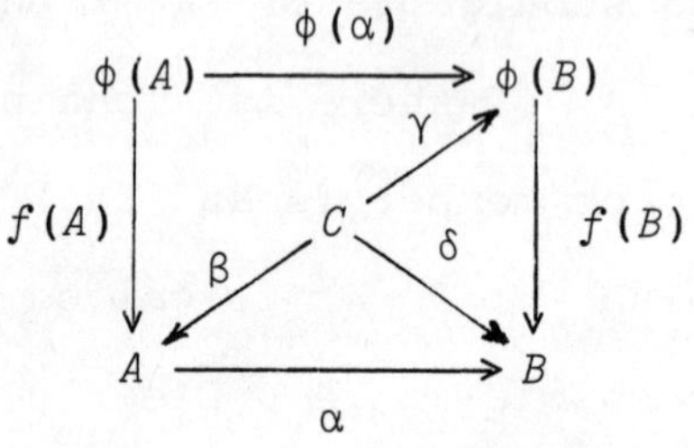

Let Θ be the radical class belonging to F . Since $\phi(B) \in \Theta$ and $\gamma: C \to \phi(B)$ is an accessible monomorphism, we have $C \in \Theta$. On the other hand, C^{β} is an accessible subalgebra of A , hence

$$C^{\beta} = F(C^{\beta}) = F(A) \cap C^{\beta} \quad , \quad C^{\beta} \subset F(A) \ ,$$

and C^{β} is an accessible subalgebra of $F(A)$. This yields an accessible monomorphism $\varepsilon: C \to \phi(A)$ which gives a commutative extension of our diagram.

It is also clear that ϕ satisfies the conditions in Proposition 1.2 and that the radical function corresponding to ϕ is F .

Some remarks should be made here. Firstly, if ϕ is a subfunctor of the identity of a category K then for every object A the image $F(A)$ of the map $f(A)$: : $\phi(A) \to A$ is a characteristic subobject of A . In particular, if K is the category of groups then $F(A)$ is a normal subgroup of A . However, we cannot in general state that $F(A)$ is an ideal of A even if ϕ is a radical. Therefore Proposition 1.2 does not apply directly to all radicals. For this reason we have to modify the definition of radicals.

In what follows our category K will be supposed to have zero objects and zero morphisms. Next we define normal monomorphisms.

Let $\mu: A \to B$ be a monomorphism and $\nu: X \to B$ be a morphism in K such that whenever $\mu\varphi = 0$ for some $\varphi: B \to C$ then $\nu\varphi = 0$. μ is said to be a *normal monomorphism* if for all ν as above, there is a $\nu': X \to A$ such that $\nu = \nu'\mu$:

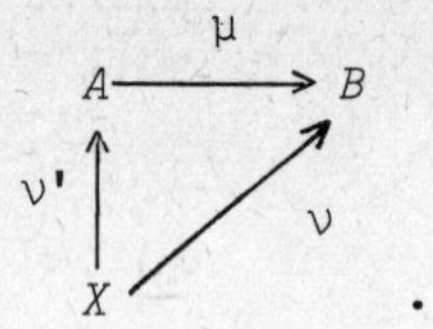

The following statement is easy to prove. If our category K is a variety of multioperator groups then a monomorphism $\mu: A \to B$ is normal in K if and only if A^{μ} is an ideal of B .

Now we define a subfunctor ϕ of the identity in K to be a *normal subfunctor* if every $f(A): \phi(A) \to A$ is a normal monomorphism.

Finally, a radical ϕ is defined now as a *normal* subfunctor of the identity which satisfies the conditions of the previous definition. This concerns also radicals in bicategories.

Now we have:

THEOREM 1.5. *Let K be a variety of multioperator groups considered as a bicategory with the accessible homomorphisms as morphisms. Then there is a* 1-1 *correspondence between radical functions in the variety K and radicals in the bicategory K , up to equivalence of radicals.*

Proof. Let ϕ be a radical in the bicategory K . By definition, the function F belonging to ϕ assigns an ideal to each object, and $f(A): \phi(A) \to A$ is always a monomorphism. By Proposition 1.2, F is then a radical function. Let now ϕ' be the radical determined by F according to Proposition 1.4. For any object A, the map $f(A): \phi(A) \to A$ induces in an obvious way a map $\varphi(A): \phi(A) \to \phi'(A)$, and the function φ establishes equivalence of the radicals ϕ and ϕ'. In view of Proposition 1.4, this proves the theorem.

Dually to normal monomorphisms we define normal epimorphisms, and we can consider normal factor functors of the identity. Then verbals could be defined, dually to radicals, as normal factor functors of the identity. However, this approach would yield nothing new.

2. DISCUSSION AND SOME PROBLEMS

The picture that has just been presented becomes especially simple and natural in the category of groups. Here we need not make use of bicategories and need not distinguish normal subfunctors or normal factor functors of the identity. The same applies to the category of modules, too.

Notice that in our approach radicals and verbals in categories are defined as functors while in varieties of algebras they are based on radical and verbal functions, respectively. If a variety is considered as a category then we have a natural correspondence of the two definitions. This approach can be found in [3] already. That paper also treated the duality between radicals and verbals but there it was assumed that the notion of exact sequence makes sense in the given category. Here we make practically no assumptions on the category.

The notions of radical and verbal variety can be carried over to categories also from another point of view. For instance, in the book [1] several conditions are imposed on the categories in order to develop a general radical theory in them - this leads to special categories which imitate the case of multioperator groups. This approach enables one not only to formulate definitions but also to prove some theorems in the categorical language. This is sometimes useful.

These two approaches can be related, say, as we did for radicals in a category which is a variety of multioperator groups and for radical functions in the variety. Instead of this variety we now consider our special category. In doing so, one has to proceed parallelly to the considerations in [1] and carry over to categories the general radical theory without supposing radical classes to be closed under extensions.

Let us have a look at this condition of extension closedness. It means that the factor algebra by the radical is semisimple, and this is needed for applications to structure theories in the algebras. As is well known, these applications led to the development of general radical theory. However, other applications have also arisen in the meanwhile.

Radicals are used in proving various theorems, the idea of a radical class is one of the fundamental organizational ideas - together with that of varieties, it is a helpful tool in classifying algebraic structures. And this concerns not only extension closed (for short: closed) radicals. For instance, as we have noticed already, the locally nilpotent radical in groups or Lie algebras is not closed. It is possible that the consideration of radicals which need not be closed is more important in groups or Lie algebras than in associative algebras.

One of the advantages of considering not necessarily closed radicals is the following. The product of two radical classes is again a radical class, but the product of two closed radicals need not be closed. Because of this fact, there are interesting problems involving properties of this product which are meaningless for closed radicals. For instance, free semigroups arising in connection with products of radical classes are investigated in several papers on groups and group representations. This product distinguishes closed radicals for they are exactly the idempotents.

In the case of modules radicals are connected with radical filters. The definition of such filters is much simpler and easier to visualize if the radicals in question need not be closed. Similarly to the product of radicals we can also define the product of filters. Then the idempotent filters will be just the ones corresponding to idempotent (i.e., closed) radicals.*)

Finally, notice that every radical admits an idempotent closure in the natural way. For instance, the closure of the locally nilpotent radical in groups is the class of those groups which have an ascending normal series with locally nilpotent factors.

Of course, with all this we do not want to say that

---

*) Speaking about modules, let us mention that our notion of radical has already been introduced for them, under a different name, in works of the Prague radical school, but their investigations go in a completely different direction as compared with ours.

one should never impose the condition of extension closedness. Closed radicals have their advantages and special features. For example, the lattice of radical classes seems to be more suited for investigations under the presence of extension closedness.

Let us exhibit now some special properties of radicals in groups and rings.

A class Θ of groups is called a radical class if it is closed under homomorphic images and normal subgroups and if every group generated by invariant subgroups belonging to Θ is itself in Θ. For groups this definition is equivalent with the one formulated above for multioperator groups. If Θ is a radical class of groups and $\Theta' = F$ is the function which assigns to each group A the sum of its invariant Θ-groups, then this function enjoys the following important property: if B is a normal subgroup in A then $F(B)$ is also. It is due to this fact that the radical classes of groups defined in this way share the necessary additional properties. Moreover, for any class of groups the smallest radical class containing it can be described by sufficiently simple closure operators.

For rings everything is more complicated. Should we take the same definition as for groups, it would not follow that B being an ideal in A implies that $F(B)$ is also. This is true if the corresponding ring theoretical condition of being characteristic is satisfied. For various concrete radicals the validity of this condition can be proved or disproved. It is not clear how this condition could be included in the axioms of radicals in a decent way, having in mind that one should have a good procedure to compute the smallest radical class containing an arbitrary class of rings.

And now some words about the connections of radical classes and varieties. We have seen that they are dual to each other. In the case of modules every variety is also a radical class, and this radical class is connected with a principal filter - a filter generated by one two-sided ideal.

For groups the picture is just the opposite. If a variety is also a radical class then it is either the trivial class or the class of all groups. For rings similar investigations have been made for closed radicals.

If a radical class is at the same time a variety then it can be axiomatized in a first order language. What can be said about axiomatizable radical classes in general? Answering a question of mine, G. M. Bergman constructed a series of radical classes of groups which are axiomatizable by first order formulas. All these classes consist of periodic groups. It is not clear how important the condition of periodicity is here. If a radical class is axiomatizable then it is also closed under Cartesian products. Now the following question arises naturally. Consider the smallest radical class of groups closed under Cartesian products and containing the infinite cyclic group. Is this class the class of all groups? - For closed radicals of rings, problems of axiomatizability have been investigated quite recently.

Finally, let us make some remarks on the philosophy of radicals (again we mean not necessarily closed radicals). What is a radical? Now it is clear that in algebraic structures it is at the same time a radical class and a radical function. In categories it is also a special functor. All these viewpoints on radicals are related to each other in a natural way. The same can be said also about verbals and varieties. The situation with varieties is simpler. They are defined by identities, and the verbal functions can be determined for arbitrary universal algebras. Radicals do not have such a universal property, they can be defined only in nice algebras, and even in that case, not everything is so clear from the definition as for verbals. Here we made the following attempt. We defined verbals in a functorial way and then radicals by duality, having in mind that such a duality could be observed in groups and modules. The result led to the related notion of torsions in rings. However, the locally nilpotent radical in Lie algebras is not conform to our definition. Maybe it would be worth thinking over such a functorial definition of radicals which does not exclude this important example.

The functorial approach to radicals can be used to construct radicals also for manysorted structures other than modules, e.g., for automata.

REFERENCES

[1] V. A. Andrunakievič, Ju. M. Rjabuhin, *Radicals of algebras and structure theory* (Russian), Moscow, 1979.

[2] N. M. Kahn, Epimorphisms, dominions, and varieties of semigroups, *Semigroup Forum* 25 (1982), 331-337.

[3] B. I. Plotkin, On functorials, radicals and coradicals in groups (Russian), *Mat. Zapiski Ural'sk. Gos. Univ.* 7 (1970), no. 3, 150-183.

[4] B. I. Plotkin, Radicals in groups, operators on classes of groups and radical classes (Russian), *Selected Questions of Algebra and Logic*, Novosibirsk 1973, 205-244.

[5] B. Terlikowska-Osłowska, Category with a self-dual set of axioms, *Bull. Acad. Polon. Sci. Sér. Sci. Math. Astronom. Phys.* 25 (1977), 1207-1214.

B. I. Plotkin
226024 Riga 24,
al. Vidzemes 8/35.
USSR

COLLOQUIA MATHEMATICA SOCIETATIS JÁNOS BOLYAI
38. RADICAL THEORY, EGER (HUNGARY), 1982.

A RELATIVE JACOBSON RADICAL WITH APPLICATIONS

T. PORTER

There is a very successful method often used in localisation theory which goes as follows: one has a subcategory T, of "torsion modules" of some module category Mod-A and one can form the quotient category, (Mod-A)/T, which is a Grothendieck category. Now, if P is a classical property of modules (e.g. Noetherian, Artinian, finite length, etc.), try if possible, to translate this into a property of objects in a general Grothendieck category and say a module M has the property P relative to T if the object corresponding to M in (Mod-A)/T has property P.

An alternative and often equivalent method is to use the lattice, $C_{\mathbb{F}}(M)$ of submodules N of M such that M/N is torsion free. Then if a property P is expressible in terms of the lattice of submodules or of finitely generated submodules of a module, one can say M has "P rel. T" if $C_{\mathbb{F}}(M)$ has this property P . Some of the clearest examples of uses of this method are to be found in the work of Albu and Năstăsescu ([1], [13] and [14]) but many other authors have also used it and indeed it is basic to much of modern localisation theory.

(Actually it is more useful for us to express the structure in terms of the additive topology (topologising and idempotent filter), $\mathbb{F}$, associated with the localisation than in terms of the torsion theory, thus one speaks of "P rel. $\mathbb{F}$".)

This lattice $C_{\mathbb{F}}(M)$ is isomorphic to the lattice of subobjects of image of M in (Mod-A)/T and as such need not have proper maximal elements (Zorn's lemma fails in general in a topos and (Mod-A)/T is exactly an abelian subcategory of sheaves on the site, with underlying category, the (additive) category with one object, A , and with Grothendieck topology $\mathbb{F}$). To be able to adapt classical proofs one therefore has to ensure that at least finitely generated objects have maximal subobjects. More exactly we introduce a condition $\mathbb{F}$-Max which states that $C_{\mathbb{F}}(M) - \{M\}$ has maximal elements for all finitely generated M . In [17] where this condition was introduced, a relative form of Nakayama's Lemma was proved. This result uses the relative form of the Jacobson radical of M namely the intersection of the maximal elements of $C_{\mathbb{F}}(M) - \{M\}$. The image of this in (Mod-A)/T is the radical of the image of M hence one can hope to generalise many useful notions of ring theory using this tool.

The motivation for this study is primarily an attempt by the author to generalise a result of Jensen [11]. This is fully discussed in Section 8 so suffice it to say here that one possible generalisation involves a relative form of semi-local ring. If such a generalisation is to be fruitful, these semi-local rings mod $\mathbb{F}$ must have nice properties other than those studied in that generalisation, hence it seems logical to devote some time and effort to investigating these relative semi-local rings and related relative versions of semi-simple, regular, semi-primary etc. rings in order to give some context and perspective to the conjectured generalisation. The discussion of these relative properties takes up the greater part of the paper. However in the last section we apply the theory of flat localisations to these relative properties and show that they induce the (absolute) property on the localised rings concerned.

These latter results seem to have "non empty intersection" with classical localisation results of Goldie and others on maximal semi-simple rings of quotients etc.

I would like to thank Jonathan Golan for the following two remarks (i) The relative radical of a ring is introduced in his book [21] on page 190. (ii) The use of the term "$\mathbb{F}$-saturated", for those submodules N of a module M for which M/N is $\mathbb{F}$-torsion free is not completely standard. Popescu [15] uses it but other authors use "$\mathbb{F}$-pure" or "$\mathbb{F}$-closed". Each term has its advantages and disadvantages and is used in slightly different senses by various authors. This may lead to some confusion.

In this paper we only scratch the surface of possible applications of the study of relative radicals. Our results are for the most part relative analogues of classical results; clearly there are many other similar results. The author hopes, however, that the suggested links with homological algebra given in Section 8 will cause the reader to search for more links between the general area of radical theory/torsion theories and that of homological dimension.

1. DEFINITIONS, NOTATION AND PRELIMINARY RESULTS.

All rings considered have non zero identity elements,

are associative but not necessarily commutative. All modules are unitary and are right modules over the ring in question. In this section we shall fix notation and terminology for localisations and make some preliminary definitions.

Let A be a ring and Mod-A the category of right A-modules.

An *additive topology* (of right ideals) on A is a non empty set, $\mathbb{F}$ of right ideals satisfying the following conditions:

(1) If $I \in \mathbb{F}$, $x \in A$ then $(I:x) \in \mathbb{F}$ where $(I:x) = \{\lambda | \lambda \in A, x\lambda \in I\}$

(2) If I, J are right ideals of A such that $J \in \mathbb{F}$ and for every $x \in J$, $(I:x) \in \mathbb{F}$, then $I \in \mathbb{F}$.

The notion of an additive topology on A coincides with the notion of a topologising and idempotent system of right ideals of A as defined by Gabriel ([6] Ch.5 p.411).

If $\mathbb{F}$ is an additive topology on A, we will denote by

$$T = \{M | M \in \text{Mod-}A,\ x \in M,\ x \neq 0 \Rightarrow \text{Ann}(x) \in \mathbb{F}\}$$

and

$$F = \{M | M \in \text{Mod-}A,\ x \in M \text{ and } \text{Ann}(x) \in \mathbb{F} \Rightarrow x = 0\}$$

The pair (T, F) is a hereditary torsion theory on Mod-A. The modules in T are called $\mathbb{F}$-*torsion modules* and those in F, $\mathbb{F}$-*torsion free*. We denote by $\tau(M)$ the largest $\mathbb{F}$-torsion submodule of a module M. τ is the associated torsion radical of the torsion theory (cf. Stenström [20]).

The canonical quotient functor associated with $\mathbb{F}$ will be denoted

$$T_{\mathbb{F}} : \text{Mod-A} \longrightarrow (\text{Mod-A})/T$$

$T_{\mathbb{F}}$ is exact but its right adjoint $S_{\mathbb{F}}$ is generally only left exact.

Recall that $L_{\mathbb{F}} = S_{\mathbb{F}}T_{\mathbb{F}}$, the localisation functor associated with $\mathbb{F}$, is such that $L_{\mathbb{F}}(A)$ is a ring and the natural transformation.

$$\psi : 1 \longrightarrow L_{\mathbb{F}}$$

is such that $\psi(A)$ is a ring homomorphism and

$$\tau(M) = \text{Ker}(\psi(M) : M \longrightarrow L_{\mathbb{F}}(M)) .$$

The quotient category (Mod-A)/T is a Grothendieck category. By the theorem of Gabriel-Popescu, all Grothendieck categories arise in this way (cf. Popescu [15] p.180) and so our results can be interpreted as being results about certain classes of Grothendieck categories.

Apart from using techniques from the theory of Grothendieck categories we also use lattice theoretic information on the lattice of subobjects of an object in (Mod-A)/T . This lattice can be extracted from the lattice $C_{\mathbb{F}}(M)$ of $\mathbb{F}$-saturated submodules of M to which it is isomorphic.

If M is any right A-module, $L \subseteq M$ a submodule, then L^c will denote the submodule of M defined by

$$\frac{L^c}{L} = \tau\left(\frac{M}{L}\right)$$

The $\mathbb{F}$-*saturated submodules* are those satisfying: $L^c = L$ (cf. Stenström [20] Section 11). $C_{\mathbb{F}}(M)$ is a complete modular lattice with inclusion as order and if $(L_i)_{i \in I}$ is a family of elements of $C_{\mathbb{F}}(M)$,

$$\bigvee_{i \in I} L_i = \left(\sum_{i \in I} L_i\right)^c$$

and

$$\bigwedge_{i \in I} L_i = \bigcap_{i \in I} L_i$$

We shall write $\mathbb{F}(M) = \{N \subseteq M \mid N^c = M\}$,

$\mathrm{Max}_{\mathbb{F}}(M)$ = set of maximal elements of $C_{\mathbb{F}}(M) - \{M\}$. This latter can be empty as maximal subobjects of $T_{\mathbb{F}}(M)$ need not exist. To enable classical module theoretic proofs to be adapted to this setting it is important that one has maximal subobjects or equivalently $\mathbb{F}$-maximal submodules, that is elements of $\mathrm{Max}_{\mathbb{F}}(M)$. Thus we introduce the following condition:

$\mathbb{F}$-Max : *If* M *is finitely generated then* $\mathrm{Max}_{\mathbb{F}}(M) \neq \emptyset$.

If A is $\mathbb{F}$-*Noetherian* (i.e. if $T_{\mathbb{F}}(A)$ is a Noetherian object or equivalently if $C_{\mathbb{F}}(A)$ is a Noetherian lattice) then clearly $\mathbb{F}$-Max is satisfied. As yet it has not proved possible to prove $\mathbb{F}$-Max if A is $\mathbb{F}$-Artinian without recourse to the Năstăsescu-Miller-Teply result that $\mathbb{F}$-Artinian implies $\mathbb{F}$-Noetherian.

One trivial consequence of $\mathbb{F}$-Max is that if N is a submodule of a finitely generated M, there is an $\mathbb{F}$-maximal L contained in M containing N since $\mathrm{Max}(M/N) \neq \emptyset$.

We shall say that an A-module M is $\mathbb{F}$-*simple* if $C_{\mathbb{F}}(M) = \{0, M\}$ and $\tau(M) = 0$.

Thus our $\mathbb{F}$-simple modules are the "$\mathbb{F}$-cocritical" modules of other authors, likewise our $\mathbb{F}$-maximal subobjects would be "$\mathbb{F}$-critical". This departure from existing terminology is intended to show how nearly classical our proofs are. (Note $T_{\mathbb{F}}(M)$ may be simple without M being $\mathbb{F}$-simple; $T_{\mathbb{F}}(M)$ is simple if and only if $M/\tau(M)$ is $\mathbb{F}$-simple.)

The $\mathbb{F}$-*radical* of an A-module is defined to be

$$\begin{aligned} \mathrm{rad}_{\mathbb{F}}(M) &= \cap\{\mathrm{Ker}(h : M \xrightarrow{\text{onto}} S) \mid S\ \mathbb{F}\text{-simple}\} \\ &= \cap\{H \mid H \in \mathrm{Max}_{\mathbb{F}}(M)\} \end{aligned}$$

If $\mathrm{Max}_{\mathbb{F}}(M) = \emptyset$, $\mathrm{rad}_{\mathbb{F}}(M) = M$ of course. In Raynaud ([19] p.21), one finds the lemma that $\mathrm{Max}_{\mathbb{F}}(A) \neq \emptyset$ if and only if (Mod-A)/T has simple objects. Thus if A is $\mathbb{F}$-Artinian, $T_{\mathbb{F}}(A)$ is Artinian, hence has simple subobjects and $\mathrm{Max}_{\mathbb{F}}(A) \neq \emptyset$. However the extension of this to $\mathrm{Max}_{\mathbb{F}}(M) \neq \emptyset$ for all M of finite type does not seem to be easy.

In [17] we have proved that $\mathbb{F}$-Max suffices to prove Nakayama's lemma module $\mathbb{F}$ (i.e. in (Mod-A)/T for ideals $J \subseteq \mathrm{rad}_{\mathbb{F}}(A)$). The statement and proof are fairly clear from the context and so are omitted here.

PROPOSITION 1.1

(i) *Let* $g : N \to M$ *be a morphism in* Mod-A, *then*

$$g(\mathrm{rad}_{\mathbb{F}} N) \subseteq \mathrm{rad}_{\mathbb{F}}(M)$$

(ii) $\mathrm{rad}_{\mathbb{F}}(M)$ *is a fully invariant submodule of* M

(iii) $\mathrm{rad}_{\mathbb{F}}(A)$ *is a two sided ideal of* A

(iv) *For any* M *in* Mod-A, $M.\mathrm{rad}_{\mathbb{F}}(A) \subseteq \mathrm{rad}_{\mathbb{F}}(M)$.

Proof

(i) $\Longrightarrow$ (ii) $\Longrightarrow$ (iii) trivially.

(i) is proved as in the classical case (see Bass [3]). This same reference provides a proof which can be easily adapted to give one of (iv).

Given a surjection $\phi : A \to B$, and an additive topology $\mathbb{F}$, there is an induced topology $\phi(\mathbb{F})$ having a particularly nice description,

$$\phi(\mathbb{F}) = \{\phi(I) : I \in \mathbb{F}\}.$$

PROPOSITION 1.2

Let $\phi : A \to A/\mathrm{rad}_{\mathbb{F}}(A)$ *be the natural morphism. There is an induced lattice homomorphism*

$$C(\phi) : C_{\phi(\mathbb{F})}(A/\mathrm{rad}_{\mathbb{F}}(A)) \to C_{\mathbb{F}}(A)$$

given by

$$C(\phi)(\phi(I)) = I + \mathrm{rad}_{\mathbb{F}}(A)$$

$C(\phi)$ *is a lattice monomorphism and sends* $\mathrm{Max}_{\phi(\mathbb{F})}(A/\mathrm{rad}_{\mathbb{F}}(A))$ *bijectively onto* $\mathrm{Max}_{\mathbb{F}}(A)$.

The proof is straightforward.

COROLLARY 1.3

$$\mathrm{rad}_{\phi(\mathbb{F})}(A/\mathrm{rad}_{\mathbb{F}}(A)) = (0)$$

One can equally well show that if $J \subseteq \mathrm{rad}_{\mathbb{F}}(A)$ is a two sided ideal and $\phi : A \to A/J$ is the canonical surjection,

$$\mathrm{rad}_{\phi(\mathbb{F})}(A/J) = \frac{\mathrm{rad}_{\mathbb{F}}(A)}{J}$$

as A/J-modules. Thus for instance

$$\mathrm{rad}_{\phi(\mathbb{F})}(A/\tau(A)) = \frac{\mathrm{rad}_{\mathbb{F}}(A)}{\tau(A)}$$

gives the passage from the radical of an arbitrary ring to that of its torsion free quotient.

Note as before that these results only have substance if A actually has some proper $\mathbb{F}$-maximal ideals e.g. if $(A, \mathbb{F})$ satisfies $\mathbb{F}$-Max . Otherwise they reduce to trivialities.

We next list some results, analogues of results on the Jacobson radical. As the standard proofs (found for instance in Bourbaki [5]) go over to this more general situation almost word for word, the proofs are left to the diligent reader.

PROPOSITION 1.4

(i) $\mathrm{rad}_{\mathbb{F}}(\oplus M_i) = \oplus \mathrm{rad}_{\mathbb{F}}(M_i)$ *for any family of modules* (M_i) .

(ii) *If* $\mathrm{rad}_{\mathbb{F}}(M) = \tau(M)$, *(the smallest it can be), then there is an* $\mathbb{F}$*-monomorphism*

$$M \longrightarrow \Pi S_i$$

where (S_i) *is a family of* $\mathbb{F}$*-simple modules.*

The result that all nilpotent ideals are contained within $\mathrm{rad}(A)$ has an analogue in this situation.

Firstly we shall say that $J \subset A$ is $\mathbb{F}$*-nilpotent* if there is some n , $J^n \subseteq \tau(A)$.

PROPOSITION 1.5

If $(A, \mathbb{F})$ *satisfies* $\mathbb{F}$-Max, *and the ideal* J *is* $\mathbb{F}$*-nilpotent then* $J \subseteq \mathrm{rad}_{\mathbb{F}}(A)$

Proof

We have from Nakayama's lemma modulo $\mathbb{F}$ ([17]) that $J \subseteq \mathrm{rad}_{\mathbb{F}}(A)$ if and only if for all M of $\mathbb{F}$-finite type, $M/MJ \in T \Longrightarrow M \in T$.

Let M be any module and assume $M/MJ \in T$, then

clearly $MJ/MJ^2 \in T$, so examination of the short exact sequence

$$0 \longrightarrow \frac{MJ}{MJ^2} \longrightarrow \frac{M}{MJ^2} \longrightarrow \frac{M}{MJ} \longrightarrow 0$$

shows $M/MJ^2 \in T$. Continuing one obtains $M/MJ^n \in T$ so $M/M\tau(A) \in T$. Finally since $M\tau(A) \subset \tau(M)$, $M/\tau(M) \in T$ so $M = \tau(M)$. Hence $J \subseteq \mathrm{rad}_{\mathbb{F}}(A)$ as required.

We finish this section with some remarks on the interpretation of $C_{\mathbb{F}}(M)$, $\mathrm{Max}_{\mathbb{F}}(M)$, etc.

If $L \in C_{\mathbb{F}}(M)$, then $T_{\mathbb{F}}(L) \subseteq T_{\mathbb{F}}(M)$.

If X is a subobject of $T_{\mathbb{F}}(M)$, then by standard arguments of localisation theory, one can find an $L \subset M$ such that

$$\begin{array}{ccc} X & \xrightarrow{\;\cong\;} & T_{\mathbb{F}}(L) \\ & \searrow \quad \swarrow & \\ & T_{\mathbb{F}}(M) & \end{array}$$

commutes (cf. Popescu 15 4.3.10, p.172). Now $T_{\mathbb{F}}(L) \to T_{\mathbb{F}}(M)$ factors as $T_{\mathbb{F}}(L) \xrightarrow{\cong} T_{\mathbb{F}}(L^c) \to T_{\mathbb{F}}(M)$ so each subobject X of $T_{\mathbb{F}}(M)$ determines a unique $L^c \in C_{\mathbb{F}}(M)$ and conversely.

Under this correspondence the lattice structure of $C_{\mathbb{F}}(M)$ corresponds to the subobject lattice of $T_{\mathbb{F}}(M)$. $\mathrm{Max}_{\mathbb{F}}(M)$ corresponds to the set of maximal subobjects of $T_{\mathbb{F}}(M)$ - both may be empty, of course - and

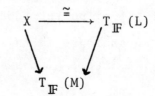

$$T_{\mathbb{F}}(\mathrm{rad}_{\mathbb{F}}(M)) = \wedge\{T_{\mathbb{F}}(H) \mid H \in \mathrm{Max}_{\mathbb{F}}(M)\}$$
$$= \wedge\{H' \mid H' \in \mathrm{Max}(T_{\mathbb{F}}(M))\}$$

so $\mathrm{rad}_{\mathbb{F}}(M)$ corresponds to the Jacobson radical of $T_{\mathbb{F}}(M)$ in (Mod-A)/T .

Finally we note the following:

PROPOSITION 1.6

For any $M \in$ Mod-A , *there are lattice isomorphisms*

$$C_{\mathbb{F}}(M) \longrightarrow C_{\mathbb{F}}(M/\tau(M))$$
$$L \longmapsto L/\tau(M)$$

and

$$C_{\mathbb{F}}(L_{\mathbb{F}}(M)) \longrightarrow C_{\mathbb{F}}(M/\tau(M))$$
$$L \longmapsto L \cap (M/\tau(M))$$

Proof

The first is easy; the second is a consequence of Stenström [20] 11.2 if one identifies $M/\tau(M)$ with its natural image in $L_{\mathbb{F}}(M)$.

2. $\mathbb{F}$-SEMISIMPLICITY AND THE $\mathbb{F}$-RADICAL.

We shall say that M in Mod-A is $\mathbb{F}$-*semisimple* if $T_{\mathbb{F}}(M)$ is a semisimple object in (Mod-A)/T .

PROPOSITION 2.1

If M *is an* $\mathbb{F}$-*semisimple* $\mathbb{F}$-*closed module, then*

(i) *every* $\mathbb{F}$-*closed submodule of* M *is a direct summand,*

(ii) *the endomorphism ring of* M *is von Neumann regular.*

Proof

M is $\mathbb{F}$-closed if $\psi(M) : M \to L_{\mathbb{F}}(M)$ is an isomorphism (cf. Stenström [20]). Under these conditions $L \subseteq M$ is in $C_{\mathbb{F}}(M)$ if and only if L is itself $\mathbb{F}$-closed. M being $\mathbb{F}$-semisimple, $C_{\mathbb{F}}(M)$ is a complemented lattice (by Stenström [20] 11.5) so the result follows from [20] 11.6.

Remark

If one assumes M is merely $\mathbb{F}$-semisimple, then one has problems of "$\mathbb{F}$-direct summands":

If L_1, L_2 are two submodules of a module M, we shall say that M is the $\mathbb{F}$-*direct sum* of L_1 and L_2 if

$$\text{(i)}\quad L_1 \cap L_2 \subseteq \tau(M)\ , \quad \text{(ii)}\quad L_1 + L_2 \in \mathbb{F}(M)\ .$$

L_1 is an $\mathbb{F}$-direct summand of M if there is such an L_2. Clearly one can in general state

COROLLARY 2.2

If M *is* $\mathbb{F}$-*semisimple, each submodule is an* $\mathbb{F}$-*direct summand.*

Proof

One has that $C_{\mathbb{F}}(M) \cong C_{\mathbb{F}}(L_{\mathbb{F}}(M))$ and is a complemented lattice i.e. given $L \subset M$, one has $L^c \in (C_{\mathbb{F}}(M))$, and there is an $N \in C_{\mathbb{F}}(M)$

$$L^c \wedge N = 0^c \text{ , the zero of the lattice}$$

whilst

$$L^c \vee N = M$$

Thus $L^c \cap N = \tau(M)$, whilst

$$(L^c + N)^c = M$$

The result follows.

We shall say that A is $\mathbb{F}$-*semisimple* if it is $\mathbb{F}$-Artinian and as a module over itself is $\mathbb{F}$-semisimple.

Remark

If A is $\mathbb{F}$-semisimple and $\mathbb{F}$-torsion free then $\mathbb{F}$ must contain the Goldie Topology (generated by the essential right ideals).

PROPOSITION 2.3

If A *is* $\mathbb{F}$-*semisimple, then* (Mod-A)/T *is a semisimple Grothendieck category (also called a discrete spectral Grothendieck category), and hence is locally finite.*

Proof

Let $M \in$ Mod-A and let

$$A^{(I)} \longrightarrow M$$

be an epimorphism then $T_{\mathbb{F}}(M)$ is a quotient of the semisimple $T_{\mathbb{F}}(A)^{(I)}$ and so is semisimple. The rest follows from the general theory of spectral Grothendieck categories (cf. [15]).

Remark

A may be $\mathbb{F}$-semisimple as a right A-module without being $\mathbb{F}$-Artinian (cf. Albu and Năstăsescu [1]).

We next turn to $\mathbb{F}$-nilpotent two sided ideals in $\mathbb{F}$-Artinian rings.

THEOREM 2.4

If A *is* $\mathbb{F}$-*Artinian and* $\mathbb{F}$-Max *holds then* $\mathrm{rad}_{\mathbb{F}}(A)$ *is the largest* $\mathbb{F}$-*nilpotent two sided ideal of* A .

Proof

We already have that if J is $\mathbb{F}$-nilpotent, then $J \subseteq \mathrm{rad}_{\mathbb{F}}(A)$.

Set $\mathrm{rad}_{\mathbb{F}}(A) = R$ and assume A is $\mathbb{F}$-Noetherian, then there is a p such that $R^p/R^{p+1} \in T$.

Let $R^p = Q$, then $Q/RQ \in T$. If $Q \in T$ then we ar finished, so assume $Q \notin T$. Then $RQ \notin T$ and there exist right ideals P such that $PQ \notin T$. If $PQ \notin T$ then $P^cQ \notin T$ so we may assume $P \in C_{\mathbb{F}}(A)$. Let M be an $\mathbb{F}$-minimal such ideal (i.e. if $M' \subseteq M$ and $M'Q \notin T$ then $M/M' \in T$). One has $(MR)Q = M(RQ)$ and $(MQ)/M(RQ) \in T$ and since $MR \subseteq M$, $\mathbb{F}$-minimality gives us $M/MR \in T$

If we can show that M is of $\mathbb{F}$-finite type then we are finished since the relative form of Nakayama's lemma [17] will show that $M \in T$, hence that $MQ \in T$, contradicting the choice of M . As $MQ \notin T$, there must be some $x \in M$, $xQ \notin T$. Therefore $(xA)Q \notin T$ (since $Q = R^P$ is two sided). However this implies $M/xA \in T$ by $\mathbb{F}$-minimality of M i.e. M is of $\mathbb{F}$-finite type.

Remark

This proof is a direct transcription mod $\mathbb{F}$ of Bourbaki's proof of the classical result.

We shall say that M is of $\mathbb{F}$-*finite length* if $T_{\mathbb{F}}(M)$ is an object of finite length in (Mod-A)/T .

This is at variance with some other notions of relative finite length, for instance in Goldman where M must be torsion free and each quotient in the filtration must be torsion free. The notion here is weaker but any object of $\mathbb{F}$-finite length in this sense is $\mathbb{F}$-isomorphic to one which is of "σ-finite length" in Goldman's sense (i.e. for $\sigma = \tau$, cf. Miller-Teply [12]).

THEOREM 2.5

Let M *be an* A-*module*. M *is* $\mathbb{F}$-*semisimple of* $\mathbb{F}$-*finite length if and only if* M *is* $\mathbb{F}$-*Artinian and* $rad_{\mathbb{F}}(M) = \tau(M)$.

Proof

Since $T_{\mathbb{F}}(M)$ being of "finite length" is defined by stating that it is both Artinian and Noetherian, it is clear that if M is of $\mathbb{F}$-finite length then it is $\mathbb{F}$-Artinian. If M is $\mathbb{F}$-semisimple then $L_{\mathbb{F}}(M)$ splits as a direct sum of $\mathbb{F}$-closed $\mathbb{F}$-simple submodules by 2.1. Hence $\text{rad}_{\mathbb{F}}(L_{\mathbb{F}}(M)) = 0$. By 1.1, applied to

$$\psi(M) : M \longrightarrow L_{\mathbb{F}}(M)$$

we have $\text{rad}_{\mathbb{F}}(M) \subseteq \text{Ker}\psi(M) = \tau(M) \subseteq \text{rad}_{\mathbb{F}}(M)$.

Conversely suppose M is $\mathbb{F}$-Artinian and $\text{rad}_{\mathbb{F}}(M) = \tau(M)$. Consider the set of finite intersections of elements of the non-empty $\text{Max}_{\mathbb{F}}(M)$. Since M is $\mathbb{F}$-Artinian, there is an $\mathbb{F}$-minimal element, R , of this set (i.e. if R' is smaller finite intersection of $\mathbb{F}$-maximal subobjects then $R/R' \in T$. In fact here since R' is $\mathbb{F}$-saturated, $R = R'$).

If $N \in \text{Max}_{\mathbb{F}}(M)$, $N \cap R$ is thus equal to R by the minimality of R , so $R \subseteq N$. We therefore have $R = \text{rad}_{\mathbb{F}}(M) = \tau(M)$. Thus there is a finite family of $\mathbb{F}$-maximal submodules of M having intersection $\tau(M)$. We have a morphism

$$g : M \longrightarrow \prod_{\text{finite}} M/M_i = \bigoplus_{\text{finite}} M/M_i = P \text{ , say .}$$

where $\{M_i\}$ is the finite family of $\mathbb{F}$-maximal submodules concerned. $\text{Ker } g = \tau(M)$ so g is $\mathbb{F}$-monic.

$T_{\mathbb{F}}(g)$ is therefore monic; $T_{\mathbb{F}}(P)$ is semi-simple of finite length and so $T_{\mathbb{F}}(M)$ must also be semisimple of finite length.

COROLLARY 2.6

If M *is* $\mathbb{F}$*-Artinian,* $M/\text{rad}_{\mathbb{F}}(M)$ *is* $\mathbb{F}$*-semisimple.*

COROLLARY 2.7

If A *is* $\mathbb{F}$*-Artinian then,* A *is* $\mathbb{F}$*-semisimple if and only if* $\mathrm{rad}_{\mathbb{F}}(A) = \tau(A)$.

COROLLARY 2.8

If A *is* $\mathbb{F}$*-Artinian and* $\phi : A \to A/\mathrm{rad}_{\mathbb{F}}(A)$ *is the natural map then* $A/\mathrm{rad}_{\mathbb{F}}(A)$ *is* $\phi(\mathbb{F})$*-semisimple.*

Proof

Follows from 1.2, 1.3 and 2.7.

THEOREM 2.9

Let $(A, \mathbb{F})$ *be given and suppose* A *has an* $\mathbb{F}$*-nilpotent two sided ideal* N *such that, if* $\phi : A \to A/N$ *is the natural map* A/N *is* $\phi(\mathbb{F})$*-semisimple. Then for an* A*-module* M *, the following are equivalent:*

(a) M *is of* $\mathbb{F}$*-finite length,*

(b) M *is* $\mathbb{F}$*-Artinian,*

(c) M *is* $\mathbb{F}$*-Noetherian.*

Proof

By definition (a) implies (b) and (c).

Assume M is $\mathbb{F}$-Artinian (resp. $\mathbb{F}$-Noetherian) and that $p \in \mathbb{N}$ is such that $N^p \subseteq \tau(A)$. It will suffice to prove that all of the modules $M/MN, MN/MN^2, \ldots, MN^{p-1}/MN$ are of $\mathbb{F}$-finite length, since $MN^p \subset M\,(A) \subseteq \tau(M)$.

Each module in this list is naturally an A/N-module and is $\phi(\mathbb{F})$-Artinian precisely because it is $\mathbb{F}$-Artinian.

As A/N is $\phi(\mathbb{F})$-semisimple, each MN^i/MN^{i+1} is of $\phi(\mathbb{F})$-finite length. It is now a simple matter to check that any A/N module S such that $T_{\phi(\mathbb{F})}(S)$ is simple, also satisfies $T_{\mathbb{F}}(S)$ is simple when S is considered as an A-module. Hence each MN^i/MN^{i+1} is of $\mathbb{F}$-finite length. (The case when M is $\mathbb{F}$-Noetherian is similar.)

COROLLARY 2.10

If A *is* $\mathbb{F}$*-Artinian and* M *is of* $\mathbb{F}$*-finite type, then* M *is of* $\mathbb{F}$*-finite length. In particular* A *is itself of* $\mathbb{F}$*-finite length.*

COROLLARY 2.11

If $\mathbb{F}$-Max *holds and* A *is* $\mathbb{F}$*-Artinian then* A *is* $\mathbb{F}$*-Noetherian.*

Remark.

This result holds without $\mathbb{F}$-Max, by Năstăsescu [14] and Miller-Teply [12], however the general non-commutative case is considerably more difficult to prove than this special case. It follows from their result that if A is $\mathbb{F}$-Artinian, then $\mathbb{F}$-Max holds, but as mentioned before, the author has been unable to prove this directly.

3. $\mathbb{F}$-REGULARITY

We shall say that given $(A, \mathbb{F})$, A is an $\mathbb{F}$*-regular ring* if given any element $a \in A$, $(aA)^c$ is an $\mathbb{F}$-direct summand of A.

Remarks

(a) If $(aA)^c$ is an $\mathbb{F}$-direct summand, then so is aA itself since one has an ideal I such that $aA \cap I \subseteq (aA)^c \cap I \subseteq \tau(A)$ and $((aA)^c + I)^c = A$; thus

$$((aA)^c + I) \subseteq (aA + I)^c \subseteq A$$

and $A = ((aA)^c + I)^c \subseteq (aA + I)^c \subseteq A$

i.e. $(aA + I) \in \mathbb{F}$.

(b) We can define a morphism

$$aA + I \longrightarrow A/\tau(A)$$

as follows

$$ax + i \longrightarrow ax + \tau(A) .$$

Note that since $aA \cap I \subsetneq \tau(A)$, this is well defined. Since $aA + I \in \mathbb{F}$, this defines an element e , of $L_{\mathbb{F}}(A)$ (constructible as equivalence classes of maps from ideals in $\mathbb{F}$ to $A/\tau(A)$). We claim: $e^2 = e$.

To see this we merely note that

$$e^{-1}\left(\frac{aA+I+\tau(A)}{\tau(A)}\right) = aA + I$$

and that $e^2(ax+i) = ax + \tau(A)$.

This corresponds to the fact that in (Mod-A)/T $T_{\mathbb{F}}(A) \cong T_{\mathbb{F}}(aA) \oplus T_{\mathbb{F}}(I)$ and e is the endomorphism of $T_{\mathbb{F}}(A)$ corresponding to the projection. Of course

$$\mathrm{End}(T_{\mathbb{F}}(A)) = \mathrm{Hom}(T_{\mathbb{F}}(A), T_{\mathbb{F}}(A))$$

$$= \mathrm{Hom}(A, S_{\mathbb{F}} T_{\mathbb{F}} (A))$$
$$= \mathrm{Hom}(A, L_{\mathbb{F}} (A))$$
$$= L_{\mathbb{F}} (A)$$

as a ring. Thus e is the idempotent corresponding to a , more or less as in the classical theory.

PROPOSITION 3.1

If A *is* $\mathbb{F}$*-regular, then for any* $a_1, \ldots, a_n$, $a_1A + a_2A + \ldots + a_nA$ *is an* $\mathbb{F}$*-direct summand of* A .

Proof

One can easily adapt the usual proof.

THEOREM 3.2

Let $(A, \mathbb{F})$ *be given. The following are equivalent statements:*

(a) A *is* $\mathbb{F}$*-semisimple*

(b) A *is* $\mathbb{F}$*-Artinian and* $\mathbb{F}$*-regular*

(c) A *is* $\mathbb{F}$*-Noetherian and* $\mathbb{F}$*-regular.*

Proof

Clearly (a) ⇒ (b) by 2.2; (b) ⇒ (c) from 2.11.

For (c) ⇒ (a) : A is $\mathbb{F}$-Noetherian so all ideals are of $\mathbb{F}$-finite type. Hence given any ideal I , there are elements $a_1, \ldots, a_n \in I$ such that $a_1A + a_2A + \ldots + a_n$ $\subseteq I \subseteq (a_1A + \ldots + a_nA)^c$. Again I is an $\mathbb{F}$-direct summand of A . The result now follows easily.

Remark

Năstăsescu ([14] II Proposition 1.3) shows that if A is $\mathbb{F}$-Noetherian then A is $\mathbb{F}$-Artinian if one of the additional conditions that follow is satisfied:

(a) A is semi-Artinian (in particular for a perfect ring)

(b) A is regular (in the sense of von Neumann)

(c) Every torsion free injective module has singular radical zero. (In fact (c) implies A is $\mathbb{F}$-semi-simple).

Our condition (c) in 3.2 is a refinement of Năstăsescu's (b).

Another way of proving (b) implies (a) in the classical case of 3.2 is to use the result that von Neumann regular rings have zero radical. This result also has its relative analogue.

PROPOSITION 3.3

If A *is* $\mathbb{F}$*-regular and* $\mathbb{F}$-Max *holds then* $\mathrm{rad}_{\mathbb{F}}(A) = \tau(A)$.

Proof

Suppose $a \in \mathrm{rad}_{\mathbb{F}}(A)$, $a \notin \tau(A)$ and A $\mathbb{F}$-regular. Then aA $\mathbb{F}$-splits off i.e. there is an $N \subset A$ with $aA \subset N \subset \tau(A)$ and $aA + N \in \mathbb{F}$. Then $N \notin \mathbb{F}$ since otherwise one would have $a \in \tau(A)$, so $N^c \in C_{\mathbb{F}}(A)$.

Thus there is a $\mathbb{F}$-maximal M, $N \subseteq M$. But then $aA \subseteq M$ and $N \subseteq M$ so $aA + N \subseteq M$ which is silly.

Remark

This result suggests that it may be possible to handle the family of ideas "perfect" "semi-perfect" "T-nilpotent" etc. in a relative form. Although some of these results certainly generalise, parts of Bass [2] depend on homologic arguments which are not immediately able to be relativised. However if his results do relativise, their relative versions would be extremely useful in the study of homologic dimensions.

4. $\mathbb{F}$-COPERFECT RINGS

The nearest we have been able to get, so far, to a relative version of the Bass theory is in relativising semi primary and coperfect rings.

As usual we fix $(A, \mathbb{F})$. We say A is $\mathbb{F}$-*semi-primary* if $\text{rad}_{\mathbb{F}}(A)$ is proper and $\mathbb{F}$-nilpotent and $A/\text{rad}_{\mathbb{F}}(A)$ is $\phi(\mathbb{F})$-semisimple.

Examples

Any $\mathbb{F}$-Artinian ring is $\mathbb{F}$-semi primary.

PROPOSITION 4.1

If A *is* $\mathbb{F}$-*semi primary and* $\mathbb{F}$-*Noetherian, then* A *is* $\mathbb{F}$-*Artinian.*

Proof

Let $J = \mathrm{rad}_{\mathbb{F}}(A)$ and suppose $J^n \subseteq \tau(A)$. (We adapt the proof of the classical case to be found on page 98 of Stenström [20] to this relative case. Very little needs to be changed!)

We use induction on the smallest n such that $J^n \subseteq \tau(A)$. If $n = 1$ then A is $\mathbb{F}$-semisimple and so is certainly $\mathbb{F}$-coperfect.

Next suppose that the assertion is true for all pairs $(A, \mathbb{F})$ with A $\mathbb{F}$-semi primary and $J^m \subseteq \tau(A)$ for $m < n$ and suppose that our given $(A, \mathbb{F})$ is such that $J^n \subseteq \tau(A)$ but $J^{n-1} \not\subseteq \tau(A)$.

Let $I_1 \supset I_2 \supset \ldots$ be a descending chain of $\mathbb{F}$-saturated ideals of $\mathbb{F}$-finite type. The natural map $\psi : A \to A/J^{n-1}$ takes this chain into a chain.

$$\bar{I}_1 \supset \bar{I}_2 \supset \bar{I}_3 \supset \ldots$$

of $\psi(\mathbb{F})$-pure ideals of $\psi(\mathbb{F})$-finite type in A/J^{n-1}. $\mathrm{rad}_{\psi(\mathbb{F})}(A/J^{n-1})$ is J/J^{n-1}, so this induced chain is stationary in $C_{\psi(\mathbb{F})}(A/J^{m-1})$ i.e. there is some k such that $\bar{I}_r/\bar{I}_{r+1}$ is $\psi(\mathbb{F})$-torsion for all $r \geq k$ and thus that $\bar{I}_k/\bar{I}_r$ is $\psi(\mathbb{F})$-torsion for all $r \geq k$. It follows that

$$\frac{I_k + J^{n-1}}{I_r + J^{n-1}}$$

is $\mathbb{F}$-torsion for all $r \geq k$ and hence that $\frac{I_k J}{I_r J}$ is $\mathbb{F}$-torsion (because $J^n \subseteq \tau(A)$).

Now look at the descending chain

$$\frac{I_k}{I_k J} \supset \frac{I_{k+1}}{I_k J} \supset \ldots$$

of A/J-submodules of the A/J-module I_k/I_kJ . A/J is $\phi(\mathbb{F})$-semisimple and I_k/I_kJ , being of $\phi(\mathbb{F})$-finite type, is $\phi(\mathbb{F})$-semisimple and of $\phi(\mathbb{F})$-finite length. Therefore there exists an $s > k$ such that

$$(I_s/I_kJ)/(I_t/I_kJ) \text{ is } \phi(\mathbb{F})\text{-torsion}$$

for all $t > s$. But hence

$$(I_s/I_kJ)/((I_t + I_kJ)/I_k) \text{ is } \phi(\mathbb{F})\text{-torsion}$$

so

$$I_s/(I_t + I_kJ) \text{ is } \mathbb{F}\text{-torsion.}$$

A simple use of Nakayama's Lemma mod $\mathbb{F}$ now shows that

$$I_s/I_t \text{ is } \mathbb{F}\text{-torsion for all } t \geq s$$

so A has descending chain condition on $\mathbb{F}$-pure ideals of $\mathbb{F}$-finite type.

We shall say that A is $\mathbb{F}$-*semi-Artinian* if $(\text{Mod-A})_{\mathbb{F}}$ is a semi-Artinian category (i.e. every non zero object has a simple subobject).

An ideal N in A is an $\mathbb{F}$-nil ideal if for each $a \in N$, there is a positive integer n such that

$$a^n \in \tau(A) .$$

An element such as a above is said to be $\mathbb{F}$-*nilpotent*.

PROPOSITION 4.3

If A *is* $\mathbb{F}$-*coperfect, then*

(i) A *is* $\mathbb{F}$-*semi-Artinian*

(ii) $\mathrm{rad}_{\mathbb{F}}(A)$ *is an* $\mathbb{F}$-*nil ideal.*

Proof

(i) A is $\mathbb{F}$-semi-Artinian if and only if for each ideal I, $(A/I)/\tau(A/I)$ if non zero contains an $\mathbb{F}$-s m le submodule.

But there is an ideal I' $\mathbb{F}$-minimal amongst the $\mathbb{F}$-finitely generated $\mathbb{F}$-saturated ideals of A, not contained in I then $S = (I' + I)/I$ is such that $T_{\mathbb{F}}(S)$ is simple. Dividing out by $\tau(S)$ therefore gives an $\mathbb{F}$-simple submodule of $(A/I)/\tau(A/I)$ as required.

(ii) The argument given in Bass [2] p.470 generalises on replacing the socle by the $\mathbb{F}$-socle, that is the sum of the $\mathbb{F}$-pure $\mathbb{F}$-simple submodules of a module M (this is the largest $\mathbb{F}$-semisimple submodule of M).

Any element in $\tau(J) = \tau(A) \cap J$ has h-value zero.

Remark

Combining 3.2, 4.1 and Năstăsescu's result [14] we have the following conditions together with "A is $\mathbb{F}$-Noetherian" imply "A is $\mathbb{F}$-Artinian" :

(i) A is semi-Artinian (is it possible with $\mathbb{F}$-semi

-Artinian or $\mathbb{F}$-perfect?)

(ii) A is $\mathbb{F}$-regular.

(iii) Every torsion free injective module has zero singular radical.

(iv) A is $\mathbb{F}$-semi primary or more generally, A contains a two sided $\mathbb{F}$-nilpotent ideal such that A/N is $\phi(\mathbb{F})$-semisimple.

5. PREPARATORY REMARKS FOR COMMUTATIVE RINGS WITH ADDITIVE TOPOLOGY

As usual $(A, \mathbb{F})$ will be fixed but now A will be commutative. We shall need various relative analogues of classical results to be able to prove results on $\mathbb{F}$-semi-local rings later.

Firstly we shall say ideals I , J are $\mathbb{F}$-*comaximal* if $I + J \in \mathbb{F}$.

LEMMA 5.1

Suppose $M \in$ Mod-A , *and* I , J $\mathbb{F}$-*comaximal ideals in* A , *then*

$$\frac{MI \cap MJ}{MIJ} \in T .$$

Proof

If $x \in MI \cap MJ$, $x(I + J) \in MIJ$. As $I + J \in \mathbb{F}$ this means $\mathrm{Ann}(x + MIJ) \in \mathbb{F}$. As x was arbitrary we find $\mathrm{Ann}((MI \cap MJ)/MIJ) \supset I + J$ and the result follows.

THEOREM 5.2 (CHINESE REMAINDER THEOREM)

Suppose $(A, \mathbb{F})$ *is such that* $\mathrm{Max}_{\mathbb{F}}(A) \neq \emptyset$ *and let*

$(I_i \mid 1 \le i \le n)$ *be a finite family of pairwise* $\mathbb{F}$*-comaximal ideals in* A . *For any* M *in* Mod-A , *one has*

$$M \,.\, (\Pi I_i) \rightarrowtail M \,.\, (\cap I_i)$$

is $\mathbb{F}$*-surjective and*

$$M \longrightarrow \bigoplus_{1 \le i \le n} M/MI_i$$

is $\mathbb{F}$*-surjective with kernel* $\cap MI_i$.

Proof

The case $n = 1$ is trivial.

Assume the result holds for all M and all families of pairwise $\mathbb{F}$-comaximal ideals having fewer than n members. Consider a family $(I_i \mid 1 \le i \le n)$ and set

$$I_i^! = \prod_{j \ne i} I_j$$

If $I_i + I_i^! \notin \mathbb{F}$ then there is a $J \in \mathrm{Max}_{\mathbb{F}}(A) \ne \emptyset$, $J \supseteq I_i + I_i^!$, $\mathrm{Max}_{\mathbb{F}}(A) \subseteq \mathrm{Spec}(A)$ (see Albu and Nastasescu [1] Prop. 2.4) so a classical argument shows that there must be some j with $I_i + I_j \in \mathbb{F}$ contradiction. Hence $I_i + I_i^! \in \mathbb{F}$ and

$$M(\Pi I_i) = MI_i I_i^! \longrightarrow MI_i \cap MI_i^!$$

is an $\mathbb{F}$-epimorphism by the lemma. As $MI_i \cap MI_i^! \longrightarrow MI_i \cap M(\bigcap_{j \ne 1} I_j)$ is an $\mathbb{F}$-epimorphism by the induction hypothesis we are done for the first part.

The second part follows similarly by induction and reduction to the case $n = 1$.

6. $\mathbb{F}$-SEMI LOCAL RINGS.

If A is commutative, an obvious property to generalise to the relative case is that of semi-locality. In fact it was in order to understand what such rings migh look like in general that this study was undertaken. As the concept is closely related to semi-simplicity it was deemed necessary to understand to what extent that concept relativised before embarking on the key concept of semi local rings. The intervening chapters are intended to do two things: firstly the theorems presented there give information that directly and indirectly aids the understanding of the $\mathbb{F}$-semi local rings and secondly in the case of semi-Noetherian rings and topologies from the Krull-Gabriel filtration, these rings would seem to be of great interest in applications similar to that dealt with in Section 8 for A an $\mathbb{F}$-semi local ring.

We say a commutative ring A with fixed additive topology $\mathbb{F}$ is $\mathbb{F}$-*semi local* if

(i) A is Noetherian

(ii) $A/\mathrm{rad}_{\mathbb{F}}(A)$ is $\phi(\mathbb{F})$-semisimple where

$$\phi : A \rightarrow A/\mathrm{rad}_{\mathbb{F}}(A) .$$

THEOREM 6.1

(A, $\mathbb{F}$) *being given,* A *is* $\mathbb{F}$-*semi local if and only if* A *is Noetherian and* $\mathrm{Max}_{\mathbb{F}}(A)$ *is finite.*

Proof

We have that A is $\mathbb{F}$-semi local, $\mathrm{Max}_{\phi(\mathbb{F})}(A/\mathrm{rad}_{\mathbb{F}}(A))$ is finite. The description of $\phi(\mathbb{F})$ now makes it clear that $\mathrm{Max}_{\mathbb{F}}(A)$ is also finite.

Conversely if $\mathrm{Max}_{\mathbb{F}}(A)$ is finite then the Chinese Remainder theorem mod $\mathbb{F}$ gives an $\mathbb{F}$-epimorphism

$$A/\mathrm{rad}_{\mathbb{F}}(A) \longrightarrow \bigoplus_{i=1}^{n} A/M_i$$

where the M_i, $i = 1, \ldots, n$ are the $\mathbb{F}$-maximal ideals. Hence $A/\mathrm{rad}_{\mathbb{F}}(A)$ is $\mathbb{F}$-semisimple as a module and $\phi(\mathbb{F})$-semisimple as a ring.

In the application treated in Section 8 we shall need the following definition.

An $\mathbb{F}$-semi local ring A is called $\mathbb{F}$-*complete* if the canonical morphism

$$A \longrightarrow \hat{A}$$

is an $\mathbb{F}$-isomorphism where $\hat{A}$ is the completion of A in the $\mathrm{rad}_{\mathbb{F}}(A)$-adic topology.

PROPOSITION 6.2

If A *is* $\mathbb{F}$*-semi local, then*

$$\theta : A \longrightarrow \hat{A}$$

is an $\mathbb{F}$*-monomorphism. In fact if* E *is any* A*-module of finite type*

$$\theta : E \longrightarrow \hat{E}$$

is an $\mathbb{F}$*-monomorphism.*

Proof

(In [17], sketched here for completeness.) We treat the second case. $\operatorname{Ker}\theta = \bigcap_{n=1}^{\infty} J^n E$ where we have written $J = \operatorname{rad}_{\mathbb{F}}(A)$; by Krull's theorem $x \in \bigcap_{n=1}^{\infty} J^n E$ if and only if there is some $j \in J$ with $jx = j$ or $(1 - j)x = 0$.

By the relative form of Nakayama's lemma, "$1 - j$ is a unit mod $\mathbb{F}$" (approximately $A(1 - j) \in \mathbb{F}$). Thus $\operatorname{Ann}(x) \in \mathbb{F}$ and $x \in \tau(E)$ as required.

Knowledge about $\operatorname{Coker}\theta$ may be gleaned in the case o $\mathbb{F}$ being the additive topology associated to Krull-Garbiel dimension n .

$$\operatorname{Coker}\theta = \varprojlim_k{}^{(1)} J^k E$$

and if $\text{KG-dim}\, A = n$, $\varprojlim{}^{(1)} J^k E$ is approximately a limit of modules of KG dimension $\leq n - 1$. It seems likely tha a close analysis of this result would yield conditions tha would imply that $\text{KG dim}\operatorname{Coker}\theta \leq n - 1$.

Remark

What would here be called $\mathbb{F}$-*local rings* have been studied in detail in the non-commutative case by Jacques Raynaud ([18]). He does not however attempt to characteri those pairs $(A, \mathbb{F})$ for which $L_{\mathbb{F}}(A)$ is complete in the natural topology.

The condition that $\operatorname{Max}_{\mathbb{F}}(A)$ be finite is also considered by Albu and Năstăsescu [1] and is linked there wit properties of $L_{\mathbb{F}}(A)$.

7. $\mathbb{F}$-ZARISKI RINGS?

In the previous section, we briefly examined completions of $\mathbb{F}$-semi-local rings. In some sense that process is contrary to the main drift of this paper, namely to work in (Mod-A)/T . In this section we discuss the difficulties involved in studying completions in (Mod-A)/T .

As one cannot form $a + N$ for N a subobject of an object M in (Mod-A)/T , it is necessary to replace I-adic topologies by I-adic filtrations.

Suppose $I \subset A$, we can define a filtration on $T_{\mathbb{F}}(A)$ (resp $T_{\mathbb{F}}(M)$) by $\{T_{\mathbb{F}}(I^n)\}_{n\in\mathbb{N}}$ (resp $\{T_{\mathbb{F}}(MI^n)\}_{n\in\mathbb{N}}$. If $N \subset M$ is a sub module then define

$$\overline{T_{\mathbb{F}}(N)} = \bigwedge_n (T_{\mathbb{F}}(N) \vee T_{\mathbb{F}}(MI^n))$$

Here we are working in the subobject lattice of $T_{\mathbb{F}}(M)$. This gives a closure operation on that lattice and hence also on $C_{\mathbb{F}}(M)$. We say N is *closed relative* to $\mathbb{F}$ if $\bar{N} = N$.

We also say that the filtration is *discrete* if $T_{\mathbb{F}}(I^n) = 0$ for some n and is *Hausdorff* if $\{\bar{0}\} = \{0\}$ i.e. 0 is closed mod $\mathbb{F}$.

PROPOSITION 7.1

Let A *be a commutative Noetherian ring,* $\mathbb{F}$ *-an additive topology on* A *and* I *an ideal in* A *. Consider* *the following statements:*

(a) $I \subseteq \mathrm{rad}_{\mathbb{F}}(A)$

(b) *If* X *in* (Mod-A)/T *is finitely generated, then the* $T_{\mathbb{F}}(I)$*-adic filtration,* $(T_{\mathbb{F}}(MI^n))_{n \in N}$, *(for* $X = T_{\mathbb{F}}(M)$) *is Hausdorff.*

(c) *If* X *in* (Mod-A)/T *is finitely generated, and* Y *is a subobject of* X *then* Y *is closed rel.* *in* X .

(d) *If* $M \in \mathrm{Max}_{\mathbb{F}}(A)$, $T_{\mathbb{F}}(M)$ *is closed rel.* $\mathbb{F}$ *in* $T_{\mathbb{F}}(A)$.

Then (b) *and* (c) *are equivalent and imply* (d).

(d) *implies* (a) *and if* $T_{\mathbb{F}}$ *preserves intersections of nested sequences of ideals, then* (a) *implies* (b) *and al the statements are equivalent.*

Remark

$T_{\mathbb{F}}$ will preserve these intersections, for instance, when it has a left adjoint i.e. T is a bilocalising subcategory of Mod-A (cf. Popescu 15). In such a case there is an idempotent ideal B such that

$$\mathbb{F} = \{J \subseteq A \mid J \supseteq B\}$$

and since A is supposed here to be commutative Noetheria $\mathbb{F}$ is determined by a set of prime ideals $p_1, \ldots, p_n$ such that

$$\mathbb{F} = \{J \subseteq A \mid J \supseteq p_1 \ldots p_n\} .$$

Thus if T is to be bilocalising, one must have a very strong restriction on the structure of the ring A. (For a detailed study of bilocalising subcategories see Goblot [7].)

Proof of 7.1

The equivalence of (b) and (c) is almost classical. (c) implies (d) trivially.

Assume (d) therefore and assume also that $I \nsubseteq \mathrm{rad}_{\mathbb{F}}(A)$ then there is some $M \in \mathrm{Max}_{\mathbb{F}}(A)$ with $M + I^n \in \mathbb{F}$ for all $n > 0$. Thus

$$\begin{aligned} \wedge_n (T_{\mathbb{F}}(M) \vee T_{\mathbb{F}}(I^n)) &= \wedge_n T_{\mathbb{F}}(M + I^n) \\ &= T_{\mathbb{F}}(A) \end{aligned}$$

and so $T_{\mathbb{F}}(M)$ is not closed rel $\mathbb{F}$.

Finally assume $I \subseteq \mathrm{rad}_{\mathbb{F}}(A)$, by the Krull theorem and the relative form of Nakayama's Lemma ([17]), one has $\cap I^n \subseteq \tau(A)$. If $T_{\mathbb{F}}$ satisfies the intersection condition of the statement of the theorem, then

$$\wedge_n T_{\mathbb{F}}(I^n) = T_{\mathbb{F}}(\cap I^n) = 0$$

and 0 is closed rel $\mathbb{F}$ in $T_{\mathbb{F}}(A)$.

Clearly one should call such a ring A with given additive topology, $\mathbb{F}$, an $\mathbb{F}$-*Zariski ring* (Bourbaki [4]). One expects following the classical theory that the properties of such rings should be essential for "good" properties following localisations. One would also hope that $\mathbb{F}$-semi local rings would be $\mathbb{F}$-Zariski. However the intersection condition is a block. Essentially the problem would seem to be to link a completion process in (Mod-A)/T with some completion like process in Mod-A itself. At the moment it is not clear how to do this as the complete filtered objects of (Mod-A)/T are as yet lacking a characterization.

8. APPLICATIONS TO THE DERIVED FUNCTORS OF LIM

Jensen proves the following theorem in [11]:

If A *is a commutative Noetherian ring, the following are equivalent:-*

(a_0) A *is a finite direct product of complete local rings (or equivalently a complete semi-local ring*

(b_0) *for all projective systems* (M_i) *of* A-*modules of finite type,* $\underleftarrow{\lim}^{(i)} M = 0$ *for all* $i > 0$.

(Here $\underleftarrow{\lim}^{(i)}$ is the ith derived functor of the limit functor see Jensen's notes [11] for other connections between the behaviour of these derived functors and the structure of the rings concerned.)

Jensen and Gruson ([8] and [9]) proved that if A is of Krull-Gabriel dimension $\leq n$ and the M_i are of finite type over A, $\underleftarrow{\lim}^{(i)} M = 0$ for $i > n$. Of course since commutative Artinian rings are semi local and have nil-potent radical, they are trivially complete in the rad(A)-

adic topology. Hence this latter result of Jensen and Gruson includes as the special case $n = 0$, a particular case of "(a_0) *implies* (b_0)". *Is there, for each* n, *a pair of statements* (a_n), (b_n) *with* (a_n) *equivalent to* (b_n) *for commutative Noetherian rings with*

(b_n):*for all projective systems* (M_i) *of* A-*modules of finite type,* $\varprojlim^{(i)} M = 0$ *for all* $i > n$,

if so what is the n-*dimensional version of* (a_0)?

In [16] the author proved the Gruson-Jensen result using localisation techniques. Approximately the proof used the following idea. One can prove the case $n = 0$ independently and in an arbitrary Grothendieck category, so assume that one has the result for dimension $n - 1$ and localise at dimension n, throwing everything into the relevant quotient category. One proves an easy lemma that the inductive hypothesis implies that this does not disturb the vanishing of the $\ell im^{(i)}$ for $i > n$ and one thus pulls back the initial case in the quotient category to the module category and dimension n. This raises a second version of the problem mentioned above: *Is Jensen's result true in a commutative locally Noetherian Grothendieck category and, if the answer is yes, can one pull back that theorem into the module category to find our conjectures* "(a_n) $\Longleftrightarrow$ (b_n)"?

The "inspiration" for the study undertaken in this paper is thus to find a "relative" version of "complete semi local" which can hopefully fit the bill for (a_n). "Relative" means relative to $\mathbb{F}_{n-1}$ the $(n-1)^{th}$ additive topology in the Krull-Gabriel filtration sequence. We shall assume that the reader is conversant with the theory of Krull-Gabriel dimension as to summarise it here would take too long. (cf. Popescu [15]) (For $n = 0$, notions rel $\mathbb{F}_{n-1}$ are of course classical, thus "$\mathbb{F}_{-1}$-Artinian" is precisely "Artinian" and so on.)

The following result is the best indication that this direction of study may be fruitful.

THEOREM 8.1

Let A *be a commutative Noetherian ring and* $J = J_n$ *its* $\mathbb{F}_{n-1}$*-radical. Suppose* A *is* $\mathbb{F}_{n-1}$*-semi local and complete in the* J*-adic topology, then for any projective system* (M_i) *of* A*-modules of finite type,* $\ell im^{(i)} M = 0$ *for all* $i > n$.

Proof

Firstly we note that

$$M = \varprojlim M/MJ^r$$

in the category of projective systems and that each M/MJ^r is finitely generated module over A/J^r . A/J is $\mathbb{F}_{n-1}$-semi simple hence clearly is of KG-dimension exactly n .

Thus KG dim $M/MJ^r \le n$ by an easy argument using short exact sequences.

We have a doubly indexed projective system

$$\bar{M} : I \times \mathbb{N} \to \text{Mod-}A$$
$$\bar{M}(i, r) = M/MJ^r$$

and hence by results of Roos (cf. Jensen [11]) a spectral sequence

$$E_1^{p,q} = \begin{cases} \varprojlim^{(p)} (\varprojlim_{\mathbb{N}}{}^{(q-p)} M/MJ^r & p \ge 0,\ q \ge 0 \\ 0 & \text{otherwise} \end{cases}$$

converging to $\varprojlim_{I\times\mathbb{N}}{}^{(t)} \bar{M}$

Each $\varprojlim^{(q-p)} M(i)/M(i)J^r = 0$ for $q \ne p$ since $(M(i)/M(i)J^r)$ is epimorphic projective sequence (cf. Jensen [11]) and so the spectral sequence collapses giving only

$$E_1^{p,p} = \varprojlim^{(p)} (\varprojlim_r M/MJ^r)\ ,\ p \ge 0$$
$$= \varprojlim^{(p)} M\ , \qquad p \ge 0$$

as possibly non-zero terms. By the results of Gruson and Jensen [9] (cf. Porter [16]) mentioned earlier $\varprojlim_{I\times\mathbb{N}}{}^{(t)} \bar{M} = 0$ if $t > n$. Hence $\varprojlim^{(p)} M = 0$ if $p > n$, as required.

Thus if (a_n) is taken as: A is a $\mathbb{F}_{n-1}$-semi local ring complete in the J_n-adic topology, one has (a_n) implies (b_n) . However the author suspects that the true (a_n) is actually somewhat weaker. Clearly if A is $\mathbb{F}_{n-1}$-semi local and $\hat{A}$ is its J_n-adic completion then a slight adaption of the proof given above will work if

$$A \longrightarrow \hat{A}$$

is an $\mathbb{F}_{n-1}$-isomorphism. (We know it always to be an $\mathbb{F}_{n-1}$ monomorphism the problem is therefore its cokernel.) Thus one can weaken the above to have A being $\mathbb{F}_{n-1}$-complete in the terminology suggested in Section 6. Finally should one really not require that $T_{\mathbb{F}_n}(A)$ be a complete semi local generator in $(\text{Mod-}A)/T_n$?

Remark

Jensen's proof in [11] that (b_0) implies (a_0) can be simplified somewhat, but one clearly has a problem (in the interaction of $T_{\mathbb{F}}$ and $\underleftarrow{\lim}$ or of $L_{\mathbb{F}}$ and $\underleftarrow{\lim}$) in finding a proof which can generalise to give (b_n) implies (a_n) .

9. APPLICATIONS WHEN $\mathbb{F}$ IS FLAT

Hacque [10] has defined an additive topology $\mathbb{F}$ on a ring A (now not necessarily commutative) to be *flat* if it satisfies the following condition.

(P) For every $M \in \mathbb{F}$, there is an $I \in \mathbb{F}$ and finite families $(x_i)_{1 \le i \le n}$ of elements of M and $(f_i)_{1 \le i \le n}$ of homomorphisms from I to $A/\tau(A)$ such that

$$\sum_{i=1}^{n} f_i . \rho(x_i) = \rho_I$$

where $\rho : A \to A/\tau(A)$ is the natural homomorphism and

ρ_I is ρ/I , its restriction to I .

He proves that $\mathbb{F}$ being flat is equivalent to $L_{\mathbb{F}}$ being a flat functor and $\psi(A) : A \to L_{\mathbb{F}}(A)$ being a flat epimorphism of rings, so that $T = \operatorname{Ker} \psi(A)^*$ where $\psi(A)^*$ is the change of rings functor

$$B \to B \otimes_A L_{\mathbb{F}}(A)$$

In this case the image of $\mathbb{F}$ in $L_{\mathbb{F}}(A)$ is the trivial topology $\{L_{\mathbb{F}}(A)\}$ (Hacque [10] Lemma 3.3) and by Albu and Năstăsescu ([1] Proposition 0.7) we get the following.

PROPOSITION 9.1

If $\mathbb{F}$ *is flat, then* $C_{\mathbb{F}}(M)$ *is canonically isomorphic to the lattice of* $L_{\mathbb{F}}(A)$*-submodules of* $L_{\mathbb{F}}(M)$ *for any right module* M *over* A .

This together with our previous results implies the following:

THEOREM 9.2

Let A *be a ring and* $\mathbb{F}$ *a flat topology on* A .

(a) *If* M *is any* A*-module,* M *is* $\mathbb{F}$*-semi-simple if and only if* $L_{\mathbb{F}}(M)$ *is semi-simple.*

(b) A *is* $\mathbb{F}$*-regular (resp.* $\mathbb{F}$*-semi primary; resp.* $\mathbb{F}$*-coperfect) if and only if* $L_{\mathbb{F}}(A)$ *is regular (resp. semi primary; resp. right coperfect).*

If in addition A *is commutative Noetherian then*

(c) A *is* $\mathbb{F}$*-semi local if and only if* $L_{\mathbb{F}}(A)$ *is semi local. In this case* $L_{\mathbb{F}}(A)$ *is a ring of fractions of* A .

The only part not clear from 9.1 is the last remark which is a result of Albu and Năstăsescu [1].

Remark

This final case includes as special cases certain of the results of Raynaud [18]. Moreover (a) and the regular case of (b) intersect non trivially with the famous theorem of Goldie in the case when $\mathbb{F}$ is the Goldie topology (cf. Stenström [20]). In fact if $\mathbb{F} = \mathbb{G}$ and A is $\mathbb{G}$-Noetherian, $\mathbb{G}$ is flat so $\mathbb{G}$-regular (which is sufficient by 3.2 to give $\mathbb{G}$-semi simple) implies $L_{\mathbb{G}}(A)$ is semi simple.

REFERENCES

[1] T. Albu and C. Năstăsescu, Décompositions primaires dans les catégories de Grothendieck commutatives, *J. Reine Angew. Math.*, 280 (1976), 172-196 and 282 (1976), 172-185.

[2] H. Bass, Finitistic dimension and a homological generalization of semi-primary rings, *Trans. Amer. Math. Soc.*, 95 (1960), 466-488.

[3] H. Bass, *Algebraic K-theory*, W.J. Benjamin Inc., New York, 1968.

[4] N. Bourbaki, *Algèbre commutative, Ch.3 Graduations, Filtrations et topologies*, Hermann, Paris, 1961.

[5] N. Bourbaki, *Algèbre, Ch.8 Modules et Anneaux Semi simples*, Hermann, Paris, 1958.

[6] P. Gabriel, Des catégories abéliennes, *Bull. Soc. Math. France*, 90 (1962), 325-448.

[7] R. Goblot, *Sur deux classes de catégories de Grothendieck*, Thesè d'état. Lille, 1971.

[8] L. Gruson and C.U. Jensen, Modules algèbriquement compacts et foncteurs $\varprojlim^{(i)}$, *Comptes Rendus Acad. Sci. Paris*, 276 (1973), 1651-1653.

[9] L. Gruson and C.U. Jensen, Dimensions cohomologiques relieés aux foncteurs $\varprojlim^{(i)}$, *Kobenhavns Universitet Mathematisk Institut, Preprint Series*, 1980, No. 19.

[10] M. Hacque, Localisations extates et localisations plates, *Publ. Dépt. Math. Lyon*, (1969), 97-117.

[11] C.U. Jensen, *Les foncteurs dérivés de lim et leurs applications en théorie des modules*, Lecture Notes in Math. 254, Springer 1972.

[12] R.W. Miller and M.L. Teply, The descending chain condition relative to a torsion theory, *Pacific J. Math.*, 83 (1979), 207-219.

[13] C. Năstăsescu, La structure des modules par rapport a une topologie additive, *Tôhuku Math. J.*,

[14] C. Năstăsescu, Conditions de finitude pour les modules, *Rev. Roum. Math. Pures et Appl.*, 24 (1979), 745-758 and 25 (1980), 615-630.

[15] N. Popescu, *Abelian categories with applications to rings and modules*, London Math. Soc. Monographs, 3. Academic Press, 1973.

[16] T. Porter, Essential properties of pro objects in Grothendieck categories, *Cahiers Top. Géom. Diff.*, 20 (1979), 3-57.

[17] T. Porter, The kernels of completion maps and a relative form of Nakayama's Lemma, *Journal of Algebra* (to appear).

[18] J. Raynaud, *Sur la théorie de la localisation*, Thèse (3eme cycle) Lyon I, 1971.

[19] J. Raynaud, *Localisations et Spectres d'Anneau*, Thèse d'état, Lyon I, 1976.

[20] B. Stenström, *Rings and Modules of Quotients*, Lecture Notes in Math. 237, Springer 1971.

[21] J.S. Golan, *Localisation of noncommutative rings*, Marcel Dekker, New York, 1975.

T. Porter
School of Mathematics and Computer Science
University College of North Wales
BANGOR
Gwynedd, LL57 2UW
Wales, U. K.

COLLOQUIA MATHEMATICA SOCIETATIS JÁNOS BOLYAI
38. RADICAL THEORY, EGER (HUNGARY), 1982.

BEHAVIOUR OF RADICAL PROPERTIES OF RINGS UNDER SOME ALGEBRAIC CONSTRUCTIONS

E. R. PUCZYŁOWSKI

In this paper we describe the behaviour of properties of associative rings under some typical algebraic constructions (polynomial rings, power series rings, complete direct products of rings, etc.). Our considerations are restricted mainly to properties related to "classical" radicals: the prime radical P, the locally nilpotent radical L, the upper nil radical N, the Jacobson radical J, and the Brown-McCoy radical U.

Investigations of connections between rings can be frequently reduced to investigations of some extensions of finite type and graded rings. Some problems of this type are discussed in § 1. In § 2 we investigate radicals of polynomial rings in non-commutative indeterminates. It appears that many radicals of such rings take a very regular form. In § 3 and § 4, using power series rings and complete direct products, we describe some connections between the radicals P and N.

The results presented here show that all the "classical" radicals, except the radical U, are connected with the Jacobson radical via some algebraic constructions. This fact reduces many problems to the ones relating to the Jacobson radical.

All rings in the paper are associative and not necessarily with unity. To simplify some considerations, however, we sometimes embed a ring R into the usual extension R^1 of R by the ring of integers to obtain a ring with unity.

All results on radicals used in the paper can be found in [5, 23].

1. RADICALS OF NORMALIZING EXTENSIONS AND GRADED RINGS

Let $A \subseteq B$ be an extension of rings with the same unity. We recall that B is said to be a finite normalizing extension with a normalizing set $\{x_1, \ldots, x_n\}$ of elements of B if $B = \sum_1^n Ax_i$ and $Ax_j = x_jA$ for all j. Thus $x_1, \ldots, x_n$ are generators of B as a left A-module. If these elements are free generators of the left A-module B then we say that $\{x_1, \ldots, x_n\}$ is a basis for $A \subseteq B$.

A deep analysis of the connection between the prime ideals of A and B done in [8, 17] gives in particular

1.1. Let $A \subseteq B$ be a finite normalizing extension with a normalizing set $\{x_1,\ldots,x_n\}$ and let $S = P, J$ or U . Then

a) ([8]) $S(B) \cap A = S(A)$;

b) ([17]) If $\{x_1,\ldots,x_n\}$ is a basis for $A \subseteq B$, $S=P$ or J and $S(A) = 0$ then $(S(B))^n = 0$.

Using the connection between N and P given in [19] or the connection between prime and primitive rings done in [17] one can show (see § 3) that 1.1 for P is a consequence of the result for J . The relation between L and J given in [19] extends 1.1 to L .

Regarding 1.1 for the nil radical N we prove

1.2 PROPOSITION. *The assertion 1.1 for* N *is equivalent to the Koethe problem: does the nil radical of a ring contain all one-sided nil ideals?*

Proof. It is well known ([2, 11]) that the Koethe problem is equivalent to each of the following:

i) if R is a nil ring then for any n the ring R_n of all $n \times n$ -matrices over R is nil;

ii) the Jacobson radical of the polynomial ring

$R[x]$ in one indeterminate x over R is equal to $N(R)[x]$.

Now let us assume that 1.1 holds for N and let R be a nil ring. Identifying R^1 with the diagonal of R_n^1 we obtain a finite normalizing extension $R^1 \subseteq R_n^1$. Thus $R = N(R^1) \subseteq N(R_n^1)$, so $N(R_n) = R_n$ and we obtain i).

Now we prove that ii) implies 1.1. If $A \subseteq B$ is a finite normalizing extension with a normalizing set $\{x_1, \ldots, x_n\}$ then so is $A[x] \subseteq B[x]$. Now by ii) and 1.1 a) for J, $N(A)[x] = J(A[x]) \subseteq J(B[x]) = N(B)[x]$. Hence $N(A) \subseteq N(B) \cap A$. The converse inclusion is clear.

If $N(A) = 0$ then $J(A[x]) = 0$. So by ii) and 1.1 b) $(N(B)[x])^n = J(B[x])^n = 0$, which leads to $(N(B))^n = 0$.

The property 1.1 has many applications (c.f. [4, 13, 14, 18]). The following result was proved in [12] in a different way.

1.3. THEOREM. *If* R *is a* U*-algebra over a field* F *with* $\operatorname{card} F > \dim_F R$ *then the polynomial algebra* $R[x]$ *in one indeterminate* x *over* R *is* U*-radical.*

Proof. If $\operatorname{card} F \leq \aleph_0$, the result is clear as in this case $\dim_F R$ is finite and finite dimensional U-algebras are nilpotent. Thus let us assume that $\operatorname{card} F > \aleph_0$ and $U(R(x)) \neq R[x]$. Then there exists a homomorphism f of

$R[x]$ onto a simple ring P with unity. Let R^* be the usual extension of R by the field F to obtain an F-algebra with unity. Since $R[x]$ is an ideal of $R^*[x]$, the homomorphism f can be extended to a homomorphism $\bar{f}:R^*[x]\to P$. Obviously $\bar{f}(F)$ is contained in the center of P. Thus, identifying F with $\bar{f}(F)$, we may consider the ring P as an F-algebra in the natural way. Then the homomorphism $\bar{f}$ is F-linear. If $\bar{f}(x)$ is algebraic over F then P is a finite normalizing extension of the subalgebra P_1 of P generated by $\bar{f}(R)$ and by the unity of P. But then by 1.1 the U-radical of P is non-zero as, obviously, $\bar{f}(R)$ is a nonzero U-radical ideal of P_1. This is impossible. Hence $\bar{f}(x)$ is transcendental over F. Now the elements $1/(\bar{f}(x)-a)$ for $a\in F$ are linearly independent over F, so $\dim_F P \geq$ $\geq \operatorname{card} F$. This is also impossible as P is a homomorphic image of $R[x]$ and $\dim_F R[x]=\max(\dim_F R, \aleph_0)<\operatorname{card} F$.

Remark. The only fact we have made use of in the above is that if $A\subseteq B$ is a finite normalizing extension with a central normalizing set then $U(A)\subseteq U(B)$. This result appears essentially easier to prove than 1.1 is (cf. [15]).

From 1.1 it follows immediately:

1.4 [4, 15]. If $R[\omega]$ is the ring obtained by adjoining a primitive n th root of unity to a ring R with unity and if $S = P,L,J$ or U then

$$S(R[\omega]) \cap R = S(R) .$$

The foregoing result is used in [4] to investigate the Jacobson radical of graded rings. Practically the same method is applied in [3] to study the Jacobson radical of skew polynomial rings. All the main results of [4], except Theorem 10, hold for any radical satisfying 1.4. Modifying the arguments from [4] we extend Theorem 10 to the Brown-McCoy radical as well.

Let Z be the ring of integers and, for a prime p, let $Z[C(p^{\infty})]$ be the ring obtained by adjoining the group $C(p^{\infty})$ of all p^n th roots of unity to Z. From 1.4 it follows easily that for any ring R and $S = P,L,J$ or U,

1.5. $S(R\otimes_Z Z[C(p^{\infty})]) \cap R = S(R)$.

Now we prove

1.6. THEOREM. *If a ring* R *is graded by a torsion-free abelian group* G *and if* S *is a radical satisfying* 1.5 *then* $S(R)$ *is homogeneous.*

Proof. Since any ideal of a G-graded ring contains a largest homogeneous ideal and the factor ring of a G-

graded ring by a homogeneous ideal is a G-graded ring in a natural way, it suffices to prove that if $S(R)\neq 0$ then $S(R)$ contains a non-zero homogeneous element. Let $\{R_g\}_{g\in G}$ be the set of homogeneous components of R. If $S(R)\neq 0$ then let $r=r_1+\ldots+r_n\in S(R)$, where $0\neq r_i\in R_{g_i}$ for $1\leq i\leq n$. If $n=1$ then we are done, so let $n\geq 2$. Since $r_n\neq 0$, there exists a prime p with $p^m r_n\neq 0$ for any natural number m. By the assumption $r\in S(R\otimes_Z Z[C(p^\infty)])$. We may grade $P=R\otimes_Z Z[C(p^\infty)]$ by putting $P_g=R_g\otimes_Z Z[C(p^\infty)]$. Since $g_1^{-1}g_n\neq 1$, G is torsion-free abelian and the group $C(p^\infty)$ is divisible, there exists a group homomorphism $f_{g_1}:G\to C(p^\infty)$ such that $\omega=f_{g_1}(g_1^{-1}g_n)$ is a p th root of unity. Now let $\bar{f}:P\to P$ be defined by $\bar{f}(\Sigma a_g)=\Sigma f_{g_1}(g)a_g$. Obviously $\bar{f}$ is a $Z[C(p^\infty)]$-automorphism of P. Thus $\bar{f}(S(P))=$ $=S(P)$. In particular, $b=r-(1/f_{g_1}(g_1))\bar{f}(r)=$ $=(1-f_{g_1}(g_1^{-1}g_2))r_2+\ldots+(1-\omega)r_n\in S(P)$. Since $1+\omega+\ldots+$ $+\omega^{p-1}=0$, we have $p=p-(1+\omega+\ldots+\omega^{p-1})=1-\omega+1-\omega^2+\ldots+$ $+1-\omega^p=\varepsilon(1-\omega)$ for some $\varepsilon\in Z[C(p^\infty)]$. Consequently, $\varepsilon b=b_2+\ldots+b_n\in S(P)$, where $b_i\in P_{g_i}$ for $2\leq i\leq n$ and $b_n=pr_n\neq 0$. If $b_i\neq 0$ for some $2\leq i\leq n-1$, we repeat the argument with f_{g_i}. After a finite number of these steps we obtain that for some m, $0\neq p^m r_n\in S(P)\cap R=S(R)$, proving that $S(R)$ contains a non-zero homogeneous element.

Remark. In [10] it is proven that the radicals P and L of rings graded by torsion-free non-abelian groups are homogeneous and a characterization of the Jacobson radical of such rings is obtained.

The following results are simple consequences of 1.6.

1.7. Let $S = P, L, J$ or U. Then

a) [1, 12] if $R[X]$ denotes the polynomial ring in a set X of commutative indeterminates then $S(R[X]) = (S(R[X]) \cap R)[X]$;

b) if $R[G]$ is the group ring of a torsion-free abelian group G then $S(R[G]) = (S(R[G]) \cap R)[G]$;

c) [3] if $R_\sigma[x]$ $(R_\sigma\langle x\rangle)$ is a skew polynomial (group) ring then $S(R_\sigma[x]) = \{\Sigma_i r_i x^i \mid r_0 \in I \cap S(R),\ r_i \in I$ for $i \geq 1$, where $I = \{r \in R \mid rx \in S(R_\sigma[x])\}\}$ $(S(R_\sigma\langle x\rangle) = (S(R_\sigma\langle x\rangle) \cap R)_\sigma\langle x\rangle)$.

The following result, proved in [18] by another method, is also deduced from 1.6.

1.8. COROLLARY. *Let* $R\{X\}$ *be the polynomial ring in a set* X *of non-commutative indeterminates commuting with the coefficients from* R . *If* $\operatorname{card} X \geq 2$ *then*

$J(R\{X\}) = L(R)\{X\}$.

Proof. Clearly $L(R)\{X\} \subseteq J(R\{X\})$, so it suffices to prove that $J(R\{X\}) \neq 0$ implies $L(R) \neq 0$. By 1.6, $J(R\{X\})$ is homogeneous under the grading of $R\{X\}$ given by the total degree of a polynomial. Let $a = a_1w_1 + \ldots + a_nw_n \in J(R\{X\})$, where $0 \neq a_i \in R$ and the w_i are monomials of the same total degree. We will show that the right ideal of R generated by $a_1, \ldots, a_n$ is locally nilpotent. To obtain this it is enough to prove that for any $b_1, \ldots, b_m \in R$ the set $A = \{a_{i_k}b_k \mid 1 \leq i_k \leq n , k=1,\ldots,m\}$ is nilpotent. Since $\operatorname{card} X \geq 2$, there exist m different monomials $p_1, \ldots, p_m$ of the same degree. Now $b = a(b_1p_1 + \ldots + b_mp_m) \in J(R\{X\})$ is homogeneous. Obviously, the set of coefficients of b coincides with A . It is not difficult to check (see [21]) that b is nilpotent. This and the fact that w_ip_j , $1 \leq i \leq n$, $1 \leq j \leq m$ are different monomials of the same degree, imply that A is nilpotent.

Corollary 1.8 gives the possibility to use properties of the Jacobson radical in investigations of the locally nilpotent radical. In that way we can obtain for instance the following well-known result.

1.9. For any ring R , $L(R) = \cap\{I \mid I$ is a prime ideal of R with $L(R/I) = 0\}$.

Proof. $L(R)\{X\} = J(R\{X\})$ is the intersection of all primitive ideals P_α of $R\{X\}$. Hence $L(R) = \cap(P_\alpha \cap R)$ and the $P_\alpha \cap R$ are prime ideals of R. If for some α, $L(R/P_\alpha \cap R) = J/P_\alpha \cap R \neq 0$ then $J\{X\}+P_\alpha/P_\alpha$ is a non-zero L-ideal of $R\{X\}/P_\alpha$, which is impossible. Hence $L(R)$ is the intersection of a set of prime ideals I with $L(R/I) = 0$. But if I is an ideal of R such that $L(R/I) = 0$, then $L(R) \subseteq I$, and this completes the proof.

Combining 1.8 and 1.1 applied to the Jacobson radical one can get easily

1.10 (cf. [18]). The radical L satisfies 1.1.

Some other applications of 1.8 are given in [18].

A ring R with unity is called right strongly prime [7] if for any non-zero element $t \in R$, there exists a finite subset $\{r_1, \ldots, r_k\} \subseteq R$ (depending on t) such that $tr_i r = 0$ for all i $(r \in R)$ implies $r=0$. In [7] the set $\{r_1, \ldots, r_k\}$ is called an insulator for t. It is obvious that R is right strongly prime if every non-zero two-sided ideal of R contains a finite subset the right annihilator of which is zero. Also strongly prime rings do not contain non-zero locally nilpotent ideals. The following result related to 1.8, generalizes Lemma 1.2 of [17] a little and will be proved almost in the same way.

1.11. THEOREM. *If* R *is a right strongly prime ring then for any infinite set* X *with* $\operatorname{card} X \geq \operatorname{card} R$, $R\{X\}$ *is right primitive.*

Proof. Clearly $\operatorname{card} X = \operatorname{card} R\{X\}$, so to each $f \in R\{X\} \setminus \{0\}$ we can associate a distinct indeterminate $x_f \in X$. Let $\{r_1,\ldots,r_n\}$ be an insulator for the coefficient a_f at a monomial p of the largest degree occurring in f and let $\bar{f} = r_1x_1+\ldots+r_nx_n \in R\{X\}$, where the $x_i \in X$ are distinct indeterminates. We prove that the right ideal $N = \sum_{f\neq 0} (1+x_f f\bar{f})R\{X\}$ is proper. If not, then there exists a finite set of $g_f \in R\{X\}$ such that $1 = \sum_{f\neq 0} (1+x_f f\bar{f})g_f$. Let f be the element of the sum such that $\deg g_f$ is maximal and let b_f be the coefficient at a monomial q of the largest degree occurring in g_f . Then the coefficient at $x_f p x_i q$ in the sum is equal to $a_f r_i b_f$. By the definition of an insulator, $a_f r_i b_f \neq 0$ for some i . Thus $N \neq R\{X\}$.

Now let M be a maximal right ideal of $R\{X\}$ containing N . Then for any non-zero ideal I of $R\{X\}$, $M+I = R\{X\}$. Indeed, if $f \in I \setminus \{0\}$ then $1+x_f f\bar{f} \in M$, so $1 \in M+I$. In consequence $R\{X\}$ is right primitive.

In [7] it is asked whether there exists a right

strongly prime ring with non-zero nil radical. Now we present an example of such a ring.

1.12. *Example.* Let A be a finitely generated nil ring which is not nilpotent [6]. Since for any natural number n , A^n is a non-zero finitely generated ring, by Zorn's Lemma we can find a maximal ideal I of A^1 such that $A^n \not\subset I$ for all n . It is easy to check that A^1/I is prime, so for every n the right annihilator of $A^n + I/I$ is zero. Now if J/I is a non-zero ideal of A^1/I then $J \supset A^n$ for some n . Let $\{a_1, \ldots, a_k\}$ be a set of generators for A^n . Then the right annihilator of $\{a_1+I, \ldots, a_k+I\}$ in A^1/I is equal to the right annihilator of $A^n + I/I$, so it is zero. In consequence each non-zero ideal of $A^1 + I/I$ contains a finite set whose right annihilator is zero. Hence A^1/I is right strongly prime. Obviously, the nil radical of A^1/I equals $A+I/I$, which is not zero.

Questions. 1. Is the nil radical of a ring graded by a torsion-free abelian group homogeneous? By 1.6 it is so if the condition 1.5 is satisfied for N . In view of 1.2 this condition seems to be easier to prove than the Koethe problem. Let us remark that in special oases for polynomial rings in commutative and non-commutative indeterminates the answer is yes.

2. From 1.6 it follows immediately that if $S = P, L, J$ or U and $R[G]$ is the semigroup ring of a torsion-free abelian semigroup G with unity then $S(R[G])$ is G-graded, i.e., if $\sum a_p p \in S(R[G])$ then $a_p p \in S(R[G])$. For $S=P$ or L this easily implies $S(R[G]) = S(R)[G]$. Does, for $S = J, U$, $S(R[G]) = (S(R[G]) \cap R)[G]$ hold or, equivalently, does $a_p p \in S(R[G])$ imply $a_p \in S(R[G])$?

3. Is the Jacobson radical of a ring graded by a free (non-abelian) semigroup locally nilpotent? If it is so then by [10] the radical is graded.

4. In [7] an example is given of a right strongly prime ring which is not left strongly prime. Does there exist a right strongly prime ring R such that for some infinite set X with $\operatorname{card} X \geq \operatorname{card} R$, $R\{X\}$ is not left primitive?

2. RADICALS OF POLYNOMIAL RINGS IN NON-COMMUTATIVE INDETERMINATES

In this section we investigate radicals of polynomial rings $R\{X\}$ in a set X of non-commutative indeterminates commuting with the coefficients from R.

Using the fact that $J(R\{X\})$ is homogeneous under the grading of $R\{X\}$ given by the total degree of a polynomial we proved in 1.8 that $J(R\{X\}) = L(A)\{X\}$. By 1.6, $U(R\{X\})$ is homogeneous under this gradation, too; this does not mean that $U(R\{X\}) = I\{X\}$ for some ideal I of R though. In the present sectior we prove this for infinite X . Giving a proof which is independent of 1.6, we obtain a bit stronger result.

For a ring R let R_n denote the ring of $n\times n$ - matrices over R .

2.1. THEOREM. *Let* S *be a radical such that if* $R_n \in S$ *for some* n *then* $R\in S$. *Then for any infinite set* X , $S(R\{X\}) = (S(R\{X\}) \cap R)\{X\}$.

To prove the theorem it suffices to show (s. [12]) that if $S(R\{X\}) \neq 0$ then $S(R\{X\}) \cap R \neq 0$. So let us assume that $S(R\{X\}) \neq 0$. Since $R\{X\}$ is an ideal of $R^1\{X\}$, $S(R\{X\})$ is a non-zero ideal of $R^1\{X\}$.

Throughout the section "homogeneous polynomial" means homogeneous under the total degree gradation.

2.2. LEMMA (cf. [20]). $S(R\{X\})$ *contains a non-zero homogeneous polynomial.*

Proof. Let $f = f_1+\ldots+f_n$ be a non-zero polynomial of $S(R\{X\})$ with the minimal number n of non-zero homogeneous components $f_1,\ldots,f_n$. We will show that $n=1$. If $n\geq 2$ and $x\in X$ is an ideterminate which does not occur in f then $f_1xf_2 \neq f_2xf_1$. Indeed, if not, then for some monomials p_1 , p_2 occurring in f_1 and some q_1, q_2 occurring in f_2 , $p_1xq_1 = q_2xp_2$. This is impossible as $\deg p_1 \neq \deg p_2$ and x occurs neither in p_1 nor in q_2 . Now $g = fxf_1 - f_1xf \in S(R\{X\})$ and $g = f_2xf_1 - f_1xf_2 + \ldots + f_nxf_1 - f_1xf_n$. Since $f_ixf_1 - f_1xf_i$ for $2 \leq i \leq n$ are homogeneous, by the minimality of n , $g=0$. This is impossible as $f_2xf_1 - f_1xf_2 \neq 0$.

Remarks. i) A bit more complicated arguments show [20] that the lemma holds for any set X with $\operatorname{card} X \geq 2$.

ii) Lemma 2.2 does not say that every radical of $R\{X\}$ is homogeneous (see [20]).

For a polynomial f , let $\operatorname{supp} f$ denote the set of monomials occurring in f with non-zero coefficients.

2.3. LEMMA. *$S(R\{X\})$ contains a non-zero homogeneous polynomial of the form $a \cdot w$, where $a \in R$ and w is a homogeneous polynomial with coefficients equal to* ± 1 .

Proof. By Lemma 2.2, $S(R\{X\})$ contains a non-zero homogeneous polynomial $f = a_1p_1 + \ldots + a_np_n$, where $0 \neq a_i \in R$ and the p_i are monomials. We say that the components a_ip_i and a_jp_j are similar if $a_i = \pm a_j$. Reducing similar components we can uniquely express f in the form $b_1w_1 + \ldots + b_mw_m$, where $0 \neq b_i \in R$ and the w_i are homogeneous polynomials of the same degree with coefficients ± 1, $\operatorname{supp} w_i \cap \operatorname{supp} w_j = \emptyset$ and $b_i \neq b_j$ for $i \neq j$. Let f be a non-zero homogeneous polynomial in $S(R\{X\})$ with minimal m. It is enough to prove that $m=1$. If not, then let $g = fw_1 - w_1 f$. Obviously, $g \in S(R\{X\})$ and $g = b_2(w_2 w_1 - w_1 w_2) + \ldots + b_m(w_m w_1 - w_1 w_m)$. Since the polynomials w_i are of the same degree and $\operatorname{supp} w_i \cap \operatorname{supp} w_j = \emptyset$ for $i \neq j$,

1° $\operatorname{supp} w_i w_1 \cap \operatorname{supp} w_1 w_i = \emptyset$ for $2 \leq i \leq m$,

2° $\operatorname{supp}(w_i w_1 - w_1 w_i) \cap \operatorname{supp}(w_j w_1 - w_1 w_j) = \emptyset$ for $2 \leq i \neq j \leq m$.

By 1° and the fact that $\operatorname{supp} w_1 \cap \operatorname{supp} w_i = \emptyset$ for $i \neq 1$, we obtain that for any $2 \leq i \leq m$ the coefficients in $w_i w_1 - w_1 w_i$ are equal to ± 1 and $g \neq 0$. Now by 2° we infer that g is a non-zero polynomial of $S(R\{X\})$ shorter (in the above sense) than m.

2.4. LEMMA. *$S(R\{X\})$ contains a non-zero polynomial of the form aw, where $a \in R$ and w is a multilinear polynomial with coefficients ± 1.*

Proof. The arguments in the proof are based on well known ones used in the theory of PI-algebras. By Lemma 2.3, $S(R\{X\})$ contains a non-zero polynomial of the form $aw(x_1, \ldots, x_n)$, where $a \in R$ and $w(x_1, \ldots, x_n)$ is a homogeneous polynomial in indeterminates $x_1, \ldots, x_n$

with coefficients ± 1. Let us assume that $w(x_1,\ldots,x_n)$ is of degree $d_i > 1$ in x_i. Let $x,y \in X \setminus \{x_1,\ldots,x_n\}$. Since X is infinite, there exists a homomorphism $f: R^1\{X\} \to R^1\{X\}$ such that $f|_R = id$, $f(x_i) = x+y$, $f(x_j) = x_j$ for $j \neq i$ and $f(X) \supseteq X$. All these conditions guarantee that $f(R\{X\}) = R\{X\}$ and $f(aw(x_1,\ldots,x_n)) = aw(x_1,\ldots,x_{i-1},x+y,x_{i+1},\ldots,x_n)$. Thus $f(S(R\{X\})) \subseteq S(R\{X\})$ and $aw(x_1,\ldots,x_{i-1},x+y,x_{i+1},\ldots,x_n) \in S(R\{X\})$. Similarly, $aw(x_1,\ldots,x_{i-1},x,x_{i+1},\ldots,x_n) \in S(R\{X\})$ and $aw(x_1,\ldots,x_{i-1},y,x_{i+1},\ldots,x_n) \in S(R\{X\})$. In consequence, $aw_1(x_1,\ldots,x_{i-1},x,y,x_{i+1},\ldots,x_n) = a(w(x_1,\ldots,x_{i-1},x+y,x_{i+1},\ldots,x_n) - w(x_1,\ldots,x,x_{i+1},\ldots,x_n) - w(x_1,\ldots,x_{i-1},y,x_{i+1},\ldots,x_n)) \in S(R\{X\})$. By the lemma on page 224 of [9], $w_1(x_1,\ldots,x_{i-1},x,y,x_{i+1},\ldots,x_n)$ is a non-zero polynomial such that: i) the degree of $w_1 \leq$ degree of w, ii) the degree of w_1 in x_j, $j \neq i$, is $\leq$ degree of w in x_j, iii) the degree of w_1 in x and y is $d_i - 1$. It is clear that the coefficients of w_1 are ± 1. If w_1 is of degree >1 in one of the x_j, x, y then we repeat the argument. After a finite number of these steps we obtain a multilinear polynomial of the needed form.

In [12] it has been proved:

2.5. For any ring R and any $n \geq 1$, $S(R_n) = I_n$ for some ideal I of R.

Proof of the theorem. Let $S(R\{X\}) \neq 0$. Then by Lemma 2.4, $S(R\{X\})$ contains a non-zero polynomial of the form $aw(x_1,\ldots,x_n)$, where $a \in R$ and $w(x_1,\ldots,x_n)$ is a multilinear polynomial in indeterminates $x_1,\ldots,x_n$ with coefficients ± 1. We may, obviously, assume that the coefficient at $x_1 \ldots x_n$ in $w(x_1,\ldots,x_n)$ is 1. Let Z denote the ring of integers and let e_{ij} be the matrix of Z_{n+1} with 1 in the (i,j)-place and 0's elsewhere. Since the set X is infinite, we can find a homomorphism f mapping $Z\{X\}$ onto $Z_{n+1}\{X\}$ such that $f(x_i) = e_{ii+1}$ for $i = 1,\ldots,n$. Observe that $f(w(x_1,\ldots,x_n)) = e_{1n+1}$. Now let $g: R\{X\} \to R \otimes_Z Z\{X\}$ and $h: R \otimes_Z Z_{n+1}\{X\} \to (R\{X\})_{n+1}$ be the natural isomorphisms. Then $\bar{f} = h \circ (id \otimes f) \circ g$ is a homomorphism mapping $R\{X\}$ onto $(R\{X\})_{n+1}$. Hence $\bar{f}(S(R\{X\})) \subseteq S((R\{X\})_{n+1})$. But by Lemma 2.5 and the assumption on S, $S((R\{X\})_{n+1}) \subseteq$ $\subseteq (S(R\{X\}))_{n+1}$, so $\bar{f}(S(R\{X\})) \subseteq (S(R\{X\}))_{n+1}$. By the foregoing $\bar{f}(aw(x_1,\ldots,x_n)) = ae_{1n+1} \in (S(R\{X\}))_{n+1}$, so $a \in S(R\{X\})$. Thus $0 \neq a \in S(R\{X\}) \cap R$ and the result follows.

It is well known that the Brown-McCoy radical satisfies the assumption of Theorem 2.1. Thus we have

2.6. COROLLARY. *If the set* X *is infinite then* $U(R\{X\}) = (U(R\{X\}) \cap R)\{X\}$.

Also, radicals which are hereditary with respect to subrings satisfy the assumption of 2.1. But for these radicals it is proved by another method [20] that 2.1 holds for any set X with $\mathrm{card}\, X \geq 2$.

It is known [12] that if S is a radical then so is $S_X = \{R \mid R\{X\} \in S\}$. Corollary 1.8 describes S_X for $S = L, N, J$ and X with $\mathrm{card}\, X \geq 2$. Namely then $S_X = L$. Obviously $P_X = P$. For $S = U$ we know only (by 2.5) that if X is infinite then $U(R\{X\}) = U_X(R)\{X\}$. Obviously $L \subseteq U_X$. Now we describe another subclass of U_X .

Set $\bar{U} = \{R \in U \mid \text{for any } a \in R,\ a \in aR + Ra + RaR\}$. It is easy to see that $R \in \bar{U}$ if and only if for any $a \in R$ there exists a finite subset $\{a_1, \ldots, a_n, b_1, \ldots, b_n\} \subseteq R^1$ such that $a = \sum a_i a b_i$ and $\sum a_i b_i \in R$. One can check that no non-zero locally nilpotent ring is contained in $\bar{U}$. Clearly, the class $\bar{U}$ contains those U-rings whose ideals are idempotent. In particular, $\bar{U}$ contains the simple idempotent rings without unity. Some subclasses of $\bar{U}$ were discussed in [22].

The following result extends a little Proposition 3 of [20] (cf. also [12], Theorem 10).

2.7. THEOREM. *If* $R \in \bar{U}$ *then for any* X , $R\{X\} \in U$.

Proof. If $R\{X\} \notin U$ then there exists a homomorphism

f of $R\{X\}$ onto a simple ring P with unity 1. Let $a = a_0 + a_1 p_1 + \ldots + a_n p_n$, where $a_0 \in R$, $0 \neq a_i \in R$ for $1 \leq i \leq n$ and the $p_i \neq 1$ are monomials of different degrees, be an element of $R\{X\}$ such that $f(a) = 1$ and n is minimal. Let us notice that $n \geq 1$. If not, then $f(a_0) = 1$ and $f(R)$ contains a unity. This is impossible as $R \in U$. Now by the assumption on R, $a_n = \sum b_i a_n c_i$ for some $b_i, c_i \in R^1$ with $\sum b_i c_i \in R$. But then $\sum b_i c_i - \sum b_i a c_i + a =$
$= (a_0 + \sum b_i c_i - \sum b_i a_0 c_i) + (a_1 - \sum b_i a_1 c_i) p_1 + \ldots + (a_{n-1} - \sum b_i a_{n-1} c_i) \times$
$\times p_{n-1} \in R\{X\}$ and $f\left(\sum b_i c_i - \sum b_i a c_i + a\right) = 1$. This contradicts the minimality of n.

Now we present an example which shows that the assumption on S in Theorem 2.1 cannot be dropped.

2.8. *Example*. Let F be a field and let S be the upper radical determined by F, i.e., $S = \{R \mid R$ cannot be mapped homomorphically onto $F\}$. Obviously, $F_2 \in S$ and $F \notin S$, so S does not satisfy the assumption of 2.1. Now let x, y be different indeterminates of X and let I be the ideal of $F\{X\}$ generated by $xy - yx$. We claim that $I \in S$. If not, then there exists a homomorphism f mapping I onto F. Since F is a ring with unity, f can be extended to a homomorphism $\bar{f}: F\{X\} \to F$. Now $f(xy - yx) = \bar{f}(xy - yx) = \bar{f}(x)\bar{f}(y) - \bar{f}(y)\bar{f}(x) = 0$. Thus $f(I) = 0$

a contradiction. In consequence $S(F\{X\}) \neq 0$. Since also $F\{X\} \notin S$, we have $(S(F\{X\}) \cap F)\{X\} \neq S(F\{X\})$.

3. NIL IDEALS OF POWER SERIES RINGS

Let $R\{\{X\}\}$ denote the power series ring in a set X of non-commutative indeterminates commuting with the coefficients from R . If $\operatorname{card} X \geq 2$ then

3.1 [20] a) $R\{\{X\}\}$ is nil if and only if R is nilpotent;

b) $N(R\{\{X\}\}) \neq 0$ if and only if $P(R) \neq 0$.

As an easy consequence of 3.1 and 1.8 we obtain 1.1 for the prime radical. The following lemma is clear.

3.2. LEMMA. *Let $A \subseteq B$ be a finite normalizing extension with a normalizing set $\{x_1, \ldots, x_n\}$. Then*

i) *$A\{\{X\}\} \subseteq B\{\{X\}\}$ is a finite normalizing extension with the same normalizing set $\{x_1, \ldots, x_n\}$;*

ii) *if $\{x_1, \ldots, x_n\}$ is a basis for $A \subseteq B$ then $\{x_1, \ldots, x_n\}$ is a basis for $A\{\{X\}\} \subseteq B\{\{X\}\}$;*

iii) *for any ideal I of A , $B\{\{X\}\}I\{\{X\}\}B\{\{X\}\} = (BIB)\{\{X\}\}$.*

3.3. COROLLARY. *Let $A \subseteq B$ be a finite normalizing extension of rings. If I is a nilpotent ideal of A then so is the ideal BIB of B generated by I .*

Proof. Obviously $I\{\{X\}\}$ is a nilpotent ideal of

$A\{\{X\}\}$. Hence by 1.10 and 3.2, $B\{\{X\}\}I\{\{X\}\}B\{\{X\}\} = (BIB)\{\{X\}\}$ is a locally nilpotent ideal of $B\{\{X\}\}$. Now 3.1 a) implies that BIB is nilpotent.

3.4. COROLLARY (cf. [8, 17, 18]). *Let* $A \subseteq B$ *be a finite normalizing extension of rings with a normalizing set* $\{x_1,\ldots,x_n\}$. *Then*

a) $P(B) \cap A = P(A)$;

b) *if* $\{x_1,\ldots,x_n\}$ *is a basis for* $A \subseteq B$ *and* $P(A) = 0$ *then* $(P(B))^n = 0$.

Proof. a) Since P is hereditary on subrings, we have $P(B) \cap A \subseteq P(A)$. Now $A/P(B) \cap A \subseteq B/P(B)$ is a finite normalizing extension with a normalizing set $\{x_1+P(B), \ldots, x_n+P(B)\}$. Since $B/P(B)$ is semiprime, so is $A/P(B) \cap A$ by 3.3. As a consequence $P(A) = P(B) \cap A$.

b) By 3.1 b), $N(A\{\{X\}\}) = 0$, so $L(A\{\{X\}\}) = 0$. Since, by 3.2 ii), $A\{\{X\}\} \subseteq B\{\{X\}\}$ is a finite normalizing extension with a basis $\{x_1,\ldots,x_n\}$, by 1.10 we have $(I\{\{X\}\})^n = 0$ for any nilpotent ideal I of B . Thus the indexes of all nilpotent ideals of B are $\leq n$, so $(P(B))^n = 0$.

In [17] it is proven that for any prime ring R and for sufficiently large X the ring $R\{\{X\}\}$ is strongly prime. This and 1.11 give a functor sending prime rings into primitive ones. This functor was very successfully used in [17] to investigate prime ideals of normalizing extensions.

Let ON be the lower strong radical determined by the nil radical N [2]. The following result proved in [20] gives a bit more than 3.1.

3.5. If $\mathrm{card}\, X \geq 2$ then for any ring R the following conditions are equivalent:

i) $P(R) = 0$,

ii) $P(R\{\{X\}\}) = 0$,

iii) $N(R\{\{X\}\}) = 0$,

iv) $ON(R\{\{X\}\}) = 0$.

If the set X is uncountable then

3.6 [20]. $a \in ON(R\{\{X\}\})$ if and only if the ideal of R generated by the coefficients of a is nilpotent. As a consequence, $ON(R\{\{X\}\}) = N(R\{\{X\}\}) =$ the sum of all nilpotent ideals of $R\{\{X\}\}$.

For a ring R and for a set X let $R^s\{\{X\}\} = \sum A\{\{X\}\}$, where A runs over all finitely generated subrings of R . Obviously $R^s\{\{X\}\}$ is a subring of $R\{\{X\}\}$. From 3.1 a) it is clear that if $\mathrm{card}\, X \geq 2$ then

3.7. $R^s\{\{X\}\}$ is nil if and only if R is locally nilpotent.

In [15] the following ideal of a ring R was defined: $N^*(R) = \{a \in R \mid aA$ is nilpotent for any finitely generated subring A of $R\}$.

Using 3.6 we can find a connection between the nil radical of $R^s\{\{X\}\}$ and $N^*(R)$. To obtain this we need the following.

3.8. LEMMA. *If* $\langle a_1,\dots,a_n\rangle$ *is the subring of* R *generated by the elements* $a_1,\dots,a_n \in N^*(R)$ *then for any finitely generated subring* B *of* R , $\langle a_1,\dots,a_n\rangle B$ *is nilpotent.*

Proof. We proceed by induction on n . Let $B = \langle b_1,\dots,b_k\rangle$ for some $b_1,\dots,b_k \in R$. If $n=1$ then $\langle a_1\rangle B \subseteq a_1 B_1$, where $B_1 = \langle a_1, b_1,\dots,b_k\rangle$. Since $a_1 \in N^*(R)$, $a_1 B_1$ is nilpotent. Thus $\langle a_1\rangle B$ is nilpotent. Now let $a_1,\dots,a_{n+1} \in N^*(R)$ and, as before, let $B = \langle b_1,\dots,b_k\rangle$. Put $B_2 = \langle a_1,\dots,a_{n+1}, b_1,\dots,b_k\rangle$. To see that $(\langle a_1,\dots,a_{n+1}\rangle B)^r = 0$ it is enough to show that $x_1y_1\dots x_ry_r = 0$ whenever $x_i \in \langle a_1,\dots,a_{n+1}\rangle$ and $y_i \in B_2$ for $1 \le i \le n$. But by induction there exists a k such that $(a_1B_2)^k = 0$ and $(\langle a_2,\dots,a_{n+1}\rangle B_2)^k = 0$. This implies $x_1y_1\dots x_ry_r = 0$ for $r = 2k$. The lemma follows.

3.9. COROLLARY. *The ideal* $N^*(R)$ *is locally nilpotent.*

Proof. Let $a_1,\dots,a_n \in N^*(R)$ and $B = \langle a_1,\dots,a_n\rangle$. By 3.8 $\langle a_1,\dots,a_n\rangle B = B^2$ is nilpotent.

3.10. COROLLARY. *If* X *is uncountable then* $N(R^s\{\{X\}\})=0$ *if and only if* $N^*(R)=0$.

Proof. Let $N^*(R)\neq 0$ and let A be a finitely generated subring of $N^*(R)$. We prove that the ideal of $R^s\{\{X\}\}$ generated by $A\{\{X\}\}$ is nil. To come to the point we show that for any finitely generated subrings B_1, B_2 of R , $B_1\{\{X\}\}\, A\{\{X\}\}\, B_2\{\{X\}\}$ is nilpotent. But to obtain this it suffices to prove that B_1AB_2 is nilpotent. Since B_2B_1 is contained in the ring generated by $B_1\cup B_2$, which is a finitely generated subring of R , AB_2B_1 is nilpotent by 3.8. The rest follows from the fact that $(B_1AB_2)^n = B_1(AB_2B_1)^{n-1}AB_2$.

Now let $0\neq s\in N(R^s\{\{X\}\})$ and let A be a finitely generated subring of R . A being countable, we can write $A=\{a_i \mid i=1,\dots\}$. Let $u=\sum a_ix_i$, where the x_i for $i=1,2,\dots$ are different indeterminates from X . By the definition of $R^s\{\{X\}\}$ there exists a finitely generated subring B of R such that $A\subseteq B$ and $s\in B\{\{X\}\}$. Thus $su\in N(B\{\{X\}\})$. It is easy to see that the set of coefficients of su is equal to CA , where C is the set of coefficients of s . Now by [19], CA is nilpotent. This proves that $C\subseteq N^*(R)$, so $N^*(R)\neq 0$.

Questions. 1. Is it true that for any X with

$\operatorname{card} X \geq 2$, $s \in ON(R\{\{X\}\})$ if and only if the ideal of R generated by the set of coefficients of s is nilpotent?

2. Let $R\{\{x\}\}$ be the power series ring in the indeterminate x commuting with the coefficients from R . It is known that if $R\{\{x\}\}$ is nil then R is nil of bounded index [19].

a) Does the converse hold? In other words, is $R\{\{x\}\}$ nil if R is nil of bounded index? Let us observe that if R is an algebra over an infinite field then the polynomial ring $R[x]$ is a subdirect sum of copies of R . Thus in this case if R is nil of bounded index then so is $R[x]$. But then, clearly, $R\{\{x\}\}$ is nil of bounded index, too.

b) Describe the structure of the nil radical of $R\{\{x\}\}$.

c) Does $ON(R\{\{x\}\}) = R\{\{x\}\}$ imply that R is nil of bounded index?

4. NIL IDEALS OF COMPLETE DIRECT PRODUCTS

In this section $\prod_{\alpha\in\Omega} A_\alpha$ and $\bigoplus_{\alpha\in\Omega} A_\alpha$ denote the complete direct product and the discrete direct sum of rings $\{A_\alpha\}_{\alpha\in\Omega}$, respectively. If $a = (a_\alpha) \in \Pi A_\alpha$ then $\bar{a}$ denotes the image of a by the natural surjection of ΠA_α onto

$A = \Pi A_\alpha / \oplus A_\alpha$.

4.1. LEMMA. *The right ideal of* A *generated by* $\bar{a}$ *is nil if and only if there exists an integer* n *such that for almost all* α , *the right ideal of* A_α *generated by* a_α *is nil of bounded index* $\le n$.

Proof. If such an n does not exist then there exist different $\alpha_1, \alpha_2, \ldots \in \Omega$, $a_i \in A_{\alpha_i}$, $x_i \in A^1_{\alpha_i}$ such that $(a_i x_i)^i \ne 0$. Let $x = (x_\alpha)$, where $x_\alpha = x_i$ for $\alpha = \alpha_i$ and $x_\alpha = 0$ otherwise. Now $\overline{ax}$ is an element of the right ideal of A generated by $\bar{a}$. For any integer $k \ge 1$, $(\overline{ax})^k \ne 0$ as $(a_i x_i)^k \ne 0$ for $i \ge k$. This contradicts the assumption.

The converse is clear.

4.2. COROLLARY. $ON(A) = P(A)$.

Proof. It is enough to prove that a nil right ideal of A generated by a singleton element is contained in $P(A)$. But by 4.1 such an ideal is nil of bounded index, so it is P-radical. Now the strongness of P ends the proof.

4.3. COROLLARY. $ON(A) = A$ *if and only if there exists a natural number* n *such that almost all* A_α *are nil of bounded index* $\le n$.

Proof. Let $ON(A) = A$. If the condition does not

hold then there exist different $\alpha_1, \alpha_2, \ldots \in \Omega$ and $a_i \in A_{\alpha_i}$ such that $a_i^i \neq 0$. Let $a = (a_\alpha)$, where $a_\alpha = a_i$ for $\alpha = \alpha_i$ and $a_\alpha = 0$ otherwise. By 4.1 $\bar{a}$ is not nil. This is impossible as by 4.2, $N(A) = A$.

4.4. COROLLARY. *If the set* Ω *is infinite then* $ON(A) \neq 0$ *if and only if* $P(A_\alpha) \neq 0$ *for almost all* α .

Proof. If $ON(A) \neq 0$ then by 4.1 almost all A_α contain non-zero nil right ideals of bounded index. All these ideals are P-radical. By the strongness of P, the P-radicals of all these A_α are non-zero.

Conversely, let $\Omega_1 = \{\alpha \in \Omega \mid P(A_\alpha) \neq 0\}$. All A_α for $\alpha \in \Omega_1$ contain non-zero ideals I_α such that $I_\alpha^2 = 0$. Now if $\hat{I}_\alpha = I_\alpha$ for $\alpha \in \Omega_1$ and $\hat{I}_\alpha = 0$ for $\alpha \in \Omega \setminus \Omega_1$, then $I = \Pi \hat{I}_\alpha$ is an ideal of ΠA_α such that $I^2 = 0$. Since the set Ω is infinite, $I \not\subset \oplus A_\alpha$. Hence $I + \oplus A_\alpha / \otimes A_\alpha$ is a non-zero nilpotent ideal of A , so $ON(A) \neq 0$.

4.5. COROLLARY. *Let, for* $n = 1, 2, \ldots,$ A_n *denote the ring of* $n \times n$ *-matrices over a ring* A . *The ring* $\Pi A_n / \oplus A_n$ *is* ON *-radical if and only if* A *is nilpotent.*

Proof. If $\Pi A_n / \oplus A_n$ is ON-radical then by 4.3 there exists an integer m such that almost all A_n are nil of bounded index $\leq m$. But since A_k is isomorphic to a subring of A_l for $l \geq k$, all A_n are nil of bounded index $\leq m$. Now let $a_1,\ldots,a_m \in A$ and let $a = a_1e_{12}+$ $+a_2e_{23}+\ldots+a_me_{mm+1}$, where e_{ij} denotes the matrix of A^1_{m+1} with 1 in the (i,j)-position and the zero element in all other positions. Observe that $a^m = a_1\ldots a_me_{1m+1}+$ $+\sum_{i\geq 2} a_{ij}e_{ij}$ for some $a_{ij} \in A$. But $a^m=0$, so $a_1\ldots a_m = 0$. In consequence $A^m=0$.

The converse is clear.

Now, for a ring R , let ΠR denote the complete direct product of $\aleph_0$ copies of R .

By 4.3 we conclude immediately:

4.6. COROLLARY. *$ON(\Pi R) = \Pi R$ if and only if the ring R is nil of bounded index.*

The following lemma is clear.

4.7. LEMMA. a) *If $A \subseteq B$ is a finite normalizing extension of rings with a normalizing set $\{x_1,\ldots,x_n\}$ then $\Pi A \subseteq \Pi B$ is a finite normalizing extension with a normalizing set $\{x_1^*,\ldots,x_n^*\}$, where $x_i^* = (x_\alpha)$ and for all α $x_\alpha = x_i$;*

b) *For any ideal I of A , $\Pi(BIB) = \Pi B\ \Pi I\ \Pi B$.*

As an application of 4.6 and 4.7 we obtain a result analogous to 3.3.

4.8. COROLLARY. *Let $A \subseteq B$ be a finite normalizing extension. If I is a nil ideal of bounded index in A*

then so is the ideal BIB *of* B *generated by* I .

Proof. The ideal ΠI of ΠA is nil of bounded index. Thus $\Pi I \subseteq P(\Pi A)$, so by 1.1 or 3.4, $\Pi B \Pi I \Pi B \subseteq$ $\subseteq P(\Pi B)$. In particular, $\Pi B \Pi I \Pi B$ is a nil ideal of ΠB . Now by 4.7 b) $\Pi B \Pi I \Pi B = \Pi(BIB)$. This and 4.6 imply that the ring BIB is nil of bounded index.

For any ring R , the ring R^1_n of $n \times n$ -matrices over R^1 is a finite normalizing extension of R^1 in the natural way. Thus from 4.8 we obtain:

4.9. COROLLARY. *If a ring* R *is nil of bounded index then so is the ring* R_n .

Let, for a ring R , $\bar{R} = \Pi R / \oplus R$. By 4.3 and 4.4 we obtain

4.10. a) $ON(\bar{R}) = \bar{R}$ if and only if R is nil of bounded index;

b) $ON(\bar{R}) = 0$ if and only if $P(R) = 0$.

The results analogous to 4.7 hold, too. Therefore we can use the functor $\bar{\ }$ to obtain results similar to some of those given in the preceding section as well as to obtain 4.8.

REFERENCES

[1] S. A. Amitsur, Radicals of polynomial rings, *Canad. J. Math.* 8 (1956), 355-361.

[2] S. A. Amitsur, Nil radicals. Historical notes and some new results, *Rings, modules and radicals (Proc. Conf. Keszthely, 1971)*, 47-65. Colloq. Math. Soc. J. Bolyai, 6, North-Holland, Amsterdam, 1973.

[3] S. S. Bedi and J. Ram, Jacobson radical of skew polynomial rings and skew group rings, *Israel J. Math.* 35 (1980), 327-338.

[4] G. M. Bergman, On Jacobson radicals of graded rings, preprint.

[5] N. Divinsky, *Rings and radicals*, London-Ontario, 1965.

[6] E. S. Golod, On nil algebras and finitely approximable p-groups (Russian), *Izv. Akad. Nauk SSSR, Mat. Ser.* 28 (1964), 273-276.

[7] D. Handelman and J. Lawrence, Strongly prime rings, *Trans. Amer. Math. Soc.* 211 (1975), 209-233.

[8] A. G. Heinicke and J. C. Robson, Normalizing extensions, prime ideals and incomparability, *J. Algebra* 72 (1981), 237-268.

[9] N. Jacobson, *Structure of rings, A. M. S. Colloquium Publications*, Providence, R. I., 1964.

[10] E. Jespers, J. Krempa and E. R. Puczyłowski, On radicals of graded rings, *Comm. Algebra*, to appear.

[11] J. Krempa, Logical connections among open problems of non-commutative rings, *Fund. Math.* 76 (1972), 121-130.

[12] J. Krempa, Radicals of semi-group rings, *Fund. Math.* 85 (1974), 57-71.

[13] J. Krempa, On Passman's problem concerning nilpotent free algebras, *Bull. Acad. Polon. Sci.* 27 (1979), 645-648.

[14] J. Krempa and A. Sierpińska, The Jacobson radical of certain group and semigroup rings, *Bull. Acad. Polon. Sci.* 26 (1976), 963-967.

[15] C. Năstăsescu and F. Van Oystayen, Jacobson radicals and maximal ideals of normalizing extensions applied to Z-graded rings, preprint.

[16] D. S. Passman, *The algebraic structure of group rings*, New York, 1977.

[17] D. S. Passman, Prime ideals in normalizing extensions, *J. Algebra* 73 (1981), 556-572.

[18] E. R. Puczyłowski, Radicals of polynomial rings, power series rings and tensor products, *Comm. Algebra* 8 (1980), 1699-1709.

[19] E. R. Puczyłowski, Nil ideals of power series rings, *Austral. J. Math.*, to appear.

[20] E. R. Puczyłowski, Radicals of polynomial rings in non-commutative indeterminates, to appear.

[21] A. Sierpińska, Radicals of rings of polynomials in non-commutative indeterminates, *Bull. Acad. Polon. Sci.* 21 (1973), 805-808.

[22] F. A. Szász, An almost subidempotent radical property of rings, *Rings, modules and radicals (Proc. Conf. Keszthely, 1971)*, 483-499. Colloq. Math. Soc. J. Bolyai, 6, North-Holland, Amsterdam, 1973.

[23] R. Wiegandt, *Radical and semisimple classes of rings*, Queen's Papers in Pure and Applied Mathematics, 37, Queen's University, Kingston, 1974.

Edmund R. Puczyłowski
Institute of Mathematics
University of Warsaw
PKiN
00-901 Warsaw, Poland

COLLOQUIA MATHEMATICA SOCIETATIS JÁNOS BOLYAI

38. RADICAL THEORY, EGER (HUNGARY), 1982.

RADICALS GENERATED BY HEREDITARY CLASSES

R. F. ROSSA

1. Introduction. This paper considers lower radical classes generated by hereditary, homomorphically closed classes of rings. The notation $I \triangleleft R$ (or $R \triangleright I$) will mean that I is an ideal of R. A class M of rings is hereditary if $R \in M$ and $I \triangleleft R$ together imply $I \in M$. A. E. Hoffman and W. G. Leavitt [3] showed that if M is a hereditary class of rings, then the lower radical LM is also hereditary. This is valid for nonassociative rings in general.

For every hereditary radical class P there is at least one hereditary, homomorphically closed class M such that LM = P, namely, M = P itself. A hereditary radical P may determine a minimal hereditary, homomorphically closed class M such that LM = P, which we shall call a *minimal generator* of P. If M_1 and M_2 are hereditary and homomorphically closed, then $L(M_1 \cap M_2) = LM_1 \cap LM_2$ [4, Theorem 4]: hence a minimal generator of a hereditary radical P is uniquely determined, provided it exists. We do not know whether $L(\bigcap M_i) = \bigcap LM_i$ if the number of hereditary, homomorphically closed classes M_i is infinite.

Let M be a hereditary, homomorphically closed class and let R be any nonzero ring in M. In this paper we characterize the largest hereditary, homomorphically closed subclass d(R, M) of M obtained by removing R from M, and then all other rings which must be removed as well to restore heredity and homomorphic closure. We are interested in the case that the lower radicals LM and Ld(R, M) differ. In particular, if $R \in M$, R may not be in Ld(R, M). We examine the implications of $R \notin Ld(R,M)$ if R has either dcc on ideals or acc on one-sided ideals. In the one case nonzero images of R, and the other nonzero ideals of R, must all have been deleted from M. We use the characterization of d(R, M) to show that if M_1 and M_2 are minimal generators of hereditary radicals, then so is $M_1 \cup M_2$. However, $M_1 \cap M_2$ need not be a minimal generator of a hereditary radical.

We shall use the notation I $\lhd\lhd$ R (or R $\rhd\rhd$ I) to mean that I is an accessible subring of R. The proofs of Theorems 5, 6 and 7 each require that if I $\lhd\lhd$ R and I' is the ideal of R generated by I, then $I'^n \subseteq I$ and $I'^n \lhd R$ for some n. Hence these results require that the universal class W of nonassociative rings under consideration satisfy an Andrunakievich lemma and that the rings in W be s-rings for some s. For more information on these points, see the paper of T. Anderson and B. J. Gardner [1].

We shall also use a result of Krempa [5, Lemma 5] which can be stated as follows. Let M be any homomorphically closed class of alternative rings. Then an alternative ring R is in LM if and only if every nonzero homomorphic image of R contains a nonzero accessible subring in M. This result will be valid in any variety of nonassociative rings in which the Kurosh lower radical construction terminates at the first infinite ordinal and in which radical accessible subrings of R generate radical ideals of R (essentially property (a) of Leavitt [6]). Thus although some of our results are stated for alternative rings, they will be valid in other universal classes as well.

2. <u>Hereditary classes and radicals</u>. Let M be a hereditary, homomorphically closed class of rings and let $R \in M$. We begin by determining the largest hereditary, homomorphically closed subclass of M that does not contain R. We shall use the notation $A \to B$ (or $B \leftarrow A$) to mean that B is a homomorphic image of A.

LEMMA 1. *Let T_1, T_2 and R be nonassociative rings such that $R \triangleleft T_2 \leftarrow T_1$. Then we can write $R \leftarrow R_1 \triangleleft T_1$ for some ideal R_1 of T.*

Proof. Let $T_1/K \cong T_2$. Then $R \cong R_1/K \triangleleft T_1/K$ for some $R_1 \triangleleft T$, and so we have $R \leftarrow R_1 \triangleleft T_1$.

LEMMA 2. *Let M be a homomorphically closed class of nonassociative rings and let $G = \{A : A \triangleright\triangleright T \to R\}$ where R is a fixed nonzero ring in M. Let $X = M - G$. Then X is also homomorphically closed.*

Proof. Let $S \in X$ and $K \triangleleft S$. We are to show that $S/K \in X$. Since $S \in M$ we have $S/K \in M$. Now $S/K \in X$ would imply that S/K has an accessible subring with image R, so that $S \to S/K \triangleright\triangleright T \to R$. But $S \triangleright\triangleright T_1 \to T \to R$ by Lemma 1, so $S \triangleright\triangleright T_1 \to R$, that is, $S \notin X$, a contradiction.

LEMMA 3. *Let M be a hereditary class of nonassociative rings and let G, X be as in Lemma 2. Then X is also hereditary.*

Proof. Let $S \in X$, $K \triangleleft S$. Now $K \triangleright\triangleright T \to R$ would imply $S \triangleright K \triangleright\triangleright T \to R$, i.e., $S \triangleright\triangleright T \to R$, so that $S \notin X$. This is a contradiction.

Combining Lemma 2 and Lemma 3, we obtain

THEOREM 4. *Let M be a hereditary, homomorphically closed class of nonassociative rings. Let R be a fixed nonzero ring in M and let* $G = \{S : S \triangleright\triangleright T \to R\}$. *Let* $d(R, M) = M - G$. *Then d(R, M) is also hereditary and homomorphically closed.*

The class d(R, M) is thus the hereditary, homomorphically closed class obtained by deleting R from M, in the sense that d(R, M) is the largest hereditary, homomorphically closed subclass of M that does not contain R. The lower radical Ld(R, M) may or may not still contain R. Let S be a simple ring and let $M = \{0, S \oplus S\}$. Then M is hereditary and homomorphically closed and $d(S \oplus S, M) = \{0, S\}$. $S \oplus S$ is still a member of $Ld(S \oplus S, M)$. On the other hand we have the following two examples, which will be referred to in the sequel.

Example 1. Let M_1 consist of Z^0, the zero ring on the group of integers, together with all its homomorphic images. Note that each ideal of Z^0 is isomorphic to Z^0. Then $d(Z^0, M_1)$ consists of all finite cyclic zero rings and $Z^0 \notin Ld(Z^0, M_1)$. In this example Z^0 is noetherian.

Example 2. Let p be a prime and let $Z(p^\infty)$ denote the zero ring on the additive group p^∞ (see [2], p. 14). Let M_2 consist of $Z(p^\infty)$ and all its ideals. Note that $Z(p^\infty)$ is isomorphic to each of its homomorphic images and is an artinian ring. The class $d(Z(p^\infty), M_2)$ consists of all zero rings on the finite cyclic group of order p^k, k integral. Then $Z(p^\infty) \notin Ld(Z(p^\infty), M_2)$.

The next two results show that these two examples exhibit some typical features.

THEOREM 5. *Let M be a hereditary, homomorphically closed class of associative rings. Let $S \in M$ be left or right Artinian and suppose $S \notin Ld(S, M)$. If $K \triangleleft S$ and $K \neq S$, then $S/K \notin d(S, M)$ and $K \in d(S, M)$. Also either S is simple or S is a zero ring.*

Proof. If T is any ideal of S, either $T \notin d(S, M)$ or $S/T \notin d(S, M)$, for otherwise S would be in $Ld(S, M)$ since radical classes are closed under homomorphic extensions.

Suppose first that S has the (classical) radical $J \neq S$. Then J is nilpotent and S/J is a direct sum of simple rings $\{S_i : i = 1, . . ., k\}$. Since S is not nilpotent, $J \in d(S, M)$, for it cannot happen that $J \triangleright\triangleright T \to S$. Hence $S/J \notin d(S, M)$. Thus $S/J \triangleright\triangleright T \to S$, so that S is a direct sum of rings S_i and $J = 0$. Indeed, $k = 1$, for otherwise $S_i \in d(S, M)$ and hence we would have $S \in Ld(S, M)$.

For the remaining case, let $S \neq 0$ be radical. Then S is nilpotent with index of nilpotency h. Suppose $h > 2$. Then S^2 and S/S^2 would both have index of nilpotency smaller than h, so that both must be in $d(S, M)$ (otherwise, e.g., $S^2 \triangleright\triangleright T \to S$ would imply that S^2 has an index of nilpotency $\geq$ that of S). This is impossible again by the closure of $Ld(S, M)$ under extensions. Thus $h = 2$, so that $S^2 = 0$.

Let K be a nonzero ideal of S such that $K \notin d(S, M)$ and $K \neq S$. Then $K \triangleright\triangleright U \to S$ for some U by Theorem 4. Using this isomorphism to obtain subrings of K corresponding to K and L, we derive the diagram

$$\begin{array}{ccccccccccccc} S & \triangleright & K & \triangleright\triangleright & U_1 & \triangleright & K_1 & \triangleright\triangleright & U_2 & \triangleright & K_2 & \triangleright\triangleright & U_3 \cdots \\ & & & & \triangledown & & & & \triangledown & & & & \triangledown \\ & & & & L_1 & & \subseteq & & L_2 & & \subseteq & & L_3 \cdots \end{array}$$

where each $U_i/L_i \simeq S$ and, since $K \neq S$, the K_i must form a properly descending chain. But since $S^2 = 0$, all of the subrings of S are ideals of S, so no such chain is possible. This contradiction shows that $S/K \notin d(S, M)$.

THEOREM 6. *Let M be a hereditary, homomorphically closed class of associative rings. Let $S \in M$ be a ring with ascending chain condition on left ideals and suppose $S \notin Ld(S, M)$. If T is a nonzero ideal of S, then $T \notin d(S, M)$ and $S/T \in d(S, M)$. Also S is either nil-radical or nil-semisimple.*

Proof. As in the proof of Theorem 5, if T is any ideal of S, then either $T \notin d(S, M)$ or $S/T \notin d(S, M)$. Let T be a maximal ideal of S such that $S/T \notin d(S, M)$. Then $T \neq S$. We can write $S/T \triangleright\triangleright W/T \to W/U \simeq S$ by Theorem 4. Hence we obtain sequences of subrings of S, $S \triangleright\triangleright W_1 \triangleright\triangleright W_2 \triangleright\triangleright \ldots$ and $T \subseteq U_1 \subseteq T_1 \subseteq U_2 \subseteq T_2 \subseteq \ldots$ with the properties that, for each i, $T_i/U_i \simeq T$, $S \simeq W_i/U_i \to W_i/T_i \triangleright\triangleright W_{i+1}/T_i \to W_{i+1}/U_{i+1} \simeq S$, and T_i properly contains U_i (since $T \neq 0$). For each i let U_i' be the ideal of S generated by U_i. Then $U_1' \subseteq U_2' \subseteq \ldots$ so that, by the acc, $U_k' = U_{k+1}'$ for some k. Now $U_k'^n \subseteq U_k$ for some n by the Andrunakievich lemma. Since $U_k'^n \subseteq U_k \subseteq T_k \subseteq U_{k+1} \subseteq T_{k+1} \subseteq U_{k+1}' = U_k'$, we have $T \simeq T_k/U_k \subseteq U_{k+1}/U_k \subseteq T_{k+1}/U_k \subseteq W_k/U_k \simeq S$, and since $U_k'/U_k'^n$ is nilpotent, $T \simeq T_k/U_k \subset T_{k+1}/U_k$ are nilpotent. Therefore T is properly contained in the nil radical N of S.

Thus $S/N \in d(S, M)$ by the maximality of T, so that $N \notin d(S, M)$. Then we can write $N \triangleleft\triangleleft W \to S$, so that S must be nil. Hence S is nilpotent since S is a nil ring with acc on one-sided ideals.

If $S^2 \notin d(S, M)$, then $S^2 \triangleright\triangleright W \to S$, so that S has an index of nilpotency no greater than that of S^2. Hence $S/S^2 \notin d(S, M)$ instead, so that $T \supseteq S^2$. But then S is a zero ring, i.e., an abelian group, with acc on subgroups. Hence we cannot have $S/T \triangleright\triangleright W/T \to S$ for any $T \neq 0$.

Finally, if $N \neq 0$ is the nil radical of S, then $N \not\subseteq d(S, M)$, so that we can write $N \triangleright\triangleright W \to S$. Hence S is nil if $N \neq 0$. Therefore S is either nil-radical or nil semisimple.

We shall say that a hereditary, homomorphically closed class M is a *generator* of a hereditary radical class P if LM = P. We say that M is a *minimal generator* of P if no hereditary, homomorphically closed class properly contained in M is a generator of P. As remarked in the introduction, a minimal generator M of P must be unique if it exists. Let us use Theorem 4 to see that minimal generators of hereditary radicals (if they exist) are well-behaved with respect to class union. On the other hand, we shall see from an example that they are not well-behaved with respect to class intersection. If $R \not\in M$, let us understand that $d(R, M) = M$.

THEOREM 7. *Let M_1 and M_2 be hereditary, homomorphically closed classes of alternative rings such that M_i is a minimal generator of $LM_i = P_i$ for $i = 1, 2$. Then $M_1 \cup M_2$ is a minimal generator of $L(M_1 \cup M_2)$.*

Proof. First note that $M_1 \cup M_2$ is a hereditary, homomorphically closed class and that therefore $L(M_1 \cup M_2)$ is a hereditary radical. Suppose some hereditary, homomorphically closed class N, properly contained in $M_1 \cup M_2$, is also a generator of $L(M_1 \cup M_2)$.

Then if $R \in M_1 \cup M_2$ but $R \notin N$, we have $N \subseteq d(R, M_1 \cup M_2) \subseteq M_1 \cup M_2$, so that $LN = Ld(R, M_1 \cup M_2) = L(M_1 \cup M_2)$. Let $X_1 = d(R, M_1)$ and $X_2 = d(R, M_2)$. Without loss of generality, suppose $R \in M_1$. Note that $d(R, M_1 \cup M_2) = X_1 \cup X_2$.

Let S be any nonzero ring in M_1. Then each nonzero homomorphic image $S\emptyset$ of S contains a nonzero accessible subring I in $X_1 \cup X_2$, since $S \in M_1 \subseteq M_1 \cup M_2 \subseteq L(M_1 \cup M_2) = L(X_1 \cup X2)$. Here we have used [5, Lemma 5]. We shall show that $I \in X_1$. Now either $R \in M_1 \cap M_2$ or $R \notin M_2$.

First suppose $R \in M_1 \cap M_2$. Again note that $d(R, M_1 \cup M_2) = X_1 \cup X_2$. If I is in X_2 but not X_1, then we can write $S\emptyset \triangleright\triangleright I \triangleright\triangleright T \to R$. Therefore I, which as an accessible subring of a homomorphic image of S must be in M_1, is not in $d(R, M_1)$. But then I cannot be in $d(R, M_2)$ either. So in this case $I \in X_1$.

On the other hand, suppose $R \in M_1$ but $R \notin M_2$. Then $d(R, M_1 \cup M_2) = X_1 \cup M_2$. If $I \in M_2$ but $I \notin X_1$ then I must have been deleted from M_1, i.e., $I \notin d(R, M_1) = X_1$. But then $I \triangleright\triangleright T \to R$ implies $R \in M_2$, a contradiction.

Thus, in any event, $I \in X_1$. Hence any $S\emptyset \neq 0$ must contain a nonzero accessible subring I in $X_1 = d(R, M_1)$. But then $LX_1 = LM_1$, contradicting the minimality of M. We have shown that if $R \in M$ can be deleted from $M_1 \cup M_2$

and the resulting class still generates $L(M_1 \cup M_2)$ then R can be deleted from M_1, and the resulting class still generates LM_1. This completes the proof.

We now use the examples given earlier to show that an intersection of minimal generators of hereditary radicals need not be a minimal generator.

Example 3. Let M_1 and M_2 be the classes defined in Examples 1 and 2. M_1 and M_2 are minimal generators of LM_1 and LM_2, respectively, but $M_1 \cap M_2$ is the class of all cyclic abelian groups of order $p^k (k = 1, 2, \ldots)$ together with the ring {0}. Then $M_1 \cap M_2$ is not a minimal generator for $L(M_1 \cap M_2)$; clearly the class consisting of {0} and the abelian group of order p is a minimal generator for this radical.

We can raise the following questions. When does a hereditary radical class P have a minimal generator? This would seem to be related to the question of when $L(\bigcap M_i) = \bigcap LM_i$ for an infinite class $\{M_i : i \in I\}$. We should also like to know whether, if $LM \neq Ld(R, M)$ for some $R \in M$, there must exist $S \in M$ such that $S \notin Ld(S, M)$.

The author wishes to thank the referee for his invaluable assistance with this paper.

REFERENCES

[1] T. Anderson and B. J. Gardner, Semi-simple classes in a variety satisfying an Andrunakievich Lemma, Bull. Australian Math. Soc. 18 (1978), 187-200.

[2] N. J. Divinsky, Rings and Radicals, Univ. of Toronto Press, 1965.

[3] A. E. Hoffman and W. G. Leavitt, Properties inherited by the lower radical, Portugal. Math, 27 (1968) 63-66.

[4] D. Kreiling and R. Tangeman, Lower radicals in non-associative rings, J. Australian Math. Soc. 14 (1972), 419-423.

[5] J. Krempa, Lower radical properties for alternative rings, Bull. Acad. Polon. Sci. 23 (1975), 139-142.

[6] W. G. Leavitt, Strongly hereditary radicals, Proc. Amer. Math. Soc. 33 (1972), 247-249.

R. F. Rossa
Division of Mathematics and Physics
Arkansas State University
State University, Arkansas 72467
USA

COLLOQUIA MATHEMATICA SOCIETATIS JÁNOS BOLYAI

38. RADICAL THEORY, EGER (HUNGARY), 1982.

A CLASS OF RINGS ASSOCIATED WITH RADICAL THEORY

A. D. SANDS

We shall work in the class of associative rings. The fundamental definitions and results in radical theory used here may be found in Wiegandt [11]. Certain classes of radicals are considered. A-radicals have been defined by Gardner [2] as radicals which depend only on the underlying additive structure of the ring. The concepts of normal radicals and N-radicals have been introduced by Jaegermann [5] and Sands [8] in connection with Morita contexts. The notation $\ell(B)$ is used for the lower radical class generated by a class B.

Radical theory is not always sufficient to distinguish one ring from another. Given an abelian group $(R,+)$ and two rings defined over it, $(R,+,\cdot)$ and $(R,+,*)$, it may be the case that for all radicals α

the underlying abelian groups of $\alpha(R,+,\cdot)$ and $\alpha(R,+,*)$ are equal. In particular there is a class of rings which radical theory does not distinguish from the underlying abelian group. If $(R,+)$ is any abelian group then multiplication may be defined trivially by $a \circ b = 0$ for all $a, b \in R$; $(R,+,\circ)$ will be used to denote this trivial ring. In Sands [10] the class C is defined to be the class of all rings $(R,+,\cdot)$ such that $\alpha(R,+,\cdot) = \alpha(R,+,\circ)$ for all radicals α . Here, and later, an equality or inclusion relation between subsets of different rings defined over the same underlying abelian group is to be interpreted as a relation on the abelian group and not as an isomorphism or monomorphism of rings. Since radical theory is concerned with the structure of rings it seems appropriate to study the class C . Some properties of C were found in [10] and we give some further results here.

In [10] a ring R was defined to be M-nilpotent if given any doubly infinite sequence $\ldots, r_{-1}, r_0, r_1, \ldots$ of elements of R there exists k such that $r_{-k} \cdots r_{-1} r_0 r_1 \cdots r_k = 0$. I am indebted to Professor S.A. Amitsur for drawing my attention to the paper [7] by Levitzki. Unfortunately I have not been able to obtain a copy of this paper but it is clear from the review that this concept has already been considered there and that there is some overlap between [7] and section 3 of [10]. It was shown in [10] that the class of M-nilpotent rings is strictly contained in the class C. These results use methods developed by Gardner [4].

We shall denote the class of lower Baer radical rings by β ; β is the lower radical generated by the class of all trivial rings. A ring R is idempotent if $R^2 = R$. The class of all idempotent rings is the upper radical generated by all trivial rings. We shall call this class the idempotent radical and denote it by I. It is shown in [10] that C is contained in $\beta \cap S_I$, where S_I denotes the semisimple class of I. An example was given there to show that this containment is strict.

One may not be interested in all radicals but only in some specified class F of radicals satisfying some property. In a similar way we define classes relative to such a family F of radicals as follows:-

$$C_1(F) = \{(R,+,\cdot) \mid \alpha(R,+,o) \subseteq \alpha(R,+,\cdot) \text{ for all } \alpha \in F\} ;$$

$$C_2(F) = \{(R,+,\cdot) \mid \alpha(R,+,o) \supseteq \alpha(R,+,\cdot) \text{ for all } \alpha \in F\} ;$$

$C(F) = C_1(F) \cap C_2(F)$.

We use C_1 and C_2 whenever F is the class of all radicals. Of course if F is a family of A-radicals one has $C(F) = A$, where A denotes the class of all associative rings. For certain other classes of radicals $C(F)$ is of interest and in some cases we are able to determine it precisely. A result is given classifying the rings in C , but it is not a completely satisfactory description.

Given any radical class α the class of rings α^+ is defined by

$$\alpha^+ = \{(R,+,\cdot) \mid (R,+,o) \in \alpha\} .$$

It has been shown by Jaegermann and Sands [6] that α^+ is an A-radical. We also have the following relation between these radicals .

THEOREM 1. *Let* α *be a radical class and let* $(R,+,\cdot)$ *be a ring and* $(R,+,o)$ *the corresponding trivial ring. Then* $\alpha(R,+,o) = \alpha^+(R,+,\cdot)$.

Proof. Let $\alpha(R,+,o) = K$ and $\alpha^+(R,+,\cdot) = L$. Then $(K,+,\cdot)$ is an ideal of $(R,+,\cdot)$. Since $(K,+,o) \in \alpha$ we have $(K,+,\cdot) \in \alpha^+$ and so $K \subseteq L$. Similarly $(L,+,\cdot) \in \alpha^+$ implies $(L,+,o) \in \alpha$ and so $L \subseteq K$. Hence $K = L$, as required .

One may use the radical α^+ to redefine the classes $C(F), C_1(F), C_2(F)$ in terms of ideals on each ring R, e.g.

$$C(F) = \{R \mid \alpha(R) = \alpha^+(R), \text{ for all } \alpha \in F\} .$$

If the class F contains any radical α such that $\beta \subseteq \alpha$ and $(R,+,\cdot) \in C_1(F)$ then from $(R,+,o) \in \beta$ it follows that $(R,+,o) \in \alpha$ and so that $(R,+,\cdot) \in \alpha$. Thus $C_1(F) \subseteq \alpha$. In particular if $\beta \in F$ we have $C_1(F) \subseteq \beta$. A radical α is said to be subidempotent if $\alpha \subseteq I$, i.e. if each ring in α is idempotent. If F contains any subidempotent radical α and $(R,+,\cdot) \in C_2(F)$ then from $\alpha(R,+,o) = 0$ it follows that $\alpha(R,+,\cdot) = 0$. Thus $C_2(F) \subseteq S_\alpha$. In particular if $I \in F$ we have $C_2(F) \subseteq S_I$.

THEOREM 2. *For any radical* α *we have*

$$\alpha^+ \cap S_\alpha \cap \beta = \{0\} .$$

Proof. Let $R \in \alpha^+ \cap S_\alpha \cap \beta$. By Lemma 5 of [6], $\beta(R) = \{x \in R \mid Rx = 0\}$. Since $\beta(R) = R$ it follows that $R^2 = 0$. $R \in \alpha^+$ then implies $R \in \alpha$ and so $R \in \alpha \cap S_\alpha = \{0\}$.

THEOREM 3. *If* F *is a class of radicals such that* $\beta \in F$ *then* $C_1(F) = \beta$.

Proof. We have seen above that $\beta \in F$ implies $C_1(F) \subseteq \beta$.

Let $R \in \beta$ and let α be any radical in $\mathcal{F}$. Let $\alpha^+(R) = K$ and let $\alpha(K) = A$. Since β is a hereditary radical we have $K \in \beta$ and so $K/A \in \beta$. Since $K \in \alpha^+$ we have $K/A \in \alpha^+$ and also $K/A = K/\alpha(K) \in S_\alpha$. By Theorem 2 we have $K/A = 0$. It follows that $K \in \alpha$ and so $\alpha^+(R) = K \subseteq \alpha(R)$. Therefore $R \in C_1(\mathcal{F})$. Hence, for all families $\mathcal{F}$, $\beta \subseteq C_1(\mathcal{F})$.

It follows that $\beta \in \mathcal{F}$ implies $C_1(\mathcal{F}) = \beta$.

COROLLARY. $C_1 = \beta$.

THEOREM 4. *For any family* $\mathcal{F}$ *of radicals* $C_2(\mathcal{F})$ *is closed under formation of ideals.*

Proof. Let $R \in C_2(\mathcal{F})$ and let M be an ideal of R. We need to show that M is in $C_2(\mathcal{F})$. Let $\alpha \in \mathcal{F}$ and let $\alpha^+(R) = A$, $\alpha^+(M) = B$, $\alpha(M) = K$. For each $k \in K$ the mapping from A to K given by $x \to x\,k$ preserves addition. $A \in \alpha^+$ implies $Ak \in \alpha^+$ and so $(Ak,+,o) \in \alpha$. Since $(AK,+,o)$ is the sum of these ideals we have $(AK,+,o) \in \alpha$ and so $AK \in \alpha^+$. Now $K \subseteq \alpha(R) \subseteq A$, since $K \in \alpha$ and $R \in C_2(\mathcal{F})$. Therefore $K^2 \subseteq AK$ and, by the above results, $AK \subseteq \alpha^+(K)$. It follows that $K/\alpha^+(K)$ is a trivial ring. Since it is in α, it is also in α^+. It follows that $\alpha^+(K) = K$. Hence $\alpha(M) = K = \alpha^+(K) \subseteq \alpha^+(M)$. Therefore $M \in C_2(\mathcal{F})$.

COROLLARY. *If* F *is a family of radicals such that* $\beta \in F$ *then* $C(F)$ *is closed under formation of ideals. In particular* C *is closed under formation of ideals.*

Proof. This follows since $C(F) = C_1(F) \cap C_2(F) = \beta \cap C_2(F)$ and β is a hereditary radical.

The following example shows that some restriction on F is needed to ensure that $C(F)$ is closed under ideals.

Let Z denote the ring of integers and let α be the lower radical generated by Z and $(Z,+,o)$. Let F consist of the single radical α. Clearly Z belongs to $C(F)$. However the ideal $2Z$ of Z does not belong to $C(F)$. $(2Z,+,o) \cong (Z,+,o) \in \alpha$. $2Z$ is not in α as $2Z$ contains no non-zero subideal which is an image of either Z or $(Z,+,o)$.

In order to determine certain classes $C(F)$ we need information about the hereditary radicals contained in β. Let p be a prime. A ring R is said to be a p-ring if the underlying abelian group is p-primary. Let P be a set of primes. A ring R is said to be a P-ring if it is a direct sum of p-rings, $p \in P$.

THEOREM 5. *Let* α *be a radical contained in* β. *If* α *is hereditary then* α^+ *is hereditary and*

$\alpha = \alpha^{+} \cap \beta$. *If* γ *is a hereditary* A-*radical then* $\alpha = \gamma \cap \beta$ *is hereditary and* $\alpha^{+} = \gamma$.

Proof. Let $\alpha \subseteq \beta$ and let α be hereditary. Let A_p denote the trivial ring defined on the cyclic group of prime order p . If $\alpha = \beta$ then $\alpha^{+} = A$ and $\alpha = A \cap \beta$, as required. If $\alpha \neq \beta$ then by the results of Armendariz [1] there exists a set P of primes such that α is the lower radical generated by $A_p, p \in P$. Let $R \in \beta$ and let R be a P-ring. Suppose $R \in S_\alpha$. If $R \neq 0$ then $R \in \beta$ implies that R contains a non-zero ideal K with $K^2 = 0$. Then K also is a P-ring in S_α . It follows that, for some $p \in P$, K contains a non-zero element b with $pb = 0$. Then the ideal of K generated by b is A_p , which is in α . This contradicts $K \in S_\alpha$. It follows that $R = 0$. Since an image of a P-ring is a P-ring it follows that every P-ring in β is in α . Since α is clearly contained in this class it follows that α consists of all P-rings in β . Clearly α^{+} consists of all P-rings and so is a hereditary A-radical. Thus $\alpha = \alpha^{+} \cap \beta$, as required.

If γ is a hereditary A-radical and $\alpha = \gamma \cap \beta$ then $\alpha^{+} = \gamma^{+} \cap \beta^{+} = \gamma \cap A = \gamma$. Also α is hereditary, since β and γ are hereditary. Note that we obtain

Gardner's classification [2] of hereditary A-radicals.

From this classification we see that hereditary radicals contained in β are left and right hereditary. Since β and all A-radicals are left and right strong so also are hereditary radicals contained in β.

THEOREM 6. *If* F *is a family of radicals such that each radical in* F *is the intersection of a hereditary radical and an A-radical then* $\beta \subseteq C(F)$. *If also* $\beta \in F$ *then* $C(F) = \beta$.

Proof. Let $\alpha \in F$. Then there exists a hereditary radical γ and an A-radical δ such that $\alpha = \gamma \cap \delta$. By Theorem 5, $\gamma \cap \beta = (\gamma \cap \beta)^+ \cap \beta$ $= \gamma^+ \cap \beta^+ \cap \beta = \gamma^+ \cap \beta$. Therefore $\alpha^+ \cap \beta$ $= (\gamma \cap \delta)^+ \cap \beta = \gamma^+ \cap \delta^+ \cap \beta = \gamma^+ \cap \beta \cap \delta = \gamma \cap \beta \cap \delta$ $= \alpha \cap \beta$. Let $R \in \beta$. Then $\alpha(R) = (\alpha \cap \beta)(R)$ $= (\alpha^+ \cap \beta)(R) = \alpha^+(R)$. Therefore $R \in C(F)$. Hence $\beta \subseteq C(F)$.

If $\beta \in F$ we have $C(F) \subseteq \beta$ and so $C(F) = \beta$.

This theorem applies to many wellknown classes of radicals, e.g. the class of all hereditary radicals or of all left (or right) hereditary radicals and, of course, many subclasses of these. It has been shown in [6] that every normal radical is an intersection of an A-radical and an N-radical. Since N-radicals are hereditary and β is normal the theorem applies to suitable classes of normal radicals. Normal radicals need not be left hereditary, but it has been shown in [9] that they are principally left hereditary. This condition is weaker than left hereditary and independent from hereditary. However Gardner [3] has proved results that imply that a principally left hereditary radical contained in β is normal. Hence, as above, it may be shown that any family F of principally left or right hereditary radicals with $\beta \in F$ satisfies $C(F) = \beta$.

THEOREM 7. *If each radical* α_i *in a family* F *satisfies* $\beta \subseteq \alpha_i$ *then* $C(F) = \alpha$, *where* $\alpha = \cap \alpha_i$. *If each radical* α_i *in a family* F *is subidempotent then* $C(F) = S_\alpha$ *where* $\alpha = \ell(U\alpha_i)$.

Proof. Let $\beta \subseteq \alpha_i$ for all $\alpha_i \in F$ and let $\alpha = \cap \alpha_i$. Then $\beta \subseteq \alpha_i$ implies $\alpha_i^+(R) = R$ for all rings R. Let $R \in \alpha$. Then $R \in \alpha_i$ for all i and so $\alpha_i(R) = \alpha_i^+(R) = R$. Therefore $R \in C(F)$. Let $R \in C(F)$. Then $\alpha_i(R) = \alpha_i^+(R) = R$, for all i. Hence $R \in \cap \alpha_i = \alpha$. Therefore $C(F) = \alpha$.

Now let every $\alpha_i \in F$ be subidempotent and let α be the smallest radical class containing all the classes α_i. Since each α_i is subidempotent $\alpha_i^+(R) = 0$ for all rings R. Let $R \in C(F)$. Then $\alpha_i(R) = \alpha_i^+(R) = 0$

for all i. It follows that $\alpha(R) = 0$. Hence $R \in S_\alpha$. Now let $R \in S_\alpha$. Then $\alpha(R) = 0$ and so $\alpha_i(R) = 0 = \alpha_i^+(R)$ for all i. Therefore $R \in C(F)$. Thus $C(F) = S_\alpha$.

Among the classes F which correspond to well-known radical properties open questions do remain. For example let F consist of all strong or left strong or right strong radicals. Then both β and I are in F and so $C(F) \subseteq \beta \cap S_I$, but the precise result is not known.

None of the above results determines the class C. We now give the result promised earlier to describe the rings which belong to C.

THEOREM 8. *A ring R belongs to C if and only if each ideal A of R belongs to the lower radical generated by A/A^2.*

Proof. Let $R \in C$ and let A be an ideal of R. By the Corollary to Theorem 4 we have $A \in C$. Let α_1 be the lower radical generated by A and let α_2 be the lower radical generated by A/A^2. Since $A \in C$ and $A \in \alpha_1$, we have $(A,+,o) \in \alpha_1$. Therefore every non-zero image of $(A,+,o)$ contains a non-zero ideal which is an image of A. Since $(A,+,o)$ and its images are trivial rings the kernels of these homomorphisms from A must contain A^2 . Hence every non-zero image of $(A,+,o)$ contains a non-zero ideal which is an image of A/A^2 . It follows that $(A,+,o) \in \alpha_2$. Since $A \in C$ we have $A \in \alpha_2$, as required.

Conversely let R be a ring such that each ideal A of R belongs to the lower radical generated by A/A^2 . Let α be a radical and let $\alpha(R) = K$, $\alpha^+(R) = M$. Then $M \in \ell(M/M^2) \subseteq \ell(M,+,o) \subseteq \alpha$, as M/M^2 is an image of $(M,+,o)$ and $(M,+,o) \in \alpha$. Therefore $M \subseteq K$. If $M \neq K$ then K/M , being a non-zero image of K , contains a non-zero subideal A/M which is an image of K/K^2 . $K/K^2 \in \alpha$ and so $K/K^2 \in \alpha^+$. Therefore $A/M \in \alpha^+$; however A/M is a non-zero subideal of the α^+- semisimple ring R/M. This is a contradiction. Therefore $M = K$ and so $\alpha(R) = \alpha^+(R)$. It follows that $R \in C$.

Our point of view has been that C consists of those rings R for which radical theory does not distinguish between the multiplication in R and trivial multiplication. So one would like to obtain a classification in terms of some property of the multiplication. It remains an open question whether some suitable generalisation of M-nilpotence or some other description of the multiplication can be found.

In [10] an example was given of a ring in C which is not M-nilpotent. This example is an extension of an M-nilpotent ring by an M-nilpotent ring. Thus it is natural to ask whether C is closed under extensions. This question remains open.

Theorem 8 does suffice to answer some questions concerning closure of C under algebraic constructions. For example we shall show that C is closed under formation of matrix rings. For any ring R we let $M_n(R)$ denote the ring of $n \times n$ matrices with entries in R. We recall that since a radical of $M_n(R)$ is an ideal of $M_n(1,R)$ it has the form $M_n(A)$, A an ideal of R. Given any homomorphism f from R to a ring S there is an induced homomorphism from $M_n(R)$ to $M_n(S)$ sending the matrix (r_{ij}) to the matrix $(f(r_{ij}))$.

THEOREM 9. *If* $R \in C$ *then* $M_n(R) \in C$.

Proof. Let $R \in C$ and let α be a radical. Let $\alpha(R) = \alpha^+(R) = B$. Since $M_n(R,+,o)$ is the direct sum of n^2 copies of $(R,+,o)$ we have $\alpha^+(M_n(R)) = M_n(B)$. Let $\alpha(M_n(R)) = M_n(A)$.

If $A \not\subseteq B$ then $A/(B \cap A)$ is a non-zero image of A and by Theorem 8 there is a non-zero homomorphism from A/A^2 onto a subideal of $A/(B \cap A) \cong (A + B)/B$.

This induces a homomorphism from $M_n(A/A^2)$ onto a non-zero subideal of $M_n((A + B)/B)$.
$M_n(A/A^2) \cong M_n(A)/M_n(A^2)$ which is a trivial ring in α and so is in α^+ . This implies that $M_n(R)/M_n(B)$ contains a non-zero subideal in α^+ , which contradicts the fact that $M_n(R)/M_n(B)$ is α^+-semisimple. Therefore $A \subseteq B$.

If $A \neq B$ then B/A is a non-zero image of B . As above there exists a non-zero homomorphism from B/B^2 onto a subideal of B/A . Thus there is induced a homomorphism from $M_n(B)/M_n(B^2)$ onto a non-zero subideal of $M_n(R)/M_n(A)$. $M_n(B)/M_n(B^2)$ is a trivial ring in α^+ and so is in α . Therefore $M_n(R)/M_n(A)$ contains a non-zero subideal in α , which contradicts the fact that it is α-semisimple. Therefore $A = B$.

Hence $M_n(R) \in C$.

REFERENCES

[1] E.P. Armendariz, Hereditary subradicals of the lower Baer radical, *Publ. Math. Debrecen* 15 (1968), 91-93.

[2] B.J. Gardner, Radicals of abelian groups and associative rings, *Acta Math. Acad. Sci. Hungar.* 29 (1973), 259-268.

[3] B.J. Gardner, Sub-prime radical classes determined by zerorings, *Bull. Aust. Math. Soc.* 12 (1975), 95-97.

[4] B.J. Gardner, Some aspects of T-nilpotence, *Pacific J. Math.* 53 (1974), 117-130.

[5] M. Jaegermann, Morita contexts and radicals, *Bull. Acad. Polon. Sci.* 20 (1972), 619-625.

[6] M. Jaegermann and A.D. Sands, On normal radicals, N-radicals and A-radicals, *J. Algebra* 50 (1978), 337-349.

[7] J. Levitzki, Contributions to the theory of nilrings, *Riveon Lematematika* 7 (1953), 50-70. (Hebrew, English summary) *Math. Reviews* 15 (1954), 677.

[8] A.D. Sands, Radicals and Morita Contexts, *J. Algebra* 24 (1973), 335-345.

[9] A.D. Sands, On normal radicals, *J. London Math. Soc.* 11 (1975), 361-365.

[10] A.D. Sands, On M-nilpotent rings, *Proc. Roy. Soc. Edin.* (to appear).

[11] R. Wiegandt, *Radical and Semisimple Classes of Rings*, Queen's University, Kingston, Ontario, 1974.

A. D. Sands
Department of Mathematical Sciences,
The University,
Dundee DD1 4HN,
Scotland.

COLLOQUIA MATHEMATICA SOCIETATIS JÁNOS BOLYAI
38. RADICAL THEORY, EGER (HUNGARY), 1982.

ON AMITSUR'S CONDITION E

M. SLATER

1. INTRODUCTION

1.1 Acknowledgements. An earlier (unpublished) version of this paper was written in summer 1968, and I should like to thank the University of Chicago for its hospitality at that time (see[14]). There is a reference in [11], and again in [18],p192, at which time it was planned to complete it jointly with Dr Watters. The final stimulus to preparing it for publication came only with the invitation to the Eger conference. It is therefore my great pleasure to thank the organisers of the conference for that stimulus as well as for everything else they did so well for the participants. I would also like to thank Maureen Woodward most warmly for all her work in preparing the camera-ready copy.

1.2 Notation. Throughout this paper $\tilde{U}$ denotes a universe of (not necessarily associative) algebras over some fixed commutative associative operator ring Φ with 1, where $1x = x$ for all $x \in A \in \tilde{U}$. We write All, Ass, Comm, Anticomm (strictly: All_Φ etc) for the universes of all Φ-algebras, all associative, all commutative and all anticommutative Φ-algebras respectively, and $\tilde{Z}$ (= $\tilde{Z}_\Phi$) for $\{A \in \text{All}: AA = 0\}$, the class of *zero-algebras*. A is *trivial* if $0 \neq A \in \tilde{Z}$. The symbol $\tilde{H}$ will always denote a homomorphically closed subclass of a given $\tilde{U}$.

In the Kurosh hierarchy $\{\tilde{H}_\alpha\}$ for given $\tilde{H}$ we start with $\tilde{H} = \tilde{H}_0$ (rather than $\tilde{H}_1$ as in [7]; see [18], 1.5), and we take unions at limit ordinals (as in [8], and unlike [16], p418). We write $\tilde{H}_*$ for $\cup_\alpha \tilde{H}_\alpha$, the radical class generated by $\tilde{H}$ (the "lower" radical in Kurosh's terminology).

The operator h associated with a given $\tilde{H} \subseteq \tilde{U}$ is defined on $\tilde{U}$ by $h(A)$ = the smallest ideal I of A such that A-I has no non-0 ideals in $\tilde{H}$. We call it a *semiradical*. We write h_α for the semiradical associated with $\tilde{H}_\alpha$, and h_* (a radical operator) for that associated with the radical class $\tilde{H}_*$.

Following Amitsur ([1], pp105 and 111) we say that $\tilde{H}$ is a D-class if $I \leq A \in \tilde{H}$ implies $I \in \tilde{H}$, a Z-class (in $\tilde{U}$) if $\tilde{Z} \cap \tilde{U} \subseteq \tilde{H}$ (thus this notion is relative to the universe within which we are working), and a D+Z class if both conditions hold. We extend this notation in a natural manner (cf. [18], 3.8): $\tilde{H}_\alpha$ is a D_α-class or a D_*-class if $\tilde{H}_\alpha$ (resp $\tilde{H}_*$) is a D-class, and similarly for $\tilde{Z}_\alpha$ and $\tilde{Z}_*$.

We write $A < B$; $A \subset B$ (as opposed to $A \leq B$; $A \subseteq B$) to mean that A is a *proper* ideal or subset of B; ie we imply $A \neq B$. When specifying an algebra A over a field F we list first a basis and then all the *non-0* basis products. In the context of homomorphisms θ we will write difference algebras rather than the usual but illogical quotients: thus $A\theta \simeq A - \mathrm{Ker}\theta$. Any further unexplained notation etc may be found in [18], Sec.1.

1.3 Aims. The original purpose of this work and its projected sequel was to investigate conditions on $\tilde{H}$ and $\tilde{U}$ (and Φ) sufficient to ensure that h_* satisfies Amitsur's condition E ([1], p105); ie that if $I \leq A \in \tilde{U}$ then $h_*(I) \subseteq I \cap h_*(A)$. This aim is dual to that of [18], Sec 5. However, a closely related question is to determine some α (if any exists) such that $\tilde{H}_* = \tilde{H}_\alpha$. For we have the following result ([1], Thm 8.1), which will be used repeatedly throughout this paper:

PROPOSITION 1.3 (Amitsur). *If* h_β *satisfies* E, *then* $h_* = h_\beta$ *and* $\tilde{H}_* = \tilde{H}_{\beta+1}$.

For this reason this paper will be concerned not only with conditions on $\tilde{H}$ and $\tilde{U}$ that will ensure E on h_β for some specified β, but also with conditions ensuring the weaker conclusion that $\tilde{H}_* = \tilde{H}_\gamma$ for some specified γ.

1.4 Summary. In the remainder of this section we will establish (Theorem A) a strong form of the thesis that for results of the above type we need conditions on $\tilde{U}$ and not merely on $\tilde{H}$ (cf [18], 5.5).

In Section 2 we discuss a fair number of arguably natural and fruitful conditions (of "Andrunakievic type") on $\tilde{U}$, and give an exhaustive account of the implications and non-implications among them.

It is these conditions that will figure in the hypotheses of the results in the main Sections 4 and 5.

The aim of Section 3 is to provide the technical prerequisites needed for parts of Section 4, but the work is deliberately done in greater generality than strictly necessary for that purpose, in the hope that some of these results may have independent interest.

Section 4 contains theorems in which D or similar is a hypothesis on $\tilde{H}$. Results with other types of hypothesis on $\tilde{H}$ will appear in a projected sequel to this paper.

In Section 5 we construct examples to show that two hierarchies of conditions used in Sec.4 do not degenerate; i.e. that no two conditions in either hierarchy are equivalent.

Note (added November 1983).

Professor Gardner has called to my attention his paper [6], in which a modification of his construction in [5] (different from the one used below) is used to prove substantially the same as 1.10a below.

The machinery we use in this section is a refinement of that developed by Ryabuhin in [13].

PROPOSITION 1.5. *Given a field* F *and algebras* S,E *over* F *with* $\dim S \geq \dim E \geq 2$, *there exist algebras* $A = A_c$ *and* $A = A_a$ *over* F *such that all of:*

(i) $A = S + E$ *(additive direct sum of subalgebras);*

(ii) $E \leq A$;

(iii) *If* T *is a right or left ideal of* A *then* $T \subseteq S$ *or* $T \supseteq E$;

(iv) *For all* $s \in S$; $e \in E$ *we have* $se = es$ *in* A_c, *and* $se = -es$ *in* A_a.

(v) *If* $\dim S \geq 3$, *then for an arbitrary prespecified* $s_0 \in S$ *(with* $s_0 \neq 0$*) we have* $s_0 e = e$ *for all* $e \in E$.

(vi) *If* S *is subdirectly irreducible then any* T *as in* (iii) *must be* 0 *or contain* E.

Proof. This is somewhat tedious, and we omit it. It is available on request.

1.6. To apply the above for algebras over an arbitrary Φ let P be any prime ideal of Φ, and F any extension field of (the quotient field of) Φ-P. Let $\alpha \mapsto \overline{\alpha}$ be the induced map of $\Phi \to F$. Given any F-algebra A, we can regard A as a (unital) Φ-algebra by defining αx to mean $\overline{\alpha}x$. Then as a corollary of 1.5 we have

PROPOSITION 1.7. *Suppose* $\Phi \to F$ *as in* 1.6; S, E *are* F-*algebras with* $3 \leq \dim S \geq \dim E \geq 2$, *and* S *is simple as a* Φ-*algebra. Then there exists an* F-*algebra* A *such that:*

(i) E *is the unique proper* Φ-*ideal of* A;

(ii) $A - E \simeq S$;

(iii) *If* E *and* S *are both in* <u>Comm</u> *or* <u>Anticomm</u>, *then so also is* A;

(iv) *If* S *has a* 1 *then* A *has a* 1.

Proof. We construct A as in 1.5, taking $A = A_c$ if $E,S \in$ <u>Comm</u>, and $A = A_a$ if $E,S \in$ <u>Anticomm</u>. We also ensure 1.5(v), taking s_o to be the 1 of S if S has a 1, and arbitrary otherwise. Then clearly (iii) and (iv) will hold if they are applicable.

From 1.5(i) and (ii) we have $E < A$ and $A - E \simeq S$. Since S is simple, E is a maximal F-ideal, and it follows by (vi) that E is the unique proper F-ideal of A. To complete the proof of (i) we show that any non-0 Φ-ideal I of A must equal E or A. Set $T = FI = \{\Sigma \alpha_i x_i : \alpha_i \in F;\ x_i \in I\}$. Then T is an F-ideal of A, so T = E or A. If T = E then $I \subseteq E$, so for $\alpha \in F$, $x \in I$ we have $\alpha x = \alpha(s_o x) = (\alpha s_o)x \in I$. Thus I = T = E. If T = A, say $s_o = \Sigma \alpha_i x_i$ with $x_i \in I$. Then for $e \in E$ we have $e = s_o e = \Sigma(\alpha_i s_o)x_i \in I$. Thus $E \subseteq I$. Since S is simple as a Φ-algebra, this gives I = (E or) A, as required.

1.8. Given $\Phi \to F$ as in 1.6, and a class $\tilde{S}_o$ of F-algebras which are simple as Φ-algebras, and such that the class $\{\dim_F S\colon S \in \tilde{S}_o\}$ of cardinals is unbounded, we can construct a transfinite chain $\{A_\alpha\colon \alpha \geq -1\}$ of F-algebras with $A_{-1} \in \tilde{S}_o$, $\dim_F A_{-1} \geq 2$; $A_{\beta+1}$ = an A as in 1.7, where $E = A_\beta$ and $S \in \tilde{S}_o$ is arbitrary subject to $\dim_F S \geq \dim_F A$, and $A_\lambda = \Sigma \bigoplus_{\mu<\lambda} A_\mu$ for λ a limit ordinal (note that for present purposes $0 = (-1) + 1$ is regarded as a successor). By 1.7(i) the unique proper Φ-ideal of $A_{\beta+1}$ is A_β. If $\tilde{S}_o \subseteq$ <u>Comm</u> or <u>Anticomm</u> we can ensure the same for $\{A_\alpha\}$ by 1.7(iii), and if every $S \in \tilde{S}_o$ has a 1 we can ensure the same for every $A_{\beta+1}$ by 1.7(iv).

PROPOSITION 1.9. *Suppose* $\Phi \to F$ *as in* 1.6, *and* $\tilde{U}$ = <u>All</u> *or* <u>Comm</u> *or* <u>Anticomm</u> *over* Φ. *Then*

(a) *There exists a class* $\tilde{S}_o$ *of* F-*algebras which are simple as* Φ-*algebras, and such that* $\{\dim_F S\colon S \in S_o\}$ *is unbounded.*

(b) *For any such* $\widetilde{S}_o$, *if* $\{A_\alpha\}$ *is constructed as in* 1.8, *then there exists* $\widetilde{H}$, *for instance* $\widetilde{H} = \widetilde{S}_o \cup \widetilde{Z}$, *such that* $\widetilde{S}_o \subseteq \widetilde{H} \subseteq \widetilde{U} \setminus \{A_\alpha;\ \alpha \geq 0\}$.

(c) *If* $\widetilde{H}$ *is as in* (b), *then for all ordinals* α, $\widetilde{H}_* \neq \widetilde{H}_\alpha$.

Proof. See the end of 1.2 for notation. If $\gamma > 2$ is any cardinal (ie. initial ordinal) we define $S_\gamma^{a} = F[x_i (i<\gamma)$: for $i>0$ $x_i x_{i+1} = -x_{i+1}x_i = x_o$; $x_o x_i = -x_i x_o = x_i] \in$ Anticomm, and $S_\gamma^{c} = F[x_i (i<\gamma)$: $x_o x_i = x_i x_o = x_i$; $x_i x_i = x_o] \in$ Comm $\subseteq$ All. We easily check that in either case S_γ is simple of dimension γ over F (the condition $\gamma > 2$ is needed only for S_γ^{a}).

(b) More generally, let $\widetilde{K}$ be any homomorphically closed subclass of $\widetilde{Z}_*$ in $\widetilde{U}$. Then $\widetilde{H} = \widetilde{S} \cup \widetilde{K}$ will do. It is clearly homomorphically closed. Next, $\widetilde{S}_o \cap \{A_\alpha: \alpha \geq 0\} = \emptyset$ since for $\alpha \geq 0$ A_α is not simple (it has proper ideal A_β for at least one $\beta < \alpha$). Finally $K \cap \{A_\alpha\} = \emptyset$. For every A_α embeds A_{-1} which is Φ-simple of dim ≥ 2 over F, and no Baer-radical ring R (ie. $R \in \widetilde{Z}_*$) has this property, as is not hard to check.

(c) We will show by induction on α that $A_\alpha \in$

$\tilde{H}_{\alpha+1} \setminus \tilde{H}_\alpha \subseteq \tilde{H}_* \setminus \tilde{H}_\alpha$, so that $\tilde{H}_* \neq \tilde{H}_\alpha$. If $\alpha = -1$ this is immediate from the construction, since $\tilde{H}_0 = \tilde{H} \supseteq \tilde{S}_0$ (take $\tilde{H}_{-1} = \{0\}$).

If $\alpha = \lambda$ is a limit ordinal, then every non-0 homomorphic image A_λ' of $A_\lambda = \Sigma_{\mu<\lambda} \oplus A_\mu$ has a non-0 ideal $I = A_\mu'$ for some $\mu < \lambda$, so $I \in \tilde{H}_{\mu+1} \subseteq \tilde{H}_\lambda$, whence $A \in \tilde{H}_{\lambda+1}$. If $A_\lambda \in \tilde{H}_\lambda$ then $A_\lambda \in \tilde{H}_\mu$ for some $\mu < \lambda$. But then $A_\mu \simeq A_\lambda - \Sigma \oplus \{A_\gamma : \gamma < \lambda; \gamma \neq \mu\}$ is also in $\tilde{H}_\mu$ contrary to inductive hypothesis.

If $\alpha = \beta + 1$ (including the case $\alpha = 0$, $\beta = -1$), then by 1.8 the only proper Φ-ideal of A_α is A_β. Thus the only non-0 homomorphic images of A_α are isomorphic to $A_\alpha - A_\beta \in \tilde{S}_0 \subseteq \tilde{H} \subseteq \tilde{H}_\alpha$, and to A_α itself, and A_α has non-0 ideal A_β in $\tilde{H}_{\beta+1} = \tilde{H}_\alpha$ by inductive hypothesis if $\beta \geq 0$, and by construction if $\beta = -1$, since $A_{-1} \in \tilde{S} \subseteq \tilde{H} = \tilde{H}_0$. Thus $A_\alpha \in \tilde{H}_{\alpha+1}$.

If γ is minimal such that $A_\alpha \in \tilde{H}_\gamma$ then γ is clearly not a limit ordinal >0, and $\gamma \neq 0$ by the hypothesis on $\tilde{H}$. Thus $\gamma = \delta + 1$ for some δ. Then A_α has a non-0 ideal, necessarily A_β, in $\tilde{H}_\delta$. So by inductive hypothesis $\delta \geq \beta + 1 = \alpha$, whence $\gamma = \delta + 1 \geq \alpha + 1$. Thus $A_\alpha \notin \tilde{H}_\alpha$.

PROPOSITION 1.10. *In the universe* $\tilde{U}$ = <u>All</u> *or* <u>Comm</u> *or* <u>Anticomm</u> *over any* Φ*, suppose the radical class* $\tilde{R}$ *is such that* r *satisfies* E. *Then*

(a) $A \in \tilde{R}$ *if and only if* $A_o \in \tilde{R}$;

(b) *If* $\tilde{Z} \subseteq \tilde{R}$ *then* $\tilde{R} = \tilde{U}$.

Proof. (a) Here A_o is obtained from A by redefining all products to be 0. In case $\tilde{U}$ = <u>All</u> this was proved by Gardner ([5], 2.6). The key construction on p68 (after the trivial change of z_2x_1 to z_1x_2 in lines 3 and 4) may be stated as follows. Define $\Gamma = \Phi[e_1,e_o,e_2\colon e_1e_1 = e_1,\ e_2e_o = e_1,\ e_2e_1 = e_o]$, and Λ similarly but with $e_2e_2 = e_2$ added. For each $A \in \tilde{U}$ define $\Gamma(A) = A \otimes_\Phi \Gamma$ and $\Lambda(A) = A \otimes_\Phi \Lambda$. Then $I_A = Ae_1+Ae_o$ is an ideal of $\Gamma(A)$ (resp. $\Lambda(A)$) isomorphic to $A \oplus A_o$, and $\Gamma(A)-I_A \simeq A_o$; $\Lambda(A) - I_A \simeq A$. Furthermore the ideal of $\Gamma(A)$ (resp. $\Lambda(A)$) generated by $Ae_1 \simeq A$ is $Ae_1 + A^2e_o$, and the ideal generated by $Ae_o \simeq A_o$ is $A^2e_1 + Ae_o$.

Everything else follows from the existence in $\tilde{U}$ of algebras $\Gamma(A)$, $\Lambda(A)$ (for all $A \in \tilde{U}$) having ideals etc. with these properties. Now if we modify Γ and Λ to be commutative; ie. we specify e_oe_2 =

e_1 and $e_1 e_2 = e_o$, then all these properties continue to hold, and in addition $\Gamma(A)$ and $\Lambda(A)$ will be in <u>Comm</u> whenever A is in <u>Comm</u>, and in <u>Anticomm</u> whenever A is in <u>Anticomm</u>. Thus Gardner's construction, and hence his proof, goes through in these two universes (and in any smaller universes closed under tensoring with Γ and Λ; eg. the finitely-generated or finitely spanned algebras in <u>Comm</u> and in <u>Anticomm</u>).

(b) Given $A \in \tilde{U}$ we have $A_o \in \tilde{Z} \subseteq \tilde{R}$, so by (a) $A \in \tilde{R}$. Thus $\tilde{U} \subseteq \tilde{R}$.

COROLLARY 1.11. *Suppose* $\tilde{U}$ = <u>All</u> or <u>Comm</u> or <u>Anticomm</u>; $\tilde{Z} \subseteq \tilde{H} \subseteq \tilde{U}$, *but not all simple algebras in* $\tilde{U}$ *are in* $\tilde{H}$. *Then* h_* *does not satisfy* E.

Proof. Take $\tilde{R} = \tilde{H}_*$ in 1.10. If $r = h_*$ satisfied E then we would have $\tilde{H}_* = \tilde{U}$. Thus there would be a simple algebra $S \in \tilde{H}_* \setminus \tilde{H}$. But this is well known to be impossible.

1.12. Following Leavitt [11] we say that the operator h is *strongly hereditary* if $I \leq A \in \tilde{U}$ implies $h(I) = I \cap h(A)$; in other words h satisfies both D and E, and in particular is a radical operator by 1.3.

THEOREM A. *In each of the universes* $\widetilde{U}$ = <u>All</u> *and* <u>Comm</u> *and* <u>Anticomm</u> *over any* Φ, *there exists a subuniverse* $\widetilde{U}_o$ *and a class* $\widetilde{R} \subseteq \widetilde{U}_o$ *such that*

(a) $\widetilde{R}$ *is a* D+Z *class;*

(b) *In* $\widetilde{U}_o$, $r = r_*$ *is strongly hereditary, and* $\widetilde{R}_* = \widetilde{R}$;

(c) *In* U, r_* *is not strongly hereditary, and for no* α *do we have* $R_* = R_\alpha$ *in* $\widetilde{U}$.

Proof. Let $\Phi \longrightarrow F$ as in 1.6, let $\widetilde{S}$ be the class of all F-algebras in U which are simple as Φ-algebras, and let $\widetilde{S}_o \subset \widetilde{S}$ be such that $\{\dim_F S: S \in \widetilde{S}_o\}$ is unbounded. This is possible by 1.9(a). Set $\widetilde{R} = \widetilde{S}_o \cup \widetilde{Z}$. Then (a) is clear. By 1.11 with $\widetilde{H} = \widetilde{R}$ we have (c). By 1.9(b) and (c) we have (d). To ensure (b) we can simply take $\widetilde{U}_o = \widetilde{R}$ (a universe by (a)). Then of course $\widetilde{R}_* = \widetilde{R}$ in $\widetilde{U}_o$, and $r_* = r$ is strongly hereditary since $r(A) = A$ for all $A \in \widetilde{U}_o$.

1.13. <u>Discussion of (b)</u>. There are more interesting possibilities for $\widetilde{U}_o$, at least if $\widetilde{U}$ = <u>Comm</u> or <u>All</u>. Thus for each cardinal γ let F_γ be an extension field of transcendence degree $\geq\gamma$ over F (as above). Then take $S_o = \{F_\gamma: \text{all } \gamma\}$;

$\tilde{H} = \tilde{Z} \cup \tilde{S}_o$, and $\tilde{U}_o$ = <u>Ass</u> $\cap$ <u>Comm</u> inside $\tilde{U}$ = <u>Comm</u> or <u>All</u>, or $\tilde{U}_o$ = <u>Ass</u> inside $\tilde{U}$ = <u>All</u>. Set $\tilde{R} = \tilde{H}_1$ inside $\tilde{U}_o$. Since $\tilde{H}$ is a D+Z class and $\tilde{U}_o \subseteq$ <u>Ass</u>, it follows, eg. from Thm. D (Sec.4) that in $\tilde{U}_o$ we have $\tilde{R} = \tilde{R}_*$ and r is strongly hereditary. Also (c) holds by 1.11. In the situation of 1.9 we can ensure $A_o \notin$ <u>Ass</u> (so for all α, $A_\alpha \notin$ <u>Ass</u>), whence 1.9b holds for $\tilde{H} = \tilde{R}$, and then (d) follows from 1.9c.

If $2 \in \Phi$ is invertible we may similarly take $\tilde{U}_o$ to be the class $\tilde{J}$ of (linear) Jordan algebras inside $\tilde{U}$ = <u>Comm</u> or <u>All</u>, taking $\tilde{R} = \tilde{H}_*$ within $\tilde{J}$ for the same $\tilde{H}$ as above. Then r is strongly hereditary in $\tilde{J}$ by a result of Slin'ko [15], (c) holds by 1.11, and (d) can be made to hold by ensuring $A_o \notin \tilde{J}$ and arguing as above.

For a 'natural' example inside $\tilde{U}$ = <u>Anticomm</u> we can take $\tilde{U}_o = \tilde{L}$, the class of Lie algebras. So long as we restrict our attention to the case where $\Phi = F$ is a field, it is easy to produce a class $\tilde{S}_o \subseteq \tilde{L}$ satisfying the condition of 1.9(a), and an $\tilde{R}$ constructed as in the Jordan case will then satisfy all the conditions with the possible exception of E in $\tilde{L}$: I do not know whether D+Z radical classes in $\tilde{L}$ necessarily also satisfy E.

There is at least one case where $\tilde{R}$ itself, as well as $\tilde{U}_o$, is an interesting class. Thus let $\tilde{U}_o$ = <u>Ass</u> over Φ, and let $\tilde{R}$ be the Brown-McCoy radical class. Given $\Phi \longrightarrow F$ as in 1.6, for each infinite cardinal γ let S_γ be the F-algebra of linear transformations of finite rank on a vector space of dimension γ over F. It is well known that S_γ is simple as an F-algebra; it is also simple as a Φ-algebra since for each $s \in S_\gamma$ there exists $e_s \in S_\gamma$ such that $se_s = s$. Set $\tilde{S}_o = \{S_\gamma : \text{all } \gamma\}$. Then $\tilde{S}_o \subseteq \tilde{R}$, and we can ensure $\tilde{R} \cap \{A_\alpha : \alpha \geq 0\} = \emptyset$ by making A_0 be not associative. Then by 1.9(c) we have (d), by 1.11 we have (c) (since for example $F.1 \notin \tilde{R}$), and (a) and (b) are well known.

It would be interesting to know whether there is a sufficiently large class of Jacobson-radical simple Φ-algebras (or at least F-algebras) to allow the same construction to go through for $\tilde{R}$ = the Jacobson radical in <u>Ass</u>.

2. ANDRUNAKIEVIC-TYPE CONDITIONS

2.1. Much the most useful possible condition on $\tilde{U}$ would be that the relation $\leq$ was transitive. For in that case if $I \leq A \leq \tilde{U}$ with $h(A) = 0$, and $V \in \tilde{H}$ with $V \leq I$, then $V \leq A$, so $V = 0$, so $h(I) = 0$, whence E holds for h, and $h = h_*$; $\tilde{H}_* = \tilde{H}_1$ by 1.3. However, while $\leq$ is transitive in the universe $\tilde{Z}$ or the universe of Boolean rings (over $\Phi = GF_2$), or more generally in the universe of hereditarily idempotent associative rings, it is not so in less specialised universes.

We define $V \leq_o A$ to mean $V = A$, and $V \leq_{n+1} A$ to mean that $V \leq_n I \leq A$ for some I. V is an *accessible subalgebra* of A if $V \leq_n A$ for some n. Then transitivity of $\leq$ says that if $V \leq_2 A$ then $V \leq A$. The conditions we study in this section weaken this by requiring that if $V \leq_2 A$ then V should at least *approximate* an ideal of A in one of two general ways.

The conditions of 2.2I and II in effect consider how tightly V sits *above* an ideal of A. If X is the sum of all ideals of A contained in V, we may interpret this as asking how closely the difference algebra $V - X = V'$ comes to being zero. On factoring out X we find that (if $V \not\leq A$) we have the *Andr* situation $0 < V' < I' < A'$ (defined below). The conditions of I and II then insist that there should be some pathology in the form of the existence of soluble ideals (I) or trivial ideals (II). For example, the Andrunakievic lemma implies that if $A \in \underline{\text{Ass}}$ then V' itself is soluble.

The single condition (a) of 2.2III provides a natural link between the notions used in I and II respectively.

The conditions of 2.2IV consider how tightly V sits *below* an ideal of A: if $J = \langle V \rangle_A$ is the ideal of A generated by V, this amounts to imposing pathology on $J - V$. For example, the Andrunakievic lemma says that if $A \in \underline{\text{Ass}}$ then $(J - V)^3 = 0$.

We will work with 'soluble-type' powers of an ideal: $I^{(0)} = I$; $I^{(n+1)} = \{I^{(n)}\}^2$. We say that I is *soluble* if I is *non-0* and $I^{(r)} = 0$ for some r.

The conditions we will define are all "local"; ie. the definitions will refer to a given algebra A, and we define the condition to hold in $\tilde{U}$ iff it holds in all $A \in \tilde{U}$.

2.2 <u>The conditions</u>. (I) The first group of conditions on A gives deductions allowable in every situation $0 < V < I < A$, where V contains no non-0 ideal of A. We call this an *Andr situation* (in A), in honour of Andrunakievic. The strict inclusion $0 < V$ is essential for the listed conditions to have the intended meanings. The other strict inclusions $V < I$ and $I < A$ are a matter of convenience: if either fails then $V < A$, so the situation is not Andr, and the conditions therefore hold vacuously.

The named condition holds in A iff for every Andr situation $0 < V < I < A$ we have that:

(va), (ia), (aa): V, I or A resp. has a soluble accessible subalgebra.

(vs), (is), (as): V, I or A resp. has a soluble ideal.

(isa) I contains a soluble ideal of A.

(vsi) V contains a soluble ideal of I.

(II) For most of the results in Sec.4 we deal with trivial rather than soluble ideals. We therefore list the analogs of the conditions of (I). Note that the first three conditions of (II) are equivalent to those of (I), since $0 \neq V$ acc A with V soluble implies that for some r we have $V^{(r+1)} = 0$

and $0 \neq V^{(r)} < V^{(r-1)} < \ldots < V$ acc A, so that $V^{(r)}$ is a trivial accessible subalgebra of A.

(va), (ia), (aa): V, I or A respectively has a trivial accessible subalgebra.

(vt), (it), (at): V, I or A respectively has a trivial ideal.

(ita) I contains a trivial ideal of A.

(vti) V contains a trivial idea of I.

(III) We list a natural condition in the presence of which the conditions of (II) are equivalent to the (otherwise weaker) corresponding conditions of (I):

(a) $I < A$ implies $I^2 < A$.

(IV) The final group of conditions on A gives deductions allowable in every situation $0 < V < J < < A$ with $J = \langle V\rangle_A$; ie. J is the ideal of A generated by V. We call this an Andr' situation in A. For given $V \subseteq A$ it occurs iff $V \leq_2 A$. For this condition is clearly necessary, and conversely, if it holds with $V < I < A$ and we set $J = \langle V\rangle_A$, then $J \subseteq I$, so that $V \leq J \leq A$. The strict inclusions are a matter of convenience: if any of them is an equality then $V = J$ and all the conditions are satisfied.

The named condition holds in A iff for every Andr' situation $0 < V < J < A$ we have that:

(f) Given $a \in A$ there exists $n = n(a,V,A)$ and $W \leq J$ with $V + aV + Va \subseteq W$ and $W^{(n)} \subseteq V$.

This condition is easily seen to be equivalent to the apparently weaker pair of conditions that there exist ideals W_e, W_r of J with $aV \subseteq W_e$; $Va \subseteq W_r$, and W_e, W_r both soluble mod V.

(b) For some $n = n(V,A)$, $J^{(n)} \subseteq V$.

(b') For some $n = n(V,A)$, $J^{(n)} \subseteq V$ and $J^{(n)} \leq A$.

Note 2.3. If (1) $\tilde{U}$ = <u>Ass</u>, and even more so if (2) $\tilde{U}$ = <u>Ass</u> $\cap$ <u>Comm</u>, the strongest of the above conditions hold in a particularly strong form. Thus (Andrunakievic lemma)

(b') (1) $A \geq J^{(2)} \subseteq V$; (2) $A \geq J^{(1)} \subseteq V$; $J^{(r)} \leq A$.

(vti) (1) V or V^2, and (2) V, is a trivial ideal of I (since $V \supseteq IVI \leq A$; resp. $V \supseteq IV \leq A$).

(ita) If J is as in (b') then (1) J or J^2, and (2) J is a trivial ideal of A contained in I.

2.4. We diagram the main implications among the conditions of (I) and (IV). There is a similar net for (II) and (IV), except that (b') does not imply any of the conditions of (II) stronger than (va). In the presence of (a) the two nets merge. Also clearly (a) + (b) $\Longleftrightarrow$ (a) + (b'). The implications are obvious except possibly for those proved in 2.5, and for (as) $\Rightarrow$ (isa), proved in Theorem B.

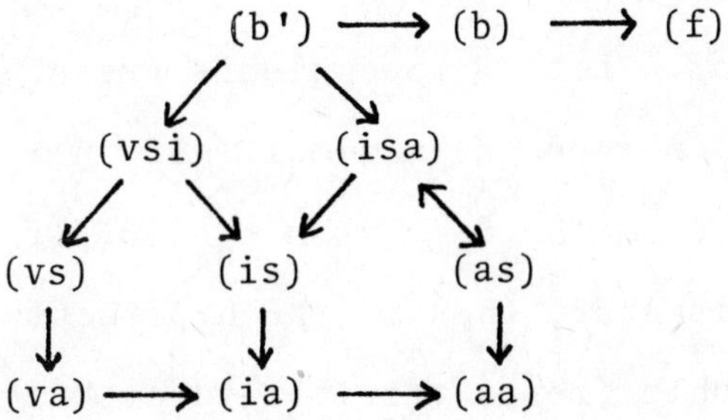

LEMMA 2.5. *For given* A, (b') $\Rightarrow$ (vsi) *and* (isa).

Proof. Given an Andr situation $V < I < A$, set $J = \langle V\rangle_A$. Then $V < J < A$ is an Andr and also an Andr' situation. By (b'), $J^{(n)} \subseteq V$ with $J^{(n)} \leq A$ for some n. So by Andr $J^{(n)} = 0$. Thus $J(\supseteq V \supset 0)$ is a soluble ideal of A contained in I. Also $V^{(n)}$ $(\subseteq J^{(n)}) = 0$, so V is itself a soluble ideal of I contained in V.

Further implications are proved in Theorem B below, and we now establish three preliminary lemmas for this.

LEMMA 2.6 *Suppose* $A \in \tilde{U}$ *has a trivial accessible subalgebra.*

(s) *If* U *satisfies* (as) *then* A *has a soluble ideal.*

(t) *If* U *satisfies* (at) *then* A *has a trivial ideal.*

Proof. Let S be soluble (resp. trivial) with $S \leq_n A$ and n minimal. If $n \geq 2$ there exist T, U with $S < T < U \leq_{n-2} A$. If $S < T < U$ $(\in \tilde{U})$ is Andr, then by the hypothesis on $\tilde{U}$, U has a soluble (resp. trivial) ideal V. If not, then S contans a non-0 ideal V of U, necessarily soluble (resp. trivial). Either way we have $V < U \leq_{n-2} A$, so $V \leq_{n-1} A$, contradicting minimality of n. Thus $n \leq 1$, and $S \leq A$.

LEMMA 2.7 *Suppose* $V \leq_n A$ *Then there exists a chain* $V = Q(0) \leq Q(1) \leq \ldots \leq Q(n) = A$ *such that* $Q(i+1) = \langle Q(i)\rangle_{Q(i+2)}$ *for* $0 \leq i \leq n-2$, *and* $Q(n-1) = \langle V\rangle_A$.

Proof. Note first that the map $P \mapsto \langle P\rangle_A$ is a closure operation (monotone, isotone, idempotent). If $P \leq I \leq A$ we have $P \subseteq \langle P\rangle_I \subseteq \langle P\rangle_A$, and it then follows from the above that $\langle\langle P\rangle_I\rangle_A = \langle P\rangle_A$.

For $n = 1$ (ie $V = Q(0) \leq Q(1) = A$) the first statement is vacuous and the second trivial. Suppose we have it for n, and are given $V \leq_{n+1} A$; say $V = R_o \leq R_1 \leq \ldots \leq R_n \leq A$. Set $Q(n) = \langle V\rangle_A$, so that $Q(n) \subseteq R_n$, and set $P_i = Q(n) \cap R_i$. Then $V = P_o \leq P_1 \leq \ldots \leq P_n = Q(n) \leq A$. By inductive hypothesis we may find $Q(0),\ldots,Q(n-1)$ with $V = Q(0) \leq Q(1) \leq \ldots \leq Q(n-1) \leq Q(n)$ (and furthermore $Q(n) \leq Q(n+1) = A$), and with the stated condition holding for $0 \leq i \leq n-2$, and $Q(n-1) = \langle V\rangle_{Q(n)}$.

Then $\langle Q(n-1)\rangle_{Q(n+1)} = \langle\langle V\rangle_{Q(n)}\rangle_A = \langle V\rangle_A$ by the remark above, $= Q(n)$ by definition. Thus the condition holds also for $i = n-1$, and the induction is complete.

As a corollary we have

LEMMA 2.8 *Suppose* $\tilde{U}$ *satisfies* (b), *and* $A \in \tilde{U}$ *is given*

(a) *If* V acc A *and* $Q = \langle V\rangle_A$ *then* $Q^{(m)} \subseteq V$ *for some* m.

(b) *Suppose* A *satisfies* (aa) *and there is an* Andr *situation in* A. *Then* A *has a soluble ideal.*

Proof. (a) If $V \leq_n A$, let $Q(i)$ be as in 2.7. Then by (b) $Q(i+1)^{(m(i+1))} \subseteq Q(i)$ for some $m(i+1)$, whence $Q^{(m)} \subseteq V$ for $m = \sum_1^n m(i)$.

(b) By (aa) A has a trivial accessible subalgebra V. If $Q = \langle V \rangle_A$ then $Q^{(m)} \subseteq V$ for some m by (a), so $Q^{(m+1)} = 0$. Also $0 \neq V \subseteq Q$, so that Q is a soluble ideal of A.

THEOREM B. *In a given universe* $\tilde{U}$:

(1) (as) ⇒ (isa); (1') (at) ⇒ (ita).

(2) (va) + (as) ⇒ (vsi); (2') (va) + (at) ⇒ (vti).

(3) (aa) + (b) ⇒ (isa).

Proof. Suppose $\tilde{U}$ satisfies the appropriate hypothesis/es, and $0 < V < I < A \in \tilde{U}$ is an Andr situation. We show that V or I contains an ideal J of the appropriate type. The argument varies somewhat according to the case considered.

Let X be an ideal of A maximal wrt (cases 2, 2') $X \cap V = 0$, and (1, 1', 3) $X \cap I = 0$. Set $\theta = \text{nat}: A \rightarrow A - X = A'$, and write V' for $V\theta$, etc. We will construct $0' \neq W' \leq A'$ with $W'^{(r)} = 0'$

for some r (r = 1 in cases 1', 2'). Assuming this done, set $W = \theta^{-1}(W')$, so that $X \subset W \leq A$, and set $J = W \cap B$, where $B = V$ in cases 2, 2', and $B = I$ in cases 1, 1', 3. Thus $J \neq O$ by the definition of X, and $J^{(r)} \subseteq W^{(r)} \cap B \subseteq X \cap B = O$. So J is the required trivial (1', 2') or soluble (1, 2, 3) ideal of I contained in V (2, 2') or ideal of A contained in I (1, 1', 3).

To construct W', suppose first $V' \leq I' \leq A'$ is an Andr situation in A'. Then A' has an ideal W' of the required type by 2.8b in case 3, and by a hypothesis in the other cases. So it remains to consider the case where V' contains a non-O ideal T' of A'. Set $\theta^{-1}(T') = T \leq A$. Then $X \subset T \subseteq X + V$, so that $T = X + K$ for $O \neq K = V \cap T \leq I$.

(Cases 2, 2') Since $K \subseteq V$, the situation $O \neq K < I < A$ is an Andr one. So K has a trivial accessible subalgebra S. But $K \cap \mathrm{Ker}\theta = K \cap X \subseteq V \cap X = O$, so $O \neq S \simeq S'$ acc $K' = (X + K)' = T' \leq A'$; ie. A' has a trivial accessible subalgebra. By 2.6 ((s) resp. (t)) it follows that A' has a soluble (resp. trivial) ideal W', and this is the required conclusion.

(Cases 1, 1', 3) We show that $K = I \cap T$. For if $i \in I \cap T = I \cap (X + K)$, say $i = x + k$ in an obvious notation. Then $x = i - k \in I + K = I$, so that $x \in I \cap X = 0$, and $i = k \in K$. Thus $I \cap T \subseteq K = V \cap T \subseteq I \cap T$, giving equality. But now $0 \neq K < A$ with $K \subseteq V$ contradicts our original assumption that $0 < V < I < A$ was an Andr situation. Thus this case cannot occur. □

For arbitrary universes the above completes the list of implications among the conditions of 2.2. However, there are two further implications in all universes $\tilde{U}$ satisfying a suitable finiteness condition. This condition holds, for example, if Φ is a field and the algebras in $\tilde{U}$ are all finite-dimensional over Φ.

THEOREM C. *Suppose* $\tilde{U}$ *is a universe of algebras all of which have acc on ideals. Then in* $\tilde{U}$ *we have*

(1) (f) ⇒ (b); (2) (isa) ⇒ (b).

Proof (1). Suppose $V \leq J \leq A$ with $\langle V \rangle_A = J$. We show that $J^{(n)} \subseteq V$ for some n, on the assumption that J has acc on ideals containing V (equivalent to acc for $J - V \in \tilde{U}$).

We construct a strictly ascending chain $\{V_i\}$ of ideals of J and $\{n(i)\}$ of integers such that $V_i^{(n(i))} \subseteq V$, starting with $V_o = V$; $n(0) = 0$. Suppose we have V_k and $n(k)$ $(k \geq 0)$. If $V_k \neq J$ we can find $a_k \in A$ such that $T_k = V_k + a_kV_k + V_ka_k \supset V_k$. Now $V \subseteq V_k \subseteq J$, so $J = \langle V_k \rangle_A$. Hence by (f) we can find $V_{k+1} \leq J$ with $V_k \subset T_k \subseteq V_{k+1}$, and s such that $V_{k+1}^{(s)} \subseteq V_k$. But then $V_{k+1}^{(n(k+1))} \subseteq V$ for $n(k+1) = n(k) + s$.

The chain must terminate, say at k, and then $J = V_k$, and $J^{(n)} \subseteq V$ for $n = n(k)$.

(2) Suppose $V \leq J \leq A$ with $J = \langle V \rangle_A$. We show that $J^{(n)} \subseteq V$ for some n, on the assumption that A has acc. on ideals contained in J.

We construct a strictly ascending chain $\{X(i)\}$ of ideals of A and a chain $\{n(i)\}$ of integers such that $X(i+1)^{(n(i))} \subseteq X(i) + V \subseteq J$, starting with $X(0) = 0$. Suppose we have $X(k)$ $(k \geq 0)$. Define $Y(k) = \Sigma\{I \leq A: I \subseteq X(k) + V\}$. Clearly $X(k) \subseteq Y(k) \subseteq X(k) + V \subseteq J$, with $Y(k) \leq A$. Let θ_k = nat: $A \longrightarrow A - Y(k) = A_k$, say. Then in an obvious notation we have $V_k \leq J_k \leq A_k$. If $I_k \leq A_k$ with $I_k \subseteq V_k$, and $I = \theta_k^{-1}(I_k) \leq A$, then

$I \subseteq V + Y(k) \subseteq V + X(k)$, so that $I \subseteq Y(k)$. Thus $I_k = 0$. If $V_k \neq 0$ it follows that $V_k \leq J_k \leq A_k$ is an Andr situation, so by (isa) we can produce a soluble ideal X_k of A_k with $X_k \subseteq J_k$. Set $X(k+1) = \theta_k^{-1}(X_k)$. Then $X(k+1) \leq A$; $X(k+1) \subseteq J + \mathrm{Ker}\theta_k = J + Y(k) = J$; $X(k+1) \supset \mathrm{Ker}\theta_k = Y(k) \supseteq X(k)$, and for some $n(k)$ we have $X(k+1)^{(n(k))} \subseteq \mathrm{Ker}\theta_k = Y(k) \subseteq X(k) + V$. This completes the construction.

By acc. the chain terminates, say at k, so that $V_k = 0$, giving $V \subseteq Y(k)$. Since $Y(k) \leq A$, this gives $J \subseteq Y(k) \subseteq X(k) + V$. Set $\phi(0) = 0$. Then we have the case $r = 0$ of the assertion that $J^{(\phi(r))} \subseteq X(k-r) + V$. Suppose we have it for $r(< n)$. Set $s = \phi(r) + n(k-r-1)$. Then, since $V \leq J$ and $X(k-r) \subseteq J$, we have $J^{(s)} \subseteq \{X(k-r) + V\}^{(n(k-r-1))} \subseteq X(k-r)^{(n(k-r-1))} + V \subseteq X(k-r-1) + V$ by the condition on the $X(i)$. Thus on setting $\phi(r+1) = s$ we have the relation for $r + 1$. Finally the case $r = k$ gives $J^{(n)} \subseteq X(0) + V = V$ for $n = \phi(k)$.

COROLLARY 2.9. *If* $\tilde{U}$ *is as in Theorem* C, *and* (aa) *holds in* $\tilde{U}$, *then* (b), (f), (as), (isa) *are all equivalent.*

Proof. By Thm. C(1) we have (b) $\Leftrightarrow$ (f);

by Thm. B(1) we have (as) $\Leftrightarrow$ (isa); by Thm. C(2) we have (isa) $\Rightarrow$ (b), and by Thm. B(3), if (aa) holds we have (b) $\Rightarrow$ (isa).

2.10. We devote the rest of Sec. 2 to showing that, even for universes of algebras over an arbitrary field F, there are no implications among the conditions of 2.2 to be added to those of 2.4, 2.5 and Thm B, and that even for universes of algebras of bounded finite dimension over F there are no further ones beyond those of Thm C.

All examples will be of the universal closure $\tilde{U}\{A\} = \tilde{U}$ of a given algebra A over F; ie. $\tilde{U}$ is the smallest universe over F to contain A. It is easily seen that $\tilde{U}\{A\}$ is the closure $\tilde{H}\tilde{A}$ under homomorphism of the set $\tilde{A}$ of all accessible subalgebras of A; more specifically it is the closure under isomorphism of the set $\{I - J: J \le I \text{ acc } A\}$ ([18], 2.4 seq.). For reasons of space we must leave the verification of each example to the reader: it starts with the successive identification of the algebras in $\tilde{A}$, and in all our examples the number of steps is bounded: C acc A implies $C \le_6 A$. It is helpful to note that the conditions of 2.2III and 2.2IV for a given algebra are preserved under homomorphism, and hence that to check them in $\tilde{U}\{\tilde{A}\}$ it suffices to check them in $\tilde{A}$. It is also useful, and easy to see, that if all proper ideals of a given $B \in \tilde{A}$ are soluble then (b') holds at least in the subclass $\tilde{U}\{\tilde{B}\}$ of $\tilde{U}\{\tilde{A}\}$.

2.11. In this paragraph we give examples of all possible non-implications where A is finite-dimensional over F, arranged in terms of the conditions of 2.2III and IV that *are* assumed. It is clear that no counterexample can satisfy more than one of: (a); (b'); (b) but not (b'); and since each of our examples satisfies one of these, they are insofar best possible. In each group we consider the possible non-implications among the remaining conditions (2.2I and II).

If (a) holds the conditions of I and II are equivalent, and (isa) $\Leftrightarrow$ (b'), since (isa) $\Rightarrow$ (aa), hence (b) by 2.9, hence (b') by (a). Examples (a1) - (a7) show that there are no further amalgamations in diagram 2.4. If (b') holds then I is automatic, and of the four remaining conditions in II we have (ita) $\Rightarrow$ (vti) by Thm. B(2'), since (va) then holds. Examples (b'1) - (b'5) show that there are no further amalgamations. Finally, if (b) holds but (b') fails, we consider I (since the (b') examples provide stronger examples for II). By 2.9 we have (aa) $\Rightarrow$ (isa), and by Thm. B(2) we have (va) $\Rightarrow$ (vsi), since (va) $\Rightarrow$ (aa) $\Rightarrow$ (isa). This leaves only (b') $\Rightarrow$ {(vsi) $\Leftrightarrow$ (vs) $\Leftrightarrow$ (va)} $\Rightarrow$ {(isa) $\Leftrightarrow$ .. $\Leftrightarrow$ (aa)}. Examples (b1) - (b3) show that there are no further implications; in (b2) and (b3) we can strengthen the hypothesis (b) to (ita) (note (ita) $\Rightarrow$ (aa), so (ita) $\Rightarrow$ (isa) $\Rightarrow$ (b) by 2.9).

In each example A has the smallest possible dimension over F.

(a1) (a) $\not\Rightarrow$ (aa). A = F[a,v,w: av=ww=w; wa=vv=v].

(a2) (a)+(aa) $\not\Rightarrow$ (ia). A = F[a,i,v,p: av=ii=pa=i; ap=p; ia=vi=vv=v]

(a3) (a)+(it) $\not\Rightarrow$ (va). A = F[a,i,j,v: aj=vv=v; ii=va=i; ij=j].

(a4) (a)+(va)+(it) $\not\Rightarrow$ (vs). A = F[a,x,i,u,v,w: xi=wa=x; ax=ii=i; av=xa=iu=uu=u; iv=uw=vv=v; vw=w].

(a5) (a)+(vt) $\not\Rightarrow$ (is). A = F[a,i,v,w: ii=aw=i; vv=wi=v; vw=w].

(a6) (a)+(vt)+(it) $\not\Rightarrow$ (vsi). A = F[a,x,i,v,w: xa=xi=x; ii=wa=i; iw=vv=v; ax=vw=w].

(a7) (a)+(vti) $\not\Rightarrow$ (as). A = F[a,i,v: ii=va=i; iv=v].

(b'1) (b')+(vt) $\not\Rightarrow$ (it). A = F[a,i,u,v: au=va=i; vi=u; uv=v].

(b'2) (b')+(it) $\not\Rightarrow$ (vt). A = F[a,i,u,v,w: aw=i; ai=u; wu=v; vw=w].

(b'3) (b')+(it)+(vt) $\not\Rightarrow$ (vti). A = F[a,i,z,v,w: av=za=i; iz=z; wi=v; vw=w].

(b'4) (b')+(vti) $\not\Rightarrow$ (ita). A = F[a,i,v: va=i; iv=v].

(b'5) (b')+(ita){+(vti)} $\not\Rightarrow$ (a). A = F[a,i,j,v: ia=i; va=j; aj=ii=v].

(b1) (b) $\not\Rightarrow$ (aa). A = F[a,i,v: av=ia=i; ii=vv=v]. This example shows that the hypothesis (aa) in Thm. B(3), and hence also in 2.9, cannot be dispensed with.

(b2) {(b)+}(ita) $\not\Rightarrow$ (va). A = F[a,w,e: ae=w; ee=e]. This example shows that the hypothesis (va) in Thm. B(2') cannot be dispensed with.

(b3) {(b)+}(ita)+(vti) $\not\Rightarrow$ (b'). A = F[a,x, v,w: aa=vv=v; wa=x; ax=vw=w].

2.12. In this paragraph we give examples of non-implications where A is compelled by Thm. C to be infinite-dimensional; ie. those where (b) fails but either (f) or (isa) holds.

If (f) holds we must assume(a) (or the (b) and (b') examples above are stronger), and therefore consider the conditions of 2.2I. Examples (f1)-(f8) show that there are no amalgamations in 2.4. If (f) fails but (isa) holds the single example (end) suffices.

To check (f) in $\tilde{U}$ (and the same applies to (b)), it is enough to consider those Andr' situations which are also Andr situations. For suppose $V < I < A \in \tilde{U}$ is Andr', and set $K = \Sigma\{T < A: T \subseteq V\}$, and θ = nat: $A \to A - K$. Then $V\theta < I\theta < A\theta$ is Andr' as well as Andr, and if (f) or (b) holds for this, then (as is easily seen) it also holds for $V < I < A$.

A subscript i in these examples ranges over all non-negative integers. Most examples exploit the algebra $X = F[x_i: x_{i+1}^2 = x_i]$ or a variant: note that all (accessible) subalgebras are ideals, these being the $X_k = F[x_i \ (i < k)]$ for $0 \leq k \leq \infty$.

Detailed varifications for all examples are available on request from the author.

(f1) (a)+(f) $\not\Rightarrow$ (aa). $A = F[a,\ x_i\colon x_i a = x_{i+1};\ {x_{i+1}}^2 = x_i;\ x_o^2 = x_o]$.

(f2) (a)+(f)+(aa) $\not\Rightarrow$ (ia). $A = F[a,p,x_i\colon pa = {x_o}^2 = x_o;\ ap = p;\ x_i a = x_{i+1};\ {x_{i+1}}^2 = x_i]$.

(f3) (a)+(f)+(ita) $\not\Rightarrow$ (va). $A = F[v,a_i,x_i\colon va_i = {x_{i+1}}^2 = x_i;\ {a_{i+1}}^2 = a_o a_i = a_i;\ a_1 a_o = v^2 = v]$.

(f4) (a)+(f)+(va)+(it) $\not\Rightarrow$ (vs). $A = F[a,w,x_i, v_i\colon ax_i = x_{i+1};\ {x_{i+1}}^2 = x_i;\ {v_{i+1}}^2 = v_i;\ wv_i = v_{i+1};\ x_1 w = w;\ v_o a = x_o]$.

(f5) (a)+(f)+(vt) $\not\Rightarrow$ (is). $A = F[a,u_i(r)$ $(0 < r < 1)\colon u_i(r)u_i(s) = u_i(r)u_{i+1}(s) = u_i(r+s)$ for $r+s < 1$; $u_i(r)u_{i+k}(s) = u_i(re^{s-k})$ for $k \geq 2$; $u_{2i}(\frac{1}{2})a = u_{2i+1}(\frac{1}{2})]$.

(f6) (a)+(f)+(vt)+(it) $\not\Rightarrow$ (vsi). $A = F[a,y,x(r), v(r)$ $(0 < r < 1)\colon ax(r) = x(\frac{1}{2}r);\ yx(r) = x(r);\ yy = x(\frac{1}{2})a = y;\ v(r)y = v(\frac{1}{2}r);\ v(\frac{1}{2})a = x(\frac{1}{2});\ x(r)x(s) = x(r+s)$ if $r+s < 1$; $v(r)v(s) = v(r+s)$ if $r+s < 1]$.

(f7) (a)+(f)+(vti) $\not\Rightarrow$ (as). $A = F[a,x_i\colon x_i a = x_{i+1};\ {x_{i+1}}^2 = x_i]$.

(f8) $(a)+(f)+(vti)+(ita) \not\Rightarrow (b)$. $A = F[a_i, x_i, v_i: a_{i+1}^2 = a_o a_i = a_i;\ v_i a_i = x_{i+1}^2 = x_i;\ x_i x_{i+1} = v_i]$.

(end) $(a)+(vti)+(ita) \not\Rightarrow (f)$. $A = F[a, x_i, v_i: v_i a = x_{i+1}^2 = x_i;\ x_i x_{i+1} = v_i]$.

In (f8) and (end) note that all proper accessible subalgebras $S \neq A^2$ of A are of the type $S = F[x_o, \ldots, x_{k-1}, x_k + w] + W$, with $w \in V$ and W a subspace of $V = F[v_i]$; thus $S^2 < A$. In (f5) such S are of type $\Sigma_o^k U_j + T$, where $T \subseteq U_{k+1} + U_{k+2}$ is in $\tilde{U}\{U_o + U_1\} \subseteq$ <u>Ass</u> (here $U_j = F[u_j(r): 0 < r < 1]$, and $k \geq -1$).

3. MACHINERY

3.1. In the first half of this section we investigate the relation of the Kurosh hierarchy for given classes $\tilde{H}$ and $\tilde{K}$ with that for $\tilde{H} \cap \tilde{K}$, and the relations between their respective semiradicals, h, k, and (as we will write it),

$h \wedge k$.

LEMMA 3.2. *Suppose* $\tilde{H}$ *is arbitrary and* $\tilde{K}$ *is a D-class. Then* $\tilde{H}_\alpha \cap \tilde{K} = (\tilde{H} \cap \tilde{K})_\alpha \cap \tilde{K}$ *for all* α, *and* $\tilde{H}_* \cap \tilde{K} = (\tilde{H} \cap \tilde{K})_* \cap \tilde{K}$.

Proof. We first do the case $\alpha=1$. If $A \in \tilde{H}_1 \cap \tilde{K}$ and A' is any non-0 homomorphic image, then A' has a non-0 ideal $I' \in \tilde{H}$. Then $I' \leq A' \in \tilde{K}$ gives $I' \in \tilde{K}$ by D for $\tilde{K}$. Thus $I' \in \tilde{H} \cap \tilde{K}$, whence $A \in (\tilde{H} \cap \tilde{K})_1$. So $A \in (\tilde{H} \cap \tilde{K})_1 \cap \tilde{K}$. The opposite inclusion is trivial.

For the general case we proceed by induction on α. If $\alpha=0$ the result is trivial, and if is a limit ordinal >0 then $\tilde{H}_\alpha \cap \tilde{K} = (\cup_{\beta<\alpha} \tilde{H}_\beta) \cap \tilde{K} = \cup_{\beta<\alpha} (\tilde{H}_\beta \cap \tilde{K}) = \cup_{\beta<\alpha} \{(\tilde{H} \cap \tilde{K})_\beta \cap \tilde{K}\} = \{\cup_{\beta<\alpha} (\tilde{H} \cap \tilde{K})_\beta\} \cap \tilde{K} = (\tilde{H} \cap \tilde{K})_\alpha \cap \tilde{K}$.

If $\alpha = \beta + 1$ then $\tilde{H}_\alpha \cap \tilde{K} = (\tilde{H}_\beta)_1 \cap \tilde{K} = (\tilde{H}_\beta \cap \tilde{K})_1 \cap \tilde{K}$ by the case $\alpha=1$, $= \{(\tilde{H} \cap \tilde{K})_\beta \cap \tilde{K}\}_1 \cap \tilde{K}$ by inductive hypothesis, $\subseteq \{(\tilde{H} \cap \tilde{K})_\beta\}_1 \cap \tilde{K} = (\tilde{H} \cap \tilde{K})_\alpha \cap \tilde{K}$. The opposite inclusion is trivial.

If $A \in \tilde{H}_* \cap \tilde{K}$ then for some α we have $A \in \tilde{H}_\alpha \cap \tilde{K} = (\tilde{H} \cap \tilde{K})_\alpha \cap \tilde{K} \subseteq (\tilde{H} \cap \tilde{K})_* \cap \tilde{K}$, and the result follows.

3.3. Given a homomorphically closed class H in a large universe, say <u>All</u>, we can consider the Kurosh hierarchy $\{\tilde{H}_\alpha\}$ in <u>All</u>, but we may also be interested in the hierarchy within a smaller universe $\tilde{U}$. Let us write $\tilde{H}_\alpha(\tilde{U})$ for the α-th step of the hierarchy *in* $\tilde{U}$ which starts with $\tilde{H}_0(\tilde{U}) = \tilde{H} \cap \tilde{U}$. Then we have the

COROLLARY 3.3. *For* $\tilde{U}$ *a subuniverse of* $\tilde{V}$, *and any* $\tilde{H}$, *we have* $\tilde{H}_\alpha(\tilde{U}) = \tilde{H}_\alpha(\tilde{V}) \cap \tilde{U}$ *for all* α, *and for* $\alpha = *$.

Proof. The general case follows from the special case $\tilde{V}$ = <u>All</u>, since we then have $\tilde{H}_\alpha(\tilde{V}) \cap \tilde{U} = (\tilde{H}_\alpha \cap \tilde{V}) \cap \tilde{U} = \tilde{H}_\alpha \cap \tilde{U} = \tilde{H}_\alpha(\tilde{U})$.

Now it is clear by induction on α from the method of construction of $\tilde{H}_\alpha(\tilde{U})$ that $\tilde{H}_\alpha(\tilde{U}) = (\tilde{H} \cap \tilde{U})_\alpha \cap \tilde{U}$, so the result now follows from 3.2 in view of the fact that any universe $\tilde{U}$ is certainly a D-class in <u>All</u>. Similarly for $\tilde{H}_*$.

PROPOSITION 3.4. *If* $\tilde{H}$ *and* $\tilde{K}$ *are* D-*classes then*

(a) $\tilde{H}_m \cap \tilde{K}_n \subseteq (\tilde{H} \cap \tilde{K})_{m+n}$ *for all integers* $m,n \geq 0$;

(b) $(\tilde{H} \cap \tilde{K})_\lambda = \tilde{H}_\lambda \cap \tilde{K}_\lambda$ *for all limit ordinals* λ;

(c) $(\tilde{H}\cap\tilde{K})_* = \tilde{H}_* \cap \tilde{K}_*$.

If $\tilde{H}$ *is arbitrary and* $\tilde{K}$ *is a radical* D-*class then*

(d) $(\tilde{H}\cap\tilde{K})_\alpha = \tilde{H}_\alpha \cap \tilde{K}$ *for all* α, *and*

(e) $(\tilde{H}\cap\tilde{K})_* = \tilde{H}_* \cap \tilde{K}$.

Proof. (a) By 4.2a below $\tilde{H}_m$ and $\tilde{K}_n$ are D-classes. So by 3.2 $\tilde{H}_m \cap \tilde{K}_n = (\tilde{H}\cap\tilde{K}_n)_m \cap \tilde{K}_n \subseteq (\tilde{H}\cap\tilde{K}_n)_m = \{(\tilde{H}\cap\tilde{K})_n \cap \tilde{H}\}_m \subseteq \{(\tilde{H}\cap\tilde{K})_n\}_m = (\tilde{H}\cap\tilde{K})_{m+n}$.

(b) We proceed by induction on λ. If $\lambda = 0$ the result is trivial. If not, then given $A \in \tilde{H}_\lambda \cap \tilde{K}_\lambda$ we can find $\alpha < \lambda$ with $A \in \tilde{H}_\alpha \cap \tilde{K}_\alpha$. If $\alpha = \beta + n$ with β a limit ordinal then $A \in (\tilde{H}_\beta)_n \cap (\tilde{K}_\beta)_n \subseteq (\tilde{H}_\beta \cap \tilde{K}_\beta)_{2n}$ by (a), $= \{(\tilde{H}\cap\tilde{K})_\beta\}_{2n}$ by inductive hypothesis, $= (\tilde{H}\cap\tilde{K})_{\beta+2n} \subseteq (\tilde{H}\cap\tilde{K})_\lambda$, since $\beta + 2n < \lambda$ for $\beta < \lambda$. The opposite inclusion is trivial.

(c) If $A \in \tilde{H}_* \cap \tilde{K}_*$ then for some (say, limit) ordinal λ we have $A \in \tilde{H}_\lambda \cap \tilde{K}_\lambda \subseteq (\tilde{H}\cap\tilde{K})_\lambda \subseteq (\tilde{H}\cap\tilde{K})_*$. The opposite inclusion is trivial. This result was obtained independently in [17] and [19].

(d) By 3.2, $\tilde{H}_\alpha \cap \tilde{K} = (\tilde{H}\cap\tilde{K})_\alpha \cap \tilde{K} = (\tilde{H}\cap\tilde{K})_\alpha \cap \tilde{K}_\alpha$ since $\tilde{K}$ is a radical class, $= (\tilde{H}\cap\tilde{K})_\alpha$.

(e) follows from (d) in just the way (c) follows from (b).

EXAMPLE 3.5. In the universe $\tilde{U}$ = <u>Ass</u> over any field F let $\tilde{H}$ be the homomorphic closure of the Zassenhaus algebra $A = F[x_s\ (0 < s < 1): x_s x_t = x_{s+t}$ for $s+t<1]$, and let $\tilde{K} = \tilde{Z}_F$. It is easily verified that $\tilde{H} \cap \tilde{K} = \{0\} = (\tilde{H} \cap \tilde{K})_*$, whereas $\tilde{H}_* \cap \tilde{K}_* \supseteq \tilde{H} \cap \tilde{K}_1 = \tilde{H}$. Thus 3.4a fails with m=0, n=1, and also 3.4b and 3.4c fail. This shows that in the first half of 3.4 it is not enough to assume that $\tilde{K}$ is a D-class.

Likewise in the second half it does not suffice for $\tilde{K}$ to be a D-class, even if $\tilde{H}$ is also a D-class. A simple example is $\tilde{H} = \tilde{K} = \tilde{Z}$ in <u>Ass</u> ∩ <u>Comm</u>, where (d) and (e) fail since $\tilde{Z} \neq \tilde{Z}_*$.

For the next result recall from 1.13 what is meant by h being strongly hereditary.

PROPOSITION 3.6. *Suppose* H *and* K *are hereditary radical classes. Then* (a) *so is* $\tilde{H} \cap \tilde{K}$.

If $A \in \tilde{U}$ *with* $h(A) = H$; $k(A) = K$,

(b) $(h \wedge k)(A) = H \cap K \subseteq h(K) = (h \wedge k)(K)$

(c) *Equality holds if and only if* $h(H+K) = H$.

(d) *Equality holds if* h *or* $h \wedge k$ *is strongly hereditary.*

(e) *If* h *and* k *are strongly hereditary then so is* $h \wedge k$.

Proof. (a) Set $\tilde{M} = \tilde{H} \cap \tilde{K}$. Clearly $\tilde{M}$ is homomorphically closed and hereditary. Also $\tilde{M}_* = \tilde{H}_* \cap \tilde{K}_* = \tilde{H} \cap \tilde{K} = \tilde{M}$ by 3.2c, so that $\tilde{M}$ is a radical class.

(b) Set $M = H \cap K$. Then $M \leq H \in \tilde{H}$ gives $M \in \tilde{H}$, and similarly $M \in \tilde{K}$, so that $M \in \tilde{M}$. Also $M \leq A$ so $M \subseteq m(A)$ for $m = h \wedge k$. Conversely $\tilde{M} \subseteq \tilde{H}$ and $\tilde{M} \subseteq \tilde{K}$ gives $m(A) \subseteq H$ and $m(A) \subseteq K$, so $m(A) \subseteq M$, giving equality. Results (a) and (b) are due to Leavitt ([10], Theorem 1).

(c) If $h(H+K) = H$, then $(H+K)-H$ has no non-0 $\tilde{H}$-ideals, whence the same holds for $K - (H \cap K)$. Thus $h(K) \subseteq H \cap K$. But $H \cap K \subseteq h(K)$ by (b), so we have equality. An exactly parallel argument using $H \subseteq h(H+K)$ gives the converse.

(d) If h is strongly hereditary then $h(H+K) = (H+K) \cap h(A) = H$, and we apply (c). If m is strongly hereditary then $m(K) = K \cap m(A) = K \cap (H \cap K) = M$, the required result.

(e) If $I \leq A \in \tilde{U}$ then by (b) $(h \wedge k)(I) = h(I) \cap k(I) = \{I \cap h(A)\} \cap \{I \cap k(A))\} = I \cap \{(h \wedge k)(A)\}$ by (b) again.

In the second half of this section we define a notion of 'p-primary component' for a class $\tilde{H}$, and in connexion with the special case $\tilde{H} = \tilde{Z}$ use it to prove a technical result vital for the theorems of Section 4.

3.7 *Definitions*. (a) Suppose P is any ideal of Φ. We define an operator u_P on $\underline{All}_\Phi$ by $A \mapsto A_P = u_P(A) = \{a \in A$: there exists $n = n(a)$ with $P^n a = (0)\}$. It is easily verified that u_P is a strongly hereditary radical operator on $\underline{All}_\Phi$: the corresponding radical class is $\tilde{U}_P = \{A \in \underline{All}_\Phi : A = A_P\} = \{A_P : A \in \underline{All}_\Phi\}$. In particular, $\tilde{U}_{(0)} = \underline{All}_\Phi$.

(b) For any $\tilde{H}$ in a given $\tilde{U}$ over Φ write $\tilde{H}_P$ for $\tilde{H} \cap \tilde{U}_P$, with corresponding operator $h_P = h \wedge u_P$. In particular $\tilde{H}_{(0)} = \tilde{H}$ and $h_{(0)} = h$ If $\tilde{H}$ is a strongly hereditary radical class then by 3.6 and (a) above we have for $A \in \tilde{U}$ that $h_P(A) = h(A) \cap A_P = h(A_P) = \{h(A)\}_P = h_P(A_P)$.

(c) For given $\tilde{U}$ let z_α be the operator for the class $\tilde{Z}_\alpha(\tilde{U})$ in the notation of 3.3, where $\tilde{Z}$ is as in 1.2, and similarly for z_*. We will say that $\tilde{U}$ *satisfies* $(z_\alpha E)$ if z_α satisfies Amitsur's condition E as in 1.3. In that case $z_\alpha = z_*$ is a strongly hereditary radical operator by Amitsur's theorem 1.3 and 4.2 below. Similarly we say that $\tilde{U}$ satisfies (z_*E) if z_* is strongly hereditary.

PROPOSITION 3.8. *Suppose* $\tilde{U}$ *satisfies* $(z_\alpha E)$. *Then*

(a) $(\tilde{Z}_P)_{\alpha+1} = (\tilde{Z}_{\alpha+1})_P = (\tilde{Z}_*)_P = (\tilde{Z}_P)_*$.

(b) *For all* $A \in \tilde{U}$ *we have* $(z_P)_\alpha(A) =$

$(z_P)_*(A) = A_P \cap z_*(A) = \{z_*(A)\}_P =$

$z_*(A_P) = (z_P)_*(A_P)$.

(c) *If* I acc $A \in \tilde{U}$ *then* $(z_P)_*(I) =$

$I \cap (z_P)_*(A) = I_P \cap z_\alpha(A) = I \cap (z_P)_\alpha(A)$.

If $\tilde{U}$ *satisfies only* (z_*E) *then those statements above that do not involve* α *remain true.*

Proof. (a) $(\tilde{Z}_P)_{\alpha+1} = (\tilde{Z} \cap \tilde{U}_P)_{\alpha+1} = \tilde{Z}_{\alpha+1} \cap$

$\tilde{U}_P$ by 3.4d and 3.7a, $= (\tilde{Z}_{\alpha+1})_P = (\tilde{Z}_*)_P$ by 1.3, $= \tilde{Z}_* \cap \tilde{U}_P = (\tilde{Z} \cap \tilde{U}_P)_*$ by 3.4e, $= (\tilde{Z}_P)_*$.

(b) $(z_P)_\alpha$ is the operator for the class $(\tilde{Z}_P)_\alpha = (\tilde{Z} \cap \tilde{U}_P)_\alpha = \tilde{Z}_\alpha \cap \tilde{U}_P$ by 3.4b, so that $(z_P)_\alpha = z_\alpha \wedge u_P = z_* \wedge u_P$ by 1.3. Now z_* and u_P are both strongly hereditary radical operators by our assumption and 3.7a, whence so also is $(z_P)_\alpha$ by 3.6e, so that $(z_P)_* = \{(z_P)_\alpha\}_* = (z_P)_\alpha$. The remaining equalities are all applications of 3.7b, taking $h = z_*$.

(c) By an obvious induction on n, where $I \leq_n A$ (see 2.1 for notation), we have $(z_P)_*(I) = I \cap (z_P)_*(A) = I \cap (z_P)_\alpha(A)$, and also $= I \cap (A_P \cap z_*(A)) = (I \cap A_P) \cap z_*(A) = I_P \cap z_\alpha(A)$.

3.9 *Definitions.* Let $\tilde{C}$ be the class of *cyclic* algebras in $\tilde{Z}$ ($\subseteq \mathrm{All}_\Phi$); ie. those of the form $Z = \Phi m$ for some $m \in Z$. If $Z = \Phi m \in \tilde{C}$, then the map $\alpha \mapsto \alpha m$ establishes an isomorphism between Φm as Φ-module and Φ-P as Φ-module, where $P = \mathrm{annih}\{m\} = \{\beta \in \Phi : \beta m = 0\} \leq \Phi$. Let us write N_P for $(\Phi\text{-}P)_o$, the Φ-algebra with all products zero and which is Φ-P *qua* Φ-module.

Thus $\Phi m \simeq (\Phi - P)_o$ if $P = \text{annih}\{m\}$. Conversely, for $P \leq \Phi$, $(\Phi - P)_o = \Phi m$ for $m = 1 + P$ (the coset), so that $\tilde{C} = \tilde{I}\{(\Phi - P)_o\colon \text{ all } P \leq \Phi\}$. We will write $\tilde{N}_P$ for the homomorphic closure $\tilde{H}\{N_P\}$ of $\{N_P\}$ in All_Φ.

PROPOSITION 3.10. *For given* $P \leq \Phi$,

(a) $\tilde{N}_P = \tilde{I}\{(\Phi - Q)_o\colon P \subseteq Q \leq \Phi\}$, *and*

(b) $\tilde{C}_P$ (ie. $\tilde{C} \cap \tilde{U}_P$) $= \tilde{I}\{(\Phi - Q)_o\colon P^n \subseteq Q \leq \Phi$ for some $n\}$.

(c) *For all* α *we have* $(\tilde{N}_P)_\alpha \subseteq (\tilde{C}_P)_\alpha \subseteq (\tilde{Z}_P)_\alpha \subseteq (\tilde{N}_P)_{1+\alpha}$.

(d) *If* $\alpha \geq \omega$, *or if* $\alpha = *$, *then* $(\tilde{N}_P)_\alpha = (\tilde{C}_P)_\alpha = (\tilde{Z}_P)_\alpha$.

Proof. (a) Write $N_P = \Phi m$ with $\text{annih}\{m\} = P$. If $N'_P \in \tilde{N}_P$, then N'_P is a homomorphic image $\Phi m'$ of N_P, and clearly $Q = \text{annih}\{m'\} \supseteq P$. Thus $N'_P \simeq (\Phi - Q)_o$ with $P \subseteq Q \leq \Phi$. Conversely, if m' is a generator of $(\Phi - Q)_o$ with $P \subseteq Q \leq \Phi$, then the map $\alpha m \mapsto \alpha m'$ of N_P onto $(\Phi - Q)_o$ is well-defined, whence $(\Phi - Q)_o \in \tilde{N}_P$.

(b) If $\Phi m \in \tilde{C}_P$, then for some n we have $P^n m = 0$, so $P^n \subseteq \text{annih}\{m\} = Q$, say, so $\Phi m \simeq$

$(\Phi\text{-}Q)_o$ with $P^n \subseteq Q$. Conversely if m is a generator of $(\Phi\text{-}Q)_o = \Phi m$ with $P^n \subseteq Q$, and $m' = \alpha m \in \Phi m$, then $P^n m' = \alpha P^n m = (0)$, so that $\Phi m \in \tilde{U}_P$. So $\Phi m \in \tilde{C}_P$.

(c) It is clear from (a) and (b) that $\tilde{N}_P \subseteq \tilde{C}_P \subseteq \tilde{Z}_P$, whence $(\tilde{N}_P)_\alpha \subseteq (\tilde{C}_P)_\alpha \subseteq (\tilde{Z}_P)_\alpha$ for all α. We will show that $\tilde{Z}_P \subseteq (\tilde{N}_P)_1$, whence $(\tilde{Z}_P)_\alpha \subseteq (\tilde{N}_P)_{1+\alpha}$ for all α. Suppose then that $A \in Z_P$, and A' is a non-0 homomorphic image. Since $A' \in \tilde{Z}_P \subseteq \tilde{U}_P$, for given $0 \neq m' \in A'$ there exists n with $P^n m' = 0$. If n is minimal such, we can find $0 \neq m \in P^{n-1}m'$. Then $\Phi m \leq A'$ (since $A' \in \tilde{Z}$), and $P \subseteq \text{annih}\{m\}$, so that $0 \neq \Phi m \in \tilde{N}_P$ by (a). Thus every $A' \neq 0$ has a non-0 ideal in $\tilde{N}_P$, so that $A \in (\tilde{N}_P)_1$.

(d) For $\alpha \geq \omega$ this follows from $1 + \alpha = \alpha$ and (c). For $\alpha = *$ we have $(\tilde{N}_P)_* = \cup_{\alpha \geq \omega}(\tilde{N}_P)$ $= \cup_{\alpha \geq \omega}(\tilde{C}_P)_\alpha = (\tilde{C}_P)_*$, which equals $(\tilde{Z}_P)_*$ similarly.

EXAMPLE 3.11. Let $\tilde{U}$ and A be as in 3.5, so that $P = (0)$. If $I \leq A$ is $F[x_s : s \geq \frac{1}{2}]$, and $C = F[x_{\frac{1}{2}}]$, then $0 < C < I < A$; $c(A) = 0$, $c(I) = I$. Thus $A \in \tilde{C}_1$; $A \notin \tilde{C}_o$. It follows easily that in $\tilde{U} = \underline{Ass}$ we have $\tilde{C}_o \subset \tilde{Z}_o \subset \tilde{C}_1 \subset \tilde{Z}_1 = \tilde{C}_2 = \tilde{Z}_* = \tilde{C}_*$. Also $c_o < z_o$ (in the sense that $c_o(A) \subseteq z_o(A)$ for all $A \in \tilde{U}$, with the inclusion strict in some cases), whereas $z_o = c_1 = z_1 = c_* = z_*$. The operator c_o does not satisfy Amitsur's condition E. In fact on A it fails in a spectacular fashion: $c_o(A) = 0$, but $c_o(I) = I$ for *every* proper ideal I of A. We omit the proof.

The final result (3.15) of this section, and the *raison d'etre* of all that goes before, requires a condition on $\tilde{H}$ which we now introduce.

DEFINITION 3.12. Given $\tilde{H} \subseteq \tilde{U}$, we say that $\tilde{H}$ is a P-*class* if given $0 \neq Z \in \tilde{H} \cap \tilde{Z}$ there exists $0 \neq C \leq \tilde{Z}$ with $C \in \tilde{H} \cap \tilde{C}$. We say that $\tilde{H}$ is a P_α-class if $\tilde{H}_\alpha$ satisfies P, and a P_*-class if $\tilde{H}_*$ satisfies P.

NOTES 3.12. (a) *$\tilde{H}$ is a P_α-class (resp. a P_*-class) if and only if every trivial algebra in $\tilde{H}$ has a non-zero cyclic subalgebra in $\tilde{H}_\alpha$ (resp. $\tilde{H}_*$). Hence a P_α-class is a P_β-class for all $\beta > \alpha$.*

(b) *If* $\tilde{H} \supseteq \tilde{Z}_U$ *(or merely* $\tilde{H} \supseteq \tilde{C}_{\tilde{U}}$ *) or if* $\tilde{H}$ *is a* D-*class (or merely* $\tilde{H} \cap \tilde{Z}$ *is a* D-*class) then* $\tilde{H}$ *is a* P-*class.*

(c) *For given* Φ EVERY *class* $\tilde{H} \subseteq All_\Phi$ *is a* P-*class if and only if* Φ *has the following property:*

(X) *Every non*-0 Φ-*module admits an endomorphism onto a non*-0 *cyclic submodule.*

Sketch proof (a) If $0 \neq A \in \tilde{H}_\alpha$ then A has a non-0 accessible subalgebra $A' \in \tilde{H}$ (eg. see [18], 4.2). Thus if the condition holds and $A \in H_\alpha \cap Z$, then A', and hence also A ($\in \tilde{Z}$), has a non-zero cyclic ideal in $\tilde{H}$. The converse is trivial. The proof for P_* is similar.

(b) Trivial.

(c) If Φ has X and $A \in \tilde{H} \cap \tilde{Z}$ then A as Φ-module has an endomorphism f with f(A) cyclic; but then $f(A) \in \tilde{H}$ also. Conversely given a Φ-module M let $\tilde{H} = \tilde{H}\{M_0\}$ and apply P to $\tilde{H}$ to obtain the desired endomorphism.

3.13. To my knowledge condition (X) has not been studied in the module-theoretic literature:

the following results were communicated to me by my colleague Dr A W Chatters.

PROPOSITION 3.13 (Chatters)

(a) *The class of* X*-rings is closed under homomorphic images and finite direct sums.*

(b) *If* Φ *has a nilpotent ideal* N *with* Φ-N *a finite direct sum of fields, then* Φ *is an* X*-ring.*

(c) *If* Φ *has a nil ideal with* Φ-N *a field, then it does* NOT *follow that* Φ *is an* X*-ring.*

(d) *If* Φ *is an* X*-ring then all prome ideals of* Φ *are maximal. Hence a Noetherian* X*-ring is necessarily also Artinian.*

Proof. Since these results are not directly relevant to the purposes of the present paper, we omit the proofs. They are avilable on request.

3.14 *Example*. Let $\tilde{U}$ = $\underline{\text{Ass}}\cap\underline{\text{Comm}}$ over the ring Φ of integers, and let $\tilde{H} = \tilde{I}\{Z(p^{\infty})_0, 0\}$. Then P fails for $\tilde{H}$, even though $\tilde{U}$ might well be regarded as a well-behaved universe.

PROPOSITION 3.15. *Given* $\tilde{H}\subseteq\tilde{U}$, *suppose* $A \in \tilde{U}$ *has a non-*0 *accessible subalgebra in* $\tilde{H}_* \cap \tilde{Z}$.

(a) *If* $\tilde{U}$ *satisfies* $(z_\alpha E)$ *and* $\tilde{H}$ *is a* P*-class then* $h_{1+\alpha}(A) \neq 0$.

(b) *If* $\tilde{U}$ *satisfies* (z_*E) *and* $\tilde{H}$ *is a* P_**-class then* $h_*(A) \neq 0$.

Proof (a) If the accessible subalgebra is M, then, arguing as in 3.12a, we can find an accessible subalgebra M' of M with $M' \in \tilde{H}$, and by condition P a non-0 cyclic subalgebra $N = \Phi m$ of M' with N also in $\tilde{H}$. Set $P = \text{annih}\{m\}$. Then $\tilde{N}_P = \tilde{H}\{N\} \subseteq \tilde{H}$. So by 3.10c $(\tilde{Z}_P)_\alpha \subseteq (\tilde{N}_P)_{1+\alpha} \subseteq \tilde{H}_{1+\alpha}$. So by 3.8c $0 \neq N = (z_P)_*(N) = N \cap (z_P)_\alpha(A) \subseteq (z_P)_\alpha(A) \subseteq h_{1+\alpha}(A)$, as required.

(b) Similar, but simpler.

3.16 *Notes* (a) In 3.11 let $\tilde{H} = \tilde{C}$. Then $\tilde{U}$ satisfies (z_o), but $h_o(A) = 0$. This shows that $1+\alpha$ in the conclusion cannot in general be improved to α.

(b) Although not all classes $\tilde{H}$ are P-classes, I conjecture that 3.15 holds for all $\tilde{H}$ without restriction, and hence that those theorems in Sec.4 whose proofs use 3.15 (F,H,J,K) remain valid with the appropriate weakenings of the hypotheses on $\tilde{H}$.

(c) The inspiration for the use of "p-primary components" for the proof of a result concerned essentially with $\tilde{Z}$ (and not primary components at

all) comes from the interesting paper [3], which considers Ass without operators (in effect taking Φ to be the ring of integers). They introduce N_P in effect for P a maximal ideal of this ring ,
and their Lemma 1 says in our notation that $(\tilde{Z}_*)_P = (\tilde{N}_P)_* = (\tilde{N}_P)_2$. We may see this as follows. Ass satisfies (z_o), so by 3.10c $(\tilde{Z}_P)_* = (\tilde{Z}_P)_1 \subseteq (\tilde{N}_P)_2 \subseteq (\tilde{N}_P)_* = (Z_P)_*$, giving equality throughout. For the application they make of this see the discussion in 4.9 below.

4. RESULTS ON HEREDITARY $\tilde{H}$

4.1 In this section we prove results of the following type. $\tilde{H}$ is a D-class (or satisfies a weaker but similar condition), and may satisfy some other condition, eg. Z (see 1.2). $\tilde{U}$ satisfies conditions of Andr type as listed in 2.2, and may also satiffy $(z_\alpha E)$ for some α, or $(z_* E)$ (see 3.7c). The conclusion to be drawn is typically that $H_* = H_\gamma$ for specified γ, and possibly also that $h_\beta = h_*$ is strongly hereditary (1.13), in which case γ will be $\beta+1$.

Since there are a large number of very similar-sounding results, the reader may find it helpful to refer to the summary at the end of this section. Our discussion of the extent to which the various results are best possible is postponed to the end (4.13): a limited number of examples there serves to deal with a much larger number of results.

We will need the following known results:

LEMMA 4.2 (a) *For any* H

(a) *We have* $\tilde{H}_* \cap \tilde{Z} = \tilde{H}_1 \cap \tilde{Z}$.

(b) *If* $\tilde{H}$ *is a* D-*class then it is also a* D_α-*class for every ordinal* α, *and for* $\alpha = *$.

Proof (a) If $A \in \tilde{H}_*$ and A' is a non-0 homomorphic image, then $A' \in H_*$ also, so A' has a non-0 accessible subalgebra $M \in \tilde{H}$ (eg [18],4.2). Now if $A \in \tilde{Z}$ also, then $A' \in \tilde{Z}$, so $M \le A'$. Thus $A \in \tilde{H}_1$.

(b) This result is due to Leavitt ([9], p15; also [7]).

The proofs of the theorems depend on one or other of the two arguments which we isolate in the next two lemmas. Recall from 2.2 the definition of an Andr situation.

LEMMA 4.3. *Suppose* $\tilde{K}$ *is a* D-*class in* $\tilde{U}$, *and for all* Andr *situations* $V < I < A \in \tilde{U}$ *with* $V \in \tilde{K}$ *we have* $k(A) \neq 0$. *Then* $k = k_*$ *is strongly hereditary, and* $\tilde{K}_* = \tilde{K}_1$.

Proof. We first show that k satisfies E; ie. given $I \leq A \in \tilde{U}$ with $k(I) \neq 0$ we show that $k(A) \neq 0$. Now I has a non-0 ideal $V \in \tilde{K}$. If V contains a non-0 ideal W of A, then $W \leq V \in \tilde{K}$ gives $W \in \tilde{K}$ by D, so that $0 \neq W \subseteq k(A)$, as required. Otherwise we have the hypothesis of the lemma, and its conclusion again gives $k(A) \neq 0$.

Since K satisfies E, it follows from 1.3 that $k = k_*$ is a radical operator, and $\tilde{K}_* = \tilde{K}_1$. Now $\tilde{K}_*$ is a D-class by 4.2b, so if $I \leq A \in \tilde{U}$ we have $k_*(I) \supseteq I \cap k_*(A)$. Since $k = k_*$, we already have $k(I) \subseteq I \cap k(A)$ by E, and we conclude that $k = k_*$ is strongly hereditary; ie. $k(I) = I \cap k(A)$ for $I \leq A \in \tilde{U}$.

LEMMA 4.4. *Suppose* $\tilde{K}$ *is a* D-*class in* $\tilde{U}$, *and for all* Andr *situations* $V < I < A \in \tilde{K}_2$ *with* $V \in \tilde{K}$ *we have* $k(A) \neq 0$. *Then* $\tilde{K}_* = \tilde{K}_1$.

Proof. $\tilde{K}_* = \tilde{K}_1$ iff $\tilde{K}_1 = \tilde{K}_2$. So it suffices to prove that if $B \in \tilde{K}_2$ then $B \in \tilde{K}_1$; ie. every non-0 homomorphic image A of B has a non-0 ideal $W \in \tilde{K}$.

Now $A \in \widetilde{K}_2$ also, so we can find a non-0 ideal I of A with $I \in \widetilde{K}_1$. Likewise we can find a non-0 ideal V of I with $V \in \widetilde{K}$. If V contains a non-0 ideal W of A then $W \leq V \in \widetilde{K}$ gives $W \in \widetilde{K}$ by D. Otherwise we have the hypothesis of the lemma, and its conclusion $k(A) \neq 0$ again gives us a non-0 ideal $W \in \widetilde{K}$ of A. □

To prove the theorems, we will assume the situation of 4.3 or 4.4 as appropriate, and for a stated $\widetilde{K}$. We then use the hypotheses on $\widetilde{U}$ to obtain the desired conclusion $k(A) \neq 0$. The conclusion of the theorem will then be automatic, being identical with that of 4.3 or 4.4 respectively, and for the $\widetilde{K}$ in question.

THEOREM D. *Suppose* $\widetilde{U}$ *satisfies* **(at)**, *and* $\widetilde{H}$ *is a* D+Z*-class in* $\widetilde{U}$. *Then* $h = h_*$ *is strongly hereditary, and* $\widetilde{H}_* = \widetilde{H}_1$.

Proof. In the situation of 4.3 with $\widetilde{K} = \widetilde{H}$, A has a non-0 ideal $W \in \widetilde{Z}$, by (at). Since $\widetilde{Z} \subseteq \widetilde{H}$, we have $0 \neq W \subseteq h(A) = k(A)$.

4.5 *Notes* (a) This result may be regarded as an axiomatization of Theorem 2 of [2], where the conclusion $H_* = H_1$ is proved in $\widetilde{U}$ = <u>Ass</u>, but using (a+b) rather than (at).

(b) Theorem D applies in particular to $\tilde{H} = \tilde{Z} \cap \tilde{U}$ which is clearly a D+Z-class in $\tilde{U}$. Thus in an (at) universe the Baer radical z_* coincides with the operator z for the class $\tilde{Z}(\tilde{U})$.

(c) If we strengthen the condition Z on $\tilde{H}$ that $\tilde{H} \supseteq \tilde{Z}(\tilde{U})$ to the condition, say S, that H contains the class $\tilde{S}(\tilde{U})$ of soluble algebras in $\tilde{U}$, then we can correspondingly weaken the condition (at) on $\tilde{U}$ to (as), for the same conclusion.

(d) We can also *weaken* Z to the condition C that $\tilde{H} \supseteq \tilde{C}(\tilde{U})$ (see 3.9)(and also D to D_1: see 1.2), with a consequent weakening of the conclusion, as in (b) below:

COROLLARY 4.6. *Suppose* $\tilde{U}$ *satisfies* (at). *If either*

(a) $\tilde{H}$ *is a* (D_1+Z_*)*-class in* $\tilde{U}$, *or*

(b) $\tilde{H}$ *is a* (D_1+C)*-class in* $\tilde{U}$,

then $h_1 = h_*$ *is strongly hereditary, and* $\tilde{H}_* = \tilde{H}_2$

Proof (a) By 4.2a $\tilde{H}$ is a Z_1-class, so the result follows on applying Thm.C to $\tilde{H}_1$ in place of $\tilde{H}$.

(b) From $\tilde{H} \supseteq \tilde{C}(\tilde{U})$ we have $\tilde{H}_1 \supseteq \tilde{C}_1(\tilde{U}) = \tilde{C}_1 \cap \tilde{U}$ by 3.3, which contains $Z_0 \cap U$ by 3.10c (taking $P = (0)$), so $\tilde{H}$ is a Z_1-class and we are back to case (a). □

We can relax the condition (at) on $\tilde{U}$ to the weakest of the Andr conditions, (aa), so long as we insist that $\tilde{Z}(\tilde{U})$ or $\tilde{C}(\tilde{U})$ is well-behaved. Thus we have

THEOREM E. *Suppose* $\tilde{U}$ *satisfies* (aa), *and* $\tilde{H}$ *is a* D-*class in* $\tilde{U}$. *Suppose further that either*

(a) $\tilde{H} \supseteq \tilde{Z}(\tilde{U})$ *and* z *is strongly hereditary, or*

(b) $\tilde{H} \supseteq \tilde{C}(\tilde{U})$ *and* c *is strongly hereditary.*

Then $h = h_*$ *is strongly hereditary, and* $\tilde{H}_* = \tilde{H}_1$.

Proof. In the situation of 4.3 with $\tilde{K} = \tilde{H}$ there exists $0 \neq N$ acc A with $\tilde{N} \in \tilde{Z}$, by (aa). Then for $m = z$ or c respectively we have
$0 \neq N = m(N) \subseteq m(A) \subseteq h(A) = k(A)$, as required.

Alternatively we can argue that in the presence of either hypothesis any (aa) universe must also be an (at) universe, and then appeal to Thm.D.

4.7 *Notes* (a) Here of course z is the operator for $\tilde{Z} \cap \tilde{U} = \tilde{Z}(\tilde{U})$, and similarly for c. The property of these that we use in the proof is the apparently weaker E. But in the case of z this is obviously equivalent to strong hereditariness, since $\tilde{Z}$, and hence by 4.2b $\tilde{Z}_*=\tilde{Z}_1$, is hereditary (cf. the proof of 4.3). It is also equivalent in the case of c, even though $\tilde{C}$ is not in general a hereditary class. For if c satisfies E then by 1.3 $c=c_*$ is the operator for the class $\tilde{C}_*(\tilde{U}) = \tilde{C}_* \cap \tilde{U}$ by 3.3, $= \tilde{Z}_* \cap \tilde{U}$ by 3.10d with P = (O), which *is* a hereditary class, whence $c_*=c$ is a hereditary operator, and so by E a strongly hereditary one.

(b) The hypotheses in (a) and (b) are incomparable. Clearly the hypothesis that $\tilde{H} \supseteq \tilde{Z} \cap \tilde{U}$ is (for all natural choices of $\tilde{U}$) strictly stronger than $\tilde{H} \supseteq \tilde{C} \cap \tilde{U}$. On the other hand the hypothesis that c is strongly hereditary is strictly stronger than the same assertion for z. For if c satisfies E, then as in (a) above we see that $c = c_1$. Now from 3.10 we have $\tilde{C}_o(\tilde{U}) \subseteq \tilde{Z}_o(\tilde{U}) \subseteq C_1(\tilde{U})$. So for every $A \in U$ we have $c(A) \subseteq z(A) \subseteq c_1(A) = c(A)$, giving equality. So z = c, whence z as well as c is strongly hereditary. The condition on c is *strictly* stronger, since for example in 3.11 it fails, while that on z holds.

(c) We may regard Thm.E(a) as saying that in an (aa) universe $\tilde{U}$ *all* D-classes containing $\tilde{Z}(\tilde{U})$ have

$h = h_*$ strongly hereditary, provided only that this holds for the single smallest such class, viz. $\tilde{Z}(\tilde{U})$ itself. To borrow a word from Polya ([12], p48), $\tilde{Z}(\tilde{U})$ is *tonangebend* for all larger hereditary $\tilde{H}$ in $\tilde{U}$, in regard to the property "$h = h_*$ is strongly hereditary .

If we relax the condition $(z_o E)$ of Thm.E(a) to $(z_\alpha E)$ for arbitrary α, or to $(z_* E)$ (see 3.7c), we find the following (see 1.2 and 3.12 for notation):

THEOREM F. *Suppose* $\tilde{U}$ *satisfies* (aa) *and* $\tilde{H}$ *is a* Z_**-class in* $\tilde{U}$.

(a) *If* $\tilde{U}$ *satisfies* $(z_\alpha E)$ *and* $\tilde{H}$ *is a* $(D_{1+\alpha}+P)$*-class, then* $h_{1+\alpha} = h_*$ *is strongly hereditary, and* $\tilde{H}_* = \tilde{H}_{1+\alpha+1}$.

(b) *If* $\tilde{U}$ *satisfies* $(z_\alpha E)$ *and* $\tilde{H}$ *is a* $D_{2+\alpha}$*-class then* $h_{2+\alpha} = h_*$ *is strongly hereditary, and* $\tilde{H}_* = \tilde{H}_{2+\alpha+1}$.

(c) *If* $\tilde{U}$ *satisfies* $(z_* E)$ *and* $\tilde{H}$ *is a* D_**-class, then* h_* *is strongly hereditary.*

Proof (a) Take $\tilde{K} = \tilde{H}_{1+\alpha}$ in 4.3, so that $\tilde{K}$ is a D-class. In the situation of 4.3 there exists $0 \neq M acc A$ with $M \in \tilde{Z}$, by (aa). Then $M \in \tilde{Z} \cap \tilde{U} \subseteq \tilde{H}_*$ gives $0 \neq M \in Z \cap H_*$. Since $(z_\alpha E)$ holds and $\tilde{H}$ is a P-class,

3.15a yields $k(A)=h_{1+\alpha}(A)\neq 0$, as required.

(b) By 4.2a $\tilde{H}_1$ is a Z-class, hence by 3.12b a P-class. So the result follows on applying (a) to $\tilde{H}_1$.

(c) With $\tilde{K}=\tilde{H}_*$ in 4.3, and M as in (a), we have $0\neq M\in\tilde{Z}\cap\tilde{H}_*$. Since $\tilde{H}_*$ is a P-class by 3.12b, we have $k(A)=h_*(A)\neq 0$ by 3.15b.

4.8 *Notes* (a) The hypotheses on $\tilde{H}$ are weaker than D throughout, in view of 4.2b and 3.12b.

(b) As stated in 3.16b, I believe that the hypothesis P in (a) is redundant. The purpose of (b) is to give the "right" result at least in case $\alpha\geq\omega$.

(c) In the language of 4.7c the result (c) says that $\tilde{Z}_*(\tilde{U})$ is "tonangebend" for all larger hereditary radical classes $\tilde{K}$ $(=\tilde{H}_*)$ in regard to the property "k is *strongly* hereditary". If $\alpha\geq\omega$ (so that $1+\alpha=\alpha$), part (b) can be read in the same way: $\tilde{Z}(\tilde{U})$ is "tonangebend" for all larger D_α-classes $\tilde{H}$ in regard to the property "h_α is strongly hereditary". Similarly for part (c), reading * in place of α. Of course in all cases $\tilde{U}$ is required to satisfy (aa).

We turn next to results in which we do not assume that $\tilde{H}$ or $\tilde{H}_*$ etc. contains $\tilde{Z}(\tilde{U})$ or $\tilde{C}(\tilde{U})$.

With the assumption (at) of Thm.D on $\tilde{U}$, all we can prove is

THEOREM G. *If* $\tilde{U}$ *satisfies* (at) *and* $\tilde{H}$ *is a* D_1*-class in* $\tilde{U}$, *then* $\tilde{H}_* = \tilde{H}_2$.

Proof. In the situation of 4.4 with $\tilde{K} = \tilde{H}_1$, A has an ideal $0 \neq W \in \tilde{Z}$, by (at). Then $W \leq A \in H_*$ gives $W \in H_*$ by 4.2b, so that $W \in H_* \cap Z$. So $W \in H_1$ by 4.2a, and $k(A) = h_1(A) \supseteq W \neq 0$.

Note 4.9. For $\tilde{U}$ = <u>Ass</u> (without operators, ie. taking Φ = the integers) this is the main result of [3]. Apart from the minor improvement of D to D_1, the present result is stronger in two ways: (1) it uses (at), whereas [3] uses the much stronger property (a+b) of <u>Ass</u>, and (2) we do not assume that h_* is strongly hereditary, a result that is automatic for <u>Ass</u>, and is heavily exploited in the proof in [3]. It is curious also that [3] uses the machinery of "p-primary components" (see 3.16c), which we do not need here (though we did need it for Thm.F).

If we wish to have h_* strongly hereditary without, as in Thm.D, assuming Z for $\tilde{H}$, then we must supplement the restriction (at) on $\tilde{U}$.

THEOREM H. *Suppose* $\tilde{U}$ *satisfies* (va) *and* (at), *and* $\tilde{H}\underline{\subset}\tilde{U}$.

(a) *If* H *is a* (D_1+P)*-class then* $h_1=h_*$ *is strongly hereditary.*

(b) *If* H *is a* D_1*-class then* $h_2(=h_*)$ *is strongly hereditary.*

Proof (a) In the situation of 4.3 with $\tilde{K}=\tilde{H}_1$, there exists $0\neq M$ acc V with $M\in\tilde{Z}$ by (va). Then M acc $V\in\tilde{H}_1$ gives $M\in\tilde{H}_1$, so $M\in H_*\cap\tilde{Z}$. Also $0\neq M$ acc A, and $\tilde{U}$ satisfies $(z_o E)$ by Thm.D. So 3.15a yields $h_1(A)\neq 0$.

(b) $\tilde{H}_1$ is a D-class, hence a D_1-class by 4.2b, and a P-class by 3.12b. So the result follows on applying (a) to $\tilde{H}_1$ in place of $\tilde{H}$.

4.10 *Notes* (a) The conclusions $H_*=H_2$, and hence also $h_*=h_2$, are known in advance from Thm.G.

(b) The hypotheses on $\tilde{H}$ are weaker than D (cf.4.8a).

(c) With reference to 3.16b, I believe that the "correct" result here is the conclusion of (a) with the hypothesis of (b).

(d) It is instructive to compare Thm.H(a) with 4.6a. The conclusions in the two cases are identical, as is the hypothesis D_1 on $\tilde{H}$, and (at) on $\tilde{U}$. The extra hypothesis Z_* on $\tilde{H}$ in 4.6a is compensated in Thm.H by the extra hypothesis (va) on $\tilde{U}$ and also (if needed at all) by the rather weak condition P on $\tilde{H}$.

(e) While the conditions on U that we actually used were (va) and (at), note that by Thm.B (1' and 2') they are equivalent to the apparently much stronger pair of conditions (vti) and (ita).

(f) It is noteworthy that the hypothesis (at) on $\tilde{U}$ was used only to ensure (z_0E). Thus Thm.H is a corollary of Thm.D and the case $\alpha=0$ of our next result.

THEOREM J. *Suppose* $\tilde{U}$ *satisfies* (va), *and* $\tilde{H}\subseteq\tilde{U}$.

(a) *If* $\tilde{U}$ *satisfies* $(z_\alpha E)$ *and* $\tilde{H}$ *is a* $(D_{1+\alpha}+P)$-*class, then* $h_{1+\alpha} = h_*$ *is strongly hereditary, and* $\tilde{H}_* = \tilde{H}_{1+\alpha+1}$.

(c) *If* $\tilde{U}$ *satisfies* (z_*E) *and* $\tilde{H}$ *is a* D_*-*class then* h_* *is strongly hereditary.*

Proof (a) Just like that of Thm.H(a), reading for 0 and $1+\alpha$ for 1.

(c) Just like that of Thm.F(c).

4.11 *Notes* (a) The notes in 4.8, insofar as they do not refer to Thm.F(b), apply verbatim here also.

(b) It is instructive to compare Thms F and J ((a) and (c)). The only differences are the weaker restriction (aa) on $\tilde{U}$ in Thm F, compensated by the extra restriction Z_* on $\tilde{H}$.

(c) If we keep only the conditions common to Thms. F(a) and J(a), we have a correspondingly weak conclusion:

THEOREM K. *Suppose* $\tilde{U}$ *satisfies* (aa) *and* $(z_\alpha E)$, *and* $\tilde{H}$ *is a* $(D_{1+\alpha}+P)$*-class in* $\tilde{U}$. *Then* $\tilde{H}_* = \tilde{H}_{1+\alpha+1}$.

Proof. In the situation of 4.4 with $\tilde{K}=\tilde{H}_{1+\alpha}$ there exists $0\neq M$ acc A with $M\in\tilde{Z}$, by (aa). Then M acc $A \in \tilde{K}_2$ gives $M \in \tilde{K}_2$ by 4.2b. So $M \in \tilde{H}_* \cap \tilde{Z}$, whence by $(z_\alpha E)$ and P on $\tilde{H}$, 3.15 yields $h_{1+\alpha}(A) = k(A) \neq 0$.

4.12 *Notes* (a) Theorem G is 'almost' a corollary of Thm.K. In the presence of (at) we have (z_0E) by Thm.D, whence $\tilde{H}_* = \tilde{H}_2$ by Thm.K, provided $\tilde{H}$ satisfies P as well as D_1.

(b) As usual, the hypothesis on $\tilde{H}$ is a weakening of D.

4.13 *Examples*. We proceed now to give examples designed to show that, at least in certain respects, the theorems above are best possible.

(a) In Thm.D, if (at) is replaced by any combination

of the Andr conditions that does not imply (at), then the conclusion $h = h_*$ may fail. Thus let $\tilde{U}$ be as in 2.11(b'4), and set $\tilde{H} = \tilde{Z}(\tilde{U})$. Then E fails for h, and $h_*=h_1\neq h$, and $\tilde{H}_*=\tilde{H}_2\neq\tilde{H}_1$. However, here the other conclusion of Thm.D, viz. that h_* is strongly hereditary, does hold. This is inevitable if $\tilde{U}$ satisfies even (f), in view of a result in Part II of this paper.

(b) For an example of a D+Z class $\tilde{H}$ in which h_* is *not* strongly hereditary, we can allow the strongest of our conditions compatible with the failure of both (at) and (f); that is, (vti)+(a). In 2.11(a7) it is easily verified that $\tilde{Z}=\tilde{Z}_*$ and $z(A)=0\neq z(A^2)$.

This example also shows that in Theorems E,F,J the assumption (c_oE) or (z_oE) or ($z_\alpha E$) or (z_*E) is independent of the assumption on $\tilde{U}$, since the latter ((va) or (aa)) is in every case implied by (vti). Likewise it shows that in these theorems the weakest common conclusion, viz. that h_* satisfies E, fails in the absence of the hypothesis on z or c, since it fails for h=z itself.

(c) We consider next an example where $\tilde{H}$ is not a Z-class. Let $\tilde{U}$ = <u>Ass</u>$\cap$<u>Comm</u> over a field F, and $\tilde{H}=\tilde{C}$, as in 3.11. Then $\tilde{H}$ is a $(D+Z_1)$-class; $h_1=h_*$ is strongly hereditary but h is not, and $\tilde{H}_*=\tilde{H}_2\neq\tilde{H}_1$. Thus the result $h_1=h_*$ in 4.6 and Thm.H(a), and $h_{1+\alpha}=h_*$ in Thms.F(a) and J(a) cannot be improved to $h_o=h_*$ or $h_\alpha=h_*$ respectively (at least for $\alpha=0$), even when $\tilde{H}$ is a $(D+Z_1)$-class and $\tilde{U}$ satisfies all the Andr-type conditions.

Similarly the result $\tilde{H}_*=\tilde{H}_2$ in 4.6 and Thm.G, and $\tilde{H}_*=\tilde{H}_{1+\alpha+1}$ in Thms.F(a), J(a) and K cannot be improved to $\tilde{H}_*=\tilde{H}_1$ or $\tilde{H}_*=\tilde{H}_{1+\alpha}$ respectively (at least for $\alpha=0$), even under strong hypotheses as above.

Finally this example shows that in Thms.D and E(a) the condition Z on $\tilde{H}$ cannot be weakened to Z_1, and that in Thm.E(b) the hypothesis (c_oE) cannot be replaced by (z_oE).

(d) For an example where h_* is not strongly hereditary let $\tilde{U}$ be as in 2.11(b2). This $\tilde{U}$ satisfies the strongest Andr-type conditions compatible with the failure of (va); viz. it satisfies (b) and (at). It is easily verified that in U we have $z_o = c_o$, and $\tilde{U}$ satisfies $(z_oE) = (c_oE)$. Set $\tilde{H}=\tilde{I}\{E,0\}$, where $E=F[e]$ is the 1-dimensional idempotent algebra. Then $h=h_*$ satisfies D but not E. This shows that if the conclusion that h_* be strongly hereditary is to survive, then the hypothesis that $\tilde{H}$ or $\tilde{H}_*$ contain $\tilde{Z}(\tilde{U})$ or $\tilde{C}(\tilde{U})$ in Thms D, E(a), E(b), F(a), F(b), F(c) and in 4.6 cannot be omitted. To put this in a slightly different way, it shows that with the hypotheses of Thm.K on $\tilde{U}$ we cannot hope to prove that h_* is strongly hereditary, even if $\tilde{H}$ is a D-class.

This example also shows that to obtain the same conclusion in Theorem J we cannot replace (va) with any combination of Andr conditions not implying (va), even if we assume that $\tilde{H}$ satisfies D. Finally, for Theorem H, it shows that the hypothesis (va) on $\tilde{U}$ additional to (at) is the weakest of the Andr conditions to ensure the desired result (that h_* be strongly hereditary), even if $\tilde{H}$ is given to satisfy D.

Key to Table 4.14. Results are arranged, so far as possible, in decreasing order of strength of hypothesis on $\tilde{U}$, and then on $\tilde{H}$. The columns are as follows:

1. Name of result.
2. Hypotheses of Andr type on $\tilde{U}$ (see 2.2)
3. Hypothesis of $(z_\alpha E)$ type on $\tilde{U}$ (see 3.7c)
4. Hypothesis of D-type on $\tilde{H}$ (see 1.2)
5. Hypothesis of Z-type on $\tilde{H}$. For Z and Z_* see 1.2, for C see 4.5d, and for S see 4.5c.

6. Hypothesis P on $\tilde{H}$ (see 3.12)

7. We display an ordinal β such that h_β is strongly hereditary, (so that $h_\beta = h_*$ by 1.3). A star here means that h_* is strongly hereditary.

8. We display an ordinal γ such that $\tilde{H}_* = \tilde{H}_\gamma$. Parentheses mean that the same result follows elsewhere (viz. Thm.G) from weaker hypotheses.

4.14 Table

Result	Hypotheses on $\tilde{U}$		Hypotheses on $\tilde{H}$			Conclusions	
	Andr	$z_\alpha E$	D	Z	P	$h_\beta E$	$\tilde{H}_*=\tilde{H}_\gamma$
1	2	3	4	5	6	7	8
Ha	va+at		D_1	P	P	1	(2)
Hb	va+at		D_1			2	(2)
D	at		D	Z		0	1
4.6a	at		D_1	Z_*		1	2
4.6b	at		D_1	C		1	2
G	at		D_1				2
4.5c	as		D	S		0	1
Ja	va	$z_\alpha E$	$D_{1+\alpha}$		P	$1+\alpha$	$1+\alpha+1$
Jc	va	z_*E	D_*			*	
Ea	aa	z_oE	D	Z		0	1
Eb	aa	c_oE	D	C		0	1
Fa	aa	$z_\alpha E$	$D_{1+\alpha}$	Z_*	P	$1+\alpha$	$1+\alpha+1$
F_b	aa	$z_\alpha E$	$D_{2+\alpha}$	Z_*		$2+\alpha$	$2+\alpha+1$
K	aa	$z_\alpha E$	$D_{1+\alpha}$		P		$1+\alpha+1$
Fc	aa	z_*E	D_*	Z_*		*	

5. TWO EXAMPLES

5.1 Among the conditions that appear in Sec.4 are the two hierarchies $\{D_\alpha\}$ and $\{(z_\alpha E)\}$ of ever-weaker conditions on $\tilde{H}$ and $\tilde{U}$ respectively, for $0 \leq \alpha \leq \infty$ (where we interpret D_∞ as D_*, etc). In this section we show that the hierarchies are strict; i.e. that for every α with $0 < \alpha \leq \infty$ $(z_\alpha E)$ may hold but $(z_\beta E)$ fail for every $\beta < \alpha$, and similarly for D . (We interpret $\beta < \infty$ for every ordinal β).

5.2 For the hierarchy $\{(z_\alpha E)\}$ we construct a chain $\{A_\alpha : -1 \leq \alpha < \infty\}$ of algebras over any Φ by the method of 1.8, but using Z_Φ in place of $\tilde{S}$ as there. Since 1.6(vi) no longer applies, we need two technical modifications:

LEMMA 5.3. *Suppose in the situation of* 1.5 *we have* $\dim_F S = \dim_F E$. *Then there exist algebras* $A=A_c$ *and* $A=A_a$ *over F satisfying* 1.5(i),(ii),(iv), (v) *and*

(vi) *Every proper right or left ideal* $T \neq 0$ *of* A *contains* E.

Proof. Available (like 1.5) on request.

5.4. For arbitrary Φ we cannot proceed with the generality of 1.6. Instead, let P be any *maximal* ideal of Φ, and set $F = \Phi\text{-}P$. Then any F-algebra may be regarded as a unital Φ-algebra as in 1.6, and furthermore it is easily verified that the F-ideals and Φ-ideals of A coincide. We therefore have the following analog of Prop 1.7:

LEMMA 5.5 *Suppose* $\Phi \to F$ *as in* 5.4, *and* S, E *are* F-*algebras with* dimE = dimS. *Then there exists an* F-*algebra* A *such that*

(i) *Every proper* Φ-*ideal of* A *contains* E;

(ii) $A - E \simeq S$;

(iii) *If* E *and* S *are both in* Comm *or* Anticomm, *then so also is* A;

((iv) *If* S *has a* 1 *then*, A *has a* 1).

We now have

THEOREM L. *Within each of the universes* $\tilde{U}$ = All *and* Comm *and* Anticomm *over any* Φ *there exists for each* α $(0 \le \alpha \le \infty)$ *a subuniverse* $\tilde{U}(\alpha) \supseteq \tilde{Z}_\Phi$ *in which* $(z_\alpha E)$ *holds but* $(z_\beta E)$ *fails for every* $\beta < \alpha$.

Proof. Let $\Phi \to F$ as in 5.4. We construct a chain $\{A_\alpha : \alpha \ge -1\}$ of F-algebras within the given $\tilde{U}$, taking $A_{-1} \in \tilde{Z}$, $\dim_F A_{-1} \ge 2$; $A_{\beta+1}$ to be an A as in 5.5 (but we do not need (iv)), where $E = A_\beta$ and $S \in \tilde{Z}_F$ is such that $\dim_F S = \dim_F E$ (such an S clearly exists), and $A_\lambda = \Sigma\oplus_{\mu<\lambda} A_\mu$ for λ a limit ordinal (>0); compare 1.8. Then we can prove almost exactly as in 1.9c that $A_\alpha \in \tilde{Z}_{\alpha+1} \setminus \tilde{Z}_\alpha$ for every α. Here $\tilde{Z}_\alpha$ means $\tilde{Z}_\alpha(\tilde{U})$ for the given $\tilde{U}$, in the notation of 3.3. If $\alpha=\beta+1$ note that every homomorphic image X of A_α is isomorphic either to A_α itself or to $A_\alpha - I$ for some I with $A_\beta \subseteq I \leq A$ (by 5.5i), in which case $X \in \tilde{Z} \subseteq \tilde{Z}_\alpha$.

Now set $\tilde{U}(\alpha) = \tilde{Z}_{\alpha+1}(\tilde{U})$. Then $z_\alpha(X) = X$ for all $X \in \tilde{U}(\alpha)$, so that z_α is strongly hereditary and (z_αE) holds. Suppose $\beta<\alpha$. Then $\beta+1\leq\alpha$, so $A_\beta < A_{\beta+1} \in \tilde{U}(\alpha)$; $z_\beta(A_\beta) = A_\beta$ since $A_\beta \in \tilde{Z}_{\beta+1}$, but $z_\beta(A_{\beta+1})=0$. For otherwise $A_{\beta+1}$ would have a non-0 ideal I in $\tilde{Z}_\beta$, but then $A_\beta \leq I \in \tilde{Z}_\beta$ by 5.5i, so $A_\beta \in \tilde{Z}_\beta$ by 4.2b (since $\tilde{Z}$ is a D-class), contrary to the result stated above. The argument works also for $\alpha=\infty$; of course $\tilde{U}(\infty)$ means $\tilde{Z}_*(\tilde{U})$.

5.6. We turn now to the construction for each α of a class $\tilde{H}(\alpha)$ in a universe $\tilde{U}(\alpha)$ for which D_α holds but D_β fails for all $\beta<\alpha$. The construction is much as sketched in [18], 6.6, where certain special cases of this result are asserted; however the detailed verifications required are by no means as routine as the authors of [18] probably thought.

We start with a rather general result. For any class $\tilde{V}$ of Φ-algebras write $\sigma\tilde{V}$ for the class of all those Φ-algebras which are isomorphic to a (weak) direct sum of $\tilde{V}$-algebras.

LEMMA 5.7 $\tilde{V} \subseteq \sigma\tilde{V} = \sigma\sigma\tilde{V}$.

Proof. $\tilde{V}$ consists precisely of the direct sums with just one summand. So $\tilde{V} \subseteq \sigma\tilde{V}$, whence $\sigma\tilde{V} \subseteq \sigma\sigma\tilde{V}$. Conversely suppose $A \in \sigma\sigma V$; say $A = \Sigma\oplus \{A_i : i \in I\}$ with $A_i \in \sigma V$. Then we may suppose that $A_i = \Sigma\oplus \{A_{\beta(i)} : \beta(i) \in B(i)\}$ with every $A_{\beta(i)}$ in $\tilde{V}$. We may suppose $B(i) \cap B(j) = \emptyset$ for $i \neq j$, eg. on replacing $B(i)$ by $B(i) \times \{i\}$ and $\beta(i)$ by $(\beta(i), i)$ for every $i \in I$. Then it is a straightforward exercise that $A \simeq \Sigma\oplus\{A_\gamma ; \gamma \in C\}$, where $C = \cup\{B(i) : i \in I\}$, so that $A \in \sigma\tilde{V}$.

PROPOSITION 5.8. *Suppose* $\tilde{V}$ *is a class of idempotent* Φ*-algebras. Then* $\sigma\tilde{V}$ *is a universe (of hereditary idempotent algebras) iff all ideals and homomorphic images of* $\tilde{V}$*-algebras are in* $\sigma\tilde{V}$.

Proof. Necessity is immediate from $V \subseteq \sigma V$ Suppose then that all the conditions on $\tilde{V}$ hold.

Then $\sigma\tilde{V}$, like $\tilde{V}$, consists of idempotent algebras, thus $T<A\in\tilde{V}$ implies $T^2=T$. Suppose $T<A\in\sigma V$; we may suppose that $A=\Sigma\oplus\{A_i: i\in I\}$ with $A_i\in\tilde{V}$. Let $T_i=T\pi_i$ be the projection of T onto A_i. Then $T_i = T_i^2\subseteq T_iA_i = TA_i\subseteq T\cap A_i\subseteq T_i$, giving equality. So $T\subseteq \Sigma\oplus\{T_i: i\in I\} = \Sigma\oplus\{T\cap A_i: i\in I\}\subseteq T$, again giving equality. Now $T_i\leq A_i\in\tilde{V}$ gives $T_i\in\sigma\tilde{V}$ by hypothesis, whence $T\in\sigma\sigma\tilde{V} = \sigma\tilde{V}$ by 5.7. Thus $\sigma\tilde{V}$ is hereditary.

Suppose $A\in\sigma V$ and A' is a homomorphic image. Then $A'\simeq A-T$ for some $T\leq A$, and in the notation above this gives $A' = \Sigma\oplus\{A_i: i\in I\}- \Sigma\oplus\{T_i: i\in I\}\simeq \Sigma\oplus\{A_i-T_i: i\in I\}$. Now $A_i\in\tilde{V}$ gives $A_i-T_i\in\sigma V$ by hypothesis, so $A'\in\sigma\sigma\tilde{V} = \sigma\tilde{V}$ by 5.7. Thus $\sigma\tilde{V}$ is homomorphically closed, and so is a universe.

Finally, if I acc $A\in\sigma V$ then $I\in\sigma V$ since σV is hereditary, so $I^2=I$. Thus every $A\in\sigma V$ is hereditarily idempotent.

5.9. We proceed now to the construction. For given Φ, and working within $\tilde{U}$ = <u>All</u> or <u>Comm</u> or <u>Anticomm</u> over Φ, let $\Phi\to F$ as in 1.4, and let $\tilde{S}$ be as in 1.9a; specifically (i) $\tilde{S}\subseteq\tilde{U}$; (ii) Every $S\in\tilde{S}$ is an idempotent F-algebra which is simple as a Φ-algebra; (iii) $\{\dim_F S: S\in\widetilde{S}\}$ is unbounded, and (for convenience) also (iv) $\widetilde{S}$ is isomorphism-closed.

Then we may construct a chain $\{A_\alpha: \alpha \geq -1\}$ as in 1.8, and based on $\widetilde{S}$. The following assertions are for $\alpha\geq 0$, but $0 = (-1)+1$ is regarded as a successor. For $0\leq\alpha\leq\infty$ let $\widetilde{S}_\alpha$ be the α-th step $\widetilde{S}_\alpha(\widetilde{U})$ in the Kurosh hierarchy within $\widetilde{U}$ that starts with $\widetilde{S}_0 = \widetilde{S}\cup\{0\}$; we take $\widetilde{S}_\infty$ to mean $\widetilde{S}_*$. Also set $\widetilde{V}_\alpha = \widetilde{S}\cup\{A_{\beta+1}: \beta+1<\alpha\}$; it is clear what this means for $\alpha=0$ and $\alpha=\infty$. Then we have

LEMMA 5.10. *For* $0\leq\alpha\leq\infty$

(a) $\widetilde{V}(\alpha)$ *is a class of idempotent algebras;*

(b) $A_\beta \in \sigma\widetilde{V}(\alpha)$ *for all* $\beta<\alpha$;

(c) $A_\alpha \in \widetilde{S}_{\alpha+1}$.

Proof. (a) It suffices to consider $A_{\beta+1} = X$, say. Now $X^2\leq X$, so if $X^2\neq X$ we have $X^2=A_\beta$ or 0 by 1.8. But then $A_{\beta+1}-A_\beta\in\widetilde{Z}$, contrary to construction and idempotence of every $S\in\widetilde{S}$ (5.9ii above).

(b) By induction on β. If β is a successor then by definition $A_\beta\in V(\alpha)\subseteq\sigma V(\alpha)$. If β is a limit ordinal then $A_\beta = \Sigma\oplus\{A_\mu: \mu<\beta\}$. For each $\mu<\beta$ we have $A_\mu\in\sigma V(\alpha)$ by inductive hypothesis, so $A_\mu\in\sigma\sigma V(\alpha) = \sigma\widetilde{V}(\alpha)$ by 5.7.

(c) Just like that of 1.9c.

PROPOSITION 5.11. *For* $0\leq\alpha\leq\infty$:

(a) $\sigma\tilde{V}(\alpha)$ *is a universe of hereditarily idempotent algebras;*

(b) $\sigma\tilde{V}(\alpha)\subseteq\tilde{S}_{\alpha+1}$.

Proof. (a) We verify the hypotheses of 5.8 for $\tilde{V}(\alpha) = \tilde{V}$. If $A\in\tilde{V}$ then $A^2=A$ by 5.10a. If $I<A\in\tilde{V}$ then trivially $I\in\tilde{V}\in\sigma\tilde{V}$ if $A\in\tilde{S}$. Suppose $A=A_{\beta+1}$. Then $I=0\in\sigma V$ (the empty direct sum) or $I=A_\beta\in\sigma\tilde{V}$ by 5.10b. Likewise if $A\in\sigma\tilde{V}$ and A' is a homomorphic image, the only non-trivial case is $A'\simeq A_{\beta+1}-A_\beta\in\tilde{S}$ $\subseteq\tilde{V}\subseteq\sigma\tilde{V}$.

(b) Suppose $A'\in\sigma\tilde{V}(\alpha)$ and A is non-0 homomorphic image. Then A also is in $\sigma\tilde{V}(\alpha)$, so has at least one non-0 ideal (direct summand) T in V(a). Then $T\in\tilde{S}\subseteq\tilde{S}_{\alpha+1}$, or $T\simeq A_{\beta+1}$ for some $\beta+1<\alpha$. But then $T\in\tilde{S}_{\beta+2}\subseteq\tilde{S}_\alpha$ in view of 5.10c and the fact that $\beta+2\leq\alpha$. Thus $A'\in\tilde{S}_{\alpha+1}$.

We now have all the prerequisites to prove

THEOREM M. *Within* $\tilde{U}$ = <u>All</u> *or* <u>Comm</u> *or* <u>Anticomm</u> *over any* Φ *there exists for each* α $(0\leq\alpha\leq\infty)$ *a subuniverse* $\tilde{U}(\alpha)$ *and a class* $\tilde{H}(\alpha)\subseteq\tilde{U}(\alpha)$ *for which* D_α *holds but* D_β *fails for every* $\beta<\alpha$.

Furthermore we can arrange for $\tilde{H}(\alpha)$ *to be a Z-class in* $\tilde{U}(\alpha)$, *for* (z_oE) *to hold in* $\tilde{U}(\alpha)$, *and for* $\tilde{Z}$ *to coincide with* $\tilde{Z}_*$ *in* $\tilde{U}(\alpha)$.

Proof. Let $\Phi \to F$ as in 1.6. There exist classes $\tilde{S}, \tilde{T} \subseteq \tilde{U}$, both satisfying (i)-(iv) of 5.9, and such that also (v) $\tilde{S} \cap \tilde{T} = \emptyset$. For example, we can decompose the class of all cardinals into disjoint cofinal subclasses P, Q, and let $\tilde{S}$ contain all Φ-algebras S satisfying (i) and (ii) of 5.9, and with $\dim_F S$ = some $\pi \in P$. There exists at least one such S for every $\pi \in P$, by 1.9a. Similarly for $\tilde{T}$.

Let $\{A_\alpha\}$ be a chain as in 5.9 and based on $\tilde{S}$. For each α let B_α be an F-algebra in $\tilde{U}$ with unique proper Φ-ideal A_α, and such that $B_\alpha - A_\alpha \in \tilde{T}$. This is possible by 1.5 in view of (i), (iii), (iv) for $\tilde{T}$. The proof is now in steps.

(1) *For all* α, β, *we have* $A_\alpha \not\simeq B_\beta$.

Proof. B_β has unique proper ideal A_β, so if $A_\alpha \simeq B_\beta$ we would have $\alpha = \gamma + 1$ for some γ, and then $\tilde{S} \ni A_\alpha - A_\gamma \simeq B_\beta - A_\beta \in \tilde{T}$ contradicts the conjunction of 5.9 (iv) and (v) above.

(2) Set $\tilde{H}(\alpha) = \tilde{S} \cup \tilde{T} \cup \tilde{I}\{B_\beta : \beta < \alpha\}$ for $0 \leq \alpha \leq \infty$. It is clear what this means for $\alpha = 0$ and $\alpha = \infty$. By the

construction of the B_β it is clear that $\tilde{H}(\alpha)$ is homomorphically closed.

(3) $A_\alpha \notin H(\infty)_\alpha$, *for all* $\alpha<\infty$.

Proof. By induction on α, much as in 1.9c. If γ is minimal such that $A_\alpha \in \tilde{H}(\infty)_\gamma$, note that $\gamma \neq 0$ since A_α is not simple, and $A_\alpha \notin \tilde{I}\{B_\beta: \beta<\alpha\}$ by (1).

Now let $\tilde{U}(\alpha)$ be any universal class with $\tilde{H}(\alpha) \subseteq \tilde{U}(\alpha) \subseteq \tilde{U}$.

(4) *If* $\beta<\alpha$ *then* D_β *fails for* $\tilde{H}(\alpha)$ *as a class in* $\tilde{U}(\alpha)$.

Proof. With notation as in 3.3 let $\tilde{H}(\alpha)_\beta(\tilde{U}(\alpha))=\tilde{X}$ be the β-th step in the Kurosh hierarchy for $\tilde{H}(\alpha)$ as constructed in $\tilde{U}(\alpha)$. We show that D fails for $\tilde{X}$. By 3.3, since $\tilde{U}(\alpha) \subseteq \tilde{U}$, we have $\tilde{X}=\tilde{H}(\alpha)_\beta(\tilde{U}) \cap \tilde{U}(\alpha) \subseteq \tilde{H}(\infty)_\beta(\tilde{U})$. So by (3) we have $A_\beta \notin \tilde{X}$. But $A_\beta < B_\beta \in \tilde{H}(\alpha) = \tilde{H}(\alpha)_o(\tilde{U}(\alpha))$ (since $\tilde{H}(\alpha)$ is homomorphically closed and $\subseteq \tilde{U}(\alpha)$), $\subseteq \tilde{H}(\alpha)_\beta(\tilde{U}(\alpha))=\tilde{X}$; i.e $B_\beta \in \tilde{X}$.

(5) To ensure that D_α holds for $\tilde{H}(\alpha)$ as a subclass of $\tilde{U}(\alpha)$ we specialise to $\tilde{U}(\alpha)=\tilde{I}\{B_\beta : \beta<\alpha\} \cup \tilde{T} \cup \sigma\tilde{V}(\alpha)$, where $\tilde{V}(\alpha)$ is as in 5.9, and $\sigma\tilde{V}(\alpha)$ is as in 5.6 for $\tilde{V} = \tilde{V}(\alpha)$.

(6) $\tilde{H}(\alpha)\subseteq\tilde{U}(\alpha)\subseteq\tilde{U}$, *and* $\tilde{U}(\alpha)$ *is a universe.*

Proof. $\tilde{T}\cup\sigma\tilde{V}(\alpha)$ is a universe in view of 5.11. If $0\neq I\leq A\simeq B_\beta$ then $I\simeq B_\beta$ or $A_\beta\in\tilde{U}(\alpha)$, and $A-I=0$ or $A-I\in\tilde{T}$. Thus $\tilde{U}(\alpha)$ is a universe. We have $\tilde{H}(\alpha)\subseteq\tilde{U}(\alpha)$ in view of $\tilde{S}\subseteq\tilde{V}(\alpha)\subseteq\sigma\tilde{V}(\alpha)$. We have $B_\beta\in\tilde{U}$ and $\tilde{T}\subseteq\tilde{U}$ by construction, and from $\tilde{V}(\alpha)\subseteq\tilde{U}$ we have $\sigma\tilde{V}(\alpha)\subseteq\tilde{U}$ from the definition of $\tilde{U}$.

(7) *If* $\tilde{Y} = \tilde{H}(\alpha)_{\alpha+1}(\tilde{U}(\alpha))$ *then* $\tilde{Y} = \tilde{U}(\alpha)$.

Proof. $\tilde{Y}\supseteq\tilde{S}_{\alpha+1}(\tilde{U}(\alpha))$ in view of $\tilde{S}\subseteq\tilde{H}(\alpha)$. But this equals $\tilde{S}_{\alpha+1}(\tilde{U})\cap\tilde{U}(\alpha)$ by 3.3, $\supseteq\sigma\tilde{V}(\alpha)\cap\tilde{U}(\alpha)$ by 5.11, $=\sigma\tilde{V}(\alpha)$. Also $\tilde{Y} = \tilde{H}(\alpha)_0(\tilde{U}(\alpha))=\tilde{H}(\alpha)$ (as in (4)), $\supseteq\tilde{T}\cup\tilde{I}\{B_\beta:\ \beta<\alpha\}$. Thus $\tilde{Y}\supseteq\tilde{U}(\alpha)$, giving equality.

(8) D_α *holds for* $\tilde{H}(\alpha)$ *as a subuniverse of* $\tilde{U}(\alpha)$.

Proof. If $X\in\tilde{V}(\alpha)$ then $X\in\tilde{H}(\alpha)_{\alpha+1}$ by (7), so $h_\alpha(X) = X$. Thus h_α is the identity function on $\tilde{U}(\alpha)$, so trivially is strongly hereditary.

(9) *Every algebra in* $\tilde{U}(\alpha)$ *is idempotent.*

Proof. If $X\in\tilde{U}(\alpha)$ is in $\sigma\tilde{V}(\alpha)$, then $X^2=X$ by 5.11a; if $X\in\tilde{T}$ then trivially $X^2=X$, and if $X\simeq B_\beta$ then $X^2=X$ by the argument of 5.10(a), in view of (ii) of 5.9 for $\tilde{T}$.

(10) *In* $\tilde{U}(\alpha)$ *we have* $\tilde{Z}=\tilde{Z}_*=\{0\}$.

For the only idempotent algebra in $\tilde{Z}(\tilde{U})$ is 0, so $\tilde{Z}(\tilde{U})\cap\tilde{U}(\alpha)=\{0\}$ by (10).

(11) Hence trivially $\tilde{H}(\alpha)$ is a Z-class in $\tilde{U}(\alpha)$ and (z_0E) holds in $\tilde{U}(\alpha)$. In case this seems 'cheating' we can redefine $\tilde{U}'(\alpha)=\tilde{U}(\alpha)\cup\tilde{Z}(\tilde{U})$ and $\tilde{H}'(\alpha)=\tilde{H}(\alpha)\cup\tilde{Z}(\tilde{U})$. The proofs of (2)-(8) are unaffected, and now in (11) we have z=0 on the subuniverse $\tilde{U}(\alpha)$ and z=id on the complementary subuniverse $\tilde{Z}$, whence again (z_0E) holds.

Note 5.12. In [18], Sec.5 a condition (k_α) weaker than D_α was introduced: viz $\tilde{H}$ satisfies (k_α) if $I<A\in\tilde{H}$ implies $I\in\tilde{H}_\alpha$. In the example of Thm.M we have in fact shown (see (4)) that (k_β) fails for every $\beta<\alpha$, while D_α holds.

Note 5.13. The result of Thm.M is not the best we could hope for: one is tempted to ask (i) does there exist for each Φ a *single* universe $\tilde{U}$ over Φ and subclasses $\tilde{H}(\alpha)$ for $0\leq\alpha\leq\infty$ such that D_β (or even (k_β)) fails for all $\beta<\alpha$, but D_α holds for $\tilde{H}(\alpha)$ as a class in $\tilde{U}$, and dually (ii) does there exist for each α a single class $\tilde{H}$ over Φ

and for every α a universe $\tilde{U}(\alpha) \supseteq \tilde{H}$ over Φ such that the above holds for $\tilde{H}$ as a subclass of $\tilde{U}(\alpha)$.

REFERENCES

[1] S. A. Amitsur, A general theory of radicals II, *American J. Math.* 76 (1954), 100-125.

[2] T. Anderson, N. Divinsky and A. Sulinski, Hereditary radicals in associative and alternative rings, *Canad. J. Math.* 17 (1965), 594-603.

[3] E. P. Armendariz and W. G. Leavitt, The hereditary property in the lower radical construction, *Canad. J. Math.* 20 (1968), 474-476.

[4] S. E. Dickson, A note on hypernilpotent radical properties for associative rings, *Canad. J. Math.* 19 (1967), 447-448.

[5] B. J. Gardner, Some degeneracy and pathology in non-associative radical theory, *Ann. Univ. Sci. Budapest Eötvös Sect. Math.* 22-23 (1970-80), 65-74.

[6] B. J. Gardner, Some degeneracy and pathology in non-associative radical theory II, *Bull. Australian Math. Soc.* 23 (1981), 423-428.

[7] A. E. Hoffman and W. G. Leavitt, Properties inherited by the lower radical, *Portug. Math.* 27 (1968), 63-66.

[8] A. G. Kurosh, Radicals of rings and algebras (in Russian), *Mat. Sbornik* 33 (1953), 13-26.

[9] W. G. Leavitt, The general theory of radicals, Mimeo notes, University of Nebraska, Lincoln, Neb., 1966.

[10] W. G. Leavitt, Sets of radical classes, *Publ. Mat. Debrecen* 14 (1967), 321-324.

[11] W. G. Leavitt, Strongly hereditary radicals, *Proc. Amer. Math. Soc.* 21 (1969), 703-705.

[12] G. Polya, Patterns of Plausible Inference, Princeton U.P., 1954.

[13] Yu. M. Ryabuhin, On lower radicals of rings (in Russian), *Mat. Zametki* 2 (1967), 239-244.

[14] M. Slater, On upper radicals and Andrunakievic classes, *Notices Amer. Math. Soc.* 15 (1968), 914.

[15] A. M. Slin'ko, On radicals of Jordan rings (in Russian), *Alg. i Logika* 11 (1972), 206-215.

[16] A. Sulinski, T. Anderson and N. Divinsky, Lower radical properties for associative and alternative rings, *J. London Math. Soc.* 41 (1966), 417-424.

[17] R. L. Tangeman and D. Kreiling, Lower radicals in non-associative rings, *J. Austral. Math. Soc.* 14 (1972), 419-423.

[18] J. F. Watters and M. Slater, On Amitsur's condition D in radical theory, *J. Algebra* 39 (1976), 175-198.

[19] Yu-Lee Lee and R. L. Propes, On intersections and unions of radical classes, *J. Austral. Math. Soc.* 13 (1972), 354-356.

Michael Slater
Department of Mathematics
University of Bristol
University Walk, Bristol
BS8 1TW
England, Great Britain

COLLOQUIA MATHEMATICA SOCIETATIS JÁNOS BOLYAI
38. RADICAL THEORY, EGER (HUNGARY), 1982.

INTERMEDIATE EXTENSIONS AND RADICALS

P. N. STEWART

Let R and S be associative rings with the same identity such that $R \subseteq S$ and suppose that there is a finite set $\{a_1,\ldots,a_n\} \subseteq S$ such that $S = \sum_{i=1}^{n} Ra_i$. If $a_i r = ra_i$ for all $r \in R$ and all $i = 1,\ldots,n$ (respectively, $a_i R = Ra_i$ for all $i = 1,\ldots,n$) , then S is a *liberal extension* (respectively, *normalizing extension)* of R . If T is a ring and $R \subseteq T \subseteq S$ where S is a liberal (respectively, normalizing) extension of R, then T is an *intermediate extension* (respectively, *intermediate normalizing extension*) of R . Liberal extensions have been studied in [15] and [19], normalizing extensions in [2], [3], [7], [8], [11], [16], [17], [21], and [23], intermediate extensions in [18] and [22], and intermediate normalizing extensions are investigated in [6].

These four types of extensions provide situations which generalize some of the classical examples of "related rings". In particular, matrix rings $M_n(R)$ and group rings $R[G]$, where G is a finite group, are examples of liberal extensions of R. Examples of normalizing extensions which are not liberal include crossed products $R*G$ where G is a finite group and certain skew matrix ring extensions of R. See, for example [7, Example 2.13]. Of course, all liberal extensions are intermediate, but intermediate extensions may not be liberal; consider, for example,

$$T = \{ae_{11} + be_{12} + ce_{22} : a,c \in R \text{ and } b \in I\}$$

where the e_{ij} are the usual matrix units and I is an ideal of R which does not have a finite set of R-centralizing generators. Similarly, intermediate normalizing extensions need not be normalizing.

In this paper we investigate some aspects of the behavior of radicals with respect to intermediate extensions. More precisely, we consider the equation

$$r(R) = R \cap r(T)$$

where r is a radical and T is an intermediate extension of R. The results of section 2 show that this equation holds for several well-known special radicals and in section 3 we show that the equation behaves well with respect to the formation of relative complements. For example, we establish the equation for the Levitzki locally nilpotent radical, the Brown-McCoy radical and the Ortiz K-primitive radical (and in [18] Robson shows that the equation holds for the Jacobson radical). Also, we prove that if T is an intermediate extension of R, then T is a Jacobson ring if and only if R is a Jacobson ring. Related results, for normalizing extensions, are given in [23].

Throughout this paper *the notation* $R \subseteq T \subseteq S$ *means that* R, T *and* S *are rings with the same identity and that* S *is a liberal extension of* R *(and hence* T *is an intermediate extension of* R). That I is an ideal of the ring A is denoted by $I \triangleleft A$, and $I \triangleleft' A$ means that I is a prime ideal of A. Also, for any ring A, mip(A) is the set of all minimal prime ideals of A. Undefined radical theoretic terms may be found in [4] or [25].

1. INTRODUCTION

We shall make frequent use of the following fundamental results concerning the prime ideal structure of intermediate extensions.

THEOREM 1.1 (Robson [18]). *Let* $R \subseteq T \subseteq S = \sum_{i=1}^{n} Ra_i$.

(a) *If* $P_r \triangleleft' R$, *then there are ideals* $P_t \triangleleft' T$ *and* $P_s \triangleleft' S$ *such that* $P_s \cap R = P_t \cap R = P_r$, $P_s \cap T \subseteq P_t$ *and* P_t *is a minimal prime over* $P_s \cap T$.

(b) *If* $P_t \triangleleft' T$, *then there are ideals* $P_r \triangleleft' R$ *and* $P_s \triangleleft' S$ *such that* $P_s \cap R = P_t \cap R = P_r$, $P_s \cap T \subseteq P_t$ *and* P_t *is a minimal prime over* $P_s \cap T$.

(c) *If* R *and* S *are prime and* $P \triangleleft' T$ *such that* $P \cap R = 0$, *then* $P \in \text{mip}(T)$ *and* $I \cap R \neq 0$ *if* $P \subsetneq I \triangleleft T$. *Also, if* $Q \in \text{mip}(T)$, *then* $Q \cap R = 0$.

(d) *If* R *and* S *are prime, then* mip(T) *is finite and* $[\cap\{P: P \in \text{mip}(T)\}]^n = 0$.

The following lemma from [22] will also be needed.

LEMMA 1.2. *Let* $R \subseteq T \subseteq S = \sum_{i=1}^{n} Ra_i$ *with* R *prime. If* $I \triangleleft T$ and $I \cap R = 0$, *then the set* $\{a_1,\ldots,a_n\}$ *can be ordered so that there is an integer* k, $1 \leq k \leq n$, *such that* $(\sum_{i=1}^{k} Ra_i + I)/I$ *is a free* R-*module and there is an ideal* D *of* R *such that* $0 \neq D \subseteq TDT \subseteq DS \subseteq F + I$ *where* $F = \sum_{i=1}^{k} Ra_i$.

We note that if $S = \sum_{i=1}^{n} Ra_i$ is a liberal extension of R we may assume that $1 \in \{a_1,\ldots,a_n\}$ and that the order referred to in 1.2 is such that $a_1 = 1$.

It is clear that the equation $r(R) = R \cap r(T)$,

where T is an intermediate extension of R, does not hold for all (or even all special) radicals r. For example, let F be a field and K a finite extension field of F, $K \neq F$. Then K is an intermediate (in fact, liberal) extension of F but $r(F) \neq F \cap r(K)$ where r is the upper radical determined by $\{K\}$. For a second example, let r be the upper nil radical; that is, the upper radical determined by the class of prime rings without proper divisors of zero. If R is a ring with identity which has no proper divisors of zero and $n > 1$, then $M_n(R)$ is a liberal extension of R but $r(R) \neq R \cap r(M_n(R))$.

We now identify a property of classes of prime rings which guarantees that the corresponding upper radical does satisfy the equation under discussion.

A class $\mathcal{P}$ of prime rings is *intermediate* if whenever $R \subseteq T \subseteq S$ with R and S prime and $P \in \mathrm{mip}(T)$, then $R \in \mathcal{P}$ if and only if $T/P \in \mathcal{P}$. If A is a ring and $\mathcal{P}$ is a class of prime rings we shall denote the intersection $\cap\{I: I \triangleleft A \text{ and } A/I \in \mathcal{P}\}$ by $p(A)$. As usual, if A has no homomorphic images in $\mathcal{P}$, $p(A) = A$. Note that if $\mathcal{P}$ is a special class of prime rings, then p is a special radical, the upper radical determined by the class $\mathcal{P}$ [1].

THEOREM 1.3. *Let* $\mathcal{P}$ *be an intermediate class of prime rings and let* T *be an intermediate extension of* R. *Then* $p(R) = R \cap p(T)$ *and if* $R \subseteq T \subseteq S = \sum_{i=1}^{n} Ra_i$, *then* $p(T)^n \subseteq p(S)$.

Proof. Assume that T is an intermediate extension of R .

Let $P_r \triangleleft' R$ such taht $R/P_r \in \mathcal{P}$. From 1.1(a) we obtain ideals $P_t \triangleleft' T$ and $P_s \triangleleft' S$ such that $R/P_r \subseteq T/(P_s \cap T) \subseteq S/P_s$, and $P_t/(P_s \cap T) \in \text{mip}(T/(P_s \cap T))$ by 1.1(c). Thus, $(T/(P_s \cap T))/(P_t/(P_s \cap T) \cong T/P_t \in \mathcal{P}$ since $\mathcal{P}$ is intermediate. It follows that $p(T) \cap R \subseteq P_t \cap R \subseteq P_r$ and hence $p(T) \cap R \subseteq p(R)$. A similar argument, using 1.1(b) instead of 1.1(a), shows that $p(R) \subseteq p(T)$.

Now assume that $R \subseteq T \subseteq S = \sum_{i=1}^{n} Ra_i$ and let $P_s \triangleleft' S$ such that $S/P_s \in \mathcal{P}$. Since S is an intermediate extension of R , $P_r = P_s \cap R \triangleleft' R$ by 1.1(b) and so $R/P_r \in \mathcal{P}$ because $\mathcal{P}$ is intermediate. Applying 1.1(d) to $R/P_r \subseteq T/(P_s \cap T) \subseteq S/P_s$ we conclude that $\text{mip}(T/(P_s \cap T))$ is finite and that $\Delta^n \subseteq P_s \cap T$ where $\Delta/(P_s \cap T) = \cap\{P/(P_s \cap T) : P/(P_s \cap T) \in \text{mip}(T/(P_s \cap T))\}$. Since $\mathcal{P}$ is intermediate and $R/P_r \in P$, $p(T) \subseteq \Delta$. Thus, $p(T)^n \subseteq P_s$ and hence $p(T)^n \subset p(S)$.

2. INTERMEDIATE CLASSES OF PRIME RINGS

In this section we show that several well-known special classes $\mathcal{P}$ of prime rings are intermediate and so the corresponding radicals p satisfy the equation $p(R) = R \cap p(T)$ whenever T is an intermediate extension of R. Of course the class of all prime rings is an intermediate class and in [18] Robson shows that the class of right primitive rings is intermediate. He also gives the equation $J(R) = R \cap J(T)$ where J is the Jacobson radical. Since matrix rings provide examples of liberal extensions, the equation $\mathcal{N}(R) = R \cap \mathcal{N}(T)$ where $\mathcal{N}$ is the nil radical would imply the correctness of the Koethe conjecture. In contrast, it is straightforward to establish the equation for the locally nilpotent radical.

PROPOSITION 2.1. *The class of prime rings without nonzero locally nilpotent ideals is intermediate.*

Proof. Let $R \subseteq T \subseteq S$ with R and S prime and let $P \in \text{mip}(T)$.

First assume that $0 \neq I/P \triangleleft T/P$ and that I/P is locally nilpotent. If F is a finitely generated subring of $I \cap R$, then $F^k \subseteq P$ for some positive integer k

and since $P \cap R = 0$ by 1.1(c) it follows that $F^k = 0$. Hence $I \cap R$ is a locally nilpotent ideal of R and $I \cap R \neq 0$ by 1.1(c).

Conversely, assume that $0 \neq J \triangleleft R$ and that J is locally nilpotent. Then JS is a locally nilpotent (two-sided) ideal of S and so $JS \cap T$ is a locally nilpotent ideal of T. Furthermore, $JS \cap T \not\subseteq P$ because $J \subseteq JS \cap T \cap R$ whereas $P \cap R = 0$. It follows that T/P contains a nonzero locally nilpotent ideal.

Two further examples of intermediate classes can be obtained from the results in [22]. First, [22, Theorem 3.3] implies that the class $\mathcal{P}$ of prime right Goldie rings is intermediate. We note that it is straightforward to verify that this class is also special. The class of prime right Johnson rings is not only special but also normal [13]. This class is intermediate as well [22, Corollary 3.5].

We now show that the equation $p(R) = R \cap p(T)$ is satisfied by the Ortiz radical [14]. This radical is special (indeed, normal [13]) and is the upper radical determined by the class $\mathcal{P}$ of (right) K-primitive rings. A prime ring A is (right) K-*primitive* [14] if A has a faithful module M_A such that $0 \neq N_A \subseteq M_A$ implies that $MI \subseteq N$ for some $0 \neq I \triangleleft A$ (equivalently, A has a right ideal K maximal with respect to not containing a nonzero (two-sided) ideal of A [9]).

PROPOSITION 2.2. *Let* $R \subseteq T \subseteq S$ *with* R *and* S *prime and let* $P \in \mathrm{mip}(T)$.

(a) R *has a prime faithful uniform module if and only if* T/P *has a prime faithful uniform module.*

(b) R *is* K-*primitive if and only if* T/P *is* K-*primitive.*

Proof. Recall that a right A-module M_A is *prime faithful* if for $m \in M$ and $J \triangleleft A$, $mJ = 0$ implies that $m = 0$ or $J = 0$ and that M_A is *uniform* if any two nonzero submodules of M_A intersect nontrivially.

(a) Assume that M is a prime faithful uniform right T/P -module. Propositions 2.2(b) and 2.3 from [22] imply that M_R is prime faithful of finite uniform dimension. Thus M_R contains a nonzero uniform R submodule which is, of course, prime faithful.

Conversely, assume that R has a prime faithful uniform right module. Then R has a cyclic module of this type so we may assume that $(R/K)_R$ is prime faithful and uniform for some right ideal K of R . Let $S = \sum_{i=1}^{n} Ra_i$ with $a_1 = 1$. In view of 1.1 and 1.2 we may assume that there is an integer k, $1 \leq k \leq n$, such that $\sum_{i=1}^{k} r_i a_i \in P$ with all $r_i \in R$ implies that $r_i = 0$ for $i = 1,\ldots,k$ and that there is an ideal D of R such that $0 \neq DS \subseteq F + P$ where $F = \sum_{i=1}^{k} Ra_i$.

Let $r \in (KT+P) \cap R$. Then $r \in KS + P$, so $r = \sum_{i=1}^{n} k_i a_i + p$ where $k_i \in K$ for all i and $p \in P$. Thus, for $d \in D$, $rd = \sum_{i=1}^{n} k_i da_i + pd$ and since $da_i \in DS$ there are elements $r_{ij} \in R$ and $p_i \in P$ such that $da_i = \sum_{j=1}^{k} r_{ij} a_j + p_i$. Hence $rd = \sum_{j=1}^{k} \sum_{i=1}^{n} k_i r_{ij} a_j + p'$ where $p' \in P$. Since $a_1 = 1$ this implies that $rd = \sum_{i=1}^{n} k_i r_{i1} \in K$ so we conclude that $rD \subseteq K$. Since R/K is a prime faithful right R-module, $r \in K$ so that $(KT+P) \cap R = K$. Thus we may choose a right ideal M of T, $M \supseteq P$, maximal with respect to $M \cap R = K$.

Since $(R/K)_R$ is uniform, the maximality of M implies that T/M is a uniform T/P-module and so the proof will be complete once we check that T/M is a prime faithful T/P-module. Let $t \in T$ and let I/P be a nonzero ideal of T/P such that $tI \subseteq P$. Then $tTI \subseteq P$, so $((tT+M) \cap R)C \subseteq K$ where $0 \neq C = I \cap R \triangleleft R$ by 1.1(c). Since $(R/K)_R$ is prime faithful, this implies that $(tT+M) \cap R \subseteq K$ and so by the maximality of M, $t \in M$ as required.

(b) Assume that M is a faithful right T/P-module and that for each nonzero T/P-submodule N of M there is an I, $P \subsetneq I \triangleleft T$ such that $MI \subseteq N$. Since the module M is necessarily prime faithful and uniform, it follows from the first part of the proof of (a) that $M_R \supseteq N_1 \oplus \dots \oplus N_m$ where each N_i is a prime faithful uniform right R-module and $\sum_{i=1}^{m} N_i$ is essential in M_R. Let Δ be a nonzero R-submodule of N_1. Then $E = \Delta \oplus N_2 \oplus \dots \oplus N_m$ is an essential submodule of M_R and so from [22, Proposition 2.4] we see that E contains a nonzero T/P-submodule X. Hence there is an I, $P \subsetneq I \triangleleft T$, such that $MI \subseteq X$. So certainly $(N_1 \oplus \dots \oplus N_m)(I \cap R) \subseteq E$ and because the sum is direct, $N_1(I \cap R) \subseteq \Delta$. Since $I \cap R \neq 0$ by 1.1(c), this shows that R is K-primitive.

Note that the above proposition shows that the class of rings with a prime faithful uniform module, as well as the class of K-primitive rings, is intermediate.

Since the Brown-McCoy radical is the upper radical determined by the class of prime simple rings with identity and the anti-simple radical β_ϕ is the upper radical determined by the class of subdirectly irreducible prime rings, the following proposition shows that the equation $r(R) = R \cap r(T)$ holds for these radicals. For any ring A, we denote the intersection of all nonzero ideals of A by $H(A)$.

PROPOSITION 2.3. *Let* $R \subseteq T \subseteq S$ *with* R *and* S *prime and let* $P \in \text{mip}(T)$.

(a) $H(R) = R \cap H_p$ *where* $H(T/P) = H_p/P$ *and so the class of prime subdirectly irreducible rings is intermediate.*

(b) R *is simple if and only if* T/P *is simple and so the class of prime simple rings is intermediate.*

Proof. (a) If $P \subsetneq I \lhd T$, then 1.1(c) implies that $0 \neq I \cap R \lhd R$. It follows that $H(R) \subseteq H_p$.

For the converse, let $S = \sum_{i=1}^{n} Ra_i$ where $a_1 = 1$ and let k, D and F be as in the proof of 2.2(a). Suppose that $0 \neq K \lhd R$. Then since $DS \subseteq F + P$, $KDS \subseteq \sum_{i=1}^{k} Ka_i + P$. Also, $KDS \cap T \lhd T$ and $KDS \cap T \not\subseteq P$ because $P \cap R = 0$ by 1.1(c) but $KD \subseteq KDS \cap T \cap R$ and $KD \neq 0$ since R is prime. Thus $H_p \subseteq KDS \cap T + P \subseteq \sum_{i=1}^{k} Ka_i + P$ so if $r \in R \cap H_p$,

$r = \sum_{i=1}^{k} k_i a_i + p$ where each $k_i \in K$ and $p \in P$.

Since $a_1 = 1$ this implies that $r = k_1 \in K$ so that $R \cap H_p \subseteq H(R)$.

(b) This follows immediately from (a) since a ring A is simple if and only if $H(A) = A$.

The class $\mathcal{P}$ of right primitive rings with nonzero socle is a normal class [13] and, since the class of right primitive rings is intermediate, the following proposition shows that $\mathcal{P}$ is intermediate. For any ring A , $\Sigma(A)$ denotes the right socle of A .

PROPOSITION 2.4. *If* $R \subseteq T \subseteq S$ *with* R *and* S *prime and* $P \in \mathrm{mip}(T)$ *, then* $\Sigma(R) = R \cap \Sigma_p$ *where* $\Sigma(T/P) = \Sigma_p/P$.

Proof. Let $S = \sum_{i=1}^{n} Ra_i$. If $r \in \Sigma(R)$, then R has minimal right ideals K_j , $j = 1,\ldots,m$, such that $r \in K = \sum_{j=1}^{m} K_j$. Now $(K_j a_i)_R$ is simple for all i and j , so $(KS)_R = \sum_{i=1}^{n} \sum_{j=1}^{m} K_j a_i$ is semisimple of finite length. Thus $(KS \cap T)_T$, and hence $((KS \cap T + P)/P)_T$, have finite length. Since T/P is (semi)prime, minimal ideals are generated by idempotents and so, as a T/P-module, $(KS \cap T + P)/P$ is semisimple of finite length. Thus $(KS\cap T+P)/P$ is contained in

$\Sigma(T/P)$ and so $\Sigma(R) \not\subseteq \Sigma_p$.

Conversely, assume that $r \in R \cap \Sigma_p$. Then $r + P \in A/P$ where A/P is semisimple of finite length as a right T/P-module. By [18, Theorem 4.5], $(A/P)_R$ is semisimple of finite length and so the same is true of $rR \cong (rR+P)/P \subseteq A/P$. Thus $r \in \Sigma(R)$.

3. RELATIVE COMPLEMENTS

THEOREM 3.1. *If P_1 and P_2 are intermediate classes of prime rings, then so is $P = \{R : R \in P_1$ and $p_2(R) \neq 0\}$.*

Proof. Let $R \subseteq T \subseteq S = \sum_{i=1}^{n} Ra_i$ with R and S prime and let $P \in \text{mip}(T)$.

If $R \in P$, then certainly $T/P \in P_1$. By 1.3, $p_2(R) = R \cap p_2(T)$ and since $p_2(R) \neq 0$, $p_2(T) \not\subseteq P$ by 1.1(c). Thus, $p_2(T/P) \neq 0$ and so $T/P \in P$.

Conversely, assume that $T/P \in P$. From 1.1(d) we know that $\text{mip}(T)$ is finite. Let $P = P_1,\ldots,P_k$ be the distinct minimal prime ideals of T and let $B_i/P_i = p_2(T/P_i)$ for $i = 1,\ldots,k$. Then $p_2(T) = \bigcap_{i=1}^{k} B_i$ because each prime ideal of T contains one of $P_1,\ldots,P_k$. Since the minimal primes P_i are incomparable, $B_i \not\subseteq P_1$ for $i > 1$, and $T/P \in P$ implies that $B_1 \neq P_1$. Thus $p_2(T) \not\subseteq P_1$. In particular, $p_2(T)^n \neq 0$ and so by 1.3, $p_2(S) \neq 0$. Using 1.3 again as well as 1.1(c) we see that $p_2(R) = R \cap p_2(S) \neq 0$. Hence $R \in \mathcal{P}$ since it is clear that $T/P \in \mathcal{P}$ implies that $R \in \mathcal{P}_1$.

We note that it is straightforward to verify that if $\mathcal{P}_1$ and $\mathcal{P}_2$ are special, then $\mathcal{P}$ is also special and the associated radical p is the p_1-relative complement of p_2 ; that is, $p = (p_1 : p_2)$ in the notation of [5], [10] and [20].

COROLLARY 3.2. *If* T *is an intermediate extension of* R *, then* T *is a Jacobson ring (strong Jacobson ring, Brown-McCoy ring) if and only if* R *is a Jacobson ring (strong Jacobson ring, Brown-McCoy ring).*

Proof. A ring A is a Jacobson ring if and only if $p(A) = A$ where $\mathcal{P} = \{R : R \in \mathcal{P}_1$ and $p_2(R) \neq 0\}$ $\mathcal{P}_1$ is the class of all prime rings and $\mathcal{P}_2$ is the class of right primitive rings. The above theorem shows that $\mathcal{P}$ is intermediate and so T is a Jacobson ring if and only if R is a Jacobson ring because $p(R) = R \cap p(T)$ by 1.3.

The proof for the other two cases is similar because

a ring A is strong Jacobson [12] if and only if $p(A) = A$ where P, P_1 and P_2 are as above except that P_2 is the class of finite simple rings, and a ring A is Brown-McCoy [24] if and only if $p(A) = A$ where P, P_1 and P_2 are as above except that P_2 is the class of all simple rings with identity.

REFERENCES

[1] V.A. Andrunakievitch, "Radicals of associative rings I", *Mat. Sbornik* 44 (1958), 179-212.

[2] J. Bit-David, "Normalizing extensions II", *Proc. Ring Theory Conference*, pp. 6-9. Lecture Notes in Mathematics, 825, Springer, Berlin, 1980.

[3] J. Bit-David and J.C. Robson, "Normalizing extensions I", *Proc. Ring Theory Conference*, pp. 1-5. Lecture Notes in Mathematics, 825, Springer, Berlin, 1980.

[4] N.J. Divinsky, *Rings and Radicals*, Univ. of Toronto Press, 1965.

[5] N.J. Divinsky and A. Sulinski, "Radical Pairs", *Canad. J. Math.* 29(1977), 1086-1091.

[6] S. Jabbour, "Intermediate normalizing extensions", preprint.

[7] A.G. Heinicke and J.C. Robson, "Normalizing extensions: prime ideals and incomparability", *J.of Algebra* 72(1981), 237-268.

[8] A.G. Heinicke and J.C. Robson, "Normalizing extensions: nilpotency", *J. of Algebra* 76(1982), 459-470.

[9] T.P. Kezlan, "On K-primitive rings", *Proc. Amer. Math. Soc.* 74(1979), 24-28.

[10] W.G. Leavitt and J. Watters, "Special closure, M-radicals, and relative complements", *Acta Math. Acad. Sci. Hung.* 28(1976), 55-67.

[11] M. Lorenz, "Finite normalizing extensions of rings", *Math. Z.* 176(1981), 447-484.

[12] M. Lorenz and G. Michler, "On maximal ideals in Ore extensions", *Ring Theory, Proc. of the 1978 Antwerp Conference,* pp. 139-151. Lecture Notes in Pure and Applied Mathematics, 51, Marcel Dekker, New York, 1979.

[13] W.K. Nicholson and J.F. Watters, "Normal classes of prime rings determined by modules", *Proc. Amer. Math. Soc.* 83 (1981), 27-30.

[14] A.H. Ortiz, "On the structure of semiprime rings", *Proc. Amer. Math. Soc.* 38(1973), 22-26.

[15] R. Paré and W. Schelter, "Finite extensions are integral", *J. Algebra* 53 (1978), 477-479.

[16] D.S. Passman, "Prime ideals in normalizing extensions", *J. Algebra* 73 (1981), 556-572.

[17] R. Resco, "Radicals of finite normalizing extensions", *Comm. Algebra* (1981), 713-725.

[18] J.C. Robson, "Prime ideals in intermediate extensions", *Proc. London Math. Soc.* (3), 44(1982), 372-384.

[19] J.C. Robson and L.W. Small, "Liberal extensions", *Proc. London Math. Soc.* (3), 42(1981), 87-103.

[20] R.L. Snider, "Lattices of radicals", *Pacific J. Math.* 40(1972), 207-220.

[21] P.N. Stewart, "Nilpotence and normalizing extensions", *Comm. Algebra,* to appear.

[22] P.N. Stewart, "Properties of intermediate extensions", *Comm. Algebra,* to appear.

[23] P.N. Stewart and J.F. Watters, "Properties of normalizing extensions and fixed rings", *Comm. Algebra*, to appear.

[24] J.F. Watters, "The Brown-McCoy radical and Jacobson rings", *Bull. Acad. Polon. Sci., Ser. Sci. Math. Astronom. Phy.* 24(1976), 91-99.

[25] R. Wiegandt, *Radical and Semisimple Classes of Rings*, Queens Papers in Pure and Applied Mathematics, No. 37, Queens University, 1974.

P. N. Stewart
Department of Mathematics, Statistics and Computing Science
Dalhousie University
Halifax, Nova Scotia
Canada B3H 4H8

COLLOQUIA MATHEMATICA SOCIETATIS JÁNOS BOLYAI
38. RADICAL THEORY, EGER (HUNGARY), 1982.

RADICALS OF RINGS GRADED BY ABELIAN GROUPS

A. SULIŃSKI AND R. WIEGANDT

Let G be an abelian group and let T_g, $g\in G$ be a family of abelian groups. The set T of all homogeneous elements of the direct sum $\bigoplus_{g\in G} T_g$ (i.e. the elements from T_g, $g\in G$) will be called a G-graded abelian group. If $0\neq x\in T_g$, $g\in G$ then g will be called the degree of x, in which case we will write

$$|x| = g .$$

Let us agree that the zero element $0\in T$ has no degree.

Let T and U be two G-graded abelian groups. An additive map $f:T\to U$ will be called a G-graded map of degree $k\in G$, if for every $g\in G$ $f(T_g)\subseteq U_{g+k}$; in this case we shall write $|f|=k$. For $x\in T$ we have $|f(x)| = |f|+|x|$ if $f(x)\neq 0$. The set $\mathrm{Hom}(T,U)$ of all G-graded maps is of course a G-graded group.

We shall say that a G-graded abelian group R is a *G-graded ring* if an associative multiplication is defined for homogeneous elements from R such that

$$R_g R_h \subseteq R_{g+h} \quad \text{for all} \quad g,h \in G$$

and this multiplication is right and left distributive with respect to the addition of elements of the same degree. Rings graded by the two element group are called 2-graded rings [4].

Let R and S be two G-graded rings. A G-graded map $f:R \to S$ will be called a G-graded homomorphism if

$$f(xy) = f(x)f(y) \quad \text{for all} \quad x,y \in R .$$

PROPOSITION 1. *Let R and S be G-graded rings and $f:R \to S$ a G-graded homomorphism of R onto S. If $|f| \neq 0$ then $S_g S_h = 0$ for all $g,h \in G$.*

Proof. Suppose that $f(x)f(y) \neq 0$ for some $x,y \in R$. Then $f(x)f(y)$ has a degree and $|f(x)f(y)| = |f(xy)|$, whence $|f|+|x|+|f|+|y| = |f|+|x|+|y|$. Therefore $|f|=0$, which is impossible.

An additive subgroup A of a G-graded ring R is a G-graded ideal of R if

$$R_g A_h \, , \; A_h R_g \subseteq A_{g+h}$$

for all $g,h \in G$. If A is a G-graded ideal of a G-graded ring R then we can construct the G-graded factor ring R/A defining the multiplication of cosets as follows:

$$(x + A_{|x|})(y + A_{|y|}) = xy + A_{|x|+|y|}$$

for $x,y \in R$.

Let π be a property of G-graded rings. A G-graded ring with property π will be called a G-graded π-ring. A G-graded ideal A of a G-graded ring R which as a G-graded ring is a π-ring, will be called a G-graded π-ideal of R. If $\pi(R)$ is a G-graded π-ideal of a G-graded ring R containing all G-graded π-ideals of R, then $\pi(R)$ is called the *G-graded π-radical of* R. And finally a G-graded ring containing no non-zero G-graded π-ideals will be called a *G-graded π-semisimple ring.*

A property π of G-graded rings will be called a *G-graded radical property* if the following conditions are satisfied:

(i) every G-graded homomorphic image of a G-graded π-ring is again a G-graded π-ring;

(ii) every G-graded ring contains a G-graded π-radical;

(iii) for every G-graded ring R the factor ring $R/\pi(R)$ is G-graded π-semisimple.

We prove the following G-graded version of the hereditariness theorem from [1] and [4].

THEOREM 1. *If A is a G-graded ideal of a G-graded ring R, then for any G-graded radical property π, $\pi(A)$ is a G-graded ideal of R. In particular, if R is π-semisimple, then so is any G-graded ideal A of R.*

Proof. Suppose that for some G-graded radical property π the radical $M=\pi(A)$ is not a G-graded ideal of R. Then there are an $x\in R$ and a $g_0\in G$ such that either $xM_{g_0}+M_{|x|+g_0}$ or $M_{g_0}x+M_{g_0+|x|}$ properly contains $M_{|x|+g_0}$, say $xM_{g_0}+M_{|x|+g_0}\neq M_{|x|+g_0}$ – the other case can be treated analogously. We shall show that the family $xM_{g-|x|}+M_g$, $g\in G$, which will be denoted shortly by $xM+M$, is a G-graded ideal of A. Indeed, for every $g,h\in G$ we have

$$A_h(xM_{g-|x|}+M_g)\subseteq A_h\cdot xM_{g-|x|}+A_hM_g\subseteq$$

$$\subseteq A_hx\cdot M_{g-|x|}+M_{h+g}\subseteq A_{h+|x|}M_{g-|x|}+M_{h+g}\subseteq$$

$$\subseteq M_{h+g}\subseteq xM_{h+g-|x|}+M_{h+g}$$

and

$$(xM_{g-|x|}+M_g)A_h \subseteq xM_{g-|x|}\cdot A_h + M_gA_h \subseteq$$

$$\subseteq x\cdot M_{g-|x|}A_h + M_{g+h} \subseteq xM_{g+h-|x|} + M_{g+h} .$$

Since $xM+M$ properly contains M, $\frac{xM+M}{M}$, i.e. the family $\frac{xM_{g-|x|}+M_g}{M_g}$, $g\in G$, is a non-zero ideal of $A/_M$.

Now define the map $f\colon M \to \frac{xM+M}{M}$ by putting

$$f(m) = xm + M_{|x|+|m|}$$

for all $m\in M$. This is of course a G-graded map of M onto $\frac{xM+M}{M}$, and $|f|=|x|$. We shall show that f is a G-graded homomorphism. Indeed, for every $m,n\in M$ we have

$$x\cdot mn = xm\cdot n \in A_{|x|+|m|}M_{|n|} \subseteq M_{|x|+|m|+|n|} ,$$

whence

$$f(mn) = x\cdot mn + M_{|x|+|m|+|n|} = 0 + M_{|x|+|m|+|n|} ,$$

which means that $f(mn)$ is the zero coset. On the other hand,

$$xm\cdot xn = xmx\cdot n \in A_{|x|+|m|+|x|}M_{|n|} \subseteq M_{2|x|+|m|+|n|} ,$$

whence

$$f(m)f(n) = (xm+M_{|x|+|m|})(xn+M_{|x|+|n|}) =$$

$$= xm\cdot xn + M_{2|x|+|m|+|n|} = 0 + M_{2|x|+|m|+|n|} ,$$

which means that $f(m)f(n)$ is again the zero coset. Therefore $f(mn) = f(m)f(n)$, i.e., f is a G-graded homomorphism.

But $\frac{xM+M}{M}$ as a G-graded homomorphic image of a G-graded π-ring M is again a G-graded π-ring. Therefore the G-graded π-semisimple ring A/M contains a non-zero G-graded π-ideal $\frac{xM+M}{M}$, which is impossible.

Next we shall prove a characterization of G-graded semisimple properties which is analogous to Sands' characterization given for associative rings in [3]. If π is a G-graded radical property, then the class of all π-semisimple G-graded rings defines the *G-graded semisimple property* σ of π . If $\pi(R)=0$ for some G-graded ring R , that is, R has property σ , then we shall write briefly $R\in\sigma$.

PROPOSITION 2. *A property σ of G-graded rings is the G-graded semisimple property of some G-graded radical property π if and only if σ satisfies the following conditions:*

(a) *if* $R\in\sigma$ *, then every non-zero* G*-graded ideal* A *of* R *has a non-zero* G*-graded homomorphic image* B *such that* $B\in\sigma$ *;*

(b) *if every non-zero* G*-graded ideal* A *of a* G*-graded ring* R *has a non-zero* G*-graded homomorphic image* B *such that* $B\in\sigma$ *, then also* $R\in\sigma$ *holds.*

The proof is the standard ring-theoretic proof (see, for instance, [2] Theorem 2).

PROPOSITION 3. *If* σ *is a* G*-graded semisimple property, then*

(i) σ *is hereditary:* $R\in\sigma$ *implies* $A\in\sigma$ *for every* G*-graded ideal* A *of any* G*-graded ring* R *,*

(ii) σ *has the coinductive property* (which is weaker than being subdirectly closed)*: if* $A_1\supseteq\ldots\supseteq A_\lambda\supseteq\ldots$ *is a descending chain of* G*-graded ideals of a* G*-graded ring* R *such that* $R/A_\lambda\in\sigma$ *for each index* λ *, then also* $R/\cap A_\lambda$ *holds,*

(iii) σ *is closed under extensions: if* A *is a* G*-graded ideal of the* G*-graded ring* R *such that* $A\in\sigma$ *and* $R/A\in\sigma$ *, then also* $R\in\sigma$ *.*

The first assertion follows immediately from Theorem 1, while the further statements can be proved exactly as their ring-theoretic versions ([5] Theorem 22.8 and

Proposition 8.5).

THEOREM 2. σ *is a* G*-graded semisimple property with respect to some* G*-graded radical property if and only if*

(i) σ *is hereditary,*

(ii) σ *has the coinductive property,*

(iii) σ *is closed under extensions.*

Proof. The necessity is the assertion of Proposition 3. In view of Proposition 2, for the sufficiency it is enough to prove the validity of condition (b), as every hereditary property satisfies condition (a). Suppose that $R\neq 0$ is a G-graded ring which satisfies the assumptions of condition (b). Then R, as a G-graded ideal of itself, has a non-zero G-graded homomorphic image $R/A\in\sigma$. In view of (ii) the G-graded ideal A of R can be chosen so that A be minimal with respect to the property $R/A\in\sigma$. If $A=0$ then we are done. Suppose that $A\neq 0$. Then by the hypothesis on R, A has a non-zero G-graded homomorphic image $A/M\in\sigma$, and as previously, M can be chosen to be minimal with respect to the property $A/M\in\sigma$. We claim that M is a G-graded ideal of R. If M is not a G-graded ideal of R, then we may confine ourselves to the case

$xM + M \neq M$ for some $x \in R$. As we have seen in the proof of Theorem 1, the map $f: M \to xM + M/M$ defined by

$$f(m) = xm + M_{|x|+|m|}$$

is a G-graded homomorphism and $xM + M/M$ is a G-graded ideal of A/M . Let us consider

$$K = \ker f = \{m \in M : xm \in M_{|x|+|m|}\} .$$

We shall show that K is a G-graded ideal of A . Indeed, for every $g,h \in G$, $A_h \subseteq A$ and $K_g \subseteq K$ we have

$$x(A_h K_g) \subseteq A_{h+|x|} K_g \subseteq M_{h+g+|x|}$$

and

$$x(K_g A_h) \subseteq K_{g+|x|} A_h \subseteq M_{g+h+|x|} ,$$

which implies

$$A_h K_g \subseteq K_{h+g} \text{ and } K_g A_h \subseteq K_{g+h} ,$$

that is, K is a G-graded ideal of A . Hence we have

$$\frac{A/K}{M/K} \cong A/M \in \sigma .$$

Since $M/K \cong xM + M/M$ and $xM + M/M$ is a G-graded ideal of A/M , the hereditariness of σ implies $M/K \in \sigma$.

Applying (iii) we get $A/K \in \sigma$, which yields $M \subseteq K$ by the minimality of M . Thus we arrived at the contradiction

$$0 = M/K \cong xM + M/M \neq 0 ,$$

therefore M is a G-graded ideal of R .

Now we have

$$\frac{R/M}{A/M} \in \sigma \quad \text{and} \quad A/M \in \sigma .$$

Hence (iii) implies $R/M \in \sigma$. Consequently, by the choice of A it follows $A \subseteq M$. Thus $A=M$, contradicting $A/M \neq 0$.

In the case of rings graded by a non-abelian group G it is not clear how to define G-graded maps, i.e. the degree of a map. One can say that an additive map f of a G-graded ring R into a G-graded ring S has the right degree $k \in G$ if $f(R_g) \subseteq S_{gk}$ for all $g \in G$. Analogously one can say that $l \in G$ is the left degree of f if $f(R_g) \subseteq S_{lg}$ for all $g \in G$. The two one-sided degrees are not necessarily equal. Indeed, let $p,q \in G$ with $pq \neq qp$. An additive map $0 = f' : R_p \to S_{pq}$ can be extended to an additive map $f: R \to S$ by setting $f(x) = 0$ for $x \in R_g$, $g \in G$, $g \neq p$. Since $S_{pq} = S_{rp}$ where

$r = pqp^{-1} \neq q$, the left degree of f is r while the right degree of f equals $q \neq r$.

REFERENCES

[1] T. Anderson, N. Divinsky and A. Suliński, Hereditary radicals in associative and alternative rings, *Canad. J. Math.* 17 (1965), 594-603.

[2] N. Divinsky, *Rings and radicals*, George Allen & Unwin Ltd, 1965.

[3] A. D. Sands, Strong upper radicals, *Quart. J. Math., Oxford*, 27 (1976), 21-24.

[4] A. Suliński, Radicals of associative 2-graded rings, *Bull. Acad. Polon. Sci.* 29 (1981), 431-434.

[5] R. Wiegandt, *Radical and semisimple classes of rings*, Queen's papers in pure & appl. math., no. 37, Kingston, 1974.

Adam Suliński
Institute of Mathematics
Warsaw University
Warsaw, Poland

Richárd Wiegandt
Mathematical Institute
Hungarian Academy of Sci.
Budapest, Reáltanoda u. 13-15
H-1053, Hungary

COLLOQUIA MATHEMATICA SOCIETATIS JÁNOS BOLYAI
38. RADICAL THEORY, EGER (HUNGARY), 1982.

ON J-EQUIRADICAL RINGS*)

A. V. TISHCHENKO

This note is motivated by the general question, in what cases the multiplicative structure determines some properties of a ring. We shall consider this question relative to the Jacobson radical. Let R be an associative ring and $J(R)$ be its Jacobson radical. Hoehnke [2] considered the notion of a faithful irreducible right S-act over a semigroup S and constructed a radical $\operatorname{rad} S$ for semigroups analogously to the ring Jacobson radical. Roiz [4] introduced the notion of J-equiradical rings as follows. Let $R^{\cdot}$ be the multiplicative semigroup of a ring R and let $J(R)_{\text{con}}$ denote the ring congruence defined by the ideal $J(R)$. For every ring R the inclusion $\operatorname{rad} R^{\cdot} \subseteq J(R)_{\text{con}}$ holds (see [4]). A ring R is called J-equiradical if

---

*) This paper was written during the author's visit in Hungary in August 1982.

rad $R^{\bullet} = J(R)_{\text{con}}$. Roiz gave an inner characterization for J-equiradical rings. In the present note we give sufficient conditions for the Jacobson radical to have a complement in the lattice of all ideals (or right ideals) of a J-equiradical ring. In Theorem 3 an example is constructed which shows that such a complement of the radical need not exist in a J-equiradical ring.

We have the following characterization of J-equiradical rings, due to Roiz [4]:

THEOREM A. *A ring* R *is* J*-equiradical iff for every* $r \in R$ *and* $Z \in J(R)$ *there exists a natural number* n *such that* $s^n Z = 0$.

This theorem has several corollaries. For example, the Jacobson radical of a J-equiradical ring is a nil-ideal. Any semisimple ring and any nil-ring are J-equiradical. Therefore we see that the class of all J-equiradical rings is wide enough.

THEOREM 1. *Let* R *be a* J*-equiradical ring and* $\bar{R} = R/J(R)$ *be generated as a right ideal by a countable set of pairwise commuting idempotents. Then* R *decomposes into a direct sum* $R = J(R) \oplus T$ *of right ideals, where* T *is a semisimple ring. Conversely,* *any ring R which decomposes into the direct sum of a nil-ideal N and a semisimple ring T both considered as right-ideals, is J-equiradical.*

To begin with, we shall prove some lemmas.

LEMMA 1. *If a ring R is J-equiradical and e is an idempotent of R then* $eR\cap J(R)=eJ(R)=0$.

Proof. This follows immediately from Theorem A.

It is known that for each pair of commuting idempotents e_1, e_2 we have $e_1R+e_2R=fR$, where $f=e_1+e_2-e_2e_1$ and $f^2=f$. Thus a right ideal of R generated by a finite set of pairwise commuting idempotents is generated by one idempotent.

LEMMA 2. *If R is a J-equiradical ring and $T=ER$ is a right ideal generated by a set E of pairwise commuting idempotents then* $T\cap J(R)=T\cdot J(R)=0$.

Proof. Let $z\in T\cap J(R)$. Then $z=e_1r_1+\ldots+e_nr_n\in e_1R+\ldots+e_nR=Q$ for some $e_1,\ldots,e_n\in E$. As we mentioned above, the right ideal Q is generated by an idempotent g . According to Lemma 1, $z\in gR\cap J(R)=0$.

LEMMA 3 (process of orthogonalization). *Let E_1 be a countable set of pairwise commuting idempotents of a ring R . Then there exists a countable set E of orthogonal idempotents such that* $ER=E_1R$.

Proof. Let $E_1 = \{e_1, e_2, \ldots, e_n, \ldots\}$. Define $g_1 = e_2$, $g_2 = e_2 - e_2 g_1, \ldots, g_n = e_n - e_n g_1 - \ldots - e_n g_{n-1}$. By induction on k we can prove that: 1) $g_k g_i = 0$ $(i<k)$; 2) $g_k^2 = g_k$; 3) $g_1 R \oplus \ldots \oplus g_k R = e_1 R + \ldots + e_k R$. Indeed, for $k=2$ the equalities 1)-3) are immediately verified. Suppose that the equalities 1)-3) are true for any $k \leq n-1$. Then $g_n g_i = (e_n - e_n g_1 - \ldots - e_n g_{n-1}) g_i = e_n g_i - e_n g_i = 0$. From the definition of g_n it follows that $g_n \in e_n R$. Hence $g_1 R \oplus \ldots \oplus g_n R \subseteq e_1 R + \ldots + e_n R$. Conversely, $e_n = g_n + g_1 e_n + \ldots + g_{n-1} e_n \in g_1 R + \ldots + g_n R$. Therefore, $g_1 R \oplus \ldots \oplus g_n R = e_1 R + \ldots + e_n R$.

Proof of Theorem 1. According to the conditions of the theorem, $\bar{R} = \bar{E}_1 \bar{R}$, where $\bar{E}_1$ is a countable set of pairwise commuting idempotents. By Lemma 3 $\bar{R} = \bar{E}\bar{R}$, where $\bar{E}$ is a countable set of orthogonal idempotents. Then $\bar{E}$ can be lifted to a countable set E of orthogonal idempotents in the ring R (see [3], §3.8). Now we have $R = ER + J(R)$. By Lemma 2 $ER \cap J(R) = 0$. Thus $R = J(R) \oplus ER$ is the direct sum of the right ideals $J(R)$ and ER . Since $ER \cong R/J(R)$, the ring ER is semisimple.

Conversely, let $R = N \oplus T$, where N is a nil-ideal and T is a semisimple ring, both of them right ideals of R . We assert that the condition in Theorem A holds in the ring R . Indeed, let $r=b+z$ be an element of the ring R , $b\in T$, $z\in N$. It is easy to verify that $(b+z)^n = b^n+zb^{n-1}+\ldots+z^{n-1}b+z^n$. There exists a natural number n such that $z^n=0$ because N is a nil-ideal. Hence for each $u\in N$ we get $(b+z)^n u = (b^n+zb^{n-1}+\ldots+ z^{n-1}b+z^n)u\in TN+RTN\subseteq T\cap N+R(T\cap N) = 0$.

COROLLARY. *Let* R *be a* J*-equiradical ring and let the factor-ring* $\bar{R}=R/J(R)$ *have an identity. Then* R *decomposes into a direct sum* $R=J(R)\oplus T$ *of the radical* $J(R)$ *and a right ideal* T *which is a semi-simple ring with identity.*

Let us remind the reader that the left sided analogue $\mathrm{rad}_l\, S$ of the radical $\mathrm{rad}\, S$ does not coincide with $\mathrm{rad}\, S$, as it takes place for the ring Jacobson radical.

Likewise one can define J_l-equiradical rings.

THEOREM 2. *Let a ring* R *be* J*-equiradical and* J_l*-equiradical and* $\bar{R}=R/J(R)$ *be generated as a right ideal by a set of pairwise commuting idempotents. Then the ring* R *decomposes into a direct sum* $R=J(R)\oplus T$, *where the ideal* T *is a semisimple ring.*

Denote by $\bar{x}$ the element $x+J(R)$ of the factor-ring $\bar{R}=R/J(R)$.

LEMMA 4. *Let a ring* R *be* J*-equiradical and* J_l*-equiradical. If* $\bar{E}$ *is a set of pairwise commuting idempotents in* $\bar{R}$ *and* $E=\{e\in R \mid e^2=e \text{ and } \bar{e}\in\bar{E}\}$ *then the elements of* E *are also pairwise commuting.*

Proof. Let $e,f\in E$. Then $\bar{e}\bar{f}=\bar{f}\bar{e}$. Hence $ef-fe\in$ $\in J(R)$. Multiplying $ef-fe$ by e on the left we obtain $z=ef-efe\in J(R)$. By Lemma 1 $z\in eR\cap J(R)=0$. Thus $ef=efe$ for any $e,f\in E$. Similarly, $fe=efe$ because the ring R is J_l-equiradical. Therefore, $ef=fe$ for all $e,f\in E$.

Proof of Theorem 2. Let $\bar{R}=\bar{E}\bar{R}$, where $\bar{E}$ is a set of pairwise commuting idempotents. Then $R=ER+J(R)$ and by Lemma 4 E is a set of pairwise commuting idempotents in R . In view of Lemma 2, $ER\cap J(R)=0$. Thus $R=ER\oplus J(R)$ is a direct sum of right ideals. From the theorem dual to Theorem A it can be easily seen that $J(R)\cdot e=0$ for any idempotent e . Therefore $J(R)\cdot ER=0$. Thus ER is a two-sided ideal.

A ring is called biregular if each of its principal ideals is generated by a central idempotent. Every biregular ring is semisimple ([3], §9,4).

COROLLARY. *If a ring* R *is* J*-equiradical and* J_l*-*

equiradical and the factor-ring $\bar{R}=R/J(R)$ *is biregular then* R *decomposes into the direct sum of the radical* $J(R)$ *and an ideal* T *which is a biregular ring.*

Finally we show in Theorem 3 that the condition of countability of a generating set of idempotents in Theorem 1 and the condition for R to be J_l-equiradical in Theorem 2 are important.

Let a right ideal T of a ring R be generated by a set E of pairwise commuting idempotents. Then $T=ER$. Without loss of generality we may assume that E is closed under multiplication because a product of any two idempotents of E is idempotent, too. Now recall that a semigroup is called a band if each of its elements is idempotent, and a commutative band is called a semilattice. The following lemma suggests the way how to construct the example in Theorem 3.

LEMMA 5. *Let* R *be a* J*-equiradical ring,* $\bar{E}$ *be a semilattice of idempotents in* $\bar{R}=R/J(R)$ *and put* $E=\{e\in R \mid e^2=e \text{ and } \bar{e}\in\bar{E}\}$. *Then* E *is a band satisfying the identity* $xyz=xzy$.

Proof. Let $e,f,g\in E$. Then $\bar{e}\bar{f}=\bar{f}\bar{e}$, i.e. $ef-fe\in J(R)$. Since $gJ(R)=0$ holds for every idempotent g in a J-equiradical ring, $gef=gfe$ for all $e,f,g\in E$.

In particular, for $g=e$ we get $efe=ef$, whence ef is an idempotent. Therefore E is closed under multiplication.

THEOREM 3. *There exists a* J*-equiradical semigroup algebra* $R=FS$ *over a field* F *such that* $\bar{R}=R/J(R)$ *is a commutative biregular ring generated by an uncountable set of idempotents and* $J(R)$ *is not a direct summand (as a right ideal) of* R.

Proof. Let F be a field, I be an index set of power continuum, and S be a band generated by a set $\{e_i | i\in I\}$ of idempotents and free in the semigroup variety $[x^2=x\ ,\ xyz=xzy]$ (see for example [1]). The elements of S are the finite products $e_{i_1}\dots e_{i_n}$ where $e_{i_k}\neq e_{i_m}$ for $k\neq m$. An equality $e_{i_1}\dots e_{i_n}= e_{j_1}\dots e_{j_m}$ holds iff $e_{i_1}=e_{j_1}$ and $\{i_1,\dots,i_n\}=\{j_1,\dots,j_m\}$. Consider the semigroup algebra $R=FS$ over the field F. The elements of R are the finite sums $\sum_{i=1}^{l} a_i s_i$ with $a_i\in F$ and $s_i\in S$. The set of all elements of S forms a basis of R considered as a linear space over F. Thus each element x of R admits a unique decomposition of the form $x=\sum a_i s_i$ with $a_i\in F_i$ and $s_i\in S$.

Consider the ideal N in R generated by the elements $e_ie_k-e_ke_i$ for all $i,k\in I$. Note that $e_m(e_ie_k-e_ke_i)=0$ for all $m\in I$. Thus $RN=0$ and, in particular, $N^2=0$. Moreover, N consists of the finite sums of elements of the kind $a(e_ie_k-e_ke_i)e_{j_1}\dots e_{j_m}$. Clearly, $N\subseteq J(R)$. On the other hand, we have $\bar{e}_i\bar{e}_k = \bar{e}_k\bar{e}_i$ in $\bar{R}=R/N$ for all $i,k\in I$.

The ring $\bar{R}=F\bar{S}$ is a semigroup algebra over the field F. Since $\bar{S}$ is a semilattice generated by $\{\bar{e}_k | k\in I\}$, $\bar{R}$ is a J-semisimple ring by [5], Theorem 1. Thus $J(R)=N$.

An element of R is called a monomial if it is of the form as with $a\in F$ and $s\in S$. Two monomials $ae_{i_1}\dots e_{i_m}$ and $be_{j_1}\dots e_{j_k}$ are called balanced if $\{i_1,\dots,i_m\}=\{j_1,\dots,j_k\}$ and similar if, in addition, $i_1=j_1$. An element x is called balanced if any two monomials a_is_i and a_js_j in its decomposition $x = \sum a_is_i$ are balanced. Every balanced element is of the type

(1) $p=a_1e_{i_1}e_{i_2}\dots e_{i_m}+a_2e_{i_2}e_{i_1}\dots e_{i_m}+\dots+a_me_{i_m}e_{i}\dots e_{i_{m-1}}$.

A balanced element p is called a balanced component of an element x if $x=p+y$ and no monomial from the decomposition of $y=\sum a_i s_i$ is balanced with monomials from p. Every element of R can be uniquely represented as the sum of its balanced components.

One can verify directly that $x \in N$ iff for every balanced component of x, written in the form (1), the equality $\sum_{k=1}^{m} a_k = 0$ holds.

As usual, let $r(b)$ denote the right annihilator of b in R, i.e., $r(b) = \{x \in R \mid bx=0\}$. We are going to show that $r(e_i) = N + \sum_{j \in I} (e_j - e_i e_j)R$ for each $i \in I$. Note that the equality $e_i e_{j_1} \ldots e_{j_k} = e_i e_{i_1} \ldots e_{i_m}$ holds iff $\{i, j_1, \ldots, j_k\} = \{i, i_1, \ldots, i_m\}$. Thus, the sets $\{j_1, \ldots, j_k\}$ and $\{i_1, \ldots, i_m\}$ coincide or differ only in that only one of them contains i. One can assume that these sets are of the form $\{j_1, \ldots, j_k\}$ and $\{i, j_1, \ldots, j_k\}$, respectively, with $i \neq j_l$ for $l=1,\ldots,k$. Let $e_i x = 0$, x containing the balanced components $p = a_1 e_{j_1} e_{j_2} \ldots e_{j_k} + \ldots + a_k e_{j_k} e_{j_1} \ldots e_{j_{k-1}}$ and $q = b_0 e_i e_{j_1} \ldots e_{j_k} + b_1 e_{j_1} e_i \ldots e_{j_k} + \ldots + b_k e_{j_k} e_i \ldots e_{j_{k-1}}$. The equality $e_i x = 0$ implies $e_i(p+q) = 0$. From the last equality it follows that $\sum_{j=1}^{k} a_j + \sum_{j=0}^{k} b_j = 0$. If $q=0$ then $\sum_{j=1}^{k} a_j = 0$ and $p \in N$. Let $q \neq 0$. Next we show that the element $p+q$ can be reduced to the form $r+p_1$, where $r \in \sum_j (e_j - e_i e_j)R+N$ and p_1 is a balanced element of type p. Firstly, $b_0 e_i e_{j_1} \ldots e_{j_k} = -b_0(e_{j_1} - e_i e_{j_1}) \cdot$ $\cdot e_{j_2} \ldots e_{j_k} + b_0 e_{j_1} \ldots e_{j_k}$, where $b_0 e_{j_1} \ldots e_{j_k}$ is a monomial of type p. Next, $b_1 e_{j_1} e_i e_{j_2} \ldots e_{j_k} =$ $= b_1(e_{j_1} e_i - e_i e_{j_1}) e_{j_2} \ldots e_{j_k} - b_1(e_{j_1} - e_i e_{j_1}) e_{j_2} \ldots e_{j_k} + b_1 e_{j_1} e_{j_2} \ldots e_{j_k}$. Here the first term belongs to N, the second to $(e_{j_1} - e_i e_j)R$, and the third is of type p. Now it is clear that by similar transformations $p+q$ can be written in the desired form. Moreover, the sum of the coefficients in the decomposition of p_1 equals zero. Therefore $p_1 \in N$ and x is a sum of elements from $N + \sum_j (e_j - e_i e_j)R$.

Now assume that $R=T \oplus N$, where T is a right ideal of R. Let $e_k = b+z$ with $b \in T$ and $z \in N$. Then $e_k = e_k^2 = be_k + ze_k$. Since T and N are right ideals and $T \cap N = 0$, we have $b = be_k$ and $z = ze_k$. In particular, if $b = \sum a_j s_j$ with $a_j \in F$ and $s_j = e_{i_1} \ldots e_{i_m} \in S$ then $k \in \{i_1, \ldots, i_m\}$. On the other hand, $e_k = e_k^2 = e_k b + e_k z = e_k b$. Thus $e_k(e_k - b) = 0$, i.e., $e_k - b \in r(e_k)$. The decomposition of $e_k - b$ cannot contain non-zero summands from $\left(\sum_m (e_m - e_k e_m)R\right) \setminus N$ because, as is shown above, each monomial in the decomposition of b contains a multiplier e_k. Hence $e_k - b \in N$, so for all $k \in I$ we have $e_k = (e_k - z_k) + z_k$ with $z_k \in N$. Since T is a right ideal and $e_k - z_k, e_m - z_m \in T$, we obtain $y_{km} = (e_k - z_k)e_m + (e_m - z_m)(-e_k) = (e_k e_m - e_m e_k) - z_k e_m + z_m e_k \in T \cap N = 0$. We shall show that this is impossible to hold simultaneously for all pairs $k, m \in I$. For this end consider the relation $\rho \subset I \times I$ defined as follows:

$(k,m) \in \rho$ iff there is a summand $ae_k e_m$ with $0 \neq a \in F$ in the decomposition of $z_k = \sum_j a_j s_j$.

It should be noted that if z_k contains a monomial $ae_k e_m$ then z_k contains a monomial $-ae_m e_k$ in its decomposition. In order that y_{km} be equal to zero, one of the elements $z_k e_m$ or $z_m e_k$ should contain a monomial $ae_k e_m$ with $0 \neq a \in F$. Since the decompositions of z_k and z_m contain only monomials whose degree is greater than or equal to 2, we conclude that z_k or z_m contains in its decomposition a monomial $ae_k e_m$ with $0 \neq a \in F$. Thus, the relation ρ possesses the following property: either $(k,m) \in \rho$ or $(m,k) \in \rho$. Since the decomposition of z_k contains only a finite number of non-zero monomials (in particular, of the kind

$ae_k e_m$), for every $k \in I$ there exists only a finite number of $m \in I$ such that $(k,m) \in \rho$. Hence, the set $A_k = \{m \mid (k,m) \in \rho\}$ is finite for all $k \in I$. Therefore the set $B_k = \{m \mid (m,k) \in \rho\}$ has finite complement C_k in I . Let $\{k_1, k_2, \ldots, k_n, \ldots\}$ be a countable subset of I . Then the set $I \setminus \bigcap_{n=1}^{\infty} B_{k_n} = \bigcup_{n=1}^{\infty} \left(I \setminus B_{k_n} \right) = \bigcup_{n=1}^{\infty} C_{k_n}$ is countable. Hence, $\bigcap_{n=1}^{\infty} B_{k_n}$ is not empty. Let j be an element contained in it. Then $(j, k_n) \in \rho$ for all natural numbers n , which contradicts the finiteness of the set A_j . Therefore the assumption that $R = T \oplus N$ for some right ideal T , is false.

If in this example we take F to be the two-element field then the factor-ring $\bar{R}$ is a Boolean ring. This completes the proof of Theorem 3.

Acknowledgement. The author thanks Prof. A. V. Mikhalev and Prof. L. A. Skornjakov for fruitful discussions.

REFERENCES

[1] T. Evans, The lattice of semigroup varieties, *Semigroup Forum* 2 (1971), 1-43.

[2] H.-J. Hoehnke, Structure of semigroups, *Canad. J. Math.* 18 (1966), 449-491.

[3] N. Jacobson, *Structure of rings*, Amer. Math. Soc., Providence, 1956.

[4] E. N. Roiz, On the Jacobson and Baer radicals of rings and their multiplicative semigroups, *Izv. Vysš. Učebn. Zaved.*, 1975, no. 10, 71-79 (Russian).

[5] M. L. Teply, E. G. Turman, A. Quesada, On semi-simple semigroup rings, *Proc. Amer. Math. Soc.* 79 (1980), 157-163.

A. V. Tishchenko
Department of Mathematics
All-Union Correspondence Institute of
Food Industry
ul. Chkalova 73
109803 Moscow
USSR

COLLOQUIA MATHEMATICA SOCIETATIS JÁNOS BOLYAI

38. RADICAL THEORY, EGER (HUNGARY), 1982.

STRUCTURE OF T -RINGS

J. TRLIFAJ and T. KEPKA

0. INTRODUCTION

This paper is devoted to the study of rings with only trivial orthogonal theories of the functor Ext. There are two reasons for this study.

Firstly, we continue the study of orthogonal theories that represent a natural generalization of torsion theories. Let us recall the basic concept. Let R be an associative ring with unit, R-mod the category of unitary left R-modules and F a bifunctor from R-mod into the category of abelian groups. If X is a non-empty class of modules, put $^{\perp}X = \{A;\ F(A,B) = 0 \text{ for each } B \in X\}$ and $X^{\perp} = \{B;\ F(A,B) = 0 \text{ for each } A \in X\}$. By [9], an ordered pair (X,Y) of non-empty classes of modules is called an orthogonal theory of F if $X = {}^{\perp}Y$ and $Y = X^{\perp}$. An orthogonal theory is said to be trivial if either $X = R$-mod or $Y = R$-mod . The orthogonal theory $({}^{\perp}Y, ({}^{\perp}Y)^{\perp})$ is called the left orthogonal theory generated by the class Y .

Orthogonal theories of Hom coincide with torsion theories introduced in [5] and playing an important role in the theory of radicals. By [8], every torsion theory for R-mod is trivial if and only if the ring R is isomorphic to a full matrix ring over a local left and right perfect ring. Similarly, by [10], if R is commutative then every orthogonal theory of the tensor product is trivial if and only if R is a local perfect ring.

Secondly, we were encouraged by the recent development of the theory of Whitehead groups. In our terminology, it is just the study of a special orthogonal theory of the functor Ext . For example, the solution of the Whitehead problem (see [13]) says that if R is the ring of integers then the assertion "the left orthogonal theory of Ext generated by $\{R\}$ is trivial" is independent of ZFC.

In view of the results mentioned above we can ask for the structure of rings with only trivial orthogonal theories of the functor Ext . These rings were introduced in [2, Appendix A], called T -rings and divided into five classes. In this paper, we proceed and get a full description of three of the five classes (see Theorem 4.4 and Theorem 6.1).

1. PRELIMINARIES

In this paper, all rings are associative with unit. If R is a ring then R^{op} denotes the opposite ring of R. If S and T are rings then $S \boxplus T$ is the ring direct sum of S and T. A ring R is said to be indecomposable if $S \boxplus T \neq R$ for any subrings S and T of R. If $\alpha \geq 1$ is an ordinal number and R is a ring then $M(\alpha, R)$ denotes the ring of row-finite matrices of type α over R.

Let R be a ring. Then R-mod (resp. mod-R) denotes the category of unitary left (resp. right) R-modules. Unitary left R-modules are simply called modules. Let M be a module and $\alpha \geq 1$ an ordinal number. Then $M^{(\alpha)}$ denotes the direct sum $M_0 \oplus M_1 \oplus \ldots \oplus M_\beta \oplus \ldots$, where $M_\beta = M$ for every $0 \leq \beta < \alpha$. Further, $E(M)$ denotes the injective hull of M. A module M is said to be Σ-injective if $M^{(\gamma)}$ is injective for every $1 \leq \gamma$.

Let R be a ring. A preradical r for R-mod is any subfunctor of the identity functor. A module M is

r -torsion (resp. r -torsionfree) if $r(M) = M$ (resp. $r(M) = 0$). For a non-empty class X of modules and a module M, let $p_X(M)$ be the sum of all $\operatorname{Im} f$, $f \in \operatorname{Hom}(A,M)$, $A \in X$, and $q_X(M)$ the intersection of all $\operatorname{Ker} g$, $g \in \operatorname{Hom}(M,A)$, $A \in X$. Then p_X and q_X are preradicals.

Put $\operatorname{Soc} = p_X$, $J = q_X$, $U = p_Y$ and $Z = p_W$ where X is the class of simple modules and zero modules, Y is the class of projective modules from X, and $W = \{M/N;\ M \in R\text{-mod and } N \text{ is an essential submodule of } M\}$.

A ring is said to be completely reducible if $R = \operatorname{Soc}(R)$. Further concepts and notation concerning rings, modules and preradicals can be found in [1], [2] and [7].

A ring R is called a left (resp. right) T -ring if every orthogonal theory of Ext for R -mod (resp. mod-R) is trivial. Hence R is a left T -ring if and only if $\operatorname{Ext}_R(M,N) \neq 0$ for each non-projective module M and each non-injective module N. Clearly, every completely reducible ring is both a left and right T -ring.

In the remaining part of this section, let R be a left T -ring such that R is not completely reducible. Then, by [2,A.2.6], all non-projective simple modules are isomorphic. Let A be a non-projective simple module and put $I = p_{\{A\}}(R)$. Then by [2,A.2], R is said to be

- of type 1 if $I \neq 0$,
- of type 2 if $I=0$ and $\mathrm{Soc}(R)$ is a left direct summand of R ,
- of type 3 if $I=0$, $\mathrm{Soc}(R)$ is not a left direct summand of R and R is not regular,
- of type 4 if $I=0$, $\mathrm{Soc}(R)$ is not a left direct summand of R , R is regular and R is not a left V -ring,
- of type 5 if $I=0$, $\mathrm{Soc}(R)$ is not a left direct summand of R and R is a regular left V -ring.

2. BASIC PROPERTIES OF T -RINGS

The main purpose of this section is to reduce the study of T -rings to the study of the indecomposable ones.

2.1. PROPOSITION. (i) *If* K *is an idempotent ideal of a left* T *-ring* R *then* R/K *is a left* T *-ring.*
(ii) *Let* $R = S \boxplus T$ *. Then* R *is a left* T *-ring if and only if both* S *and* T *are so and at least one of them is completely reducible.*

Proof. (i) The assertion is an immediate consequence of [12,§5.5.3].
(ii) Let M be a module. Then $M = SM \oplus TM$ and we can use (i).

2.2. PROPOSITION. *Every left T-ring is either regular or left artinian.*

Proof. Let R be a left T-ring. If every flat module is projective then R is left artinian by [2,A.1.4] and [1,28.4]. Thus we can assume that there is a flat non-projective module A. Denote by S the ring of integers and let C be an injective cogenerator in S-mod. Moreover, take an arbitrary $B \in$ mod-R and put $D = \mathrm{Hom}_S(B,C)$. Using the left-hand version of [3,VI.5.1] we get

$$\mathrm{Ext}_R(A,D) \simeq \mathrm{Hom}_S(\mathrm{Tor}_R(B,A),C) = 0 .$$

Hence D is an injective module and the functor $\mathrm{Hom}_S(B\otimes_R -,C)$ is exact by [1,20.6]. By [1,18.14], B is a flat right R-module and R is regular.

2.3. COROLLARY. *Every ring of type 3 is left artinian*

2.4. PROPOSITION. *Let R be a ring of type 2. Then either R is indecomposable or $R = S \boxplus T$, where S is a completely reducible ring and T is an indecomposable ring of type 2. Every indecomposable ring of type 2 is a simple left hereditary regular ring.*

Proof. Let R be an indecomposable ring of type 2. Then, by [2,A.3.5], $\mathrm{Soc}(R) = 0$ and, by 2.2, R is regular

Hence R contains an infinite set of orthogonal idempotents and R is left hereditary by [2,A.2.1]. The remaining assertions follow from [2,A.2.6 and A.3.4].

2.5. *Remark.* Similarly, rings of type 1 were shown to be either indecomposable or of the form $S \boxplus T$, where S is a completely reducible ring and T is an indecomposable ring of type 1. Moreover, indecomposable rings of type 1 are full matrix rings over local left artinian rings of type 1 (see [2,A.3.1 and A.3.2]). The following proposition provides us with examples of such rings.

2.6. PROPOSITION. *Let R be a local left artinian ring such that every left ideal is principal and every right ideal is an ideal. Then R is an indecomposable left and right T-ring of type* 1.

Proof. By [7,18.38.2], there are an element $a \in R$ and a positive integer n such that $Ra^{n-1} \neq 0$ and

$$0 = Ra^n \subseteq Ra^{n-1} \subseteq \dots \subseteq Ra \subseteq R$$

is the complete list of left (resp. right) ideals of R. Moreover, by [7,25.4.2], every module is a direct sum of cyclic modules. Since R is selfinjective, it suffices to show that for all $1 \leq r,s \leq n-1$, $\mathrm{Ext}_R(R/Ra^r, R/Ra^s) \neq 0$. Define a homomorphism f of Ra^r into R/Ra^s by

$f(a^r) = 1 + Ra^s$ if $r+s \leq n$ and $f(a^r) = a^{r+s-n} + Ra^s$ in the opposite case. It is easy to see that $f(a^r) \neq g(a^r)$ for every homomorphism g of R into R/Ra^s .

2.7. PROPOSITION. *Let* R *be a ring of type* 3, 4 *or* 5 . *Then:*

(i) *Every non-projective simple module is* Σ *-injective.*

(ii) *For every module* M , *both* $U(M)$ *and* $M/U(M)$ *are completely reducible, the submodule is projective and the factormodule injective.*

(iii) R *is left semiartinian, left hereditary and* $J(R) \subseteq \mathrm{Soc}(R) = U(R)$, $J(R)^2 = 0$ *and* $\mathrm{Soc}(R)^2 = \mathrm{Soc}(R)$.

Proof. Let A be a non-projective simple module. Then every non-projective simple module is isomorphic to A by [2,A.2.6]. Since $p_{\{A\}}(R) = 0$, we have $\mathrm{Soc}(R) = U(R)$ and every U -torsionfree module is completely reducible and injective by [2,A.2.7]. In particular, A is Σ -injective and (i) holds. Further, for every module M , the module $M/U(M)$ is U -torsionfree and (ii) holds. Since $R/\mathrm{Soc}(R)$ is completely reducible, $J(R) \subseteq \mathrm{Soc}(R) = \mathrm{Soc}(R)^2$ and hence $J(R)^2 = 0$. The remaining assertions follow from [2,A.4.1].

2.8. PROPOSITION. *Let* R *be a ring of type* 3 *(resp. type* 4 , *resp. type* 5*). Then either* R *is indecomposable* *or* $R = S \boxplus T$, *where* S *is a completely reducible ring and* T *is an indecomposable ring of type* 3 *(resp. type* 4 *, resp. type* 5*). Anyway, if* R *is indecomposable then there exist two simple modules* A *and* B *such that* A *is* Σ *-injective and non-projective,* B *is projective, and every simple module is isomorphic to one of them.*

Proof. By 2.7(i), there is a Σ -injective simple module A such that A is non-projective and every non-projective simple module is isomorphic to A . First, we shall prove that the number of isomorphism classes of simple modules is finite. Let Y_1 and Y_2 be two non-empty classes of projective simple modules such that no module from Y_1 is isomorphic to a module from Y_2 . Put $K_1 = p_{Y_1}(R)$, $K_2 = p_{Y_2}(R)$ and suppose that K_1 is not a finitely generated left ideal. Then $\mathrm{Hom}(K_1, K_2) = 0$, K_1 is not a left direct summand of R and K_2 is injective by [2,A.2.3]. In particular, K_2 is a finitely generated left ideal and the assertion is clear. However, this implies that R is a finite ring direct sum of indecomposable rings and the first part of the proposition follows from 2.1(ii). To prove the rest, suppose that R is indecomposable. By 2.7(iii), $\mathrm{Soc}(R) = U(R)$ and hence there is a projective simple module B such that $L =$

$= p_{\{B\}}(R)$ is not injective. Let V be the class consisting of zero modules and all simple modules isomorphic neither to A nor to B. Put $r = p_V$. Then $r(L) = 0$ and, by [2,I.7.1 and A.2.7], either $r(R) = 0$ or $r(R)$ is a ring direct summand of R. Since R is indecomposable, V contains only zero modules and the proof is finished.

3. COMMUTATIVE T-RINGS

3.1. THEOREM. *Let* R *be a commutative* T*-ring. Then either* $R = S$ *or* $R = T$ *or* $R = S \boxplus T$*, where* S *is a finite ring direct sum of fields and* T *is a commutative local* QF*-ring of type* 1.

Proof. With respect to 2.1(ii), 2.4, 2.8 and [2, A.3.1], we can assume that R is an indecomposable ring such that R is not completely reducible. Moreover, by 2.4 and 2.8, R is of type 1, and by [2,A.3.2] R is a local artinian ring. Let $\mathrm{Soc}(R) = Ra_0 + \ldots + Ra_m$, $n = m+1$, $a = (a_0,\ldots,a_m) \in R^{(n)}$ and $M = R^{(n)}/Ra$. The module Ra is simple and hence neither Ra nor M is projective. Let f be a homomorphism of Ra into R, $f(a) = r_0a_0 + \ldots + r_ma_m$. Define a homomorphism g of $R^{(n)}$ into R by $g(e_0) = r_0, \ldots, g(e_m) = r_m$, where $\{e_0,\ldots,e_m\}$ is the canonical basis of $R^{(n)}$. Then $g(a) = f(a)$ and

$\mathrm{Ext}_R(M,R) = 0$. Thus R is selfinjective.

3.2. *Remark*. Further results concerning rings of type 1 and 2 can be found in [15].

4. T -RINGS OF TYPES 4 AND 5

In this section, a method of construction of large non-injective modules is described. These modules are maximal submodules of the injective ones, and hence they are "likely" to generate nontrivial orthogonal theories of Ext . This method is then used to prove that rings of type 4 resp. 5 do not exist.

4.1. PROPOSITION. *Let* R *be an indecomposable ring of type* 3 *(resp. type* 4 *, resp. type* 5*). Then* R *is Morita equivalent to an indecomposable ring* S *of type* 3 *(resp. type* 4 *, resp. type* 5*) such that* $\mathrm{Soc}(S)$ *is a maximal left ideal of* S .

Proof. If R is of type 3 then, by 2.3 and 2.8, we can take any basic ring of R . Hence we can suppose that R is regular. Let A be a non-projective simple module and B a projective simple module. Consider a non-zero proper ideal K of R . By 2.7(iii), K contains a minimal left ideal L and $L = L^2 = KL$. By 2.8, L is isomorphic to B and hence $B = KB$. Further, let M be a module with $KM = 0$. Then $\mathrm{Soc}(M)$ is a direct sum of copies of A , $\mathrm{Soc}(M)$ is injective and $M = \mathrm{Soc}(M)$, by 2.8 and 2.7(iii). Consequently, R/K is a simple ring and $K = \mathrm{Soc}(R)$, since $KA = 0$. We have proved that 0 , $\mathrm{Soc}(R)$ and R are the only ideals of R . Further, there is an idempotent $e \in R$ such that the module $(Re + \mathrm{Soc}(R))/\mathrm{Soc}(R)$ is simple. Since $e \notin \mathrm{Soc}(R)$, we have $ReR = R$ and the rings R and $S = eRe$ are Morita equivalent. It is easy to see that $\mathrm{Soc}(S) = e\mathrm{Soc}(R)e$. Finally, let $a, b \in S$, $a \notin \mathrm{Soc}(S)$. Then $b \in Re + \mathrm{Soc}(R) = Ra + \mathrm{Soc}(R)$, $b = ca + d$, $c \in R$, $d \in \mathrm{Soc}(R)$ and $b = eca + ede \in Sa + \mathrm{Soc}(S)$. Hence $S = Sa + \mathrm{Soc}(S)$ and $\mathrm{Soc}(S)$ is a maximal left ideal of S .

4.2. Let R be a ring such that $T = R/\mathrm{Soc}(R)$ is a division ring. Let ϕ be an ordinal number such that $T = \{t_0, t_1, \ldots, t_\psi, \ldots\}$, $0 \le \psi < \phi$. Put $V = E(R)^{(1+\phi)}$, so that $\mathrm{Soc}(E(V)) = \mathrm{Soc}(V) = \mathrm{Soc}(R)^{(1+\phi)}$. For every ordinal number $0 \le \psi < 1+\phi$, let q_ψ designate the ψ -th natural injection of $P = E(R)/\mathrm{Soc}(R)$ into $V/\mathrm{Soc}(V)$. Further, let $\{u_0, u_1, \ldots, u_\rho, \ldots\}$, $0 \le \rho < \lambda$, be a basis of the left T -module P such that $u_0 = 1 + \mathrm{Soc}(R)$. Then either $\lambda = 1$ and R is left selfinjective, or $\lambda \ge 2$ and we can define a subset D of $V/\mathrm{Soc}(V)$ by

$$D = \{q_\psi(u\rho)\,;\ 1 \le \psi < 1+\phi \ ,\ 0 \le \rho \le \lambda \ ,\ \rho \neq 1\} \cup$$
$$\cup\ \{t_\psi q_0(u_0) + q_{1+\psi}(u_1)\,;\ 0 \le \psi < \phi\} \ .$$

LEMMA. *If* $\lambda \geq 2$, *then there is a maximal submodule* N *of* $E(V)$ *such that* $\mathrm{Soc}(V) \subseteq N$, $q_0(u_0) \notin N/\mathrm{Soc}(V)$, $D \subseteq N/\mathrm{Soc}(V)$ *and* $E(V)/\mathrm{Soc}(V) = N/\mathrm{Soc}(V) \oplus Tq_0(u_0)$.

Proof. The set $D \cup \{q_0(u_0)\}$ is T -independent and hence there is a subset D_0 of $E(V)/\mathrm{Soc}(V)$ such that $D \subseteq D_0$ and $D_0 \cup \{q_0(u_0)\}$ is a basis of the left T -module $E(V)/\mathrm{Soc}(V)$. Now it suffices to put $N = \{n \in E(V);\ n + \mathrm{Soc}(V) \in D_0\}$.

4.3. PROPOSITION. *Let* R *be an indecomposable ring of type* 3 *(resp. type* 4 , *resp. type* 5*). By* 2.8, *there are a minimal left ideal* K_0 *and a left ideal* K' *such that* $K_0 \oplus K' = R$. *Let* $\mathrm{Soc}(K') = K_1 \oplus \ldots \oplus K_\beta \oplus \ldots,\ 1 \leq \beta < \alpha$, *be a complete decomposition of* $\mathrm{Soc}(K')$. *Then there exists an injective ring homomorphism* t *of* R *into the ring* $Q = M(\alpha, S)$, $S = \mathrm{End}_R(K_0)^{\mathrm{op}}$, *such that the following three conditions are satisfied:*

(i) *The module* Q *is a copy of* $E(t(R))$.

(ii) $L = t(\mathrm{Soc}(R))$ *is a right ideal of the ring* Q *and every matrix from* L *has only finitely many non-zero columns.*

(iii) *Let* $e \in Q$ *such that* $e(0,0) = 1$ *and* $e(\beta, \gamma) = 0$ *for all* $(\beta,\gamma) \neq (0,0)$. *Then* $e \in L$.

Proof. For every $0 \le \beta < \alpha$, let p_β be the natural projection of $\mathrm{Soc}(R)$ onto K_β and q_β an isomorphism of K_β onto K_0 , $q_0 = id_{K_0}$ (see 2.8). By 2.7(iii), $Z(R) = 0$ and hence for every $x \in E(R)$ there is a unique endomorphism f of $E(R)$ with $f(1) = x$. Since $f(\mathrm{Soc}(R)) \subseteq \mathrm{Soc}(R)$, we can define a mapping g of $E(R)$ into Q by $g(x)(\beta,\gamma) = q_\gamma p_\gamma f q_\beta^{-1}$. Then g is an isomorphism of the additive groups and $g(rx) = g(r)g(x)$ for all $r \in R$ and $x \in E(R)$. Now put $t = g|R$. Then (i) holds and for every $0 \le \beta < \alpha$ and every $y \in K_\beta$ the matrix $t(y)$ is zero outside the β -th column. Thus, every matrix from L has only finitely many non-zero columns. Since $f(\mathrm{Soc}(R)) \subseteq \mathrm{Soc}(R)$ for each endomorphism f of $E(R)$, L is a right ideal of the ring Q and (ii) holds. Finally, there is an idempotent $e' \in R$ such that $K_0 = Re'$ and $K' = R(1-e')$. Put $e = t(e')$. Then e is zero outside the 0 -th column and $e(0,0) = 1$. For every $1 \le \beta < \alpha$, $K_\beta e' = 0$ and hence $e(\beta,0) = 0$ and the proof is finished.

4.4. THEOREM. *No ring is of type* 4 *(resp. type* 5*).*

Proof. Let, on the contrary, R be an indecomposable ring of type 4 (resp. type 5) such that $T = R/\mathrm{Soc}(R)$ is a division ring - see 2.8 and 4.1. Let $T = \{t_0, \dots, t_\psi, \dots\}$, $0 \le \psi < \phi$. By 4.3 we can assume that R is a subring of the ring $Q = M(\alpha, S)$ for a division ring S and an ordinal number α such that the conditions 4.3(i), (ii),(iii) are satisfied with $t = id$. Since R is not left artinian, $\mathrm{Soc}(R)$ is not finitely generated and α is infinite. The rest of the proof is divided into four parts in which we shall construct a non-injective module N and a non-projective module M such that $\mathrm{Ext}_R(M,N) = 0$, in contradiction with the premises.

(i) Consider the matrix $w \in Q$ such that $w(n,n+1) = 1 = w(n+1,n)$ for every integer $n \geq 0$ and $w(\beta,\gamma) = 0$ otherwise. Suppose that the elements $1 + \mathrm{Soc}(R)$ and $w + \mathrm{Soc}(R)$ of $P = Q/\mathrm{Soc}(R)$ are T-dependent. Since $1 + w \notin \mathrm{Soc}(R)$, we have $u = r(1+w) - 1 \in \mathrm{Soc}(R)$ for some $r \in R$. However, the matrix u has infinitely many non-zero columns, a contradiction. Hence there is a basis $\{u_0, \ldots, u_\rho, \ldots\}$, $0 \leq \rho < \lambda$, of P such that $\lambda \geq 2$, $u_0 = 1 + \mathrm{Soc}(R)$ and $u_1 = w + \mathrm{Soc}(R)$. Consider the maximal submodule N of $E = E(V)$, $V = Q^{(1+\phi)}$, corresponding to this basis by 4.2. Then $\mathrm{Soc}(E) \subseteq N$, N is essential in E and N is not injective.

(ii) For every integer $n \geq 0$, let $e_n \in Q$ be such that $e_n(0,n) = 1$ and $e_n(\beta,\gamma) = 0$ otherwise. Since $e_0 \in \mathrm{Soc}(R)$ and $\mathrm{Soc}(R)$ is a right ideal of Q, $e_n \in \mathrm{Soc}(R)$ for every n. Denote by G the submodule of $R^{(2)}$ generated by the pairs $(e_n, -e_{n+1})$, $0 \leq n$, n even. Then G is not finitely generated, and hence the module $M = R^{(2)}/G$ is not projective.

(iii) Put $H = \{(x,y) \in E^{(2)};\ e_n x = e_{n+1} y,\ 0 \leq n,\ n \text{ even}\}$ and denote by p the natural projection of $E^{(2)}$ onto $F^{(2)}$, $F = E/\mathrm{Soc}(E)$. Then H is a subgroup of $E^{(2)}$ and, for every $0 \leq \psi < 1+\phi$, (x,y) and (y,x) are contained in H, where $x,y \in V$ are such that $x_\psi = 1$, $y_\psi = w$ and $x_\xi = 0 = y_\xi$ for each $\xi \neq \psi$. Consequently, $(N/\mathrm{Soc}(E))^{(2)} + p(H) = F^{(2)}$. Since $\mathrm{Soc}(E) \subseteq N$, $N^{(2)} + H = E^{(2)}$.

(iv) We shall prove that $\mathrm{Ext}_R(M,N) = 0$. Let p and q be the natural projections of $R^{(2)}$ onto M and of E onto E/N, respectively. Let f be a homomorphism of M into E/N. We have $fp(1,0) = u+N$ and $fp(0,1) = v+N$ for some $u,v \in E$. By (iii), $x-u$, $y-v \in N$ for some $x,y \in H$. Let k be the homomorphism of $R^{(2)}$ into E with $k(1,0) = x$ and $k(0,1) = y$. Then $k(G) = 0$ and k induces a homomorphism g of M into E such that $f = qg$.

5. SPECIAL MATRIX RINGS

In this section we get a criterion for matrix rings of a special kind to be left T-rings. This criterion is used in section 6 to determine the structure of rings of type 3.

Let m be a positive integer, $n = m+1$, and let S, T be division rings such that T is a subring of $M(m,S)$. Denote by $R = U(m,S,T)$ the set of all matrices $a \in M(n,S)$ such that $a(1,0) = \ldots = a(m,0) = 0$ and $b \in T$ where $b \in M(m,S)$ and $b(i,j) = a(i+1,j+1)$ for all $0 \leq i,j < m$. Then R is a subring of $M(n,S)$. Further, for $0 \leq i \leq m$, let $e_i \in R$ be such that $e_i(0,i) = 1$ and $e_i(j,k) = 0$ otherwise. Finally, denote by A the module $R/\mathrm{Soc}(R)$, by B the left ideal Re, $e = e_0$, and by C the left ideal $R(1-e)$.

5.1. PROPOSITION. (i) *R is indecomposable, left hereditary and left artinian.*

(ii) *A is a non-projective Σ-injective simple module and B is a non-injective projective simple module.*

(iii) *Every simple module is isomorphic either to A or to B.*

(iv) $Z(R) = 0$, $\mathrm{Soc}(R) = Re_0 \oplus \ldots \oplus Re_m$ *and* $J(R) = Re_1 \oplus \oplus \ldots \oplus Re_m$.

(v) *The module $M(n,S)$ is a copy of $E(R)$.*

(vi) *The ring $R/\mathrm{Soc}(R)$ is isomorphic to T.*

Proof. This proposition is an easy generalization

of [11, ch. 13, Exercise 2].

5.2. PROPOSITION. $m=1$ *and* $T=S$ *then* $R=U(m,S,T)$ *is both a left and right* T*-ring of type* 3.

Proof. Let M be a module. Then, by 5.1, $Z(M)$ is a direct sum of copies of A, whence $M=Z(M)\oplus N$ for a module N such that $E(N)$ is a direct sum of copies of $E(B)$. Since $m=1$, C is a copy of $E(B)$. Thus both $E(N)$ and N are projective and, by [1,27.11], N is a direct sum of copies of B and C. Now, since $\mathrm{Ext}_R(A,B)\neq 0$, we see that R is a left T-ring of type 3. On the other hand, R^{op} is isomorphic to the ring $U(1,S^{\mathrm{op}},S^{\mathrm{op}})$ and hence R is a right T-ring of type 3.

5.3. PROPOSITION. *Let* $R=U(m,S,T)$. *Then* R *is a left* T*-ring if and only if* $m=1$ *and* $T=S$.

Proof. By 5.2 it suffices to prove that if $T\neq M(m,S)$, then there are a non-injective module N and a non-projective module M such that $\mathrm{Ext}_R(M,N)=0$. Suppose that $T\neq M(m,S)$. For $x\in M(m,S)$, let $x'\in Q=M(n,S)$ be such that $x'(i,j)=x(i-1,j-1)$ and $x'(0,k)=0=x'(k,0)$ for all $i\neq 0\neq j$, $0\leq i,j,k\leq m$. The mapping $x\to x'+\mathrm{Soc}(R)$ is an isomorphism of T onto the factor-ring $T'=R/\mathrm{Soc}(R)=\{t_0,\ldots,t_\psi,\ldots\}$, $0\leq\psi<\phi$. The rest of the proof is divided into four parts:

(i) Let $\{x_0,\ldots,x_\rho,\ldots\}$, $0\le\rho<\lambda$, be a basis of the T -module $M(m,S)$ such that $x_0=1$. We have $\lambda\ge 2$. Put $u_\rho=x'_\rho+\mathrm{Soc}(R)\in P=Q/\mathrm{Soc}(R)$. Then $\{u_0,\ldots,u_\rho,\ldots\}$ is a basis of the T' -module P . Consider the corresponding maximal submodule N of $V=E(V)=Q^{(1+\phi)}$, by 4.2. Then N is not injective.

(ii) Put $x=x'_1$. Then $e_1,e_1x\in J(R)$, $(e_1,e_1x)\in J(R)^{(2)}$ and the module $M=R^{(2)}/R(e_1,e_1x)$ is not projective.

(iii) We have $x\in Q$ and V is also a left Q -module. Hence xN is a subgroup of V . Denote by p the natural projection of V onto $V/\mathrm{Soc}(V)$. Then, for every $1\le$ $\le\psi<1+\phi$, we have $q_\psi(u_1)\in p(xN)$ where q_ψ is the ψ -th natural injection of P into $p(V)$. Consequently, $p(N)+p(xN)=p(V)$ and $V=N+xN$. From this, e_1N+ $+e_1xN=e_1V$ and hence for all $u,v\in V$ there are $y,z\in V$ such that $y-u$, $z-v\in N$ and $e_1y+e_1xz=0$.

(iv) We shall prove that $\mathrm{Ext}_R(M,N)=0$. Let p and q be the natural projections of $R^{(2)}$ onto M and of V onto V/N , respectively. Let f be a homomorphism of M into V/N . We have $fp(1,0)=u+N$ and $fp(0,1)=v+N$ for some $u,v\in V$. By (iii), $y-u$, $z-v\in N$ for some $y,z\in V$. Let k be the homomorphism of $R^{(2)}$ into V such that $k(1,0)=y$ and $k(0,1)=z$. Then $k(R(e_1,e_1x))=0$ and

k induces a homomorphism g of M into V such that $f = qg$.

5.4. *Remark*. By [14], there exist rings which are not left T-rings, but $\mathrm{Ext}(M,N) \neq 0$ for every finitely generated non-projective module M and every non-injective module N . On the other hand, if S and T are division rings such that $T \subseteq S$, the left dimension of S over T is 2 and the right dimension is infinite (see [4]) and if $R = U(1,S,T)$, then $\mathrm{Ext}_R(M,N) \neq 0$ for every non-projective module M and every cyclic non-injective module N .

6. T-RINGS OF TYPE 3

For a division ring S , let $U(2,S)$ designate the ring of upper triangular 2-matrices over S , i.e. $U(2,S) = U(1,S,S)$.

6.1. THEOREM. *The following conditions are equivalent for a ring* R *:*

(i) R *is a left* T*-ring of type* 3 .

(ii) R *is a right* T*-ring of type* 3 .

(iii) *Either* $R=L$ *or* $R = K \boxplus L$ *, where* K *is a completely reducible ring and* L *is Morita equivalent to the ring* $U(2,S)$ *for a division ring* S .

Proof. (i) implies (iii). According to 2.3, 2.8 and [2,A.1.2], we can assume that R is an indecomposable basic left artinian left T -ring of type 3. Moreover, by 2.8 and [1,§27] we can suppose that there is an idempotent $e_0 \in R$ such that $\{e_0, 1-e_0\}$ is a basic set of primitive idempotents of R and Re_0 is a simple module. Put $I = R(1-e_0)$ and let $\mathrm{Soc}(I) = B_1 \oplus \ldots \oplus B_m$ be a complete decomposition of $\mathrm{Soc}(I)$. By 4.3, we can assume that there is a division ring S such that R is a subring of $Q = M(n,S)$, where $n = m+1$ and the conditions 4.3(i),(ii),(iii) are satisfied for $t = id$. Using 4.3(iii), we see that $e_0(0,0) = 1$ and $e_0(i,j) = 0$ otherwise. Since $1-e_0$ is a primitive idempotent and I is a maximal left ideal, I is an ideal of R . Further, for $1 \leq i \leq m$, let $e_i \in Q$ such that $e_i(0,i) = 1$ and $e_i(j,k) = 0$ otherwise. Since $\mathrm{Soc}(R)$ is a right ideal of Q , $e_1, \ldots, e_m \in I \cap \mathrm{Soc}(R)$. Now, if $b \in R$ then $e_1 b, \ldots, e_m b \in I$ and hence $b(i,0) = 0$ for every $1 \leq i \leq m$. Using this we see that $\mathrm{Soc}(R)$ is just the set of all $c \in Q$ such that $c(i,j) = 0$ for all $1 \leq i \leq m$ and $0 \leq j \leq m$. Further, for every $x \in M(m,S)$, let $x' \in Q$ be such that $x'(i,j) = x(i-1,j-1)$ and $x'(k,0) = 0 = x'(0,k)$ for all $0 \leq i,j,k \leq m$, $i \neq 0 \neq j$. Put $T = \{x \in M(m,S);\ x' \in R\}$. Then T is a subring of $M(m,S)$ and T is isomorphic to $R/\mathrm{Soc}(R)$. Since $\mathrm{Soc}(R)$ is a maximal left ideal of R ,

T is a division ring and we have $R = U(m,S,T)$. By 5.3 , $m=1$ and $T=S$, whence $R = U(2,S)$.

(ii) implies (iii). This implication is dual to the preceding one (take into account that $U(2,S)$ is isomorphic to the lower triangular matrix ring).

(iii) implies (i) and (ii). This implication follows from 5.2, [2,A.1.2] and 2.1(ii).

6.2. *Remark*. Let S be a division ring. For all $1 \leq m < n$ let $V(m,n,S)$ be the set of all matrices $a \in M(n,S)$ such that $a(i,j) = 0$ for all $m \leq i < n$ and $0 \leq j < m$. Then this set is a subring of $M(n,S)$. Using [1,22.7 and Exercise 21.6], it is easy to see that a ring R is Morita equivalent to $U(2,S)$ if and only if R is isomorphic to $V(m,n,S)$ for some positive integers $1 \leq m < n$.

REFERENCES

[1] F. W. Anderson and K. R. Fuller, *Rings and categories of modules*, Springer-Verlag, New York/Heidelberg/ Berlin, 1974.

[2] L. Bican, T. Kepka and P. Němec, *Rings, modules, and preradicals*, M. Dekker Inc., New York/Basel, 1982.

[3] H. Cartan and S. Eilenberg, *Homological algebra*, Princeton University Press, Princeton, 1956.

[4] P. M. Cohn, Quadratic extensions of skew-fields, *Proc. London Math. Soc.* 11 (1961), 531-556.

[5] S. E. Dickson, A torsion theory for abelian categories, *Trans. Amer. Math. Soc.* 121 (1966), 223-235.

[6] V. Dlab, On a class of perfect rings, *Canad. J. Math.* 22 (1970), 822-826.

[7] C. Faith, *Algebra II. Ring theory*, Springer-Verlag, New York/Heidelberg/Berlin, 1976.

[8] B. J. Gardner, Rings whose modules form few torsion classes, *Bull. Austral. Math. Soc.* 4 (1971), 355-359.

[9] P. Jambor, An orthogonal theory of a set-valued bifunctor, *Czech. Math. J.* 23 (1973), 447-454.

[10] P. Jambor, Hereditary tensor-orthogonal theories, *Comment. Math. Univ. Carolinae* 16 (1975), 139-145.

[11] F. Kasch, *Moduln und Ringe*, Teubner, Stuttgart, 1977.

[12] J. Lambek, *Lectures on rings and modules*, Blaisdell Publ. Co., London, 1966.

[13] S. Shelah, A compactness theorem for singular cardinals, free algebras, Whitehead problem and transversals, *Israel J. Math.* 21 (1975), 319-349.

[14] J. Trlifaj, *E*-rings and differential polynomials over universal fields, *Comment. Math. Univ. Carolinae* 23 (1982), 159-166.

[15] J. Trlifaj, Ext and von Neumann regular rings (to appear).

J. Trlifaj
Čimická 15/257
182 00 - Praha
Czechoslovakia

T. Kepka
Mat.-fyz. fak.
Universitá Karlova
Sokolovská 83
18600 Praha 8
Czechoslovakia

COLLOQUIA MATHEMATICA SOCIETATIS JÁNOS BOLYAI

38. RADICAL THEORY, EGER (HUNGARY), 1982.

CONNECTEDNESSES AND DISCONNECTEDNESSES OF TOPOLOGICAL SPACES CHARACTERIZED BY PAIRS OF POINTS

S. VELDSMAN

ABSTRACT : Let X be a topological space that belongs to some connectedness or disconnectedness. Then quite some is known about continuous images and subspaces of X. What about the points of X? We look at the properties that the points (or rather, pairs of points) of such a space must posses. These properties are also used to characterize connectednesses and disconnectednesses. These ideas originated from a paper by Richard Wiegandt [5] where radicals of associative rings were defined by means of the properties of elements of such rings. From a categorical viewpoint, it is well known that the radical classes of rings and the connectednesses of topological spaces correspond — hence a similar theory for topological space can be expected.

1. DEFINITIONS AND CONSTRUCTIONS

Let P be a property that two points in a topological space may have. For example, P can be the property (1) there exists a U open such that $x \in U$ and $y \notin U$ or (2) x and y can be seperated by open sets. For each space X, P_X will denote all the pairs of points in X which satisfies the property P. Hence P_X can be considered as a subset of XxX. If, for a space X and points x and y in X the property P holds for x and y, we'll denote it by $(x , y) \in P_X$. We will however drop the subscript X in P_X and simply say $(x , y) \in P$ in X for $x , y \in X$ and if there is no confusion, just $(x , y) \in P$ We note that $(x , y) \in P$ might hold while $(y , x) \in P$ might not hold.

The ordered pair (x , y), with x and y points from a space X, is then called a *P-pair* in X if $(x , y) \in P$ in X or $(y , x) \in P$ in X holds. (x , y) is called a *~ P-pair* in X if $(x , y) \notin P$ and $(y , x) \notin P$. A space X is called a *P-space* (*~P-space*) if (x , y) is a P-pair (~ *P-pair*) in X for all x and y in X. A subspace A of X is called a *P-subspace* of X if (a , b) is a P-pair in X for all $a , b \in A$. Note that if A is a P-subspace of X, then A need not be a P-space. Also, if A is a subspace of X such that A is a P-space, then A need not be a P-subspace of X. For convenience, we will assume that (x, x) is a P-pair and also a $\sim P$-pair in X for all $x \in X$. Hence X is a P-space and a $\sim P$-space if and only if X is a one-point space. A trivial P-pair or a trivial $\sim P$-pair is a pair of the form (x, x), $x \in X$. Note that (x, y) is a P-pair if and only if (y, x) is a P-pair. If $\neg P$ is the logical negation of P, then every $\sim P$-pair is also a $\neg P$-pair but the converse need not be true. If, however, P is simmetrical, i.e. $(x, y) \in P$ if and only if $(y, x) \in P$, then the converse holds.

When P and P' are two properties such that whenever X is a P-space, then also X is a P'-space, we'll denote this by $P \leqq P'$. If $P_X \subseteq P'_X$ for all spaces X, then $P \leqq P'$. If $P \leqq P'$ and $P' \leqq P$ we'll indicate this by $P \doteq P'$. Hence, if $P \doteq P'$, then X is a P-space if and only if X is a P'-space.

Let $f : X \to Y$ be an onto continious function and let A be a subspace of X. We will consider the following conditions on a property P (these conditions correspond very much to those given by Wiegandt in [5] for rings):

(a) If (x, y) is a P-pair in X, then $(f(x), f(y))$ is

a P-pair in Y.

(a*) If (f(x), f(y)) is a non-trivial P-pair (i.e. $f(x) \neq f(y)$), in Y, then (x , y) is a P-pair in X.

(b) If (a , b) is a P-pair in A, then (a , b) is a P-pair in X.

(b*) If $a , b \in A$ and (a , b) is a P-pair in X, then (a , b) is a P-pair in A.

(c) If f is a quotient map, $f^{-1}(y)$ is a P-subspace of X for all $y \in Y$ and (f(x), f(y)) is a P-pair in Y, then (x , y) is a P-pair in X.

(c*) If f is a quotient map, $f^{-1}(y)$ contains only trivial P-pairs in X for all $y \in Y$ and (x , y) is a P-pair in X, then (f(x), f(y)) is a P-pair in Y.

(d) If (x , z) and (z , y) are P-pairs in X, then (x , y) is a P-pair in X.

(d*) If (x , z) and (z , y) are not P-pairs in X, then (x , y) is not a P-pair in X.

Obviously, if P satisfies (a*) (or (a)), then it also satisfies (c) ((c*) respectively). Also, P satisfies (x) if and only if $\sim P$ satisfies (x*) and $\sim P$ satisfies (x) if and only if P satisfies (x*) where x is any of a, b, c or d. Note also that if P satisfies (b), then A is a P-subspace of X if A is a P-space and if P satisfies (b*), then A is a P-space if A is a P-subspace of X. We will consider the following classes

$$\underline{R}_1(P) = \{X \mid X \text{ is a } P\text{-space}\}$$

$$\underline{R}_1^*(P) = \{X \mid X \text{ is a } \sim P\text{-space}\}$$

$$\underline{R}_2(P) = \{X \mid \text{ each non-trivial continious image of X has a non-trivial } \sim P\text{-pair}\}$$

and $\underline{R}_2^*(P) = \{X \mid$ each non-trivial continious image of X has a non-trivial P-pair$\}$

where a non-trivial topological space is a space with more than one point. If a space is not non-trivial, it is called trivial and the class of all trivial spaces will be denoted by Φ. Note that $\underline{R}_i(P) = \underline{R}_i^*(\sim P)$ and $\underline{R}_i(\sim P) = \underline{R}_i^*(P)$ for i = 1, 2. For convenience, we recall the following form [1] and [4]: A *connectedness* is a class $\underline{C}$ of topological spaces which satisfies any one of the following two conditions:

I $X \in \underline{C}$ if and only if whenever Y is a non-trivial continious image of X, then Y contains a non-trivial subspace which belongs to $\underline{C}$.

II (i) $\Phi \subseteq \underline{C}$.

(ii) $\underline{C}$ is continiously closed.

(iii) $\underline{C}$ is a component class (i.e. whenever $(X_i \xrightarrow{f_i} X)_I$ is a chained sink, where X_i is a subspace of X which is in $\underline{C}$ and f_i is the inclusion, then $X \in \underline{C}$ must hold).

(iv) $\underline{C}$ is closed under extensions (i.e. q-reversible).

A *disconnectedness* is a class D of topological spaces which satisfies any one of the following two conditions:

I $X \in \underline{D}$ if and only if whenever A is a non-trivial subspace of X, then A has a non-trivial continious image which is in $\underline{D}$.

II (i) $\Phi \subseteq \underline{D}$.

(ii) $\underline{D}$ is hereditary.

(iii) $\underline{D}$ is closed under subdirect embeddings (i.e. if $X_i \in \underline{D}$ for all $i \in I$ and $f : X \to \Pi X_i$ is an injective continious function such that $\pi_i \circ f$ is surjective for all i (where π_i is the i-th projection), then $X \in \underline{D}$ must hold).

(iv) $\underline{D}$ is closed under extensions.

If $\underline{B}$ is a class of spaces, then the operators R and S

on the class $\underline{B}$ are defined by:

$R\underline{B}$ = {X|X has no non-trivial continious image in $\underline{B}$}

and $S\underline{B}$ = {X|X has no non-trivial subspace that is in $\underline{B}$}.

If $\underline{B}$ is a connectedness, then $S\underline{B}$ is the disconnectedness determined by $\underline{B}$ and if $\underline{B}$ is a disconnectedness, then $R\underline{B}$ is the connectedness determined by $\underline{B}$.

We are now ready to look at the constructions.

2. CONSTRUCTIONS

2.1 THEOREM *Suppose P satisfies conditions (a),(b) (c) and (d). Then $\underline{R}_1(P)$ = {X|X is a P-space} is a connectedness. If P also satisfies (b*), then $\underline{R}_1(P)$ is a hereditary connectedness and in such a case, S $\underline{R}_1(P)$ = {X|X is a ~P-space} = $\underline{R}_1^*(P)$ is the disconnectedness determined by $\underline{R}_1(P)$. Conversely, if $\underline{C}$ is a connectedness, then there exists a property P, which is an equivalence relation and also satisfies (a), (b) and (c), such that $\underline{C} = \underline{R}_1(P)$. If $\underline{C}$ is hereditary, then P also satisfies (b*) and in such a case, if $\underline{D}$ is the disconnectedness determined by $\underline{C}$, then $\underline{D} = \underline{R}_1^*(P)$.*

PROOF Because every trivial space is a P-space, $\Phi \subseteq \underline{R}_1(P)$ holds. Obviously, in view of (a), $\underline{R}_1(P)$ is continiously closed. To see that $\underline{R}_1(P)$ is a component class, let A_i be a subspace of X with A_i a P-space, $f_i : A_i \to X$ the inclusion and such that $(A_i \xrightarrow{f_i} X)_{i \in I}$ is a chained sink. Let $x \neq y$ in X. Because the above sink is chained, it follows from [3] that there is a finite number of subspaces $A_{i_1}, A_{i_2}, \ldots, A_{i_n}$ with $x \in A_{i_1}$, $y \in A_{i_n}$ and $A_{i_k} \cap A_{i_{k+1}} \neq \phi$ for all k = 1, 2,...,n-1. Let $a_1 \in A_{i_1} \cap A_{i_2}$. Then (x, a_1) is a P-pair in A_{i_1} and from (b), also in X. Let $a_2 \in A_{i_2} \cap A_{i_3}$. Then (a_1, a_2) is a P-pair in A_{i_2} and from (b) also in X. Using (d) we then have (x, a_2) a P-pair in X. Continuing in this manner, we can find a $a_{n-1} \in A_{i_{n-1}} \cap A_{i_n}$ such that (x, a_{n-1}) and (a_{n-1}, y) are P-pairs in X. Hence (x, y) a P-pair in X follows which shows that X must be a P-space in this case. Finally, to show that $\underline{R}_1(P)$ is a connectedness, we proof that $\underline{R}_1(P)$ is closed under extensions. Let $f : X \to Y$ be a quotient map such that Y and $f^{-1}(y)$ are P-spaces for all $y \in Y$. Let $x, z \in X$. Then $(f(x), f(z))$ is a P-pair in Y and from (c), (x, z) must be a P-pair in X ((c) is applicable, because in view of (b), every $f^{-1}(y)$ is a P-subspace of X). Obviously, if P satisfies condition (b*), then $\underline{R}_1(P)$ is hereditary and $S\,\underline{R}_1(P) = \{X|X$ is a $\sim P$-space$\}$. For the converse, let $\underline{C}$ be any connectedness and let $\underline{D}$ be the disconnectedness determined by $\underline{C}$. If X is any topological space, then $\underline{C}$ partitions X in disjoint maximal $\underline{C}$-components, say X_i. If $f : X \to D$ is the maximal $\underline{D}$-image of X, then $X_i = f^{-1}(y)$ for some $y \in D$. Furthermore, if A is a subspace of X with $a \in A$ and $A \in \underline{C}$, then $A \in f^{-1}(y)$ for some $y \in Y$, i.e. A is contained in the $\underline{C}$-component of a. Let P_X be the following property:

P_X : x and y are in the same $\underline{C}$-component of X.

Then P_X is an equivalence relation on each space X and X is a P-space if and only if $X \in \underline{C}$. Hence $\underline{C} = R_1(P)$. We now show that P satisfies conditions (a), (b) and (c). Let $(x\ ,\ y)$ be a P-pair in X and let $f : X \to Z$ be any onto continious function. Suppose the $\underline{C}$-component of x and y is X_i and let g be the restriction of f to X_i. Because $\underline{C}$ is continiously closed, $f(X_i) \in \underline{C}$ holds and this yields f(x) and f(y) in the same $\underline{C}$-component of Y. Hence P satisfies condition (a). For (b), let A be a subspace of X with $a, b \in A$ such that $(a\ ,\ b)$ is a P-pair in A. If A_j is the $\underline{C}$-component of a and b in A, then $A_j \subseteq X_i$, where X_i is the $\underline{C}$-component of a in X, because $A_j \in C$. To show the validity of (c), let $g : X \to Y$ be a quotient map such that $g^{-1}(y)$ is a P-subspace of X for all $y \in Y$. Let $x_1 \neq x_2$ in X such that $(g(x_1), g(x_2))$ is a P-pair in Y. Let $f : X \to D$ be the maximal $\underline{D}$-image of X in $\underline{D}$. For any $\underline{C}$-component, say X_i, of X, we have $f^{-1}(d) = X_i = \cup \{g^{-1}(y) | g^{-1}(y) \cap X_i \neq \phi\}$ for some $d \in D$. The second equality follows if we can verify $\cup\{g^{-1}(y) | g^{-1}(y) \cap X_i \neq \phi\} \subseteq X_i$. Let $a \in g^{-1}(y)$ for some $g^{-1}(y)$ such that $g^{-1}(y) \cap X_i \neq \phi$. Let $b \in g^{-1}(y) \cap X_i$. Because $a, b \in g^{-1}(y)$ and $g^{-1}(y)$ a P-subspace of X, (a , b) is a P-pair in X. But $b \in X_i$. Hence $a \in X_i$ follows. In view of the above equality, $f(g^{-1}(y))$ is trivial for each $y \in Y$. Using the properties of a quotient map, we can find a unique continious function $h : Y \to D$ such that $h \circ g = f$. Because $(g(x_1), g(x_2))$ is a P-pair in Y, $(h(g(x_1)), h(g(x_2))) = (f(x_1), f(x_2))$ a P-pair in D follows using (a) above. But D, being in $\underline{D}$ has only trivial $\underline{C}$-components. Thus $f(x_1) = f(x_2) = d'$, say. Hence (x_1, x_2) is a P-pair in X. If $\underline{C}$ is hereditary, then P satisfies condition (b*). Indeed, let A be a subspace of X and let (a , b) be a P-pair in X, a , b $\in$ A. Let X_i be the $\underline{C}$-component of X such that a , b $\in X_i$. Because $A \cap X_i \subseteq X_i$, $X_i \in \underline{C}$ and $\underline{C}$ hereditary, $A \cap X_i \in \underline{C}$ holds. Hence $b \in A_j$ where A_j

is the $\underline{C}$-component of a in A, i.e. (a , b) is a P-pair in A. The last assertion is trivial.

COROLLARY If property P' satisfies conditions (a*), (b*), (c*) and (d*), then $\underline{R}_1^*(P') = \{X | X \text{ is a } \sim P'\text{-space}\}$ is a connectedness. If P' also satisfies (b), then $\underline{R}_1^*(P')$ is a hereditary connectedness and in such a case $S\ \underline{R}_1^*(P') = \{X | X \text{ is a } P'\text{-space}\} = \underline{R}_1(P')$.

Remembering that there are only three hereditary connectednesses, we see that if P is some property which satisfies conditions (a), (b), (c), (d) and (b*), then $\underline{R}_1(P)$ must be the class of all indiscrete spaces, or the class *Top* (i.e. the class of all topological space), or the class of all trivial spaces Φ.

2.2 LEMMA *Suppose a property P satisfies conditions (c) and (d). If X is a topological space such that every non-trivial continious image of X has a non-trivial P-pair, then every continious image of X is a P-pair. In particular, X must be a P-space.*

PROOF Let $f : X \to Y$ be an onto continious function. If Y is a trivial space, then Y is a P-space. Suppose thus Y is non-trivial. For each $y \in Y$, let $A_y = \{z \in Y|$ (z , y) is a *P*-pair in Y$\}$. For each $y \in Y$, $A_y \neq \phi$ and if $A_y \cap A_z \neq \phi$, then $A_y = A_z$ in view of (d). Let $g : Y \to Z$ be the cokernel of the sink $(A_y \to Y)_{y \in Y}$. If g is not constant, then, by our hypothesis, there must be a non-trivial *P*-pair, say (z_1, z_2), in Z. This however means that if $g(a_1) = z_1$ and $g(a_2) = z_2$, then (a_1, a_2) is a non-trivial *P*-pair in X by (c). Hence $a_1 \in A_{a_2} = g^{-1}(z)$ for some $z \in Z$ which contradicts $z_1 \neq z_2$. Thus g is constant and if $x \neq y$ in Y, then there is a finite number of subsets A_y of Y, say $A_{y_1}, A_{y_2}, \ldots, A_{y_n}$ such that $x \in A_{y_1}$, $y \in A_{y_n}$ and $A_{y_i} \cap A_{y_{i+1}} \neq \phi$ for each $i = 1,2,\ldots,n-1$. Using (d) and remembering that any two elements in A_{y_i} is a *P*-pair in X, we have that (x , y) is a *P*-pair in Y. Hence Y a *P*-space follows.

<u>*COROLLARY*</u> Suppose a property *P* satisfies conditions (c*) and (d*). Let X be a topological space such that every non-trivial continious image of X has a non-trivial ~*P*-pair. Then every continious image of X is a ~*P*-space. In particular, X is a ~*P*-space.

2.3 <u>THEOREM</u> *Suppose the property P satisfies conditions (a*), (b*) and (b). Then* $\underline{R}_1(P) = \{X|X$ *is a P-*

space} is a disconnectedness. If P also satisfies conditions (c) and (d*), then the connectedness determined by* $\underline{R}_1(P)$ *is* $R\,\underline{R}_1(P) = \{X \mid$ *every continious image of X is a ~P-space*$\} = \underline{R}_1(\sim P) = \underline{R}_1^*(P)$.

PROOF $\Phi \subseteq \underline{R}_1(P)$ and in view of (b*), $\underline{R}_1(P)$ hereditary follows. To see that $\underline{R}_1(P)$ is closed under subdirect embeddings, let $f : A \longrightarrow \prod_{i \in I} P_i$ be a monomorphism, with $\pi_i : \prod_{i \in I} P_i \longrightarrow P_i$ the i-th projection, such that $\pi_i \circ f$ is a epimorphism for all $i \in I$ and where each P_i is a P-space. Then A is a P-space, for if $a, b \in A$, $a \neq b$, then $f(a) \neq f(b)$. Hence $\pi_j(f(a)) \neq \pi_j(f(b))$ for some $j \in I$. Because $\pi_j \circ f : A \longrightarrow P_j$ is a continious surjection and because P_j is a P-space, (a , b) a P-pair in A follows from (a*). $\underline{R}_1(P)$ is closed under extensions, because, if $f : X \longrightarrow Y$ is a quotient map such that Y and $f^{-1}(y)$ are P-spaces for each $y \in Y$, then X is a P-space. Indeed, if $x_1 \neq x_2$ in X and $f(x_1) = f(x_2) = y_o$, say, then $x_1, x_2 \in f^{-1}(y_o)$ which yields $(x_1\ x_2)$ a P-pair in X using (b). If $f(x_1) \neq f(x_2)$, then (x_1, x_2) a P-pair in X follows from (a*). The last assertions follows from the Corollaries to Lemma 2.2 and Theorem 2.1.

COROLLARY Suppose the property P satisfies conditions (a), (b) and (b*). Then $\underline{R}_1^*(P) = \{X | X \text{ is a } {\sim}P\text{-space}\}$ is a disconnectedness. If P also satisfies conditions (c) and (d), then the connectedness determined by $\underline{R}_1^*(P)$ is $R\,\underline{R}_1^*(P) = \{X|$ every continious image of X is a P-space$\}$ $= \underline{R}_1^*({\sim}P) = \underline{R}_1(P)$.

We note that if P satisfies conditions (a*), (b*), (c*) and (d*), then $\underline{R}_1^*(P)$ is a connectedness by the corollary of Theorem 2.1 and the disconnectedness determined by $\underline{R}_1^*(P)$ is $S\underline{R}_1^*(P) = \{X|$ for all subspaces A of X, if $A \in \underline{R}_1^*(P)$, then $A \in \Phi\}$. If P also satisfies condition (b), then $S\underline{R}_1^*(P) = \underline{R}_1^*({\sim}P) = \underline{R}_1(P)$. Hence, if P satisfies conditions (a*), (b*), (c*), (d*) and (b), then $\underline{R}_1(P)$ is a disconnectedness with corresponding connectedness $\underline{R}_1^*(P) = \underline{R}_1({\sim}P)$. From Theorem 2.3 however, if P satisfies conditions (a*), (b*) and (b), then $\underline{R}_1(P)$ is a disconnectedness. Thus (c*) and (d*) are not necessary for $\underline{R}_1(P)$ to be a disconnectedness. But, to show that the connectedness determined by $\underline{R}_1(P)$ is $\underline{R}_1({\sim}P)$, we need (c*) and (d*) as have been seen in the above mentioned theorem.

There are only two continuously closed disconnectednesses

viz. *Top* and Φ. Hence, if P satisfies conditions (a*), (b*), (a) and (b), then $\underline{R}_1(P) = Top$ or $\underline{R}_1(P) = \Phi$.

2.4 <u>THEOREM</u> *Suppose the property P satisfies condition (b*). Then* $\underline{R}_2(P) = \{X|$ *every non-trivial continious image of X has a non-trivial ~P-pair*$\}$ *is a connectedness. If P also satisfies conditions (c*) and (d*), then* $\underline{R}_2(P) = \{X|$ *every continious image of X is a ~P-space*$\}$.

PROOF Firstly note $\underline{R}_2(P)$ is continiously closed. Indeed, if $f : X \to Y$ is a continious surjection and $X \in \underline{R}_2(P)$, then $Y \in \underline{R}_2(P)$ if $Y \in \Phi$. Suppose thus $Y \notin \Phi$ and let $g : Y \to Z$ be a continious surjection with $Z \notin \Phi$. Then Z is a non-trivial continious image of X which must contain a non-trivial ~P-pair. Hence $Y \in \underline{R}_2(P)$ follows. Suppose $X \notin \underline{R}_2(P)$. Then X has a non-trivial continious image, say Y, which has no non-trivial ~P-pairs. Hence (x, y) is a P-pair in Y for all $x, y \in Y$. Let A be any subspace of X. Then A is a P-space because of (b*); hence $A \notin \underline{R}_2(P)$. Thus $\underline{R}_2(P)$ a connectedness follows. The corollary to Lemma 2.2 yields the last part.

COROLLARY Suppose the property P satisfies condition (b). Then $\underline{R}_2^{*}(P) = \{X|$ every non-trivial continious image of X has a non-trivial P-pair} is a connectedness. If P also satisfies conditions (c) and (d), then $\underline{R}_2^*(P) = \{X \mid$ every continious image of X is a P-space$\}$.

This concludes our constructions. Lastly, before we look at examples, we discuss some relationships between P and P' and the classes $\underline{R}_1(P)$, $\underline{R}_1^*(P)$, $\underline{R}_2(P)$ and $\underline{R}_2^*(P)$. Firstly, from the definitions it is clear that $P \leqq P'$ if and only if $\underline{R}_1(P) \subseteq \underline{R}_1(P')$ and $P \leqq P'$ if and only if $\underline{R}_2(P') \subseteq \underline{R}_2(P)$. Hence $P \doteq P'$ if and only if $\underline{R}_1(P) = \underline{R}_1(P')$ and $P \doteq P'$ if and only if $\underline{R}_2(P) = \underline{R}_2(P')$.

2.5 <u>PROPOSITION</u>

(1) $\underline{R}_1(P) \cap \underline{R}_2(P) = \Phi$ *for any property* P.

(2) *(i)* *If* P *satisfies conditions (c) and (d), then* $\underline{R}_2^*(P) \subseteq \underline{R}_1(P)$.

(ii) *If* P *satisfies conditions (a), then* $\underline{R}_1(P) \subseteq \underline{R}_2^*(P)$. *Hence* $\underline{R}_1(P) = \underline{R}_2^*(P)$ *if* P *satisfies conditions (a), (c) and (d).*

(3) $\underline{R}_2(P) = R\,\underline{R}_1(P)$ *and* $\underline{R}_2^*(P) = R\,\underline{R}_1^*(P)$ *for any property* P.

(4) *If P satisfies condition (a), then S* $\underline{R}_1(P)$ *is a disconnectedness. If P satisfies conditions (b) and (b*), then S* $\underline{R}_1(P) = \underline{R}_1^*(P)$. *Hence, if P satisfies conditions (a), (b) and (b*), then S* $\underline{R}_1(P) = \underline{R}_1^*(P)$ *is a disconnectedness (compare with Theorem 2.3).*

PROOF

(1) Let $X \in \underline{R}_1(P) \cap \underline{R}_2(P)$, $X \notin \Phi$. Then X is a non-trivial continious image of X which, because $X \in \underline{R}_2(P)$, must contain a non-trivial ~*P*-pair. This contradicts $X \in \underline{R}_1(P)$. Hence $X \in \Phi$.

(2) (i) Follows from Lemma 2.2.

(ii) Because $\underline{R}_1(P)$ is continiously closed, $\underline{R}_1(P) \subseteq \underline{R}_2^*(P)$.

(3) Trivial.

(4) Because *P* satisfies condition (a), $\underline{R}_1(P)$ is continiously closed; hence $S\ \underline{R}_1(P)$ is a disconnectedness. Suppose *P* satisfies condition (b) and (b*). Let $X \in S\ \underline{R}_1(P)$ and let $x \neq y$ in X. Suppose (x , y) is a *P*-pair in X. Then {x ; y} is a *P*-subspace

of X and because of (b*), $\{x ; y\}$ is a P-space. But $X \in S\ R_1(P)$ implies $x = y$ which contradicts $x \neq y$. Hence X a $\sim P$-space follows. Conversely, suppose X is a $\sim P$-space and let A be a subspace of X. Because of (b), A is a $\sim P$-space. Hence, if A is a P-space, $A \in \Phi$ follows.

From Theorem 2.1, if P satisfies conditions (a), (b), (c) and (d), then $\underline{R}_1(P)$ is a connectedness. Hence there is an equivalence relation P' which satisfies conditions (a), (b) and (c) such that $\underline{R}_1(P) = \underline{R}_1(P')$. Then $P \doteq P$ and $S\ \underline{R}_1(P) = S\ \underline{R}_1(P') = \underline{R}_1^*(P') = \underline{R}_1(\sim P')$. Furthermore, if P also satisfies condition (b*), then so must P' and then $P_X = P_X'$ for all spaces X. Hence X is a $\sim P'$-space if and only if X is a $\sim P$-space, i.e. $\underline{R}_1^*(P') = \underline{R}_1^*(P)$.

From Theorem 2.4, if P satisfies condition (b*), then $\underline{R}_2(P)$ is a connectedness. Hence, there is an equivalence relation P' which satisfies conditions (a), (b) and (c) such that $\underline{R}_2(P) = R_1(P')$. What is the relation between P and P'? Using Proposition 2.5 (2) and (3), we have

$$R(\underline{R}_1(P)) = \underline{R}_2(P) = \underline{R}_1(P') = \underline{R}_2^*(P') = \underline{R}_2(\sim P') = (\underline{R}_1(\sim P'))$$

because $\underline{R}_1(P)$ is hereditary and because $\underline{R}_1(\sim P')$ is a disconnectedness, $\underline{R}_1(P) \subseteq \underline{R}_1(\sim P')$ follows. Hence $P \leqq \sim P'$.

3. EXAMPLES

1) Let X be a topological space and let $x, y \in X$. Let P_1 be the following property:

P_1 : $x \in U$ if and only if $y \in U$ for all U open in X.

Then $(x, y) \in P_1$ in X if and only if $(y, x) \in P_1$ in X and if P_1' is the property:

P_1' : $x \in G$ if and only if $y \in G$ for all G closed in X, then (x, y) is a P_1-pair if and only if (x, y) is a P_1'-pair in X.

It is easy to see that P_1 satisfies (a), (b), (b*) and (d). To see that P_1 satisfies condition (c), let $f : X \to Y$ be a quotient map such that $f^{-1}(y)$ is a P_1-subspace of X for all $y \in Y$. Let $x_1 \neq x_2$ in X such that $(f(x_1), f(x_2))$ is a P_1-pair in Y. Let U be open in X with $x_1 \in U$. Now $U = \cup\{f^{-1}(y) \mid f^{-1}(y) \cap U \neq \phi, y \in Y\}$. Indeed, if $u \in f^{-1}(y)$ for some $y \in Y$ such that $f^{-1}(y) \cap U \neq \phi$, let $z \in f^{-1}(y) \cap U$. Then $(u\ ,\ z)$ is a P_1-pair in X because $f^{-1}(y)$ is a P_1-subspace and because $z \in U$, $u \in U$ follows. Hence $U \supseteq \cup \{f^{-1}(y) \mid f^{-1}(y) \cap U \neq \phi,\ y \in Y\}$ and the converse is trivial. Because of this equality, $f^{-1}(f(U)) = U$, hence $f(U)$ is open in Y by the definition of a quotient map. Because $f(x_1) \in f(U)$ and because $(f(x_1), f(x_2))$ is a P_1-pair in Y, $f(x_2) \in f(U)$, i.e. $x_2 \in f^{-1}(f(U)) = U$ follows.

Hence $\underline{R}_1(P_1) = \underline{R}_2^{*}(P_1)$ is a hereditary connectedness by Theorem 2.1 and Proposition 2.5 with corresponding disconnectedness $S\ \underline{R}_1(P_1) = \underline{R}_1^{*}(P_1) = \{X \mid X \text{ is a } {\sim}P_1\text{-space}\}$. Because X is a P_1-space if and only if X is indiscrete, $\underline{R}_1(P_1) = \{X \mid X$ is indiscrete$\}$. Furthermore, $(x\ ,\ y)$ is a ${\sim}P_1$-pair if and only if there exists a $U \subsetneq X$ open such that $x \in U$, $y \notin U$ or there exists a $V \subsetneq X$ open such that $y \in V$, $x \notin V$. Hence X is a ${\sim}P_1$-space if and only if X is a T_o-space, i.e. $\underline{R}_1^{*}(P_1) = \{X \mid X$ is a T_o-space$\}$. From Theorem 2.4, $\underline{R}_2(P_1) = R(R(P_1))$ is a connectedness. Because every non-trivial topological space can be mapped continiously onto the two-point indiscrete space, $\underline{R}_2(P_1) = \Phi$ follows.

2) Let P_2 be the following property:

P_2: x ∈ U if and only if y ∈ U for all U open and closed in X

where x, y ∈ X, X some topological space.

Then P_2 is simmetrical, i.e. $(x, y) \in P_2$ in X iff $(y, x) \in P_2$ in X and P_2 satisfies conditions (a), (b), (c) and (d). Hence $\underline{R}_1(P_2) = \underline{R}_2^*(P_2)$ is a connectedness. It is easily seen that $\underline{R}_1(P_2) = \{X | X \text{ is a connected space}\}$ and the corresponding disconnectedness is given by $S\,\underline{R}_1(P_2) = \{X | X \text{ is a totally disconnected space}\}$.

Before we give the next example, let us say something about the largest and smallest proper connectednesses. It is well known that $\underline{R}_1(P_1)$ is the smallest proper connectedness and $R_1(P_2)$ is the largest proper connectedness. Hence, if $\underline{C}$ is any proper connectedness, i.e. $\underline{C} \neq \Phi$ and $\underline{C} \neq \underline{Top}$, then by Theorem 2.1 there is a property P such that $\underline{C} = \underline{R}_1(P)$. Hence $\underline{R}_1(P_1) \subseteq \underline{R}_1(P)$

$\subseteq \underline{R}_1(P_2)$, i.e. $P_1 \leqq P \leqq P_2$.

3) Let P_3 be the following property:

P_3: $f(x) = f(y)$ for every continious function $f : X \to R$, where R is the set of real numbers with the usual topology

where $x, y \in X$, X some topological space.

Clearly P_3 satisfies conditions (a), (b) and (d). To see that P_3 satisfies condition (c), let $g : X \to Y$ be a quotient map such that $f^{-1}(y)$ is a P_3-subspace of X for all $y \in Y$, and let $(f(x_1), f(x_2))$ be a P_3-pair in Y. Let $g : X \to R$ be continious. Because $f^{-1}(y)$ is a P_3-subspace of X, $g(f^{-1}(y))$ is a singleton for all $y \in Y$.

Because f is a quotient map, there exists a unique continious function $h : Y \to R$ such that $h \circ f = g$. Then $g(x_1) = h(f(x_1)) = h(f(x_2)) = g(x_2)$ because $(f(x_1), f(x_2))$ is a P_3-pair in Y. Hence (x_1, x_2) a P_3-pair in X follows.

Hence $\underline{R}_1(P_3) = \underline{R}_2^*(P_3)$ is a connectedness and $\underline{R}_1(P_3) = \{X | X$ is functionally connected$\}$ with corresponding disconnectedness $S\,\underline{R}_1(P_3) = \{X | X$ is functionally disconnected$\}$.

4) Let P_4 be the following property:

P_4: if $x \in U$ then $y \in U$ for all U open in X where $x, y \in X$ for some space X.

P_4 satisfies conditions (a), (b) and (b*). Hence $R_1^*(P_4) = S\,\underline{R}_1(P_4)$ is a disconnectedness (Theorem 2.5 (4)) with corresponding connectedness $R\,\underline{R}_1^*(P_4) = \underline{R}_2^*(P_4)$ and by Theorem 2.4, $\underline{R}_2(P_4) = R\,R_1(P_4)$ is a connectedness. It is easy to see that $\underline{R}_1^*(P_4) = \{X | X \text{ is a } T_1\text{-space}\}$ and $\underline{R}_2^*(P_4) = \{X | X \text{ is absolutely connected}\}$. Furthermore $\underline{R}_2(P_4) = R\,\underline{R}_1(P_4) = \Phi$ because the two point indiscrete space is a P_4-space.

Note that if P' is the property:

P': if $x \in G$ then $y \in G$ for all G closed in X where $x, y \in X$ for some space X, then (x, y) is a P'-pair if and only if (x, y) is a P_4-pair. Hence the property P' yields nothing new.

5) A space X is *ultraconnected* if it has no disjoint closed sets, or equivalently, if the closures of distinct points always intersect. Let P_5 be the following property:

P_5: whenever x and y are in A, $A \subseteq X$, then

$$\bigcap_{a \in A} \overline{\{a\}} \neq \phi$$

where x, y $\in$ X for some space X. Then X is a P_5-space if and only if X is ultraconnected. P_5 satisfies condition (a). Indeed, if $f : X \to Y$ is an onto continious function with (x , y) a P_5-pair in X, let $A \subseteq Y$ such that $f(x) \in A$ and $f(y) \in A$. Then $x, y \in f^{-1}(A)$ and thus $\bigcap_{z \in f^{-1}(A)} \overline{\{z\}} \neq \phi$. Let $p \in \bigcap_{z \in f^{-1}(A)} \overline{\{z\}}$. Then $p \in \overline{\{z\}}$ for all z such that $f(z) \in A$. Let $a \in A$. Because f is onto, there is a $z_a \in f^{-1}(A)$ such that $f(z_a) = a$. Thus $p \in \overline{\{z_a\}}$ and $f(p) \in f(\overline{\{z_a\}}) \subseteq \overline{\{f(z_a)\}} = \overline{\{a\}}$ follows. This means $\bigcap_{a \in A} \overline{\{a\}} \neq \phi$.

From Proposition 2.5(4) we then have $S\, \underline{R}_1(P_5) = S\{X|X$ is totally ultradisconnected$\}=\{X|X$ is a T_1-space$\}$ is a disconnectedness. Because $\underline{R}_1(P_5) \subsetneq R\, S\, \underline{R}_1(P_5) =$ $R\{X|X$ is a T_1-space$\}=\{X|X$ is absolutely connected$\}$ and because the above inclusion is strict (X = R with topology $\{R, \phi, R-\{0\}, R-\{1\}, R-\{0, 1\}\}$ is absolutely connected but not ultraconnected (cf. [2])), the class of ultraconnected spaces is not a connectedness.

6) A space X is *hyperconnected* if X has no disjoint open sets, or equivalently, if the closure of any non-empty open set is precisely X. Let P_6 be the folowing property:

P_6: if $x \in U$ then $y \in \overline{U}$ for all U open in X where $x , y \in X$ for some space X. Firstly, note that P_6 is simmetrical. It is sufficient to show that $(x , y) \in P_6$ in X implies $(y , x) \in P_6$ in X for the converse will follow likewise. Suppose thus $(x , y) \in P_6$ in X and let U be open in X with $y \in U$. If $x \notin \overline{U}$, then $x \in X-\overline{U}$ and because $(x , y) \in P_6$ holds, $y \in \overline{X - \overline{U}}$ follows. Thus $U \cap (X - \overline{U}) \neq \phi$ which obviously cannot be true. Hence $x \in \overline{U}$ must hold showing that $(y , x) \in P_6$ in X. Using this simmetric property of P_6, it is easy to see that X is hyperconnected if and only if X is a P_6-space. P_6 satisfies condition (a) and (b). Hence $S\,\underline{R}_1(P_6) = S\{X|X$ is hyperconnected$\} = \{X|X$ is totally hyperdisconnected$\}$ is a disconnectedness and $\underline{R}_2(P_6) = R\,\underline{R}_1^*(P_6) = \{X|$ every non-trivial continious image of X contains two distinct points such that the one point is in the closure of any open set that contains the other point$\}$ is a connectedness. We can just mention that the class of all hyperconnected spaces is not a connectedness (cf. [2]).

7) Let P_7 be the following property:

P_7: there is a U and V open in X such that $x \in U$, $y \in V$ and $U \cap V = \phi$

where $x, y \in X$ for some space X. Then X is a P_7-space if and only if X is a T_2-space. Because P_7 satisfies condition (b*), $\underline{R}_2(P_7) = R\ \underline{R}_1(P_7) = \{X|$ every non-trivial continious image of X is not a T_2-space$\}$ is a connectedness. The class of all T_2-spaces is not a disconnectedness.

8) Let P_8 be the following property:

P_8: there is a U and a V open in X with $x \in U$, $y \in V$ and such that $\overline{U} \cap \overline{V} = \phi$

where $x, y \in X$ for some space X. Then X is a P_8-space if and only if X is $T_{2\frac{1}{2}}$. Because P_8 satisfies condition (b*), $\underline{R}_2(P_8) = R\ \underline{R}_1(P_8) = \{X|$ every non-trivial continious image of X is not a $T_{2\frac{1}{2}}$-space$\}$ is a connectedness. The class of all $T_{2\frac{1}{2}}$-spaces is not a disconnectedness.

9) We recall, a *path (arc)* between two points x and y in a space X is a continious (injective and continious) function $f : I \to X$ with $f(0) = x$ and $f(1) = y$

where I is the unit interval with the usual topology. X is called *path (arc) connected* if there is a path (arc) from x to y in X for all x, y $\in$ X. The *path component* of x in X is the equivalence class given by the relation "there is a path from x to y". X is called *totally pathwise disconnected* if each path component is a singleton. If X is such that it has no non-trivial continious image that is totally pathwise disconnected, then X is called *weakly path connected.* Let P_9 be the following property:

P_9: there is a path from x to y in X

where x, y $\in$ X for some space X. Then X is a P_9-space if and only if X is path connected. Furthermore, P_9 satisfies conditions (a) and (b). Hence $S\ \underline{R}_1(P_9)$ is a disconnectedness and $\underline{R}_2^*(P_9)$ is a connectednesses. For $S\ \underline{R}_1(P_9)$, because I itself is arc connected and thus also path connected, we have $S\ \underline{R}_1(P_9) = \{X|X$ is totally path disconnected$\} = \underline{R}_1^*(P_9)$. Furthermore, $\{X|X$ is path connected$\} = \underline{R}_1(P_9) \subsetneq R\ S\ \underline{R}_1(P_9) = \{X|X$ is weakly path connected$\}$ and because the inclusion is strict. (cf. [2] §9 no. 5), the class of all path connected spaces is not a connectedness. For $\underline{R}_2^*(P_9)$ we have $\underline{R}_2^*(P_9) = R\ \underline{R}_1^*(P_9) = R\ S\ \underline{R}_1(P_9) = \{X|X$ is weakly path connected$\}$ which does not yield a new connectedness. Note lastly, that $S\{X|X$ is arcwise connected$\} = S\{X|X$ is weakly pathwise connected$\} = \{X|X$ is totally pathwise disconnected$\}$ (cf. [2]).

REFERENCES

[1] ARHANGEL'SKII, A.V. and R. WIEGANDT, Connectednesses and disconnectednesses in Topology, *Gen. Top and Appl*, 5, 1975, 9-33.

[2] OHLHOFF, H.J.K., Connectedness and left constants in Topology, *Research report* UP W 79/5, University of Pretoria, 1979.

[3] TILLER, J.A., Component subcategories. *Quaestiones Mathematicae*, 4, 1980, 19-40.

[4] VELDSMAN, S., On the characterization of radical and semisimple classes in categories. *Comm. in Algebra*, 10 (no. 9), 1982.

[5] WIEGANDT, R. Radicals of rings defined by means of elements. *Österreich. Akad. Wiss., math.-naturw. Kl., S.-ber.*, Abt.II, 184, 1975, 117-125.

S. Veldsman
Department of Mathematics
Rand Afrikaans University
P.O. Box 524
Johannesburg 2000

New address:

Department of Mathematics
University of Port Elizabeth
P.O. Box 1600
Port Elizabeth 6000
Republic of South Africa

COLLOQUIA MATHEMATICA SOCIETATIS JÁNOS BOLYAI
38. RADICAL THEORY, EGER (HUNGARY), 1982.

THE GALOIS CORRESPONDENCE BETWEEN RADICAL AND CORADICAL CLASSES

S. M. VOVSI

INTRODUCTION

In the present paper we are concerned with radicals in linear representations of groups and, to a lesser extent, with radicals in certain multioperator groups. It is well known that the foundations of the general theory of radicals were laid in papers by Kuroš [6] and Amitsur [1] in the early fifties. Although these papers dealt mainly with associative rings and algebras, their arguments and constructions were quite universal and turned out to be applicable to many other algebraic structures, in particular, to groups. However, it was discovered that, in contrast to the case of rings, for group theory the condition of extension-closedness of radical and coradical (=semisimple) classes in the sense of Kuroš-Amitsur proved to be too restrictive. For example, the

Hirsch-Plotkin radical, the Baer radical, the Gruenberg radical are, probably, the most popular radicals in group theory, but none of them is closed under extensions. As a result, in a number of papers on radicals in groups, published in the late fifties - early seventies, the requirement of extension-closedness was omitted. One should especially mention a series of papers by Plotkin [11, 12, 13] where this point of view has been presented in the most completed form.

Such an approach has a number of advantages. First, it covers all the known examples. Second, if we accept such an axiomatics of radical, then varieties of groups become particular cases of coradical classes, showing clearly the duality between the concepts of radical class and variety. Third, the collections of radical and coradical classes in the sense of Plotkin are "well-organised". In particular, under the usual multiplication of classes of groups they form "semigroups" (whose idempotents are just the Kuroš-Amitsur radical and semisimple classes). Finally, there arise natural applications of the theory to traditional problems of group theory connected with ascending and descending series of subgroups. For example, a well known theorem of

Mal'cev [8] on the existence of a group with derived series of arbitrary transfinite length is a special case of a general fact concerning infinite powers of coradical classes of groups [18, 10].

However, it should be noted that in passing to non-extension-closed radical and coradical classes we loose one important feature of the Kuroš-Amitsur theory. Namely, there is a natural one-to-one correspondence between radical and semisimple classes in the sense of Kuroš-Amitsur, but there is nothing of the kind if we consider *arbitrary* radical and coradical classes. This immediately leads to the question: *does there exist any "good" connection between radical and coradical classes of groups?* In the first, preliminary, section of the present paper we show that an answer to this question is easily deduced from results of Fried and Wiegandt [3]. Namely, it turns out that there is a Galois correspondence between radical and coradical classes and, moreover, the elements closed under this correspondence are just the Kuroš-Amitsur radical and semisimple classes. (Thus the Kuroš-Amitsur one-to-one correspondence is automatically obtained from the general properties of Galois connections.) We note that this fact is valid not only for groups but for arbitrary multioperator groups satisfying the condition that a characteristic ideal of an ideal be itself an ideal.

In §§ 2 and 3 we deal with the main topic of the paper - radicals in group representations. In § 2 we give some basic definitions and, using the material of § 1, prove the connection theorem for radical and coradical classes of group representations. Despite the elementary proof, this theorem seems to be of some interest because it clears up the situation completely and, on the other hand, leads to several interesting questions. These questions and some relevant results are discussed in § 3.

The author is indebted to Dr. L. Márki for his criticism and valuable remarks.

§ 1. RADICALS AND CORADICALS IN MULTIOPERATOR GROUPS

Let $\mathcal{U}$ be a class of Ω-groups of a given type Ω, satisfying the following conditions:

U1. $\mathcal{U}$ is closed with respect to homomorphic images and ideals.

U2. If $G \in \mathcal{U}$, $H \triangleleft G$ and K is a characteristic ideal of H, then $K \triangleleft G$.

In this section such a $\mathcal{U}$ will be called a *universal class.*

EXAMPLES. 1. The class of all left (right) modules over an arbitrary ring R .

2. The class of all groups.

3. The class of all lattice-ordered (ℓ-) groups.

4. The class of all finite-dimensional Lie algebras over the field of reals.

5. An arbitrary subclass of a universal class which satisfies U1.

Examples 1, 2 and 5 are trivial. Example 3 is also straightforward. The only thing which deserves to be verified is that the class of real finite-dimensional Lie algebras satisfies U2. Let L be such an algebra, $M \triangleleft L$ and N a characteristic ideal of M (i.e., $N^A = N$ for every $A \in \mathrm{Aut}\, M$). It is enough to prove that N is invariant under any derivation D of L . Consider the series

$$\exp D = E + D + D^2/2! + D^3/3! + \dots .$$

It is well known that this series converges in End L and that its sum $A = \exp D$ is an automorphism of the algebra L (see e.g. [5], p. 197). Since $M \triangleleft L$, M is invariant under D and therefore M is invariant under A. Furthermore, M is finite-dimensional, so A induces an automorphism of M. Since N is characteristic in M, it follows that $N^A = N$. Hence N is also invariant under

$$\log A = (A-E) - (A-E)^2/2 + (A-E)^3/3 - \dots$$

(the series converges only if $\|A-E\| < 1$, but it is easy to assure this condition by multiplying the original D by a suitable scalar). Since $\log A = \log \exp D = D$, we have $N^D \subseteq N$, as required.

NOTE. We did not try to present the above examples in the most general form. For instance, the class of all *operator groups* with a fixed system of operators is certainly a universal class. As special cases of the latter we obtain Examples 1 and 2. Furthermore, in Example 4 we need not restrict ourselves to the field of reals - any valuation field of characteristic 0 may be taken as well.

From now on and to the end of this section we fix a universal class $\mathcal{U}$ of Ω-groups, and every Ω-group is supposed to belong to $\mathcal{U}$. All the classes of Ω-groups (i.e., subclasses of $\mathcal{U}$) are abstract and contain the trivial Ω-group. If $\mathcal{X}$ is a class of Ω-groups then by an $\mathcal{X}$-group we mean an Ω-group belonging to $\mathcal{X}$. An ideal H of an Ω-group G is called its $\mathcal{X}$-*ideal (co-*$\mathcal{X}$*-ideal)* if $H \in \mathcal{X}$ $(G/H \in \mathcal{X})$.

The following definitions repeat the corresponding group-theoretical definitions from [12].

DEFINITIONS. A class $\mathcal{X}$ of Ω-groups is said to be *radical* if

R1. $\mathcal{X}$ is homomorphically closed.

R2. Every Ω-group has a largest $\mathcal{X}$-ideal.

A class $\mathcal{Y}$ of Ω-groups is said to be *coradical* if

C1. $\mathcal{Y}$ is closed under taking ideals.

C2. Every Ω-group has a smallest co-$\mathcal{Y}$-ideal.

If G is an Ω-group and $\mathcal{X}$ is a radical class then the largest $\mathcal{X}$-ideal of G is denoted by $\mathcal{X}(G)$ and is called the $\mathcal{X}$-*radical* of G. If $\mathcal{Y}$ is a coradical class then the smallest co-$\mathcal{Y}$-ideal of G is denoted by $\mathcal{Y}^*(G)$ and called the $\mathcal{Y}$-*coradical* of G. A variety of Ω-groups is a special case of a coradical class: the corresponding coradical is just the verbal Ω-subgroup associated with this variety.

A radical (coradical) class $\mathcal{X}$ is called a *Kuroš-Amitsur radical (semisimple) class* if it is extension-closed, i.e.,

$$(G/H \in X) \;\&\; (H \in X) \Rightarrow (G \in X)$$

for each G and each $H \lhd G$.

Our first aim is to establish a connection between radical and coradical classes. For any radical class X denote by $X^{\vee}$ the class of all Ω-groups G such that $X(G)=\{1\}$. On the other hand, for any coradical class Y denote by $Y^{\wedge}$ the class of all G such that $Y^{*}(G)=G$.

THEOREM 1. *If* X *is a radical class then* $X^{\vee}$ *is a coradical class. If* Y *is a coradical class then* $Y^{\wedge}$ *is a radical class. The maps* $X \mapsto X^{\vee}$ and $Y \mapsto Y^{\wedge}$ *form a Galois correspondence between radical and coradical classes. The closed elements under this correspondence are exactly the Kuroš-Amitsur radical and semisimple classes.*

Proof can be easily extracted from [3]. Indeed, denote by A the class of all nonzero objects from U , and for arbitrary $G,H \in A$ let $G \mapsto H$ denote that H is an epimorphic image of G , and $G \succ\!\!- H$ - that H is a subnormal Ω-subgroup of G . Then $\mapsto$ and $\succ\!\!-$ are preorders on A and, applying the arguments of [3, § 2] plus a few obvious remarks to this A , we obtain the required result. For details see [3].

From now on the Kuroš-Amitsur radical and semisimple classes will be called *closed* radical and coradical classes, respectively. Furthermore, for the sake of brevity we shall often say "radical" and "coradical" instead of "radical class" and "coradical class".

From Theorem 1 and the general properties of Galois connections we obtain

COROLLARY 1. *Let* $\mathcal{X}$ *be a radical and* $\mathcal{Y}$ *a coradical. Then*

(i) $\mathcal{X}^{\vee}$ $(\mathcal{Y}^{\wedge})$ *is a closed coradical (radical);*

(ii) $\mathcal{X}^{\vee\wedge}(\mathcal{Y}^{\wedge\vee})$ *is the smallest closed radical (coradical) containing* $\mathcal{X}(\mathcal{Y})$;

(iii) *the maps* $\mathcal{X} \mapsto \mathcal{X}^{\vee}$ *and* $\mathcal{Y} \mapsto \mathcal{Y}^{\wedge}$ *yield a one-to-one correspondence between closed radical and coradical classes.* □

For any radical $\mathcal{X}$ and coradical $\mathcal{Y}$ their Galois closures $\mathcal{X}^{\vee\wedge}$ and $\mathcal{Y}^{\wedge\vee}$ are denoted by $\tilde{\mathcal{X}}$ and $\tilde{\mathcal{Y}}$, respectively. There is a simple and well-known method of constructing these closures from the given $\mathcal{X}$ and $\mathcal{Y}$. Namely, $\tilde{\mathcal{X}}$ is the class of all Ω-groups possessing an ascending series of ideals (in general, of a transfinite length) with $\mathcal{X}$-factors, and $\tilde{\mathcal{Y}}$ is the class of Ω-groups possessing a descending series of ideals with

$\mathcal{Y}$-factors.

A radical is said to be *hereditary* if it is closed under taking ideals. A coradical is said to be a *prevariety* if it is closed under taking Ω-subgroups. It follows from the above that if $\mathcal{X}(\mathcal{Y})$ is a hereditary radical (prevariety) then $\tilde{\mathcal{X}}(\tilde{\mathcal{Y}})$ is also a hereditary radical (prevariety). Although there is sometimes a strong duality between hereditary radicals and prevarieties (see e.g. [17], [18]) the maps $^\vee$ and $^\wedge$ do not take hereditary radicals into prevarieties and vice versa.

NOTES. 1. All of the above is true for arbitrary universal classes of Ω-groups. At the same time each of these classes has its own features. For instance, a class of *groups* is both radical and coradical if and only if it is trivial (that is, either the class of all groups or the class of unit groups). For ℓ-groups the situation is quite opposite: by a theorem of Holland [4], a class of ℓ-groups is both radical and coradical if and only if it is a variety.

2. The considerations of this paper were partly stimulated by [9], where similar questions were studied in a particular case of ℓ-groups.

3. It was proved in [16] that if $\mathcal{Y}$ is a variety of groups then $\mathcal{Y}^{\wedge}$ is a closed radical. Clearly, this is a very special case of Corollary 1 (i).

§ 2. RADICALS AND CORADICALS IN LINEAR REPRESENTATIONS

We shall restrict ourselves to representations of *groups*, although it should be noted that the whole section can be literally carried over to representations of semigroups, rings, associative and Lie algebras, etc.

Let K be an arbitrarily fixed commutative ring with identity. A representation $\rho: G \to \mathrm{Aut}\, A$ of a group G in a K-module A will be denoted by $\rho = (A, G)$. All group representations over K with naturally defined morphisms form the *category* REP-*K of group representations over K*, which is the universal class of the present section. Next we recall several definitions relating to this category; for the details we refer to [15] and [20].

By a class of group representations we mean an abstract class containing the zero representation of the unit group. If $\rho = (A, G)$ is a representation, $H = \mathrm{Ker}\, \rho$ its kernel, then the naturally arising faithful representation $\bar{\rho} = (A, G/H)$ is called the *faithful image* of ρ. Two representations ρ and σ are called *equivalent* (notation $\rho\sim\sigma$) if $\bar{\rho}\cong\bar{\sigma}$. A class X of representations is said to be *saturated* if

$$(\rho\in X)\ \&\ (\rho\sim\sigma) \Rightarrow (\sigma\in X)\ .$$

Let $\rho=(A,G)$ be a representation, B a G-submodule ($=KG$-submodule) of A. Then we have two representations (B,G) and $(A/B,G)$, called a *left subrepresentation* and a *left factor-representation* of ρ, while ρ is called an *extension* of (B,G) by $(A/B,G)$. If X is a class of representations, then B is an *X-G-submodule of* A if $(B,G)\in X$, and a *co-X-G-submodule of* A if $(A/B,G)\in X$.

Definitions. A class X of group representations is said to be *radical* if

R1. X is saturated.

R2. X is homomorphically closed.

R3. For every representation $\rho=(A,G)$ there exists a largest X-G-submodule of A (called the *X-radical* of ρ and denoted by $X(\rho)=X(A,G)$).

A class Y of group representations is said to be *coradical* if

C1. Y is saturated.

C2. Y is closed with respect to left subrepresentations.

C3. For every representation $\rho=(A,G)$ there exists a smallest co-$\mathcal{Y}$-G-submodule of A (called the $\mathcal{Y}$-*coradical* of ρ and denoted by $\mathcal{Y}^*(\rho) = \mathcal{Y}^*(A,G)$).

Note that the above definition of radical class is a slight modification of that given in [14], while the notion of coradical class has not been considered earlier. It is not hard to see that there is a natural duality between radical and coradical classes. This duality becomes quite clear if one realizes that for saturated classes the condition R2 is equivalent to the following: $\mathcal{X}$ is closed with respect ot left factor-representations.

For a class $\mathcal{X}$ of group representations and a group G denote by $\mathcal{X}_G$ the G-*layer* of $\mathcal{X}$, that is, the class of right KG-modules A for which the associated representation (A,G) belongs to $\mathcal{X}$. Using this notation we shall sometimes write $A\in\mathcal{X}_G$ instead of $(A,G)\in\mathcal{X}$.

The following assertion is straightforward.

PROPOSITION 1. A class $\mathcal{X}$ of group representations is a radical (coradical) class if and only if it is saturated and for every group G the corresponding layer $\mathcal{X}_G$ is a radical (coradical) class of KG-modules.

Furthermore, X is extension-closed if and only if each X_G is extension closed. □

Radical and coradical classes can be characterized by means of certain operations on classes of representations. For a class X denote:

VX - the class of all representations possessing a right epimorphic image belonging to X ;

QX $(Q_\ell X, Q_r X)$ - the class of all (left, right) epimorphic images of X-representations;

SX $(S_\ell X, S_r X)$ - the class of all (left, right) subrepresentations of X-representations;

CX $(C_r X)$ - the class of all (right) Cartesian products of X-representations;

DX - the class of direct products of X-representations;

$D_\ell X$ - the class of representations (A,G) such that A is a direct sum of its X-G-submodules;

RX - the class of representations (A,G) such that A is the sum of its X-G-submodules;

AX - the class of subdirect products of X-representations;

$A_\ell X$ - the class of representations (A,G) such that A has a collection of co-X-G-submodules $B_i, i \in I$, with $\bigcap_{i \in I} B_i = \{0\}$;

$A_r X$ - the class of representations (A,G) where A is a direct sum of its G-submodules $A_i, i \in I$, and for each i there is a normal subgroup H of G acting identically on A and such that $(A_i, G/H_i) \in X$ and $\bigcap_{i \in I} H_i = \{1\}$.

It is easy to see that $V, Q, \ldots, A_r$ are closure operations. Almost all of them were introduced in [14] although our notation differs sometimes from that of [14].

Further, Rad X and Corad X denote the radical class and the coradical class, respectively, generated by a given X, while $\{U_1, U_2, \ldots\}X$ stands for the smallest class closed under the operations $U_1, U_2, \ldots$ and containing X.

PROPOSITION 2 (cf. [14], Theorem 1). 1. *For any class of representations* X *the following conditions are equivalent:* (i) X *is a radical class;* (ii) $X = \{V,Q,R\}X$; (iii) $X = \{V,Q,A_r\}X$; (iv) $X = \{V,Q,D_\ell\}X$.

2. *For any class of representations* X,

$$\text{Rad } X = VQA_r X = VQD_\ell VX = VQRV\,X .$$

Proof. 1. The validity of (i) ⇒ (ii) is evident. It

is also easy to see that

(4) $$D_\ell \leq A_r \leq D_\ell V \leq RV ,$$

whence (ii) ⇒ (iii) ⇒ (iv). Finally, if $X=\{V,Q,D_\ell\}X$ then X is saturated, homomorphically closed, and every representation (A,G) has a largest X-G-submodule. Hence (iv) ⇒ (i).

2. It follows from the above that Rad $X = \{V,Q,A_r\}X$. Since the inequalities

$$QV \leq VQ \ , \ A_rV \leq VA_r \ , \ A_rQ \leq QA_r$$

between closure operations are valid [14], it follows that Rad $X = VQA_rX$. The rest follows from (4). □

PROPOSITION 3. 1. *For any class of representations* Y *the following conditions are equivalent:* (i) Y *is a coradical class;* (ii) $Y=\{V,Q_r,S_\ell,A\}Y$; (iii) $Y = \{V,Q_r,S_\ell,A_\ell\}Y$.

2. *For any class of representations* Y ,

$$\text{Corad } Y = VQ_rS_\ell AY = VQ_rS_\ell A_\ell VY .$$

Proof. 1. The implications (i) ⇒ (ii) ⇒ (iii) ⇒ (i) are fairly obvious.

2. In view of the latter, Corad $= \{V,Q_\ell,S_\ell,A\}$,

therefore to prove the equality $\mathrm{Corad} = VQ_\ell S_\ell A$ one has to establish that

$$Q_r V \le VQ_r \ , \ S_\ell V \le VS_\ell \ , \ AV \le VA \ ,$$

$$S_\ell Q_r \le Q_r S_\ell \ , \ AQ_r \le Q_r A \ , \ AS_\ell \le S_\ell A \ .$$

These inequalities are more or less straightforward. Let us verify, for instance, that $AV \le VA$. Let $\rho = (B,G) \in AVX$. Then there exist normal subrepresentations $\rho_i = (B_i, G_i)$, $i \in I$, of ρ such that $\rho/\rho_i \in VX$, $\cap B_i = \{0\}$, $\cap G_i = \{1\}$. Since $\rho/\rho_i = (B/B_i, G/G_i) \in VX$, for every $i \in I$ there is a normal subgroup H_i of G such that $G_i \subseteq H_i \subseteq G$, $H_i/G_i \subseteq \mathrm{Ker}(\rho/\rho_i)$ and $(B/B_i, G/H_i) \in X$. Then $H = \bigcap_{i \in I} H_i$ acts identically on each B/B_i, so that from $\cap B_i = \{0\}$ it follows that $H \subseteq \mathrm{Ker}\,\rho$. Now the representation $(B, G/H)$ is a subdirect product of $(B/B_i, G/H_i)$, $i \in I$. Since the latter belong to X, we have $(B, G/H) \in AX$, whence $\rho = (B,G) \in VAX$.

Thus $\mathrm{Corad} = VQ_r S_\ell A$. Since $A \le A_\ell V$, it follows from the first part of the proposition that $\mathrm{Corad} = VQ_r S_\ell A_\ell V$. □

An analogue of Theorem 1 holds for radical and coradical classes of representations. Namely, if X is a radical then by $X^\vee$ we denote the class of all (A,G)

such that $X(A,G)=\{0\}$. Dually, if Y is a coradical then $Y^\wedge$ is the class of all (A,G) with $Y^*(A,G)=A$.

THEOREM 2. *If* X *is a radical then* $X^\vee$ *is a coradical. If* Y *is a coradical then* $Y^\wedge$ *is a radical. The maps* $X \mapsto X^\vee$ *and* $Y \mapsto Y^\wedge$ *establish a Galois correspondence. The closed elements under this correspondence are exactly the extension-closed radicals and coradicals.*

Proof. We could argue as in proving Theorem 1, but now this is not the best way. Note first that for every radical X , coradical Y and group G ,

$$(5) \qquad (X^\vee)_G = (X_G)^\vee \ , \ (Y^\wedge)_G = (Y_G)^\wedge \ .$$

Using these evident formulas and Proposition 2, one can directly deduce Theorem 2 from Theorem 1. Indeed, suppose one has to prove that $X^{\vee\wedge}=X \Leftrightarrow X$ is extension-closed. Using (5), Theorem 1 and Proposition 1, we have:

$$X^{\vee\wedge} = X \Leftrightarrow \forall G: (X^{\vee\wedge})_G = X_G \Leftrightarrow \forall G: (X_G)^{\vee\wedge} = X_G \Leftrightarrow$$

$$\Leftrightarrow \forall G: \ X_G \text{ is extension closed} \Leftrightarrow X \text{ is extension closed.} \ \square$$

As in § 1, the terms "closed radical" and "closed coradical" are now quite natural. For closed radicals and coradicals the analogue of Corollary 1 holds. The *closures* of a radical X and a coradical Y will be denoted, as in § 1, by $\tilde{X}$ and $\tilde{Y}$. It is easy to verify that $\tilde{X}(\tilde{Y})$ consists of representations (A,G) such that A has an ascending (descending) series of G-submodules whose factors belong to $X(Y)$.

Radicals and coradicals of representations are said to be *hereditary* if they are closed with respect to subrepresentations. Hereditary radicals were introduced in [14], hereditary coradicals (which are usually called *prevarieties*) - in [19]. If X is a hereditary radical (prevariety) of representations then for any group G the corresponding layer X_G is a hereditary radical (prevariety) of KG-modules: the converse is, in general, false. We also remark that, as in the case of Ω-groups, the maps $^\vee$ and $^\wedge$ do not take hereditary radicals into prevarieties and vice versa.

For a class of representations X denote by rad X , pvar X and var X the hereditary radical, the prevariety and the variety generated by X . It was proved in [14] and [19] that

$$\text{rad} = VQSC_r \text{ , pvar} = VQ_rSC \text{ , var} = VQSC \text{ .}$$

Therefore we have the following "diagram":

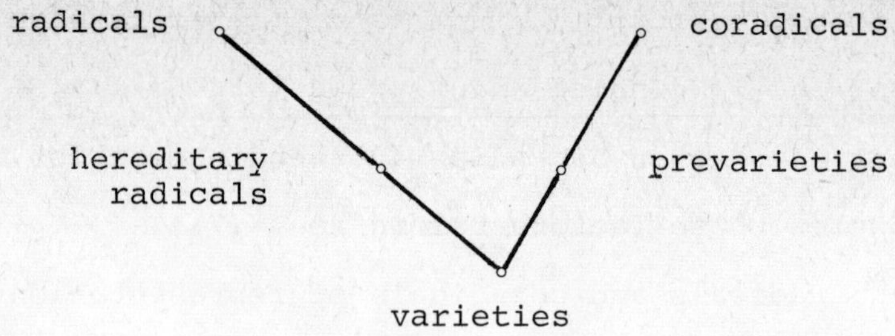

(Incidentally, as noted in § 1, the same diagram is true for lattice-ordered groups.)

§ 3. CERTAIN PROPERTIES OF THE MAPS $^\vee$ AND $^\wedge$

In this section we prove several simple assertions and pose a few problems which are closely associated with the previous considerations. Since the present interests of the author are connected with group representations, only these objects will be dealt with in what follows. However, it should be emphasized that the main body of the section is equally valid for linear representations of other algebraic structures as well as for any universal class of Ω-groups in the sense of § 1.

1. Let K be a commutative ring with 1 . Unless otherwise stated, X denotes a radical class and Y a coradical class of group representations over K . Further, O is the class of all representations over K and E the class of zero representations.

The following formulas are evident:

(6) $\quad X \cap X^{\vee} = E \; , \qquad Y \cap Y^{\wedge} = E \; ,$

(7) $\quad X^{\vee} = 0 \Leftrightarrow X = E \; , \; Y^{\wedge} = 0 \Leftrightarrow Y = E \; ,$

(8) $\quad 0^{\vee} = E \; , \qquad 0^{\wedge} = E \; .$

In contrast to (7), the formulas (8) cannot be "reversed", that is, the implications $X^{\vee} = E \Rightarrow X = 0$ and $Y^{\wedge} = E \Rightarrow Y = 0$ are not, in general, correct. Indeed, let $K=\mathbb{Z}_4$ and let E_2 be the variety of group representations over $\mathbb{Z}_4$ whose domains are abelian groups of exponent 2. Then E_2 is both a radical and a coradical and it is easy to see that $E_2^{\vee}=E_2^{\wedge}=E$. On the other hand, $E_2 \neq 0$.

The result is just the opposite if the ground ring is a field.

PROPOSITION 4. *If* K *is a field then* $X^{\vee}=E \Rightarrow X = 0$.

Proof. Let $X^{\vee}=E$. Take an arbitrary free group F and consider its right regular representation $\rho = (KF,F)$ over K. From $X^{\vee}=E$ it follows that $A = X(\rho) \neq \{0\}$, i.e., A is a nonzero right ideal of the group algebra KF. By a theorem of Cohn [2], KF is a fir (free ideal ring), so that A is a free KF-module. Since $(A,F) \in X$, it follows that X contains all representations (V,F) where V is a free KF-module, i.e., all absolutely free representations. As X is homomorphically closed, we obtain $X=0$. □

Problem 1. Let Y be a coradical class of group representations over a field. Does the implication $Y^{\wedge}=E \Rightarrow Y=0$ hold?

If Y is a hereditary coradical (i.e., a prevariety), this question is easily answered in the affirmative. For if $Y^{\wedge}=E$ then Y contains all irreducible representations, but every representation over a field can be embedded in an irreducible one. Being hereditary, Y must coincide with 0.

2. Let X be a variety of group representations over an arbitrary K. As noted at the end of § 2, X is both a radical and a coradical. Therefore we have a coradical $X^{\vee}$ and a radical $X^{\wedge}$. Can these classes of representations coincide? It is clear that over any K the equation $X^{\vee}=X^{\wedge}$ has two trivial solutions E and 0. In general, it may have nontrivial solutions, as the above example of the variety E_2 over $\mathbb{Z}_4$ shows. But if K is a field, the situation is again simpler.

PROPOSITION 5. *If K is a field then the equation $X^{\vee}=X^{\wedge}$ has only trivial solutions.*

Proof. If $X^{\vee}=X^{\wedge}$ then $X^{\vee}$ is both a closed radical and a closed coradical, i.e., a closed variety. Since the semigroup of varieties of group representations over a field is free, it does not have nontrivial idempotents, i.e., nontrivial closed varieties. Hence either $\mathcal{X}^{\vee}=\mathcal{O}$ or $\mathcal{X}^{\vee}=\mathcal{E}$. Using respectively (7) or Proposition 4, we obtain $\mathcal{X}=\mathcal{E}$ or $\mathcal{X}=\mathcal{O}$. □

Problem 2. 1) Investigate the equation $\mathcal{X}^{\vee}=\mathcal{X}^{\wedge}$ over $\mathbb{Z}$. 2) Investigate this equation over $\mathbb{Z}_n$.

3. Given two classes of representations $\mathcal{X}$ and $\mathcal{Y}$, their *product* $\mathcal{X}\mathcal{Y}$ is the class of representations (A,G) such that A possesses a G-submodule B with $(B,G)\in\mathcal{X}$ and $(A/B,G)\in\mathcal{Y}$. It is easy to verify that if $\mathcal{X}$ and $\mathcal{Y}$ are radicals (hereditary radicals, coradicals, prevarieties, varieties) then so is $\mathcal{X}\mathcal{Y}$. Since the multiplication of classes of representations is associative, one can speak of the *semigroups* of radicals and coradicals, and of their subsemigroups of hereditary radicals and prevarieties, keeping in mind that these "semigroups" are defined on classes of elements, not on sets.

Evidently, the intersection $\mathcal{X}_1\wedge\mathcal{X}_2$ of radicals is a radical, and the same holds for coradicals, hereditary radicals, etc. Further, the *joins* of radicals

$\mathcal{X}_1, \mathcal{X}_2$ and coradicals $\mathcal{Y}_1, \mathcal{Y}_2$ are defined, as usual, by

$$\mathcal{X}_1 \vee \mathcal{X}_2 = \mathrm{Rad}(\mathcal{X}_1 \cup \mathcal{X}_2) , \quad \mathcal{Y}_1 \vee \mathcal{Y}_2 = \mathrm{Corad}(\mathcal{Y}_1 \cup \mathcal{Y}_2) .$$

It is easy to see that $\mathcal{X}_1 \vee \mathcal{X}_2$ consists of all (A,G) with $A = \mathcal{X}_1(A,G) + \mathcal{X}_2(A,G)$, and $\mathcal{Y}_1 \vee \mathcal{Y}$ consists of all (A,G) such that $\mathcal{Y}_1^*(A,G) \cap \mathcal{Y}_2^*(A,G) = \{0\}$.

Under the operations $\vee$ and $\wedge$, the collections of all radicals and all coradicals of group representations over K form complete *lattices*.

Problem 3. Under which conditions on the ground ring K are the lattices of radicals and coradicals modular?

As to the distributivity of these lattices, we believe that they are *never* distributive, although this is not yet proved.

What can we say about the connections between the semigroups and the lattices of radicals and coradicals?

PROPOSITION 6. *For any radicals* $\mathcal{X}, \mathcal{X}_1$ *and* $\mathcal{X}_2$

(9) $\mathcal{X}(\mathcal{X}_1 \vee \mathcal{X}_2) = \mathcal{X}\mathcal{X}_1 \vee \mathcal{X}\mathcal{X}_2 , \quad \mathcal{X}(\mathcal{X}_1 \wedge \mathcal{X}_2) = \mathcal{X}\mathcal{X}_1 \wedge \mathcal{X}\mathcal{X}_2 .$

For any coradicals $\mathcal{Y}, \mathcal{Y}_1$ and $\mathcal{Y}_2$,

(10) $(\mathcal{Y}_1 \vee \mathcal{Y}_2)\mathcal{Y} = \mathcal{Y}_1\mathcal{Y} \vee \mathcal{Y}_2\mathcal{Y} , \quad (\mathcal{Y}_1 \wedge \mathcal{Y}_2)\mathcal{Y} = \mathcal{Y}_1\mathcal{Y} \wedge \mathcal{Y}_2\mathcal{Y} .$

Proof. To prove the first equality, suppose that $(A,G) \in X(X_1 \vee X_2)$. This means that if $B = X(A,G)$ then $(A/B,G) \in X_1 \vee X_2$, i.e., $A/B = C_1/B + C_2/B$ where $(C_i/B,G) \in X_i$. Therefore $A=C_1+C_2$ and $(C_i,G) \in XX_i$, whence $(A,G) \in XX_1 \vee XX_2$. Thus $X(X_1 \vee X_2) \subseteq XX_1 \vee XX_2$, while the reverse inclusion is obvious.

The other equalities are proved similarly. □

Problem 4. Under which conditions on the ground ring K will the radical (coradical) classes satisfy all the equalities (9) and (10)? In other words, when does the collection of all radicals (coradicals) form a lattice-ordered semigroup?

4. We return to the maps $X \mapsto X^{\vee}$ and $Y \mapsto Y^{\wedge}$. How do they interact with the multiplication, intersection and join of radicals and coradicals?

PROPOSITION 7. *For any radicals* X_1 *and* X_2 ,

$$(X_1X_2)^{\vee} = (X_2X_1)^{\vee} = (X_1 \vee X_2)^{\vee} = X_1^{\vee} \wedge X_2^{\vee} .$$

For any coradicals Y_1 *and* Y_2 ,

$$(Y_1Y_2)^{\wedge} = (Y_2Y_1)^{\wedge} = (Y_1 \vee Y_2)^{\wedge} = Y_1^{\wedge} \wedge Y_2^{\wedge} .$$

Proof. Consider, for instance, the case of coradicals. Let $\rho = (A,G)$ be an arbitrary representation, then

$$\rho \in Y_1 Y_2 \Leftrightarrow Y_1^*(Y_2^*(\rho)) = A \Leftrightarrow Y_1^*(\rho) = Y_2^*(\rho) = A \Leftrightarrow$$

$$\Leftrightarrow (\rho \in Y_1^{\wedge})\&(\rho \in Y_2^{\wedge}) \Leftrightarrow \rho \in Y_1^{\wedge} \wedge Y_2^{\wedge} .$$

Thus $(Y_1 Y_2)^{\wedge} = Y_1^{\wedge} \wedge Y_2^{\wedge}$. Furthermore, it is evident that

$$(Y_1 Y_2)^{\wedge} \subseteq (Y_1 \vee Y_2)^{\wedge} \subseteq Y_1^{\wedge} \wedge Y_2^{\wedge} .$$

The rest follows from the symmetry of the assertion. □

In view of Proposition 7 one might guess that $(X_1 \wedge X_2)^{\vee} = X_1^{\vee} \vee X_2^{\vee}$ for any radicals X_1 and X_2 (and the same for coradicals). The following example shows that this is not the case. Over the ring of integers consider the class E_2 of representations (A,G) where $\exp A = 2$, and the class P of representations (A,G) where A is a divisible abelian group. Both E_2 and P are radical classes, and it is clear that

$$(E_2 \wedge P)^{\vee} = E^{\vee} = O .$$

On the other hand, consider the unit representation $\rho = (C_{2^\infty}, 1)$ where C_{2^∞} is the quasicyclic 2-group, and suppose that $\rho \in E_2^{\vee} \vee P^{\vee}$. Then C_{2^∞} possesses subgroups A and B such that $A \cap B = \{0\}$, C_{2^∞}/A has no 2-subgroups, C_{2^∞}/B has no divisible subgroups. Since each nontrivial factor-group of C_{2^∞} is isomorphic to C_{2^∞} , it follows that C_{2^∞} cannot possess such subgroups A and B. Consequently, $\rho \notin E_2^{\vee} \vee P^{\vee}$, so that $E_2^{\vee} \vee P^{\vee} \neq 0$.

REFERENCES

[1] S. A. Amitsur, A general theory of radicals, I-III, *Amer. J. Math.* 74 (1952), 774-786; ibid. 76 (1954), 100-136.

[2] P. Cohn, Free ideal rings, *J. Algebra* 1 (1964), 47-69.

[3] E. Fried and R. Wiegandt, Connectednesses and disconnectednesses of graphs, *Algebra Universalis* 5 (1975), 411-428.

[4] W. C. Holland, Varieties of ℓ-groups are torsion classes, *Czech. J. Math.* 29 (1979), 11-12.

[5] N. Jacobson, *Lectures in Abstract Algebra*, vol. II, Van Nostrand, Princeton, 1953.

[6] A. G. Kuroš, Radicals in rings and algebras, *Mat. Sb.* 33 (1953), 13-26 (Russian).

[7] A. G. Kuroš, *Lectures in General Algebra*, 2nd ed., Nauka, Moscow, 1973 (Russian).

[8] A. I. Mal'cev, Generalized nilpotent algebras and their adjoint groups, *Mat. Sb.* 25 (1949), 347-366 (Russian).

[9] J. Martinez, The fundamental theorem on torsion classes of lattice-ordered groups, *Trans. Amer. Math. Soc.* 259 (1980), 311-317.

[10] L. M. Martynov, On attainable classes of groups and semigroups, *Mat. Sb.* 90 (1973), 235-245 (Russian).

[11] B. I. Plotkin, On the semigroup of radical classes of groups, *Sib. Mat. Ž.* 10 (1969), 1091-1108 (Russian).

[12] B. I. Plotkin, On functorials, radicals and coradicals in groups, *Mat. Zap. Ural. Gos. Univ.* 7, No. 3 (1970), 150-182 (Russian).

[13] B. I. Plotkin, Radicals in groups, operations on classes of groups, and radical classes, *Selected Questions of Algebra and Logic*, Nauka, Novosibirsk, 1973, pp. 205-244 (Russian).

[14] B. I. Plotkin, Radicals and varieties in group representations, *Latv. Mat. Yežegodnik* 10 (1972), 75-132 (Russian).

[15] B. I. Plotkin and S. M. Vovsi, *Varieties of Group Representations (general theory, connections and applications)*, Zinatne, Riga, 1983 (Russian).

[16] K. K. Ščukin, On verbal radicals of groups, *Uč. Zap. Kišinev Gos. Univ.* 82 (1965), 97-99 (Russian).

[17] S. M. Vovsi, The absolute freeness of nonclosed radicals, Certain Questions of Group Theory, *Latv. Valsts Univ. Zinatn. Raksti* 151 (1971), 14-18 (Russian).

[18] S. M. Vovsi, On infinite products of classes of groups, *Sib. Mat. Ž.* 13 (1972), 272-285 (Russian).

[19] S. M. Vovsi, The semigroup of prevarieties of linear group representations, *Mat. Sb.* 93 (1974), 405-421 (Russian).

[20] S. M. Vovsi, *Triangular Products of Group Representations and Their Applications*, Progress in Mathematics, vol. 17, Birkhäuser, Boston-Basel-Stuttgart, 1981.

S. M. Vovsi
Department of Mathematics
Riga Polytechnic Institute
Riga 226355
USSR

COLLOQUIA MATHEMATICA SOCIETATIS JÁNOS BOLYAI

38. RADICAL THEORY, EGER (HUNGARY), 1982.

ULTRACLOSED RADICAL AND SEMISIMPLE CLASSES

J. F. WATTERS

1. INTRODUCTION

In a recent paper [5] J.R. Fisher has investigated axiomatic radical and semisimple classes of rings. This paper is partly motivated by an attempt to answer some of the questions posed in [5]. We begin by showing that a radical class $\mathcal{R}$ with semisimple class $\mathcal{S}$ has both $\mathcal{R}$ and $\mathcal{S}$ (finitely) axiomatic if and only if the radical of an ultraproduct is the ultraproduct of the radicals of the component rings. This result extends the main theorem of [5] . An investigation of Question 3 in [5] leads to an extension of Proposition 2 therein and some sufficient conditions are given for $\mathcal{S}$ to be finitely axiomatic when $\mathcal{R}$ is finitely axiomatic. Using one of these sets of conditions the semisimple radical classes [14] are shown to have finitely axiomatic semisimple classes.

In the third section of this paper we show how ultraclosed semisimple classes arise from ultraclosed classes of prime rings.

To obtain exmples of the latter we follow the approach in [11] and deal with classes of prime rings having modules of some specified type preserved by the ultraproduct construction. It is well known that primitivity is an ultraclosed property [1] and we show that weak primitivity is likewise ultraclosed.

The final section of the paper gives some examples of radicals where both $\mathcal{R}$ and $\mathcal{S}$ are finitely axiomatic viz. the semisimple radical classes and the linear semiprime radicals [10], and includes some remarks on the prime radical, the Brown-McCoy radical, the class of Jacobson rings, and the weak radical which is the upper radical determined by the weakly primitive rings.

We recall the ultraproduct construction and establish some notation. Let U be an ultrafilter on a set A. Let $\{R_\alpha : \alpha \in A\}$ be a set of rings and M_α a left R_α-module for each $\alpha \in A$. Put $\bar{R} = \pi R_\alpha$ and $\bar{M} = \pi M_\alpha$. Equivalence relations are defined on $\bar{R}$ and $\bar{M}$ thus : for $(r_\alpha), (s_\alpha) \in \bar{R}$, $(r_\alpha) \sim (s_\alpha)$ if and only if $\{\alpha : r_\alpha = s_\alpha\} \in U$; a similar relation is defined on M. Then $I = \{(r_\alpha) \in \bar{R} : (r_\alpha) \sim 0\}$ is an ideal of $\bar{R}$ and $N = \{(m_\alpha) \in M : (m_\alpha) \sim 0\}$ is an $\bar{R}$-submodule of $\bar{M}$. The factor ring $R = \bar{R}/I$ is called an ultraproduct of the R_α's and we write $R = \pi R_\alpha/U$ and $[r_\alpha]$ for the coset determined by (r_α). The $\bar{R}$-module $M = \bar{M}/N$ can also be considered as an R-module since $I\bar{M} \subseteq N$. The module M is called an ultraproduct of the M_α's and we write $M = \pi M_\alpha/U$ and $[m_\alpha]$ for the coset determined by (m_α). An *ultraclosed* class is one which is closed under taking ultraproducts.

2. FINITELY AXIOMATIC CLASSES

It is well known [3, p.244] that a class $\mathcal{K}$ is finitely axiomatic if and only if $\mathcal{K}$ and $\mathcal{K}'$ are ultraclosed. We begin by observing links between $\mathcal{R}$ and $\mathcal{S}'$, and $\mathcal{S}$ and $\mathcal{R}'$.

PROPOSITION 2.1. (a) *If $\mathcal{R}$ is ultraclosed, then $\mathcal{S}'$ is ultraclosed.*

(b) *If $\mathcal{S}$ is ultraclosed, then $\mathcal{R}'$ is ultraclosed.*

Proof. (a) If $R_\alpha \in \mathcal{S}'$ then there is a non-zero ideal I_α of R_α with $I_\alpha \in \mathcal{R}$. Now $R = \pi R_\alpha/U$ contains the non-zero ideal $\pi I_\alpha/U$, which by hypothesis belongs to $\mathcal{R}$, and so $R \in \mathcal{S}'$.

(b) If $R_\alpha \in \mathcal{R}'$ then there is a non-zero homomorphic image $S_\alpha \in \mathcal{S}$. By hypothesis $S = \pi S_\alpha/U \in \mathcal{S}$ and there is an induced homomorphism from R onto S [9, p.159]. Hence $R \in \mathcal{R}'$ as required.

In section 4 we shall give examples to show that the converse statements to 2.1.(a) and (b) are false.

THEOREM 2.2. *For a radical class $\mathcal{R}$ with associated semisimple class $\mathcal{S}$ the following are equivalent :*

(1) *$\mathcal{R}$ and $\mathcal{S}$ are finitely axiomatic.*

(2) *$\mathcal{R}$ and $\mathcal{S}$ are axiomatic.*

(3) *$\mathcal{R}$ and $\mathcal{S}$ are ultraclosed.*

(4) $\mathcal{R}(\pi R_\alpha/U) = \pi\, \mathcal{R}(R_\alpha)/U$.

(5) *$\mathcal{S}$ is ultraclosed and $\mathcal{R}$ is closed under arbitrary products.*

Proof. The cycle (1) ⇒ (2) ⇒ (3) ⇒ (1) follows from [3, p.244] and 2.1, whilst their equivalence to (5) follows from Fisher's Lemma [5] and the fact that $\mathcal{R}$ is homomorphically closed. The implication (4) ⇒ (3) is immediate and we need only prove (3) ⇒ (4).

Set $J_\alpha = \mathcal{R}(R_\alpha)$ and $J = \pi J_\alpha/U$ regarded as a subring of $R = \pi R_\alpha/U$. Now $J \in \mathcal{R}$ so $J \subseteq \mathcal{R}(R)$. On the other hand $S_\alpha = R_\alpha/J_\alpha \in \mathcal{S}$ so $S = \pi S_\alpha/U \in \mathcal{S}$ and there is an induced homomorphism from R onto S with kernel K. If $[r_\alpha] \in K$ then $[r_\alpha + I_\alpha] = 0$ and $B = \{\alpha : r_\alpha \in I_\alpha\} \in U$. With $s_\alpha = r_\alpha$ when $\alpha \in B$ and $s_\alpha = 0$ otherwise, we have $[r_\alpha] = [s_\alpha] \in J$. Hence $K = J$ and $R/J \in \mathcal{S}$. Therefore $J = \mathcal{R}(R)$ as required.

COROLLARY 2.3. *If* $\mathcal{S}$ *is ultraclosed then the following are equivalent :*

(1) $\mathcal{R}$ *is finitely axiomatic.*

(2) $\mathcal{R}$ *is axiomatic.*

(3) $\mathcal{R}$ *is ultraclosed.*

(4) $\mathcal{R}$ *is closed under arbitrary products.*

Before giving examples of ultraclosed semisimple classes we look first at some circumstances under which $\mathcal{S}$ will be finitely axiomatic. The results we obtain follow the pattern set by [5, Proposition 2].

If $\mathcal{R}$ is finitely axiomatic then we can assume that there is a first—order sentence ϕ such that $\mathcal{R} = \{R : R \models \phi\}$. Moreover we can assume that ϕ is in prenex normal form [3,9]. Thus $\phi = (Q_1 x_1) \ldots (Q_n x_n)\ \theta\ (x_1, \ldots, x_n)$ where Q_i stands for either $\exists$ or $\forall$ and θ is a quantifier-free formula. From ϕ we obtain a formula $\psi = (Q_1 x_1) \ldots (Q_n x_n)\ \theta\ (xx_1, \ldots, xx_n)$, where the variable x_i in θ is replaced by the term xx_i, and a sentence $\tau = (\exists x)\psi$. The sentence τ determines a finitely axiomatic class $\mathcal{J} = \{R : R \models \tau\}$ which we prefer to think of as $\{R : \exists\ x \in R$ such that $xR \in \mathcal{R}\}$. Of course we wish to avoid $xR = 0$ but subject to modifications to ensure this we find that $\mathcal{S}'$ is largely determined by τ.

Recall that a radical $\mathcal{R}$ is *right hereditary* if every right ideal of a radical ring is radical; $\mathcal{R}$ is *right strong* if every radical right ideal of a ring R is contained in the radical of R [12]. The left-handed versions of these conditions are used in [5, Proposition 2]. As the description of $\mathcal{J}$ implies, our interest lies in the principal right ideals so we also say that a radical $\mathcal{R}$ is *principally right strong* if whenever $a \in R$ and $aR \in \mathcal{R}$, then $aR \subseteq \mathcal{R}(R)$.

THEOREM 2.4. *Let* $\mathcal{R}$ *be a right hereditary, principally right strong radical which satisfies either*

(a) $a \in R,\ aR = 0 \Rightarrow a \in \mathcal{R}(R)$; *or*

(b) $0 \neq a \in \mathcal{R}(R) \Rightarrow aR \neq 0$.

If $\mathcal{R}$ *is finitely axiomatic, then so also is* $\mathcal{S}$.

Proof. In both cases we show that $\mathcal{S}'$ is finitely axiomatic.

(a) In the above notation let ϕ' be the sentence $(\exists x)(x \neq 0 \wedge \psi)$ and $\mathcal{S}^*$ be the finitely axiomatic class $\{R : R \models \phi'\}$ which we think of as $\{R : \exists\, 0 \neq x \in R$ with $xR \in \mathcal{R}\}$.

If $R \in \mathcal{S}'$ then there is a nonzero ideal I of R with $I \in \mathcal{R}$. If $0 \neq x \in I$ then $xR \in \mathcal{R}$ from the right hereditary condition and $R \in \mathcal{S}^*$.

If $R \in \mathcal{S}^*$ so that $xR \in \mathcal{R}$ with $0 \neq x \in R$, then either $xR = 0$, in which case $\mathcal{R}(R) \neq 0$ from (a), or $xR \neq 0$ when the principally right strong condition implies that $\mathcal{R}(R) \neq 0$. Thus $R \in \mathcal{S}'$ and $\mathcal{S}^* = \mathcal{S}'$.

(b) A similar argument applies with $\mathcal{S}^*$ determined by the sentence $(\exists x)(\psi \wedge (\exists y)(xy \neq 0))$.

We remark that condition (a) is satisfied by all supernilpotent radicals and (b) by all subidempotent radicals. To deal with the semisimple radical classes we use the following variation on 2.4.

THEOREM 2.5. *Let $\mathcal{R}$ be a hereditary radical which satisfies the condition that whenever $\mathcal{R}(R) \neq 0$ in a ring R, then $\exists$ $a \in \mathcal{R}(R)$ with $aR \neq 0$ and $aR \supseteq Ra$. If $\mathcal{R}$ is finitely axiomatic, then so also is $\mathcal{S}$.*

Proof. Similar arguments to those used in 2.4. apply. The class $\mathcal{S}'$ is determined by the sentence

$(\exists x)(\psi \wedge (\exists y)(xy \neq 0) \wedge (\forall u)(\exists v)(ux = uv))$.

3. ULTRACLOSED SEMISIMPLE CLASSES

A useful property of a number of radicals, special radicals particularly, is that the semisimple class $\mathcal{S}$ is the subdirect closure of another class $\mathcal{P}$ of rings. With this in mind we prove :

THEOREM 3.1. *If the semisimple class* $\mathcal{S}$ *is the subdirect closure of an ultraclosed class* $\mathcal{P}$, *then* $\mathcal{S}$ *is ultraclosed.*

Proof. Let $R_\alpha \in \mathcal{S}$ and $R = \pi R_\alpha/U$. Suppose $0 \neq [r_\alpha] \in R$ and $B = \{\alpha : r_\alpha = 0\}$. Then $B' \in U$ and for $\alpha \in B'$ there is a nonzero ideal P_α in R_α with $R_\alpha \neq P_\alpha$ and $R_\alpha/P_\alpha \in \mathcal{P}$. For $\alpha \in B$ take P_α to be any ideal of R_α such that $R_\alpha/P_\alpha \in \mathcal{P}$. Now consider the rings $S_\alpha = R_\alpha/P_\alpha$, the elements $s_\alpha = r_\alpha + P_\alpha$, and the ultraproduct $S = \pi S_\alpha/U$. Under the natural homomorphism from R onto S, $[r_\alpha]$ is mapped to $[s_\alpha]$ which is nonzero by the choice of P_α's. If K is the kernel of this homomorphism then $[r_\alpha] \notin K$ and $R/K \in \mathcal{P}$. Hence R is a subdirect product of rings from $\mathcal{P}$ and so $R \in \mathcal{S}$.

In the light of 3.1 we wish to find ultraclosed classes One example in the literature is the class $\mathcal{P}$ of primitive rings.

Amitsur [1] showed that $\mathcal{P}$ is ultraclosed and recently Lawrence [8] has shown that $\mathcal{P}'$ is not closed under ultrapowers. Thus $\mathcal{P}$ is not axiomatic, although the corresponding semisimple class is finitely axiomatic. To show that $\mathcal{P}$ is ultraclosed we can use either the "pseudo-elementary" approach of Sabbagh [4, p.117] or two-sorted first-order logic [2, p.42]. The latter line is used in [11] and we shall use the same technique here. Briefly, the variables are taken to be of two sorts viz. ring elements and module elements, but in other respects the logic is like ordinary first-order logic.

In [11] it is noted that for a left R-module M the statements "M is faithful" and "M is irreducible" are finitely axiomatic. On this basis the class of primitive rings is seen to be ultraclosed. Other module concepts that can be dealt with in this way are :

(1) *Prime*. M is prime if $RM \neq 0$ and if $r \in R$, $m \in M$ are such that $rRm = 0$ then either $rRM = 0$ or $m = 0$.

(2) *Cyclic*.

(3) *Uniform*. This is finitely axiomatic if for $m \in M$, $Rm = 0$ implies $m = 0$. This condition holds if the module is unital or faithful and prime. M is uniform if for every nonzero $m_1, m_2 \in M$, $Rm_1 \cap Rm_2 \neq 0$.

(4) *Non-singularity*. Provided every nonzero left ideal contains a nonzero left ideal Rr, which is the case if for example $1 \in R$ or R is semiprime, nonsingularity is determined by the sentence

$(\forall m_1 \in M)(\exists r_1 \in R)(\forall r_2 \in R)(m_1 = 0 \vee (r_1 \neq 0 \wedge r_2 r_1 m_1 \neq 0 \vee r_2 r_1 = 0))$.

Some other module concepts arise in determining the class of weakly primitive rings [16] whose upper radical is the weak radical [15]. Recall that a *compressible* module M is one for which $RM \neq 0$ and M can be embedded in each of its nonzero submodules. The module M is said to be *monoform* if every partial endomorphism is either zero or monic. The *weakly primitive rings* are those with a faithful monoform compressible module. We now examine the "elementary" nature of these concepts in which module homomorphisms are involved and so would seem to be non-elementary. In a similar vein we consider the notion of a *standard prime module* M which is defined by the properties that $RM \neq 0$ and whenever $m_1, m_2 \in M$ either $m_1 = 0$ or $m_2 = 0$ or there is $\phi \in M^* = \mathrm{Hom}_R(M,R)$ such that $(m_1\phi)m_2 \neq 0$.

PROPOSITION 3.2. *The concept of a cyclic standard prime module is finitely axiomatizable in two-sorted first-order logic.*

Proof. If $M = Rm_1$ then every $\phi : M \to R$ is determined by $m_1\phi = r_1$, where $rm_1 = 0$ implies $rr_1 = 0$ for all $r \in R$, and conversely every such $r_1 \in R$ determines a homomorphism $\phi : M \to R$ via this equation. If $m_2 = r_2 m_1$ then $(m_2\phi)m_3 = r_2 r_1 m_3$ and so, apart from the condition $RM \neq 0$ which is clearly finitely axiomatic, the concept is given by the sentence

$(\exists m_1 \in M)(\forall m_2, m_3 \in M)(\exists r_1, r_2 \in R)(\forall r_3 \in R)((r_3 m_1 = 0 \to r_3 r_1 = 0)$
$\wedge r_2 m_1 = m_2 \wedge (m_2 = 0 \vee m_3 = 0 \vee r_2 r_1 m_3 \neq 0))$.

PROPOSITION 3.3 *The concept of a cyclic compressible module is finitely axiomatizable in two-sorted first-order logic.*

Proof. Let $M = Rm_1$, $m_1 \in M$. If $0 \neq m_2 \in M$ then $Rm_2 \neq 0$ (using compressibility) and M can be embedded into Rm_2. Under the embedding m_1 is mapped to $r_2 m_2$ where for all $r_1 \in R$, $r_1 r_2 m_2 = 0$ if and only if $r_1 m_1 = 0$. In this way we see that the concept is finitely axiomatic.

For monoform modules we can make a statement similar to 3.2 and 3.3 since a module M is monoform if and only if M is uniform and every partial endomorphism $\phi : Rm \to M$, $m \in M$, is either zero or monic. The following is a direct proof of ultraclosure.

PROPOSITION 3.4 *Let* M_α *be a monoform* R_α*-module such that if* $0 \neq m_\alpha \in M_\alpha$ *then* $R_\alpha m_\alpha \neq 0$. *Then the ultraproduct* $M = \pi M_\alpha / U$ *is a monoform* S*-module where* $S = \pi R_\alpha$.

Proof. Suppose N is a nonzero submodule of M and $\phi : N \to M$ is a partial endomorphism which is neither zero nor monic. Let $0 \neq [b_\alpha] \in N$ and $[b_\alpha]\phi = [c_\alpha] \neq 0$ and suppose $0 \neq [d_\alpha] \in N$ with $[d_\alpha]\phi = 0$. Set $B = \{\alpha : b_\alpha = 0\}$ with C and D similarly defined. Then B', C', and $D' \in U$.

Let $X \in U$ with $X \subseteq B' \cap C'$ and suppose that for every $\alpha \in X$ there is $s_\alpha \in R_\alpha$ such that $s_\alpha b_\alpha = 0$ but $s_\alpha c_\alpha \neq 0$. For $\alpha \notin X$ set $s_\alpha = 0$ and consider $s = (s_\alpha) \in S$. We find that $s[c_\alpha] = (s[b_\alpha])\phi = [s_\alpha b_\alpha]\phi = 0\phi = 0$. However $s[c_\alpha] = [s_\alpha c_\alpha] \neq 0$ since $\{\alpha : s_\alpha c_\alpha = 0\} = X' \notin U$. Thus we deduce that there is $\beta \in X$ such that for $r_\beta \in R_\beta$, $r_\beta b_\beta = 0$ implies $r_\beta c_\beta = 0$.

For $\alpha \in B' \cap D'$ there are $p_\alpha, t_\alpha \in R_\alpha$ such that $p_\alpha b_\alpha = t_\alpha d_\alpha \neq 0$ since M_α is uniform. If $\alpha \in B \cup D$, set $p_\alpha = t_\alpha = 0$ and write $p = (p_\alpha)$, $t = (t_\alpha) \in S$. Clearly $p[b_\alpha] = t[d_\alpha]$ and hence $(p[b_\alpha])\phi = [p_\alpha c_\alpha] = (t[d_\alpha])\phi = 0$. Thus $E = \{\alpha : p_\alpha c_\alpha = 0\} \in U$. Put $X = E \cap B' \cap D' \cap C'$.

With β as above we have an epimorphism $\phi_\beta : R_\beta b_\beta \to R_\beta c_\beta$ given by $r_\beta b_\beta \to r_\beta c_\beta$. Now $R_\beta b_\beta$ and $R_\beta c_\beta$ are nonzero submodules of M since $\beta \in B' \cap C'$ and so ϕ_β must be monic. However $\beta \in E$ so $p_\beta c_\beta = 0$ whilst $\beta \in B' \cap D'$ so $p_\beta b_\beta \neq 0$. Thus ϕ_β is not monic. This contradiction forces ϕ to be either zero or monic and M to be monoform.

COROLLARY 3.5. *The class of weakly primitive rings is ultraclosed.*

Proof. Use 3.3 and 3.4.

We remark that in addition to the prime, primitive, and weakly primitive rings the classes of (1) primitive rings with nonzero socle and (2) prime non-singular rings with a uniform left ideal are also ultraclosed. In the former case the class is finitely axiomatic [11].

4. EXAMPLES

4.1. THE LINEAR SEMIPRIME (p:q) RADICALS

Here we consider the radical class

$\mathcal{R} = \{R : (\forall r \in R)(r \in p(r)Rq(r))\} = (p:q)$ where $p, q \in \mathbb{Z}[x]$ and are products of linear polynomials with constant term 1, [10]. It is known that each such radical can be represented as a (p:1) radical where $p(x) = 1 + nx$ and n is a square-free natural number. The case $n = 1$ gives the Jacobson radical.

A result of Ortiz [10] shows that $\mathcal{R} = (p:1)$ is supernilpotent. From [10, Lemma 13] $\mathcal{R}$ is right strong and it is an easy matter to see that $\mathcal{R}$ is right hereditary. Hence $\mathcal{J}$ is finitely axiomatic from 2.4.(a).

From 2.2 we have $\mathcal{R}(\pi R_\alpha/U) = \pi\mathcal{R}(R_\alpha)/U$ for these radicals. This result for the Jacobson radical was proved by Santosuosso [13].

The case when p or q is not a product of linear polynomials with constant term 1 remains to be investigated. In particular, the question as to whether the regular $(p = q = x)$ and strongly regular $(p = x^2, q = 1)$ radicals have axiomatic semisimple classes is open.

4.2 THE SEMISIMPLE RADICAL CLASSES $\mathcal{C} = V(P,N)$

For these radicals there is a nonempty finite set $\mathcal{F}$ of finite fields, closed under taking subfields, such that $R \in \mathcal{C}$ if and only if every finitely generated subring of R is isomorphic to a finite direct product of fields in $\mathcal{F}$ [14]. It is clear that $\mathcal{C}$ is a hereditary class of commutative rings. If, for a ring R, $\mathcal{C}(R) \neq 0$ then $\mathcal{C}(R)$ contains a nonzero idempotent element e. For $r \in R$, er and re $\in \mathcal{C}(R)$ and the subring generated by e, er, and re is commutative. Hence $er = e^2r = ere = re^2 = re$ and $e \in Z(R)$. Clearly $eR \neq 0$. Finally from [6] the description of $\mathcal{C}$ in the form V(P,N), as a variety, shows that $\mathcal{C}$ is finitely axiomatic. Theorem 2.5 now applies to give us that $\mathcal{C}$'s semisimple class is finitely axiomatic.

4.3. THE p-TORSION RADICAL

Here p is a rational prime, $\mathcal{R} = \{R : \text{for each } a \in R \ \exists\, k \in N$ such that $p^k a = 0\}$, and $\mathcal{S} = \{R : (\forall r \in R)(pr \neq 0 \vee r = 0)\}$. Clearly $\mathcal{S}$ is finitely axiomatic, but $\mathcal{R}$ is not ultraclosed as the following example shows.

EXAMPLE 4.4. Let R_n be the ring of integers modulo p^n with additive generator a_n and U be a nonprincipal ultrafilter on N. Consider $a = [a_n] \in R = \pi R_n/U$. If $p^k a = 0$ then $\{n : n \leq k\} = \{n : p^k a_n = 0\} \in U$. But U contains no finite subset of N, so $a \notin \mathcal{R}(R)$. Thus $R \notin \mathcal{R}$ and this example shows that the converse to 2.1.(a) is false.

4.5. THE BROWN-McCOY RADICAL

It is an open question as to whether this radical is finitely axiomatic. Since $\mathcal{R}$ is the upper radical determined by the class of simple rings with 1, $\mathcal{R}'$ can be characterized, using Zorn's Lemma, as the class of all rings with a nonzero image in the class of rings with 1.

To express this property in first-order logic we add to the basic language of ring theory a name for I and consider structures of the form (R,I) where I is an ideal of R and $1 \in R/I$ if (R,I) is a model for the following axioms in which I is thought of as a unary relation and we write $I(x)$ instead of $x \in I$:

$I(0);$

$(\forall x,y \in R)(I(x) \wedge I(y) \rightarrow I(x+y));$

$(\forall x,y \in R)(I(x) \wedge x+y=0 \rightarrow I(y));$

$(\forall x,y \in R)(I(x) \rightarrow I(xy) \wedge I(yx));$

$(\exists e \in R)(\forall x \in R)(I(x-ex) \wedge I(x-xe) \wedge \neg I(e)).$

In this way we observe that $\mathcal{R}'$ is pseudo-elementary [4, p.116] and so ultraclosed. Thus $\mathcal{R}$ is axiomatic if and only if $\mathcal{R}$ is finitely axiomatic if and only if $\mathcal{R}$ is ultraclosed.

To see that for this radical $\mathcal{R}$ is not ultraclosed we appeal to the next example.

EXAMPLE 4.6. For each $n \in N$ let D_n be a division ring, $R_n = M_n(D_n)$, the matrix ring, and $M_n = (D_n)^n$, the space of column vectors. Thus R_n is a primitive ring with faithful irreducible module M_n and commuting ring D_n. Hence $R = \pi R_n/U$ is a primitive ring with faithful irreducible module $M = \pi M_n/U$ and commuting ring $D = \pi D_n/U$ [7]. Moreover each R_n has a nonzero socle so that R also has a nonzero socle, whence R contains nonzero transformations of M, as a vector space over D, of finite rank. Further $1 \in R$ and M is not finite dimensional over D, so the ideal F of finite rank transformations is nonzero and proper. Since F is a simple ring without 1, $F \in \mathcal{R}$, $R \in \mathcal{S}'$, but $R_n \in \mathcal{S}$.

This example shows that the converse to 2.1.(b) is false. It would be interesting to have an example to show this with $\mathcal{R}$ axiomatic.

4.7. THE RADICAL CLASS OF JACOBSON RINGS

Apart from the intrinsic interest of this class of rings we use it here to show that in the context of this paper Theorem 3.1 gives us the only property to be preserved in going from $\mathcal{P}$ to $\mathcal{S}$.

Let $\mathcal{P}$ be the class of prime rings with nonzero Jacobson radical. Then $\mathcal{P}$ is a special class and $\mathcal{R} = u\mathcal{P}$ is a special radical. In fact $\mathcal{R}$ is the class of all Jacobson rings, i.e. $R \in \mathcal{R}$ if and only if every prime image of R is semiprimitive. Since $\mathcal{P}$ is the intersection of two finitely axiomatic classes it is also finitely axiomatic. Hence by 3.1 $\mathcal{S}$ is ultraclosed but the next example shows that $\mathcal{S}'$ is not closed under taking ultrapowers whence $\mathcal{S}$ is not axiomatic.

EXAMPLE 4.8. Let U be a non-principal ultrafilter on N and let $R_n = Z$ for each $n \in N$. Let $R = \pi Z/U$ and $0 \neq r = [a_n] \in R$. Then $b = \{n : a_n = 0\} \notin U$. Put $s = [a_n^n] \in R$. Then $\{n : a_n^n = 0\} = B$ so $s \neq 0$.

Consider the ideal sR and the set $\mathfrak{a} = \{r, r^2, \ldots\}$. We show that $\mathfrak{a} \cap sR = \phi$. If $r^m \in sR$ then $[a_n^m] = [a_n^n c_n]$ for some $[c_n] \in R$. Therefore $\{n : a_n^m = a_n^n c_n\} \in U$. But $a_n^m = a_n^n c_n$ if and only if $a_n = 0$ or $n \leq m$. Hence $B \cup \{1, 2, \ldots, m\} \in U$. But $B' \in U$, so $B' \cap \{1, 2, \ldots, m\} \in U$. However, U, being non-principal, contains no finite set as a member and therefore $\mathfrak{a} \cap sR = \phi$.

Using Zorn's Lemma we can now find an ideal P of R maximal with respect to the properties that $P \supseteq sR$ and $\mathfrak{a} \cap P = \phi$. Thus P is a prime ideal of R and we now consider the ideal J of R, $J \supseteq P$, where J/P is the Jacobson radical of R/P. We claim that $r \in J$.

Let $d = [d_n] \in R$ and let $f_n \in Z$ be determined by the equation $\sum_{k=1}^{n-1} (a_n d_n)^k = - a_n f_n$. Then $a_n d_n \circ a_n f_n = a_n^n d_n^n$ (where $x \circ y = x + y - xy$) and $rd \circ rf = s[d_n^n]$, where $f = [f_n] \in R$. Therefore $rd \circ rf \in P$ and $(rR + P)/P$ is a quasiregular ideal of R/P. Hence $r \in J$ as claimed.

By construction $r \notin P$ so $J \supset P$ and $R/P \in \mathcal{P}$.

Since this argument applies to any $0 \neq r \in R$ it follows that $R \in \mathcal{S}$. Thus $\mathcal{S}'$ is not closed under taking ultrapowers.

4.9. THE WEAK RADICAL

It was noted in [5] that the prime radical is not axiomatic since it is not closed under arbitrary products. For example if $R_n = (2)/(2^n) \subsetneq \mathbb{Z}/(2^n)$ and $a_n = 2 + (2^n)$ then $a = (a_n) \in \pi R_n = R$ is not a nilpotent element and so R has a prime image. Since R is commutative this image is weakly primitive. Thus the weak radical is not axiomatic which, in some sense, answers a question posed in [15]. From 3.5 and 3.1 we see that the corresponding semisimple class is ultraclosed.

REFERENCES

1. S.A. AMITSUR, Prime rings having polynomial identities with arbitrary coefficients, *Proc. London Math. Soc.*, 17(1967), 470-486.

2. J. BARWISE, An introduction to first-order logic, *Handbook of Mathematical Logic,* North Holland Pub. Co., Amsterdam, 1977.

3. P.M. COHN, *Universal algebra,* Harper and Row, New York, 1965.

4. P. EKLOF, Ultraproducts for algebraists, *Handbook of Mathematical Logic,* North Holland Pub. Co., Amsterdam, 1977.

5. J.R. FISHER, Axiomatic radical and semisimple classes of rings, *Pacific J. Math.,* 97(1981), 81-91.

6. B.J. GARDNER & P.N. STEWART, On semisimple radical classes, *Bull. Austral. Math. Soc.,* 13(1975), 349-353.

7. I.N. HERSTEIN, *Noncommutative rings,* Math. Assoc. of America, 1968

8. J. LAWRENCE, Primitive rings do not form an elementary class, *Comm. in Algebra,* 9(1981), 397-400.

9. A.I. MAL'CEV, *Algebraic systems,* Springer, Berlin, 1973.

10. G.L. MUSSER, Linear semiprime (p:q) radicals, *Pacific J. Math.,* 37(1971), 749-757.

11. L. ROWEN, *Polynomial identities in ring theory,* Academic Press, New York, 1980.

12. A.D. SANDS, On normal radicals, *J. London Math. Soc.* (2), 11(1975), 361-365.

13. G. SANTOSUOSSO, Sul trasporto ad un ultraprodutto di anelli di proprietà dei suoi fattori, Rend. Math., 1(1968), 82-99.

14. P.N. STEWART, Semisimple radical classes, *Pacific J. Math.*, 32(1970), 249-254.

15. J. ZELMANOWITZ, Dense rings of linear transformations, *Ring Theory II* (Second Oklahaoma Conference), Dekker, New York, 1975.

16. J. ZELMANOWITZ, Weakly primitive rings, *Comm. in Algebra*, 9(1981), 23-45.

J. F. Watters
Department of Mathematics
The University
Leicester, LE1 7RH
England

COLLOQUIA MATHEMATICA SOCIETATIS JÁNOS BOLYAI
38. RADICAL THEORY, EGER (HUNGARY), 1982.

IDEALS OF WEAKLY PRIMITIVE RINGS

J. ZELMANOWITZ

The purpose of this note is to show that one-sided ideals of weakly primitive rings induce weakly primitive rings.

The basic theory of weakly primitive rings was described in [3]. A ring R is (*left*) *weakly primitive* if it has a faithful monoform compressible module ${}_RM$. (Recall that *monoform* means that partial endomorphisms $f \in \mathrm{Hom}_R(N,M)$ with ${}_RN \subseteq M$ are monomorphisms and *compressible* means that $\mathrm{Hom}_R(M,N)$ contains a monomorphism for every $0 \neq {}_RN \subseteq M$.) Such a ring is characterized by the following density property [3].

THEOREM. *Let* ${}_R M$ *be a faithful monoform compressible R-module, let* $\overline{M}$ *denote the R-quasi-injective hull of* M, *and set* $\Delta = \mathrm{End}_R \overline{M}$. *Then* Δ *is a division ring,* $\overline{M} = M\Delta$, *and (regarding* R *as a subring of* $\mathrm{End}\ \overline{M}_\Delta$) *given any elements* $v_1, \ldots, v_k \in \overline{M}$ *linearly independent over* Δ, *there exists* $0 \neq a \in \Delta$ *such that given*

$m_1, \ldots, m_k \in M$ *there exists* $r \in R$ *with* $rv_i = m_i a \in M$ *for each* i.

For convenience, we call a triple (M, V, Δ) an *R-lattice* if Δ is a division ring, M is a left R-module contained in the R-Δ-bimodule V, $M\Delta = V$, and ${}_R M$ is faithful (so that we can regard $R \subseteq \mathrm{End}\ V_\Delta$). The converse to the preceding theorem then reads as follows [3].

THEOREM. *Suppose that* (M, V, Δ) *is an R-lattice with the property that given any elements* $v_1, \ldots, v_k \in V$ *linearly independent over* Δ *there exists* $0 \neq a \in \Delta$ *such that for any* $m_1, \ldots, m_k \in M$ *there exists* $r \in R$ *with* $rv_i = m_i a \in M$ *for each* i. *Then each non-zero cyclic R-submodule of* M *is a faithful monoform compressible module.*

When the hypothesis of this theorem is satisfied we

say that R is a *weakly dense ring of linear transformations (on the R-lattice* (M,V,Δ)*)*. Observe that the element $a \in \Delta$ does not depend on the m_i; we say that a *is chosen relative to* $v_1,\dots,v_k$. In this terminology, the preceding theorems may be summarized by saying that the weakly primitive rings are precisely the weakly dense rings of linear transformations.

The key to studying one-sided ideals of weakly primitive rings is the "double annihilator" lemma which is already featured in Jacobson's treatment of primitive rings [1; page 27]. We state it but slightly revised, in a form suitable for our purpose, and include a proof for the convenience of the reader. For N a subset of a module M, $\ell_R(N)$ or simply $\ell(N)$ will denote $\{r \in R \mid rN=0\}$; we also write $\ell(m_1,\dots,m_k)$ for $\ell(\{m_1,\dots,m_k\})$. Similarly, for $I \subseteq R$, $r_M(I) = \{m \in M \mid Im=0\}$.

LEMMA [1]. *Suppose that* ${}_RV$ *is a quasi-injective module. Let* J *be a left ideal of* R *and set* $\Delta = End_R V$. *Then for any* $v_1,\dots,v_k \in V$, $\ell(v_1,\dots,v_{k-1}) \cap J \subseteq \ell(v_k)$ *if and only if* $v_k \in \sum_{i=1}^{k-1} v_i\Delta + r_V(J)$.

Proof. If $v_k \in \sum_{i=1}^{k-1} v_i\Delta + r_V(J)$ then clearly $(\ell(v_1,..,v_{k-1}) \cap J)v_k = 0$, so one half of the conclusion is obvious. Conversely, suppose that $\ell(v_1,\dots,v_{k-1}) \cap J \subseteq \ell(v_k)$. This means that the mapping $f: J(v_1,\dots,v_{k-1}) \to V$ defined by $(r(v_1,\dots,v_{k-1}))f = rv_k$ is a well-defined R-homomorphism; here we regard $J(v_1,\dots,v_{k-1})$ as a submodule of $V^{(k-1)} = V\oplus\dots\oplus V$. Since ${}_RV$ is quasi-injective, f extends to an element of $\mathrm{Hom}_R(V^{(k-1)},V)$, which we continue to denote by f.

Let e_i and p_i denote the canonical injections and projections for $V^{(k-1)}$; thus $\sum_{i=1}^{k-1} p_ie_i = 1$ on $V^{(k-1)}$. For any $r \in J$, $rv_k = (r(v_1,\dots,v_{k-1}))f = \sum_{i=1}^{k-1}(rv_i)(e_if) = r\left(\sum_{i=1}^{k-1} v_id_i\right)$ where each $d_i = e_if \in \mathrm{End}_RV$. Then, because $r \in J$ is arbitrary, we have that

$$J\left(v_k - \sum_{i=1}^{k-1} v_id_i\right) = 0,$$

which is the desired outcome.

We can now state our main result. Part (3) was first proved, in a somewhat different form, by K. Koh and A.C. Mewborn in [2].

THEOREM. *Assume that* M *is a faithful monoform compressible left* R*-module, so that* R *is a weakly dense ring of linear transformations on the* R*-lattice* (M,V,Δ) *where* V *is the* R*-quasi-injective hull of* M *and* $\Delta = End_R V$.

(1) If J *is a non-zero left ideal of* R *then* $\bar{J} = {}^{J}/J \cap r_R(J)$ *is a weakly dense ring of linear transformations on the* $\bar{J}$*-lattice* $({}^{M}/r_M(J), {}^{V}/r_V(J), \Delta)$.

(2) If I *is a non-zero right ideal of* R *then* $\bar{I} = {}^{I}/I \cap \ell_R(I)$ *is a weakly dense ring of linear transformations on the* $\bar{I}$*-lattice* (IM, IV, Δ).

(3) If A *is a non-zero ideal of* R *then* A *is a weakly dense ring of linear transformations on the* A*-lattice* (M,V,Δ).

Proof. We leave it to the reader to check that each of the above triples does in fact define a lattice. Inasmuch as a weakly primitive ring is prime the induced rings $\bar{I}$ and $\bar{J}$ will be non-trivial.

In order to prove (1), let $v_1, \ldots, v_k \in V$ be such that $\bar{v}_1, \ldots, \bar{v}_k$ are linearly independent elements of $\bar{V} = {}^{V}/r_V(J)$. By the preceding lemma, $\bigcap_{i \neq j} \ell(v_i) \cap J \not\subseteq \ell(v_j)$ for each j. Set $A_j = \bigcap_{i \neq j} \ell(v_i) \cap J$; then for each j,

$A_jv_j \cap M$ is a non-zero submodule of ${}_RM$. Since a monoform module is uniform, $N = \bigcap_{j=1}^{k} A_jv_j \cap M \neq 0$. Since ${}_RM$ is compressible, there exists a monomorphism $0 \neq a \in \mathrm{Hom}_R(M,N)$. We extend a to an element of $\Delta = \mathrm{End}_RV$, also denoted by a.

Now let $m_1,\ldots,m_k$ be arbitrary elements of M. Since $Ma \subset N$, there exists for each $i = 1,\ldots,k$ an element r_i of A_i such that $m_ia = r_iv_i \in M$. Set $r = \sum_{i=1}^{k} r_i \in J$. Then $m_ia = rv_i \in M$ for each i, and hence $\overline{m}_ia = \overline{r}\overline{v}_i \in \overline{M} = {}^M/r_M(J)$ with $\overline{r} \in \overline{J}$.

For (2), let $v_1,\ldots,v_k \in IV$ be given, linearly independent over Δ. We use the fact that R is weakly dense on (M,V,Δ) to choose $0 \neq a \in \Delta$ relative to $v_1,\ldots,v_k$. If $m_1,\ldots,m_k \in IM$ are given, write each $m_i = \sum_{j=1}^{t} s_{ij}n_j$ for some choice of $s_{ij} \in I, n_1,\ldots,n_t \in M$. For each choice of i and j, $1 \leq i \leq k$, $1 \leq j \leq t$, we apply the density condition to choose elements $r_{ij} \in R$ with the property that $r_{ij}v_i = n_ja \in M$ and $r_{ij}v_h = 0a = 0$ for $i \neq h$. Set $r = \sum_{i,j} s_{ij}r_{ij}$. Then $r \in I$ and for each $h \in \{1,\ldots,t\}$,

$rv_h = \sum_{i,j} s_{ij} r_{ij} v_h = \sum_j s_{hj} n_j a = m_h a \in IM$. Thus $\overline{I}$ is a weakly dense ring of linear transformations on the $\overline{I}$-lattice (IM, IV, Δ).

Since $A \cap r_R(A) = 0$ for an ideal A of a weakly primitive ring, (3) is an immediate consequence of (1).

The weak radical $W(R)$ of a ring R was first defined in [2]. In the terminology of this paper, $W(R)$ is the intersection of the annihilators of the monoform compressible (left) R-modules. In [2] it was shown that $W(R)$ has the properties usually associated with a radical of a ring.

COROLLARY. Let R *be a weakly primitive ring*

(1) If J *is a left ideal of* R *then* $W(J) = J \cap r(J)$ *and* $^J/W(J)$ *is a weakly primitive ring.*

(2) If I *is a right ideal of* R *then* $W(I) = I \cap \ell(I)$ *and* $^I/W(I)$ *is a weakly primitive ring.*

(3) If A *is an ideal of* R *then* A *is a weakly primitive ring.*

For A an ideal of a ring R, the relationship between the monoform compressible R-modules and the monoform compressible A-modules can be explicitly described.

LEMMA. *If* M *is a compressible R-module with* $AM \neq 0$ *for* A *a right ideal of* R, *then* $r_M(A) = \{m \in M \mid Am=0\} = 0$.

Proof. If $r_M(A) \neq 0$ then there exists a monomorphism $f\colon M \longrightarrow r_M(A)$. But then $A \subseteq \ell(r_M(A)) = \ell(M)$, contradicting the hypothesis $AM \neq 0$.

THEOREM. *Let* A *be an ideal of a ring* R.

(1) If M *is a monoform compressible R-module such that* $AM \neq 0$ *then* M *is also a monoform compressible A-module.*

(2) If M' *is a monoform compressible A-module such that* $AM' \neq 0$ *then* AM' *defines a monoform compressible R-module.*

Proof. For (1), let M be given as stated, and we study the action of A on M. Let N be a non-zero A-submodule of M and suppose that $0 \neq f \in \mathrm{Hom}_A(N,M)$. Then f induces, via restriction to the R-submodule AN of N, an element of $\mathrm{Hom}_R(AN,M)$ which we denote by f'. $f' \neq 0$ because of the preceding lemma. Since ${}_RM$ is monoform, f' is a monomorphism. Since a monoform module is uniform, we conclude that f is also a monomorphism. Thus M is a monoform A-module.

To check that M is a compressible A-module, let N be a non-zero A-submodule of M. As above, AN is a non-zero R-submodule of N. Hence there exists a monomorphism $f \in \mathrm{Hom}_R(M, AN)$ which is of course also an A-monomorphism of M into N.

To prove (2), first observe that AM' has a natural structure of R-module defined via

$$r\left(\sum_{i=1}^{k} a_i m_i\right) = \sum_{i=1}^{k} (ra_i)m_i \quad \text{where } r \in R,\ a_i \in A,\ m_i \in M'.$$

This action is well-defined. For if $\sum_{i=1}^{k} a_i m_i = 0$ and $r \in R$ then $A\left(\sum_{i=1}^{k} (ra_i)m_i\right) = 0$, and hence $\sum_{i=1}^{k} (ra_i)m_i = 0$ by the preceding lemma.

Now suppose that $0 \neq f \in \mathrm{Hom}_R(N, AM')$ with N an R-submodule of AM'. Then, of course, $f \in \mathrm{Hom}_A(N, AM')$, so f is a monomorphism and AM' is a monoform R-module. To check compressibility, let N be a non-zero R-submodule of AM'. There exists a monomorphism $f \in \mathrm{Hom}_A(M', N)$. Then one checks that $f|_{AM'}$ induces an R-monomorphism of AM' into N.

We have the following immediate consequence, which was originally proved in [2;p,556] .

COROLLARY. *If* A *is an ideal of a ring* R *then* $W(A) = W(R) \cap A$.

Proof. With intersections taken over all monoform compressible A-modules M', $W(A) = \cap \ell_A(M') = \supseteq \cap \ell_A(AM') = \cap \ell_R(AM') \cap A \supseteq W(R) \cap A$. And, in the opposite direction, with intersections taken over all monoform compressible R-modules M, $W(R) \cap A = \cap \ell_R(M) \cap A = \bigcap_{AM \neq 0} \ell_R(M) \cap \bigcap_{AM=0} \ell_R(M) \cap A = \bigcap_{AM \neq 0} \ell_R(M) \cap A = \bigcap_{AM \neq 0} \ell_A(M) \supseteq W(A)$.

REFERENCES

[1] N. Jacobson, *Structure of Rings*, Amer. Math. Soc. Colloq. Publication 37, Providence, R.I., 1964.

[2] K. Koh and A.C. Mewborn, The weak radical of a ring, *Proc. Amer. Math. Soc. 18* (1967), 554-559.

[3] J. Zelmanowitz, Weakly primitive rings, *Comm. in Algebra 9* (1981), 23-45.

J. Zelmanowitz
Department of Mathematics
University of California
Santa Barbara, California 93106
USA
and
McGill University
Montreal, Canada

COLLOQUIA MATHEMATICA SOCIETATIS JÁNOS BOLYAI

38. RADICAL THEORY, EGER (HUNGARY), 1982.

PROBLEMS

B. J. Gardner

1. Find necessary and/or sufficient conditions on a universal class W for all hereditary subclasses of W to define hereditary lower radicals.

2. Find necessary and/or sufficient conditions on a universal class W for all homomorphically closed subclasses C of W to define a smallest semisimple class $S(C)$ which is homomorphically closed.

3. In a variety W of rings etc. let V be a subvariety such that

(i) all subdirectly irreducible objects in V are simple, and

(ii) all objects in V have distributive normal subobject lattices.

When is V extension-closed? When is V a semisimple

radical class? (Not always.)

4. Let S be a simple associative ring whose lower radical $L(\{S\})$ is an atom in the lattice of all radicals of associative rings. Must S satisfy the following condition? $S \triangleleft A$, $A/S \cong S \Rightarrow A \cong S \oplus I$ for some I .

J. S. Golan

1. A lattice L is said to be arithmetic if the meet of two compact elements of L is compact. Lattice-theoretic considerations suggest that this is an important class of lattices. Characterize those rings R for which the lattice R-tors is arithmetic.

2. A topological space X is called a T_D-space if every singleton subset of X is the intersection of an open subset and a closed subset of X . One easily proves that $T_1 \Rightarrow T_D \Rightarrow T_0$ and that these implications are strict. Characterize the rings R such that R-sp (topologized with any reasonable topology) is a T_D-space.

3. A topological space X is said to be Alexandroff discrete if each of its elements has a unique minimal open neighbourhood. Characterize those rings R for which R-sp (topologized with the BO-topology) is Alexandroff discrete.

4. Characterize the rings R such that every prime

torsion theory in R-sp is basic. (This problem is related to the previous one.)

J. Krempa

1. Construct a semiprime associative ring in which every prime ideal P is the intersection of all prime ideals properly containing P. (This problem is connected with characterizing atoms in the lattice of special radicals.)

2. Let K be a finite field and $K(t)$ be the field of rational functions over K. Is it possible to construct a radical property R in the class of all associative K-algebras such that for some K-algebra A

$$R(A \otimes_K K(t)) \neq (R(A \otimes_K K(t)) \cap A) \otimes_K K(t) ?$$

L. M. Martynov

1. Characterize the radical-semisimple classes of associative algebras over an arbitrary commutative unital ring.

2. Characterize those semisimple classes of associative rings (algebras over a field) which are quasivarieties.

For background on these problems see the author's paper in the present volume.

K. McCrimmon

Let Φ be a unital, commutative, associative ring of characteristic $p>0$ prime, V a Φ-module, $V[X]$ the space of polynomials in a set of indeterminates X with coefficients from V. Find the space $\mathrm{Fix}_G(V[X])$ of all polynomials fixed by all translations $\sigma_{x,\alpha} \to f(x_1,\ldots,x,\ldots,x_n) \to f(x_1,\ldots,x+\alpha,\ldots,x_n)$ for all $x\in X$, $\alpha\in\Phi$. (If Φ contains an infinite subfield then $\mathrm{Fix}_G(V[X])=V$ consists only of constants. Always $\mathrm{Fix}_G(V[X]) \supset \sum^{\infty} V_n[X_n]$ where $V_n = \{v\in V \mid F_p^n(\Phi)v=0\}$, $X_n=F_p^n(X)$ for $F_p(t) = t^p-t$ (example: for $\Phi = \mathbb{Z}_p = V$, $\mathrm{Fix}_G(V[x]) = V[x^p-x]$) but in general $\mathrm{Fix}_G(V[X])$ contains more.) Or find all $f \in \mathrm{Fix}_G(V[X])$ which in addition have $f(x_1,\ldots,\alpha x,\ldots,x_n) = \alpha^d f(x_1,\ldots,x,\ldots,x_n)$ for all $x\in X$, $\alpha\in\Phi^*$ (units in Φ).

K. McCrimmon, Ju. Medved'ev, M. Racine, E. Zel'manov

A sample of Jordan problems:

1. Describe the Amitsur shrinkage of the nil and Jacobson radicals, in particular the "strictly nil" radical. ($R(J[X]) = R_X(J)[X]$ for a radical $R_X \subset R$; describe R_X for finite and infinite X, hopefully in an explicit elemental manner.)

2. Construct a Baer radical for Jordan algebras so that either (i) J/B has no trivial ideals or (ii) J/B has no nilpotent ideals. Are these two radicals the same? How are they related to the McCrimmon radical?

3. In the free Jordan algebra $FJ(X)$ in infinitely many indeterminates X, is there a nilpotent ideal? (The answer is "yes" for finitely generated X, by results of Medvedev.) Are there analogues of Shestakov's alternative results (the Jacobson radical $J(FJ(X))$ is nilpotent, or a sum of nilpotents - perhaps only in characteristic 0 - when X is finite)?

4. Shirshov problem (analogue of Nagata-Higman for associative algebras): does nil of bounded index m plus characteristic $>m$ imply solvable? (The answer is yes for special Jordan algebras, with characteristic $>2m$.)

5. If J is finitely generated and satisfies a polynomial identity, is the Jacobson radical $J(J)$ nilpotent? (The answer again is "yes" for special algebras, via results of Razmyslov, Kemer, and Shestakov.)

6. Is the McCrimmon radical locally nilpotent? (By Zelmanov's work, the answer is "yes" for linear Jordan algebras; the passage to quadratic Jordan algebras in characteristic 2 is the main obstacle to extending Zel-

manov's structure theory of Jordan algebras to the quadratic case.)

R. Mlitz

1. Let R be a Kurosh-Amitsur radical in a variety (or a more general class) of Ω-groups. Then R can be written in the form

$$R(G) = \phi^{-1}(R(G/I)) \cap K(I)$$

where $I = \ker \phi$ and $K(I)$ is an ideal of G depending on I. Characterize properties of R by those of $K(I)$. In particular: characterize

a) near-ring radicals with hereditary semisimple classes;

b) subidempotent near-ring radicals.

2. Give necessary and sufficient conditions on a near-ring radical R in order that $R(I)$ be an N-subgroup of N for every $I \triangleleft N$.

M. Slater

Given any homomorphically closed subclass H of the universe R of rings, the corresponding (Amitsur) operator $h_H\colon R \to R$ is defined by $h_H(R) = \cap\{I \triangleleft R\colon R/I$ has no non-zero ideals from $H\}$.

1. Give necessary and sufficient conditions on the

operator $h:R \to R$ for h to equal h_H for some H. (I am informed by Prof. Kegel that this question was frequently raised by R. Baer.)

2. (See J. Algebra 39 (1976), p. 176.) Granted that $h=h_H$ for some H, construct one such H.

3. Granted that $h=h_H$ for some H, construct the largest such H. (This always exists.)

P. Stewart

Let S be a liberal extension of R, T an intermediate extension of R, and r a radical. When is $r(S)=Sr(R)$? When is $r(R)=R\cap r(S)$? When is $r(R)=$ $=R\cap r(T)$? In particular, what is the situation when r is the weak radical or the Blair hereditarily idempotent radical?

S. M. Vovsi

It is well known that neither the semigroup of hereditary group radicals nor that of prevarieties of groups is free. Now we consider only those radicals and prevarieties which are products of finitely many indecomposables; they are called completely decomposable.

1. (Plotkin) Is the semigroup of completely decomposable radicals free?

2. (Plotkin) Is the semigroup of completely decom-

posable prevarieties free? - Special case: we consider only the bounded prevarieties; they form a subsemigroup. Is this semigroup free?

3. For a group radical X, we denote by $\bar{X}$ the Kurosh-Amitsur radical generated by X. X is called self-closed if $\bar{Y}=X$ implies $Y=X$. Characterize self-closed radicals. Conjecture: Every local, closed radical is self-closed.

For group representations all these problems are solved in the positive.

4. The same problem as the previous one, this time for ℓ-groups.

A. E. Zalesskiĭ

Let $F=P[x_1,x_2,\ldots]$ be the free algebra in countably many indeterminates over a field P. A T-ideal of F is an ideal consisting of all the identities of some fixed algebra over P or, equivalently, an ideal which is invariant under all endomorphisms of F. If L is a T-ideal of F then $\sqrt{L}$ denotes the sum of all the T-ideals J of F such that J/L is nilpotent.

1. Is $\sqrt{L}/L$ always a nilpotent ideal?

L is called semiprime if $\sqrt{L}=L$ and prime if $AB\subset L$ implies $A\subset L$ or $B\subset L$ for any T-ideals A,B in F.

2. Is it true that every semiprime T-ideal is the intersection of finitely many prime T-ideals?

3. Determine the prime T-ideals of F.

These problems occurred to me when reading A. R. Kemer's abstract in the Theses of the Leningrad algebraic conference, 1981. Prof. Bokut' informed me that Kemer had solved all these problems for characteristic zero. Some information on the case of prime characteristic can be found in the paper A. E. Zalesskiĭ - M. B. Smirnov: Lie solvable associative rings, Izv. Akad. Nauk BSSR, Ser. Fiz.-Mat. Nauk, 1982, no. 2, 15-20.